Lasers with Nuclear Pumping

S.P. Melnikov • A.N. Sizov • A.A. Sinyanskii •
George H. Miley

Lasers with Nuclear Pumping

Springer

S.P. Melnikov
The Russian Federal Nuclear Center
Sarov
Russia

A.N. Sizov
The Russian Federal Nuclear Center
Sarov
Russia

A.A. Sinyanskii
The Russian Federal Nuclear Center
Sarov
Russia

George H. Miley
University of Illinois
Urbana
USA

This Work was originally published in 2008 as Lazery s yadernoy nakachkoy (in Russian) by RFNC-VNIIEF (Russian Federal Nuclear Center – All-Russian Research Institute of Experimental Physics) ISBN 5-9515-0044-3

ISBN 978-3-319-08881-5 ISBN 978-3-319-08882-2 (eBook)
DOI 10.1007/978-3-319-08882-2
Springer Cham Heidelberg New York Dordrecht London

Library of Congress Control Number: 2014953012

Printed on acid-free paper

Springer is part of Springer Science+Business Media (www.springer.com)

Preface

The majority of the credit for this book on nuclear-pumped lasers (NPLs) goes to its original Russian authors. I (George H. Miley) became the contributing American co-author when I learned of the Russian version of the book and proposed to develop a version in English. After correspondence with the Russian authors, I obtained permission from them to work on the English version. Subsequently, I managed to get a translation into English, which was not a simple task in view of the unique technical terminology and equations. While I made some minor changes and clarifications, these chapters generally follow the original Russian version. My additional contribution has been to add Chap. 13, a brief summary discussing work on this subject in the United States. Chapter 13 is somewhat similar to the Russian view of American work provided in the early chapters, but brings in some new perspectives.

The field of NPLs was born before the Cold War and the "Star Wars" program in the United States, when Karl Thom at NASA headquarters took an interest in NPLs for space power beaming. That led to research programs at NASA's Langley Research Center, and the University of Florida. About this time I obtained support from the DOE for research at the University of Illinois, and later received some additional support from NASA. Other laboratories with programs at the time included Sandia National Laboratory (where David McArthur achieved the first NPL in the United States, using a cooled CO lasing medium and the pulsed nuclear reactor located at Sandia), and also Los Alamos National Laboratory where the pulsed Godiva reactor was employed. These programs were relatively small, however, and focused on basic research about radiation-induced plasmas as well as NPLs. But NPLs became entangled in "Star Wars" and became a part of the cat and mouse game between Russia and the United States. This provided greatly expanded funding for classified programs in several of our National Laboratories (such as Lawrence Livermore National Laboratory, Sandia National Laboratory, and the Idaho National Engineering Laboratory), intended to compete with the classified work in the "secret science cities" in Russia. This competition led to great strides forward in both countries. However, the end of the Cold War era also abruptly ended the flow of money to NPL research in the United States. Thus very few American researchers work in this area at the present, and the national laboratory programs on NPLs have all stopped. In contrast, the Russian laboratories have managed to maintain a reasonably vigorous program, as is discussed in this book.

Because I had one of the few unclassified early programs on NPLs in the United States, I was one of the few American scientists free to discuss NPLs with Russian researchers. These conversations started in September of 1981 when I met several of their scientists during Tenth European Conference on Controlled Fusion and Plasma Physics in Moscow, where I spoke about radiation-induced plasmas. Alex Filyukov, a Russian scientist, approached me during the meeting and struck up a conversation about my NPL work. This was during the height of the Cold War, so these conversations were often vague and guarded on their part. I could speak more freely because I knew little of the classified work occurring within the United States. Alex's questions made it clear that he had studied all of my papers on the subject in great detail—he even knew some details that I had forgotten! It became obvious to me that he had been asked by their KGB or some official agency to interrogate me. After several discussions over the period of the conference, Alex stated that I was on "the right track" with my NPL research, but that he and Russian colleagues had some great concepts they could not yet disclose. This led to continued discussions and invitations for me to visit and give seminars at several places such as Moscow State University. Later after the fall of the Iron Curtain, I was able to visit the Russian "secret science cities" where the classified NPL research was going on. During those trips I met many more NPL researchers, including the co-authors of this book.

My work on NPLs led to memorable collaborations and visits to Russia even prior to the collapse of the Berlin Wall. My recollections from these remarkable times are recalled in the text. Some memories are shown here via photos. In my first trip to Russia after the end of the Cold War, the Russian NPL scientists disclosed that they had held "All Russian" NPL conferences each year, rotating among locations at Sarov (Arzamas-16) and Snezhinsk (Chelyabinsk-70) two main secret science cities in the USSR. I was presented with this photograph of attendees at the first such meeting. The senior scientists who attended were some of the top laser and nuclear scientists in Russia and included my co-authors of this book.

On those trips to Russia, I visited five specially built research reactors designed for NPL research in the "secret" laboratories. Shown above (reactor VIR-2M, Sarov) is a pulsed system with large laser pump regions located on top of the structure standing on posts in the foreground. The laboratory room containing the NPL and reactor incorporated special openings through a shielding wall with diagnostics behind. This allowed good detector sensitivity with minimum radiation interference. These facilities and NPL research on them are discussed in detail in this book.

I made four trips to the USSR prior to the fall of the Berlin Wall. These trips were for discussions of NPLs, at the invitation of the Russians, hosted by Alex Filyukov and scientists from the Lebedev Physics Institute in Moscow. At the time I learned that Alex Miskevich had been doing leading experiments at a laboratory somewhere north of Moscow, but I was not permitted to visit it. A few months after the fall of the Berlin Wall, I was invited to come again, and was hosted by Alex Miskevich (second from left—I am to his left, wearing glasses), and observed his NPL facilities at the Moscow Engineering and Physics Institute (MIFI). He is shown holding one of his laser cells. During this visit I learned that there were major programs in the Russian Laboratories in the "secret cities" of Obninsk, Chelyabinsk-70, and Arzamas-16. I was invited to visit them, and shortly after that I made trips to the laboratories in these secret cities (where my present co-authors are located).

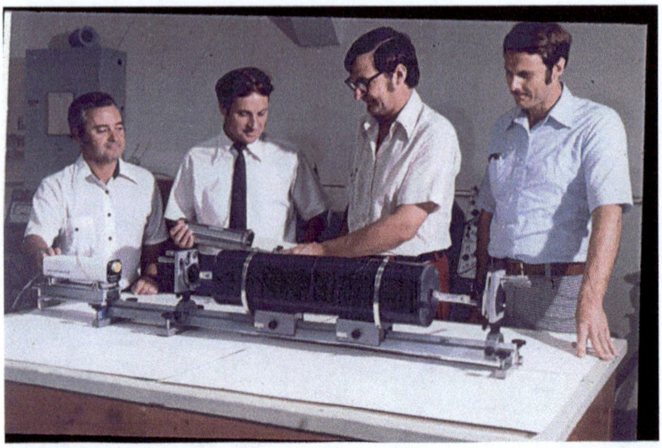

This photo shows U.S. scientists from the NASA Langley Research Center preparing a NPL experiment for the Fast Burst Reactor at the Army Aberdeen Test Laboratory in Maryland. Jack Fryer (technician), Frank Hohl, Nelson Jalufka, and Russell De Young are shown with the experimental setup.

Russell De Young at the Aberdeen Fast Burst reactor where nuclear laser experiments were carried out. The reactor is shown between two test laser setups. One is the Box Laser and the other is polyethylene (neutron moderator) covered cylindrical gas tube laser.

Much of the early NPL work in the United States was done by scientists at NASA Langley Research Laboratory, who made use of the excellent pulsed reactor facilities at the Army Aberdeen Maryland Laboratory. Dr. Frank Hohl, who was in charge of this work, and Dr. Russell De Young, who did his thesis on NPLs with me at the University of Illinois, were major contributors to the research. The photo above shows Frank Hohl and Russ De Young along with another NPL scientist Nelson Jalufka. The pulsed reactor at Aberdeen is shown in the second photo.

Here Dr. Mark Prelas and Dr. Fred Boody are being greeted by the director of the Arzamas 16 Laboratory upon their arrival for a workshop in 1991.

Another University of Illinois graduate, Dr. Mark Prelas (a Professor at the University of Missouri Columbia), along with Dr. Fred Boody (MS at University of Illinois and PhD from University of Missouri) were heavily involved in early NPL work. In particular, they focused on the concept of using nuclear pumped "flash-lamps" to pump the laser medium. Following the fall of the Berlin Wall, they made several visits to the Russian NPL laboratories and participated in meetings there, much as I did.

These photographs are intended to provide the reader some feeling for the flavor of the early NPL research. They certainly do not provide nor are they intended to provide a full picture of all the people and facilities that have been utilized in this work.

The association of NPLs with "Star Wars" was unfortunate and deleterious to NPL research in the United States. As the reader will discover, there are many very important civilian applications for NPLs that have been overlooked due to the associate with military applications. An additional problem was that just as the technology began to emerge, nuclear reactors fell into disfavor due to fears initiated by the Chernobyl and Three Mile Island reactor accidents. Practical NPLs would require the design of special types of nuclear reactors (both for terrestrial and space applications) and as a result of public concerns such reactors were "not in the cards." However, the situation is now slowly changing, as many feel nuclear reactors should play an important role in the world's future energy economy. The most unfortunate recent accident is the Fukushima Daiichi nuclear disaster in Japan represents a new setback in acceptance. But many people believe this will pass as lessons learned improve reactor safety and the Fukushima event was an extreme example. Consequently, the vision of a reactor-based laser system for various applications cannot be ruled out. This is particularly true in view of the importance of potential applications such as power beaming, inertial confinement fusion, chemical and materials processing, and deflection of asteroids and other space debris. Hopefully this book, the first on the subject, will introduce a new generation of scientists and engineers to this exciting and very important field of science. The ultimate applications for NPLs may well be something not yet envisioned, and the basic science remains challenging and intriguing. My Russian co-authors have, in my view, provided a unique insight into the status of the field and the potential now open to move into important applications. Thus I am extremely pleased to have played a role in helping bring this information to a much wider scientific audience.

I want to thank Autumn West, Robyn Bachar, and Maria Lipson for their diligent work on the translation and proofreading. However, I must take ultimate responsibility for this version of the Russian book, including any mistakes or omissions that have crept in. I also wish to thank my many colleagues and students who contributed to the advancement of NPLs in the United States and whose work I discuss here.

The first 12 chapters in this book are derived from the original version in Russian. Rather than attempting to discuss them myself, I felt it best to use the Russians' description. Thus, the following introduction is adapted from the original Russian edition. In this section, "authors" refers to my Russian co-authors.

The use of nuclear radiation to pump active media and create nuclear-pumped lasers (NPL) on this basis is a comparatively new scientific–technical area. It is at the interface of two disciplines—quantum electronics and nuclear physics. This area has rapidly evolved over the last 40 years, from the first proposals on the use of nuclear energy sources for pumping lasers to the creation of diverse NPLs. Now NPL research has reached the point where engineering design development of continuous and pulsed nuclear laser units for various applications—integrated devices based on the achievements of nuclear physics and technology, quantum electronics, gas dynamics, optics, etc.—has become possible.

The authors of this book have participated in studies into NPL-related issues since the late 1960s. At that time, NPL research was in its incipient phase. In a number of laboratories (primarily in the United States), attempts were made to pump various active media with nuclear radiation and thus prove the fundamental possibility of direct conversion of nuclear energy into laser radiation. To search for NPL active media and study their characteristics, the All-Union Scientific Research Institute of Experimental Physics (VNIIEF) formed a science team consisting primarily of graduates of the Moscow Engineering and Physics Institute (L.Ye. Dovbysh, V.M. Karyuk, M.F. Kostenko, V.N. Krivonosov, S.P. Melnikov, A.N. Sizov, A.A. Sinyanskiy), which initially was directed by A.M. Voinov and A.T. Kazakevich, and later on by A.A. Sinyanskiy. Academicians Yu.B. Khariton and A.I. Pavlovskiy devoted great attention to the development of these studies. In the first phase of the exploratory research, significant assistance was provided by I.V. Podmoshenskiy (S.I. Vavilov State Optical Institute [GOI])—a virtual scientific consultant of the team of young researchers. The decision on organization of the work to study the problems of NPLs specifically at VNIIEF was a natural one, since in that period (and at present), VNIIEF was one of the few organizations with an inventory of diverse, powerful pulsed reactors. It was this circumstance that made it possible to implement the first successful experience in laser pumping with nuclear radiation in 1972, and then to continue research into various NPL problems.

The authors have striven to show the most important results and phases of NPL development not only at VNIIEF, but also at other Russian and foreign laboratories. The book may not be entirely free of subjectivity, because as a rule a scientific publication is the result of a compromise of an author (or authors) desiring to present the modern state of research in a given field with adequate completeness while displaying the achievements of his or her own laboratory.

The basic reason for writing this book was the lack of a scientific study that reflected in adequate detail the status of the work on NPL problems and prospects for their development. The authors worked to systematize information accumulated over decades, which had frequently been published in scientific publications or digests of conference materials that were available only in limited circulation.

This book may also be seen as a reference work, because it contains information about virtually all studies published in the last approximately 40 years related to NPLs.

This book consists of 13 chapters: Chaps. 1–6 (except for Sections 1 and 2 in Chap. 6), 11, and 12 were written by S.P. Melnikov; Chaps. 7–9 were written by A.N. Sizov; and Chap. 10 and Sections 1 and 2 in Chap. 6 were written by A.A. Sinyanskiy. Chapter 13 has been added by our American co-author, George H. Miley.

Chapter 1 provides the chronology of basic events associated with the advent and study of NPLs. The specifics of pumping lasers with nuclear radiation are also discussed.

Chapter 2 provides data on the characteristics of the pulsed reactors that were used to conduct the bulk of the studies of NPLs in Russia and abroad. It examines lasers and the experimental procedures, and discusses the specific features of experimentation with pulsed reactors associated with the problem of radiation resistance of optical materials and photodetectors.

Chapter 3 systematizes the extensive material on the results of experimental studies of spectral, energy, and threshold characteristics of various types of NPLs radiating at the transitions of Xe, Kr, Ar, Ne, C, N, O, Hg, Cd, and I atoms, the ions Cd+, Zn+, and Hg+, the CO molecule, and the molecular ion N_2+. The results of studies of a number of active media for which no generation was detected during nuclear radiation pumping are discussed. And finally, Chap. 3 looks at a method of laser pumping with fast neutrons, and cites the results of studies of NPLs based on a He-Xe mixture in experiments with the BR-1 pulsed reactor.

Chapter 4 is dedicated to a discussion of processes that occur in a low-temperature nuclear-excited plasma: from the initial processes of ionization and excitation of the medium by nuclear particles to subsequent plasma-chemical reactions leading to the populating of laser levels. Data are cited on the energies of formation of primary particles of plasma (ions and excited atoms). In addition, Chap. 4 provides a survey of experimental work studying the parameters of nuclear-excited plasma and spectral luminescence characteristics of gas media.

Chapter 5 discusses the kinetics of plasma-chemical processes and lasing mechanisms of all currently known gas NPLs, and looks at NPL kinetic models published in the literature and the results of computations of their characteristics.

Chapter 6 contains information on the results of investigations of the more complex nuclear laser devices, which are structural elements of multichannel reactor-lasers operating in fixed or pulsed modes. This chapter also gives the parameters of certain variations of designed pulsed reactor-lasers.

Chapter 7 briefly examines information about the methods of calculating the specific energy deposition of ionizing particles. The influence of non-uniformities of uranium-containing layers on the efficiency of the energy contribution of uranium nuclear fission fragments to the gas is studied. The dependence of the energy deposition and its effectiveness on the thickness of the layers and the density of the gas are considered. A dimensionless parameter of optimization of the energy deposition is introduced as a function of the reduced thickness of the uranium layers and the ratio of the transverse dimension of the excited volume to the length of the free path of the average fragment in the gas. It is demonstrated that the functional dependence of this parameter is in good accord with experimental data on the output power for lasers pumped by fission fragments. Experimental data to determine the energy deposition, the causes of their mutual discrepancy, and deviation from theoretical results are analyzed.

Chapter 8 is dedicated to studies of optical non-uniformities arising in hermetically sealed gas lasers excited by fission fragments. Formation and development of these non-uniformities are caused by the specifics of distribution of the specific energy deposition of fission fragments and relaxation thermal and gasdynamic processes. It is shown that owing to heat exchange of the excited gas with the walls and the substrates of laser cells, a wall zone is formed with large positive values of the refraction index gradient. These values are so great that lasing can be carried out only outside the zone limits. The size of the wall zone increases over time: for pumping pulse durations of ~ 0.1 s, it encompasses virtually the entire volume of the laser cell. Estimates are cited regarding the influence of optical non-uniformities on laser beam divergence, and results of calculations of the distribution of non-uniformities are compared with experimental results. The possibilities for improving optical characteristics of hermetic lasers by varying the initial parameters of the gas and the conditions of energy input are studied. The optical non-uniformities in the quasi-stationary and stationary (with external cooling of the cell) modes of irradiation are investigated.

Chapter 9 examines the specific features of flowing-gas lasers. The drawbacks of longitudinal circulation of gas mixtures are discussed. The influence of turbulent pulsations on laser optical characteristics, based on which the constraints on gas pumping speed are found, is evaluated. The advantages of transverse gas circulation, with subsequent release of excess heat downstream of each laser channel to radiators, are underscored. The derivation of equations describing the action of the radiator is provided, and their solution is cited. The structure of gas flow in the laser channel with transverse circulation is described. A model for forming optical non-uniformities in the flowing-gas channel, one that concurs with the experiment, is introduced. Methods are proposed for approximate calculation of gas density distribution in such a channel. Possibilities of combined heat removal are discussed, in which the heat is removed simultaneously both by flowing the laser-active gas itself, and by surrounding the channel with an external coolant.

Chapter 10 is dedicated to a discussion of problem issues associated with development of multichannel stationary reactor lasers and their possible applications.

Chapter 11 provides a survey of the works investigating the possibility of pumping solid-state and liquid laser media with nuclear radiation and creating different versions of pulsed reactor lasers based on them.

Chapter 12 is a review of articles about the results of investigations of nanosecond pulsed NPLs. These lasers are usually pumped by radiation of the most powerful source of energy, a nuclear explosion. Such lasers can be used to resolve one of the problems of inertial thermonuclear fusion determination of the levels of energy that laser drivers must have to obtain specific amplification gains of the target.

Chapter 13, as noted earlier, has been added to the English version by George Miley to provide some added insight into research in the United States.

The authors would like to thank VNIIEF associates A.M. Voinov, L.Ye. Dovbysh, V.F. Kolesov, M.I. Kuvshinov, V.N. Krivonosov, B.V. Lazhintsev, V.A. Nor-Arevyan, and V.T. Punin for their many years of fruitful collaboration, as well as E.P. Magda (All-Russia Scientific Research Institute of Technical Physics [VNIITF]), A.I. Miskevich (Moscow Engineering and Physics Institute [MIFI]), and V.F. Tarasenko (Russian Academy of Sciences Institute of High-Current Electronics, Siberian Division [ISE SO RAN]). The authors thank A.P. Morovov for proofreading the book, and for valuable comments.

The authors are grateful to Yu.N. Deryugin, V.Yu. Matyev, and Ye.V. Prikhodko for participating in the calculations of energy deposition, spatial non-uniformities, and the stability dynamics of cavities of gas NPLs, as well as to A.A. Pikulev, V.M. Tsvetkov, S.L. Turutin, A.N. Korzenev, S.V. Patyanin, and P.V. Sosnin for investigations of various NPL characteristics.

Numerous investigations of NPLs performed at VNIIEF on the pulsed reactors VIR-2, VIR-2M, TIBR, BR-1, and BIGR would have been impossible without the help of the personnel of these units. The authors would like to express their gratitude to all associates of the reactor teams and especially note the great contribution to successful conduct of the experiments by V.N. Bogdanov, A.S. Koshelev, and S.F. Melnikov.

The authors would also like to express their gratitude to the associates of the scientific–technical library of the Institute of Nuclear and Radiation Physics at VNIIEF for assistance in finding information, and to S. Yu. Pikuleva for providing the figures in Chap. 4.

Updated to add: Finally, we are most grateful to our U.S. co-author, George H. Miley, who made this English version of the book possible.

Champaign, IL, USA George H. Miley

Contents

Chapter 1
General Issues

1.1 Chronology of Main Events. Initial Phases of Research

This section briefly examines the last 50 years of research in the field of transformation of the kinetic energy of products of nuclear reactions into laser radiation, that is, nuclear-pumped lasers (NPL). NPLs include lasers in the optical spectral range, which are excited directly by nuclear radiation or by using intermediate devices, for example, nuclear-optical converters (nuclear-excited plasma, scintillators).

The discussion of the problem of direct conversion of nuclear energy into laser radiation began in the 1960s, immediately after creation of the first lasers. Interest in this problem was prompted by the possibility of using powerful, compact, high-energy sources of nuclear energy (nuclear reactors, nuclear explosions) for laser pumping, and accordingly, the fundamental possibility of creating powerful lasers. At present, the concept of the reactor-laser (RL) is the most common embodiment of this new technology. The core of the RL consists of fissionable material and the laser medium. There the direct conversion of the escaping nuclear energy into laser radiation occurs, bypassing the intermediate stage of thermal energy.

Early experimental and theoretical studies of NPLs were primarily carried out in the United States and Russia. Table 1.1 shows the basic phases of NPL development. In some cases, the dates of events are provided (if they are reliably known to the authors), but not the dates of articles about these events. Articles sometimes appeared considerably later, since roughly until the end of the 1980s, because research on NPLs in certain laboratories was carried out in accordance with classified programs. The sequence by which individual articles reached the open press may be traced from the list of references.

In Russia, research on problems of NPLs began in the late 1960s virtually simultaneously at three institutes—VNIIEF (Sarov), the Kurchatov IAE (Moscow), and the MGU Institute of Nuclear Physics (Moscow). The neutron sources were pulsed reactors of VNIIEF and IAE. Roughly from the start of the 1970s,

S.P. Melnikov et al., *Lasers with Nuclear Pumping*,
DOI 10.1007/978-3-319-08882-2_1

Table 1.1 Basic phases of NPL development

Date	Event	Scientific organization (authors)	References
1964	Calculated estimation of a nuclear-laser device using a mixture of He-Ne ($\lambda = 633$ nm)	UAC (L.O. Herwig)	[1]
1965–1972	US experiments to find active NPL media (condensed media, mixtures of He-Ne, CO_2-N_2-He, Ne-O_2, Ar, Ne, Xe, Kr, etc.)	ANL, NL, UI, UF, NASA, NARC, GA, MC et al. (J.A. De Shong, J.W. Eerkins, J.R. Rusk, G.H. Miley, J.C. Guyot, R.T. Schneider, E. Matovich, D.M. Compton, V.E. Derr, et al.)	[2–5]
1968–1971	Investigations at VNIIEF to find laser media for NPLs (condensed media, CO_2-N_2-He, CO) and assessment of different variants of nuclear laser devices	VNIIEF (A.M. Voinov, A.T. Kazakevich, V.N. Krivonosov, M.F. Kostenko, A.N. Sizov, V.M. Karyuk, L.M. Pavlovskaya, A.A. Sinyanskiy, et al.)	[6–8]
1970	Attempt to achieve lasing in a mixture of ^{3}He-Hg ($\lambda = 615$ nm)	MGU, IAE (V.M. Andriyakhin, V.D. Pismennyy, V.Ye. Khvostionov, et al.)	[9]
1970	Achievement of lasing at transitions of Ne and Xe atoms with pumping of atmospheric-pressure gas mixtures with an electron beam	IYaF SO AN (G.G. Dolgov-Savelyev, B.A. Knyazev, Yu. L. Kozminykh, V.V. Kuznetsov, A.M. Orishich)	[10]
1971	First experiments to pump lasers with γ- radiation of a nuclear explosion (liquid xenon medium)	VNIIEF (O.V. Korniyenko, I.G. Zhidov, V.V. Varaksin, A.P. Morovov, Ye.K. Bonyushkin, et al.)	[un-published]
1972	Achievement of super-luminescence with excitation of a laser based on europium chelate with n,γ-radiation of a pulsed reactor	VNIIEF (T.A. Babicheva, A.M. Voinov, L.Ye. Dovbysh, A.A. Sinyanskiy, L.M. Pavlovskaya, et al.)	[6, 7]
1972	First successful experiments to develop an NPL (mixture of He-Xe, $\lambda \approx 3$ μm)	VNIIEF (A.A. Sinyanskiy, A.M. Voinov, A.T. Kazakevich, S.P. Melnikov, L.Ye. Dovbysh); GOI (I.V. Podmoshenskiy)	[6, 11]
1973	Experiments on pumping a mixture of SF_6-C_2H_6 (LANL) and gaseous xenon (LLNL) with γ-radiation of a nuclear explosion	LANL (P.B. Lyons, J.S. Clarke, D.S. M. Metzger); LLNL (P.J. Ebert, J.L. Ferderber, et al.)	[12–14]

(continued)

Table 1.1 (continued)

Date	Event	Scientific organization (authors)	References
1974	Statement in open press on possibility of creating a reactor-laser	FIAN (L.I. Gudzenko, S.I. Yakovlenko)	[15]
1974–1976	Cycle of experimental investigations of NPLs at IR transitions of atoms Xe, Kr, and Ar (acquiring an efficiency of up to 2 %)	VNIIEF (A.M. Voinov, A.A. Sinyanskiy, V.N. Krivonosov, S.P. Melnikov, et al.); GOI (I.V. Podmoshenskiy)	[6, 11, 16]
1975	First successful experiments in the United States using pulsed reactors to develop NPLs: CO laser ($\lambda = 5.1$–5.6 μm) and laser with mixture of He-Xe ($\lambda = 3.51$ μm)	Sandia (D.A. McArthur, P.B. Tollefsrud); LANL, UF (H.H. Helmick, J.L. Fuller, R.T. Schneider)	[13, 17, 18]
1975	Proposal to use mixtures of rare gases with vapors of metals (Mg, Hg, Cd, etc.) as prospective NPL media	VNIIEF (A.A. Sinyanskiy)	[16]
1976–1977	Creation of NPLs at transitions of atoms N and C in near IR spectral range	UI (G.H. Miley, M.A. Prelas, R.J. DeYoung, F.P. Boody, W. E. Wells, et al.)	[19, 20]
1977	First NPL in visible range (mixture of He-Hg, $\lambda = 615$ nm)	Sandia (M.A. Akerman, D.A. McArthur); UI (G.H. Miley)	[21]
1977–1984	Development and investigations of powerful HF and XeF lasers, excited by γ-radiation of a nuclear explosion	VNIIEF (Ye.K. Bonyushkin, A.I. Pavlovskiy, A.P. Morovov, B.V. Lazhintsev, V.V. Varaksin, A.Ye. Lakhtinov, et al.)	[22, 23]
1979	Achieving lasing in the mixture He-Cd ($\lambda = 534$ and 538 nm) pumped by $^3He(n, p)^3H$ reaction products	MIFI (A.I. Miskevich, B.S. Salamakha, V.A. Stepanov, et al.)	[24]
1979	Start of computational and theoretical investigations of the appearance and evolution of optical non-uniformities in NPLs	VNIIEF (A.N. Sizov)	[25]
1981	Creation of NPLs pumped by fast neutrons (He-Xe mixture)	VNIIEF (S.P. Melnikov, A.M. Voinov, A.A. Sinyanskiy, A.S. Koshelev)	[6, 11]
1981	Achieving lasing in a solid-state neodymium NPL pumped by the radiation of a CsI scintillator	VNIIEF (A.M. Voinov, L.Ye. Dovbysh, A.A. Sinyanskiy, et al.)	[6]

(continued)

Table 1.1 (continued)

Date	Event	Scientific organization (authors)	References
1982	Achieving lasing in a He-Cd mixture ($\lambda = 442$, 534, and 538 nm) pumped by uranium fission fragments	VNIITF (E.P. Magda, V.A. Kryzhanovskiy, L.V. Semkov, et al.); VNIIEF (A.A. Sinyanskiy, S.P. Melnikov)	[26]
1982	First experimental investigations of optical non-uniformities in gas NPLs	VNIIEF (V.V. Borovkov, B.V. Lazhintsev, S.P. Melnikov, A.A. Sinyanskiy, V.A. Nor-Arevyan, et al.)	[27]
1983	Achieving lasing in a He-Zn mixture ($\lambda = 748$ nm)	VNIIEF (A.M. Voinov, S.P. Melnikov, A.A. Sinyanskiy, et al.); MIFI (A.I. Miskevich, B.S. Salamakha, et al.)	[6, 16, 28]
1985	Achieving lasing in the visible spectral range at transitions of the Ne atom ($\lambda = 585$, 703 and 725 nm)	VNIIEF (S.P. Melnikov, A.M. Voinov, A.A. Sinyanskiy, et al.); MIFI (A.I. Miskevich)	[11, 16]
1985	Achieving a minimal lasing threshold of NPLs with a thermal-neutron flux density of around 2×10^{12} $cm^{-2} \cdot s^{-1}$ (0.02 W/cm^3)	VNIIEF (A.M. Voinov, S.P. Melnikov, A.A. Sinyanskiy, A.I. Konak, V.G. Zobnin, I.N. Mochkayev	[11, 16]
1985–1990	Development of kinetic models of NPLs with various active media	IOFAN (S.I. Yakovlenko, A.V. Karelin, et al.); VNIIEF (S.P. Melnikov, A.A. Sinyanskiy, A.M. Voinov)	[28, 29]
1986	Formulation of concept of a laser system with pumping from a pulsed reactor (nuclear pumped laser amplifier, OKUYaN)	FEI (A.V. Zrodnikov, P.P. Dyachenko, A.V. Gulevich, et al.)	[30]
1987–2007	Experimental and theoretical investigations of optical non-uniformities in gas-flowing NPLs	VNIIEF (V.V. Borovkov, A.N. Korzenev, B.V. Lazhintsev, V.Yu. Matyev, A.N. Sizov, A.A. Sinyanskiy, et al.)	[31–33]
1988–1993	Computational and theoretical investigations in the United States of optical non-uniformities in NPLs	Sandia (J.R. Torczynski, D.R. Neal, et al.)	[34, 35]
1989	Obtaining of an intrinsic efficiency of ~3 % for NPLs using mixtures of Ar-Xe and He-Ar-Xe ($\lambda = 1.73$ and 2.03 μm)	Sandia (WJ. Alford, G.N. Hays)	[36]

(continued)

Table 1.1 (continued)

Date	Event	Scientific organization (authors)	References
1989–1994	US development of kinetic models of NPLs operating on transitions of atoms Xe, Ar, and Ne	UI (M.J. Kushner, M. Ohwa, J.W. Shon, T. J. Moratz, R.L. Rhoads)	[37–39]
1991	Creation of a photodissociation iodine laser ($\lambda = 1.315$ μm), initiated by radiation of XeBr* molecules	UI (W.H. Williams, G.H. Miley)	[40]
1992	Creation of NPLs based on the transition of the Hg atom in a He-Xe-Hg-H2 mixture ($\lambda = 546$ nm)	VNIITF (A.V. Bochkov, E.P. Magda, V.A. Kryzhanovskiy, S.L. Mukhin, et al.)	[41]
1991–1993	A series of experimental investigations of xenon NPLs in the United States	Sandia (W.J. Alford, G.N. Hays, G.A. Hebner)	[42–44]
1994	Activation of the LM-4/BIGR setup, obtaining of cw lasing in gas flowing mode	VNIIEF (A.M. Voinov, A.A. Sinyanskiy, S.L. Turutin, B.V. Lazhintsev, V.V. Porkhayev, A.N. Sizov, A.N. Pokalo, et al.)	[45]
1996	First NPL in UV spectral range (He-N_2-H_2 mixture, $\lambda = 391$ nm)	VNIITF (N.M. Barysheva, E.P. Magda, A.V. Bochkov, V.A. Kryzhanovskiy, et al.)	[46]
1999	Start-up of demonstration model of OKUYaN reactor-laser system	FEI (A.V. Zrodnikov, P.P. Dyachenko, A.V. Gulevich, et al.)	[30]
1986–2007	Development of reactor IKAR—model of stationary reactor-laser	VNIIEF (A.M. Voinov, A.N. Sizov, A.A. Sinyanskiy, V.N. Krivonosov, I.A. Nikitin, V.I. Turutov et al.)	[47]

Note: Nomenclature of scientific organizations
UAC United Aircraft Corporation (United States), *ANL* Argonne National Laboratory (United States), *NL* Northrop Laboratories (United States), *UI* University of Illinois (United States), *UF* University of Florida (United States), *NASA*, NASA Langley Research Center (United States), *NARC* North American Rockwell Corporation (United States), *GA* General Atomic (United States), *MC* Martin Company (United States), *VNIIEF* All-Russian Scientific Research Institute of Experimental Physics (Sarov), *MGU* Moscow State University (Moscow), *IAE* Kurchatov Institute of Atomic Energy, *IYaF SO AN* Institute of Nuclear Physics, Siberian Department, Academy of Sciences (Novosibirsk), *GOI* Vavilov State Optical Institute (St Petersburg), *LLNL* Lawrence Livermore National Laboratory (United States); Sandia: Sandia National Laboratories (United States), *LANL* Los Alamos National Laboratory (United States), *FIAN* Lebedev Physics Institute, USSR Academy of Sciences (Moscow), *MIFI* Moscow Engineering and Physics Institute (Moscow), *VNIITF* All-Russian Scientific Research Institute of Technical Physics (Snezhinsk), *IOFAN* General Physics Institute, USSR Academy of Sciences (Moscow), *FEI* Physics and Power Engineering Institute (Obninsk)

investigations into the problems of NPLs began to be conducted at MIFI (Moscow), from the end of the 1970s at VNIITF (Snezhinsk), and from the start of the 1980s at FEI (Obninsk). Some experiments using a stationary reactor and calculations of the kinetics of certain types of NPLs were carried out at the Nuclear Physics Institute of the Academy of Sciences of Kazakhstan (Alma-Ata).

In parallel with the experiments, theoretical investigations of NPLs were carried out, in which FIAN (Moscow) and IOFAN (Moscow) participated, in addition to the institutes mentioned previously. Experimental investigations were carried out with the use of electron beams at FIAN (Moscow), the Institute of High Current Electronics of the Siberian Division of the Russian Academy of Sciences (ISE SO RAN) (Tomsk), the Scientific Research Center for Technological Lasers (Moscow), and the Institute of Electrophysics of the Ural Division of the Russian Academy of Sciences (Yekaterinburg).

Elsewhere, experimental and theoretical investigations of NPLs were mainly carried out in the United States (see Table 1.1). The first proposals regarding NPLs and the first experimental investigations began somewhat earlier than in Russia. The most substantial contribution to the development and creation of NPLs was made by investigations carried out at the Sandia National Laboratories, University of Illinois, NASA Langley Research Center, and University of Florida. In addition to the laboratories and universities indicated in Table 1.1, a number of NPL investigations were carried out in the United States at the University of Missouri-Columbia, the Idaho National Engineering Laboratory, Oak Ridge National Laboratory, North Carolina State University, and Vanderbilt University. Some individual investigations of NPLs were carried out in France (Laboratoire de Physique des Gaz et des Plasmas), China (Institute of Nuclear Physics and Chemistry of CAEP), and Japan (University of Tokyo, Tokyo Institute of Technology).

To model the conditions that occur in gas media of NPLs, in joint experiments carried out in Germany (Technische Universität München) and in the United States (Rutgers University), beams of heavy ions ($^4He^+$, $^{32}S^+$, $^{132}Xe^+$, etc.) were used. A number of interesting studies (basically of a high-pressure xenon laser) were also carried out using fast electron beams at the University of Twente (The Netherlands), Stanford University (United States), Naval Research Laboratory (United States), and Science Research Laboratory (United States).

The majority of published data relate to the search for NPL media and the study of their characteristics. Basic efforts were directed toward obtaining maximal efficiency of conversion of deposited nuclear energy into laser radiation (ηl) and low lasing thresholds (q_{th}). The maximal values of ηl and the lowest lasing thresholds were registered for a laser based on IR transmissions of the atom Xe: $\eta_l = 2–3$ % (VNIIEF, 1976 [6, 11, 16]; Sandia Laboratory, 1989 [36]), $q_{th} \sim 0.02$ W/cm^3 (VNIIEF, 1985 [6, 11, 16]).

Several key issues were uncovered and addressed in the problem of searching for and studying active NPL media. These included specific features of gas medium pumping with heavy charged particles and the problem of severe optical non-uniformities. Computational and theoretical investigations [25, 48–50], and then later also experimental investigations of this type were begun at VNIIEF in

1970. Later on there were reports about the conduct of analogous investigations in the United States [34, 35].

Apart from the work related to the search for NPLs and study of their various characteristics, one should also note other issues studied relating to development of powerful NPLs:

- Investigations into the properties of nuclear-excited plasma (track structure of plasma, kinetics of plasma processes, luminescence characteristics).
- Selection of radiation-resistant optical and structural materials for NPLs.
- Development and creation of various nuclear-laser devices.
- Development of the fabrication method of thin-film uranium fuel (radiator of fission fragments and fuel for the reactor-laser (RL).
- Selection of the optimal design for a RL and creation of its experimental model.
- Evaluation of possible applications of nuclear-laser devices of various types. Many of the studies done in these areas are also considered in this book.

In the history of development of NPLs, some special scientific conferences have been organized to discuss the results of research into their characteristics, prospects for development, and possible applications. The most significant of these were:

1. Second Symposium on Uranium Plasmas: Research and Applications, Atlanta (USA), 15–17 November 1971.
2. First International Symposium on Nuclear Induced Plasmas and Nuclear Pumped Lasers, Orsay (France), 23–25 May 1978.
3. The specialist conference Physics of Nuclear-Excited Plasma and Problems of Nuclear-Pumped Lasers, Obninsk (Russia), 26–29 May 1992.
4. The second international conference Physics of Nuclear-Excited Plasma and Problems of Nuclear-Pumped Lasers, Arzamas-16 (Russia), 26–30 September 1994.
5. The third international conference Problems of Nuclear-Pumped Lasers and Pulsed Reactors, Snezhinsk (Russia), 16–20 September 2002.
6. The fourth international conference Problems of Nuclear-Pumped Lasers and Pulsed Reactors, Obninsk (Russia), 18–21 September 2007.

The widest international representation was involved in the last four conferences, in which scientists from Russia, the United States, Germany, France, China, and Japan participated.

Despite the large volume of research carried out with regard to NPL problems, the number of survey papers is comparatively small, namely: K. Thom, R.T. Schneider [2] (1972); V.Ya. Pupko [51] (1981); N.W. Jalufka [3] (1983); R.T. Schneider, F. Hohl [52] (1984); D.A. McArthur [13] (1991); A.I. Miskevich [53] (1991); G.H. Miley [40] (1992); E.G. Batyrbekov [54] (1994); A.A. Sinyanskiy [6] (1995); A.V. Karelin, A.A. Sinyanskiy, S.I. Yakovlenko [28] (1997); A.A. Sinyanskiy, S.P. Melnikov, [11] (1998); E.P. Magda [55] (1998). Most of the surveys were published some time ago; many of them contain information only about research by individual laboratories. The most complete survey

Table 1.2 Experimental research to find gas media for NPLs performed before 1972

Scientific organization, reference	Pumping method	Laser medium	Results
University of Illinois [2, 3] (G.H. Miley, J.C. Guyot, et al. 1967)	TRIGA reactor, α particles (layer ^{10}B)	He-Ne (≤0.13 atm)	In the opinion of the authors, amplification was observed at λ = 3.39 μm
Northrop Laboratories [2, 3, 57] (J.R. Rusk, J.W. Eerkins, J.A. DeJuren, et al. 1968)	TRIGA reactor, α particles and uranium fission fragments (layer ^{235}U and ^{10}B)	1. $Ne\text{-}O_2$ (~0.1 atm) 2. $CO_2\text{-}N_2\text{-}He$ 3. Ne, Ar, Kr, Xe (≤0.4 atm)	An increased level of luminescence was observed in rare gases, which the authors attribute to stimulated radiation
MGU and IAE [2, 3, 58] (V.M. Andriyakhin, V.D. Pismennyy, et al. 1968)	Proton beam with energy of 3 MeV	$CO_2\text{-}N_2\text{-}^3He$ (0.013 atm)	A threefold increase in lasing power was observed upon irradiation with a proton beam of a gas-discharge CO_2 laser
MGU and IAE [2, 9] (V.M. Andriyakhin, V.D. Pismennyy, et al. 1970)	IIN pulsed reactor, nuclear reaction products $^3He(n,p)^3H$	$^3He\text{-}Hg$ (~0.5 atm); cell temperature 150 °C	Observed radiation power at λ = 615 nm substantially exceeds the level of spontaneous radiation, in the opinion of the authors
University of Florida, NASA Langley Research Center [2, 3, 59] (F. Allario, H.S. Rhoads, R.T. Schneider, 1971)	Stationary reactor, nuclear reaction products $^3He(n,p)^3H$	$CO_2\text{-}N_2\text{-}^3He$ (0.008 atm)	A twofold increase in lasing power and efficiency was observed with irradiation of a gas-discharge CO_2 laser
University of Illinois [60] (T. Ganley, J.T. Verdeyen, G.H. Miley, 1971)	TRIGA reactor, α particles (layer ^{10}B)	$CO_2\text{-}N_2\text{-}He$ (≤1 atm)	An increase in lasing power was observed with irradiation of gas-discharge CO_2 lasers at both low and high pressures
VNIIEF (A.A. Sinyanskiy, A.T. Kazakevich et al. 1971)	VIR-2 reactor, fission fragments (layer ^{235}U)	Ar, Xe (≤0.5 atm) $CO_2\text{-}(N_2)\text{-}He$ (≤1 atm)	No lasing

[28] is dedicated to the energy and threshold characteristics only gas NPLs excited with the use of pulsed reactors, and to certain problems of the RL.

To create effective NPLs, it is necessary to use high pressures of the gas medium, close to atmospheric, because it is at such pressures that effective absorption of the kinetic energy of nuclear particles (uranium fission fragments, α particles) becomes possible at the path length of several centimeters, which is comparable with the characteristic transverse dimensions of laser cells. This is why, in the initial stages of research into NPLs, when atmospheric pressure gas lasers were still unknown, primary attention was given to liquid- and solid-state laser media. Studies dealt primarily with the condensed media known at the time: ruby and neodymium lasers, liquid lasers based on organic solvents activated with europium, and inorganic aprotonic solvents activated with neodymium. Attempts to achieve lasing during pumping of condensed media with nuclear radiation did not yield positive results, although in VNIIEF's experiments with the pulsed TIBR reactor, superluminescence was registered ($\lambda \sim 610$ nm) with the use of a solution of $Eu(BTFA)_4$ HDPhH in acetone [6, 7]. The principal reason for the absence of lasing with the use of condensed media was their radiation damage: radiation defects of the crystal lattice in solid-state lasers, radiolysis and the formation of gas bubbles on the tracks of nuclear particles in liquid lasers. The problems of such NPLs are examined in detail in Chap. 11.

Interest in the study of gas NPLs grew in the late 1960s, when powerful gas-discharge CO_2 lasers ($\lambda = 10.6$ μm) were developed with a gas pressure of around 1 atm (for example, see monograph [56]) and lasing was obtained in the IR spectral range with excitation of atmospheric-pressure rare-gas mixtures with an electron beam [10]. The results of experiments performed before 1972, in which attempts were made to pump gas media with nuclear radiation, are shown in Table 1.2. As is noted in the review paper [2], none of these experiments unequivocally demonstrated the presence of lasing, although indirect evidence was cited. Table 1.2 also includes experiments on the study of the influence of nuclear radiation on the parameters of gas-discharge CO_2 lasers, which can be viewed as a preliminary phase on the path to finding gas NPLs.

The initial unsuccessful attempts at direct pumping of gas media with nuclear radiation did not stop studies to find gas media for NPLs, and soon after, in 1972 [6, 11] (Russia, VNIIEF, He-Xe mixture) and in 1974 [12, 13] (United States, LANL, SF_6-C_2H_6) lasing was achieved when gas media were pumped with uranium fission fragments and with γ-radiation from a nuclear explosion, respectively.

1.2 Methods of Excitation. Pumping Sources

As noted above, NPLs include lasers with active media that are excited directly using nuclear radiation, or with the use of intermediate nuclear-optical converters. There is also another method of pumping gas lasers using nuclear radiation, in which this radiation is used only for pre-ionization of the laser medium. In this case, the nuclear radiation performs an auxiliary function (creating the conditions for uniform volume pumping the gas medium), while direct pumping occurred through gas discharge. Such lasers (sometimes they were called electroionization lasers), which radiate at transitions of CO_2, CO, XeF molecules and of Xe and Ar atoms, are not classified as NPLs, and are not considered in this book. Information about them may be found, for example, in the survey papers [54, 56, 61].

There are three basic sources of nuclear radiation which can be used to pump NPLs:

1. Nuclear explosive devices.
2. Radioactive isotopes.
3. Laboratory neutron sources.

Nuclear Explosive Devices

Thanks to the high fluxes of nuclear radiation and the short pulse duration (~10 ns), when nuclear explosive devices are used it is possible to obtain high specific power depositions ($q \sim 10^9$ W/cm^3 for gas NPLs). Laser media are pumped in this case basically by electromagnetic radiation in the X-ray and γ-ray spectral range. Studies to develop NPLs using nuclear explosive devices were carried out up until 1987, when a ban was introduced on underground testing. The results of these experiments and possible applications of such NPLs are examined in Chap. 12.

Radioactive Isotopes

Laser media are not yet pumped by the radiation of radioactive isotopes. Radioactive isotopes were previously used for an auxiliary purpose—pre-ionization of the active media of CO_2 lasers [62], as well as lasers at transitions of Xe [63] and Ne [64] atoms.

In the case of the use of radioactive isotopes directly for pumping of laser media, the specific power deposition will be higher for isotopes with short half-lives, which naturally reduces the lifespan of such a laser. As data from study [65] show, the most suitable isotopes for pumping gas NPLs are the α active isotopes ^{210}Po, ^{242}Cm, and the spontaneously fissionable isotope ^{52}Cf. Some characteristics of these isotopes are shown in Table 1.3.

Calculations of specific power deposition were made in [65] for cylindrical laser cells with internal diameters of 1–8 cm, on the internal surface of which layers of isotopes were deposited with a thickness equal to one-half the particle path length in the layer material. Calculations showed that depending on the cell diameter and the argon pressure (0.25 and 0.5 atm), for the isotopes ^{210}Po and ^{242}Cm, the specific power depositions vary in a range of $q = 0.06$–0.1 W/cm^3, and for ^{52}Cf, $q < 0.02$ W/cm^3. With such small specific power depositions, it is possible to achieve lasing using only the most low-threshold active media of NPLs, for example, Ar(Kr)-Xe mixtures, radiating at the infrared transitions of the Xe atom (see Chap. 3, Sect. 3.1).

Study [66] examined the possibility of radioisotope pumping of a neon laser with a He-Ne-H_2 mixture ($\lambda = 585.3$ nm). When a ^{210}Po layer 65 mg/cm^2 thick used, and is deposited to the internal surface of a cylinder cell with a diameter of 15 mm, it is possible to obtain $q \leq 0.6$ W/cm^3. However, the conclusion of the authors of [66] regarding the possibility of achieving lasing under such conditions appears

Table 1.3 Characteristics of radioactive isotopes

Isotope (half-life)	Particle energy, MeV	Specific activity, Ci/g	Specific energy release, W/g
^{210}Po (138 days)	5.3 (α particle)	4,490	140
^{242}Cm (163 days)	6.1 (α particle)	3,300	120
^{252}Cf (2.65 years, α decay); (85 years, spontaneous fission)	6.1 (α particle)	540	20
	180 (two fragments)	17	11

extremely optimistic, since NPLs with He-Ne-H_2 mixture have a rather high lasing threshold (see Chap. 3, Sect. 3.2).

Apart from the thin radioisotope layers, for excitation of gas media it is possible to use gaseous isotopes, for example, β radiators ^{42}Ar, ^{85}Kr, etc. In this case, one should consider gas media having Ar or K as their basic components (for example, Ar-Xe or Kr-Xe). Calculations [54, 63] show that the isotopes ^{42}Ar and ^{85}Kr can for now be used only for pre-ionization of active media of gas-discharge lasers, since in this case the specific power deposition does not exceed 0.001 W/cm^3.

Laboratory Neutron Sources

Of the powerful neutron sources, the most widespread are the nuclear reactors, which were used to perform the bulk of investigations to find active NPL media and to study their characteristics. Nuclear reactors are sources not only of neutron radiation, but also of γ radiation; however, for pumping NPLs, neutrons are used, since in this case it is possible to obtain a specific energy deposition to the laser medium approximately two orders of magnitude greater than the energy deposition from γ radiation.

In experiments investigating NPLs with help of reactors, as a rule direct pumping of active media is carried out using not neutron radiation, but the products of exothermal nuclear reactions, which take place during interaction of neutrons with the nuclei of ^{235}U, ^{10}B and ^{3}He (Table 1.4).

For effective excitation of the gas medium, it is necessary for the isotopes that interact with the neutrons to be in immediate contact with the laser medium. When nuclear reactors are used as neutron sources, two basic types of laser-medium excitation are utilized (Fig. 1.1): (1) a gaseous isotope or compound thereof (^{3}He, $^{235}UF_6$) is a component part of the laser medium; (2) the internal surface of the gas-filled laser cell is coated with a thin layer of isotope (^{10}B, ^{235}U) or compound thereof ($^{235}UO_2$, $^{235}U_3O_8$). In studies published up to this time, both methods are used. From the comparison made in the first section of Chap. 7 of the efficiency of pumping gas NPLs using the isotopes ^{3}He, ^{235}U, and ^{10}B, it follows that approximately identical energy contributions to gas media can be obtained using the

Table 1.4 Nuclear reactions for NPL pumping

Isotope (energy of reaction, MeV)	Natural composition of isotope	Cross-section of reaction for thermal neutrons, barns	Reaction products	Kinetic energy of reaction products, MeV	Path length of reaction products in air at 1 atm pressure, cm
^{3}He (0.76)	^{4}He (100 %)+ +^{3}He (0.00014 %)	5,400	^{1}H ^{3}H	0.57 0.19	1.0 0.2
^{10}B (2.3)	^{11}B (80.4 %)+ ^{10}B (19.6 %)	3,800	^{4}He ^{7}Li	1.5 0.8	0.9 0.4
^{235}U (167)	^{238}U (99.28 %)+ ^{235}U (0.72 %)	580	Light fragment Heavy fragment	99 68	2.3 1.8

gaseous isotope ^{3}He and a thin layer of ^{235}U. However, in the first case, it is necessary to use only ^{3}He as the buffer gas; this substantially limits the possibilities of gas mixture selection. When thin uranium layers are used, the maximal specific power deposition ($q \sim 5 \times 10^3$ W/cm^3 for gas media) is achieved in experiments with pulsed reactors with a minimal pulse duration of ~100 μs. This pumping method is interesting because on its basis, it is possible to create powerful nuclear-laser units (reactor-lasers) in the core of which uranium layers are used not only to excite the laser medium, but also as nuclear fuel.

The specific power deposition and uniformity of excitation of laser media depend on the magnitude and duration of the neutron flux, the pumping method, the geometry and dimensions of the laser cell, the type of nuclear particles, and the gas pressure.

In the case of a volumetric source of pumping using ^{3}He, the non-uniformity of pumping comes from the absorption of slow neutrons in ^{3}He and from the reduction of the energy contribution in the region near the wall owing to the removal of reaction products (^{1}H, ^{3}H) to the walls of the cell. Results of computation of the total energy deposition and spatial distribution of the deposited energy depending on the ^{3}He pressure and the diameter of the cylindrical cell are given in [67, 68], while [69] shows the results of computation of the energy deposition for a $^{235}UF_6$-He mixture. The use of the gaseous (at comparatively low temperatures) compound $^{235}UF_6$ makes it possible in principle to obtain an efficiency of nuclear energy deposition in the gas mixture of up to 100 %, but the search for nuclear-pumped gas media based on $^{235}UF_6$ to date has not yielded a positive result owing to the high rates of "quenching" of excited atoms by molecules of UF_6.

In the case of a surface pumping source, the area of uniform excitation is determined by the transverse dimension of the laser cell and the path length of fission fragments or α particles, which for various gases at atmospheric pressure is 1–10 cm. Computations of the total energy deposition and its spatial distribution for laser cells in the shape of a cylinder and a rectangular parallelepiped with layers of ^{235}U applied to the internal surface, depending on the thickness of the uranium

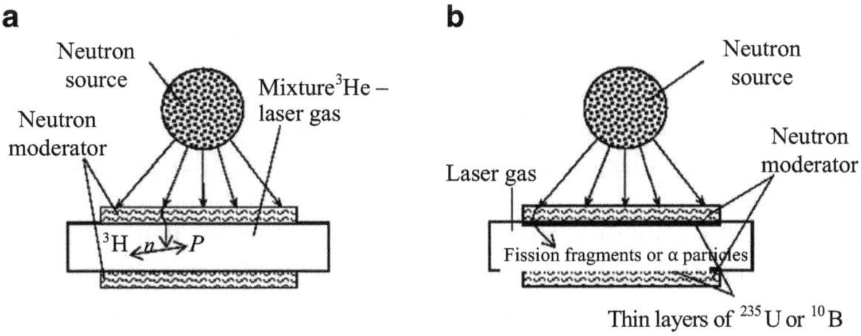

Fig. 1.1 Methods of exciting NPL gas media: (**a**) volumetric method of pumping, (**b**) surface method of pumping

layer, dimensions of the laser cell, and the gas pressure (as well as experimental methods of determining the energy deposition and the results of its measurement) are provided in Chap. 7. Here we note that when uranium layers are used, the efficiency of energy deposition from fission fragments in the gas medium does not exceed 50 % (an infinitely thin layer) and as a rule is 15–20 % when a thickness of uranium layer equal to one half of fragment path length in layer material (~5 mg/cm^2).

As was noted above, basically nuclear reactors were used as neutron sources in experiments with NPLs. Among nuclear reactors, pulsed aperiodic reactors [70, 71] with a pulse duration ranging from ~50 μs to ~10 ms possess the maximal possible neutron fluxes; they provide repeated and controlled fission bursts of uranium nuclei. The organization of experiments with NPLs using pulsed reactors is considered in Chap. 2.

Apart from aperiodic pulsed reactors, for the pumping of NPLs it is also possible to use periodic pulsed reactors [72], although the neutron fluxes which they supply are roughly 10–50 times less than in the case of aperiodic reactors. The review [28] mentions an experiment conducted in 1985 by associates of MIFI and IOFAN using the pulsed periodic reactor IBR-30 (Joint Institute of Nuclear Research (OIYaI, Dubna)). This experiment, carried out using gas mixtures He-Ne-Ar and He-Ar-Xe, did not yield a positive result, which may be explained, the authors believe, by the low power deposition.

Stationary nuclear reactors have substantially lower neutron fluxes than pulsed reactors. In stationary research reactors such as the IRT-2000 and the VVR, the thermal-neutron flux densities are ~10^{13} cm^{-2} s^{-1} (specific power deposition of gas media up to ~1 W/cm^3), which is not sufficient for researching most NPLs, especially in the stage of the search for new laser media. Therefore, experiments with stationary reactors were directed chiefly at studying the spectral-luminescent plasma characteristics and electroionization lasers (see, for example, [54]). It should be noted that there are special SM-type stationary reactors [73], in which thermal-neutron flux densities in the central cavity of the reactor core reach

2×10^{15} cm^{-2} s^{-1}. However, experiments with NPLs with an SM reactor are hampered owing to the limited volume of space with such a high neutron flux.

Of the other possible laboratory neutron sources for NPL pumping, it was proposed that the neutron radiation from tokomaks [74] could be used. Devices based on high-density plasma (plasma focus, Z-pinch) [75] could also be used for this purpose.

The options for pumping NPLs using neutron sources that were mentioned previously use nuclear reactions that take place with the interaction of nuclei of certain isotopes with thermal neutrons. To increase the flux of thermal neutrons, laser cells were surrounded by a fast-neutron moderator (Plexiglas, polyethylene, graphite). The maximal pressures of gas media do not exceed 5–6 atm, which is due to the appearance of excessive non-uniformity of pumping with a growth of pressure because of the shortening of the path length of the nuclear reaction products or weakening of the thermal neutron flux. One method of uniform pumping of gas media (at pressures of tens and hundreds of atmospheres) and of condensed media is the use of the elastic scattering of slow neutrons on atoms (nuclei) of the medium. In this case, ionization and excitation of the gas medium are carried out by recoil nuclei. This NPL pumping method was first used at VNIIEF [76] and is considered in the third section of Chap. 3 of this book.

1.3 Problems in the Search for Laser Media

Thus the maximal specific power deposition of gas NPLs is no more than 5×10^3 W/cm^3 and is achieved in experiments using pulsed reactors at the maximal possible thermal-neutron flux densities of ~10^{17} cm^{-2} s^{-1}. Such a pumping level is substantially lower than when electron and ion beams are used, or in a pulsed gas discharge. [Note that while the pump power density is low, the NPL has a tremendous opportunity to achieve enormous energy densities if proper metastable species can be found. This led some later searches to consider lasers involving states like singlet-delta oxygen, $O_2(^1\Delta)$]. This circumstance hampers the search for laser transitions for NPLs, especially in the visible and ultraviolet spectral ranges, because the unsaturated gain of the laser medium is directly proportional to λ^2. If one considers that the frequency of realization of experiments with pulsed reactors as a rule does not exceed one pulse per day (~100 pulses/year), and pulsed reactors are unique and potentially dangerous, then experiments to find and study active media for NPLs are a complicated and expensive proposition.

Significant assistance in the preliminary selection and study of active media for NPLs is provided by experiments using other, more accessible and safer sources of ionizing radiation operating at high frequency: high current electron and ion accelerators [77, 78], and accelerators with large cross-section beams that have been developed for laser pumping [79]. The kinetics of plasma processes in laser media excited by various types of ionizing radiation (γ radiation, fission fragments, and other products of nuclear reactions, fast electrons, and ions) are practically identical, so that laser characteristics depend not on the type of ionizing particles, but on the energy deposition to the laser medium.

At present, electron and ion accelerators operate in a wide range of pulse durations from ~10 ns to stationary mode with electron and ion energies from ~0.1 MeV to ~100 MeV. The specific power depositions of gas media in pulsed mode can reach 10^9 W/cm^3 [80], and in stationary mode 10 W/cm^3 [81], which makes it possible to model NPL pumping in a wide range of conditions, from nuclear explosions to stationary nuclear reactors.

From investigations with the use of electron and ion beams related to NPLs, one should note the study [10], in which pumping of lasers with mixtures of He-Ne, He-Xe, Ne-Xe, and Ar-Xe at atmospheric pressure by ionizing radiation was carried out for the first time (in this case with an electron beam), as well as studies performed at FIAN [82–84], the Institute of High Current Electronics of the SO RAN [85–87], the Scientific Research Center for Technological Lasers [88, 89], and the Institute of Electrophysics of the Ural Division of the Russian Academy of Sciences [90, 91]. Abroad, the most interesting research was carried out in the United States [92–95], Germany [81, 96], and The Netherlands [97, 98]. A detailed description of gas lasers excited by electron and ion beams is not the purpose of this book. Nonetheless, in the following chapters, in the discussion of NPL characteristics, data acquired using electron and ion beams will be cited as needed.

References

1. L.O. Herwig, Concepts for direct conversion of stored nuclear energy to laser beam power. Trans. Am. Nucl. Soc. **7**(1), 131 (1964)
2. K. Thom, R.T. Schneider, Nuclear Pumped Gas Lasers. AIAA J. **10**(4), 400–406 (1972)
3. N.W. Jalufka, Direct nuclear-pumped lasers. *NASA Technical Paper,* 2091, 1983
4. E. Matovich, In pursuit of a pulsed homogeneous nuclear laser. IEEE J. Quantum Electron. **QE-4**(5), 379 (1968)
5. V.E. Derr, G.T. McNice, P.M. Rushworth, Application of nuclear radiation to the pumping of lasers, in *Radioisotopes for Space. Part 2: Systems and Applications*, (Plenum Press, 1966), pp. 309–346
6. A.A. Sinyanskiy, *Research on continuous-action nuclear-laser devices at VNIIEF* (Issledovaniya po sozdaniyu yaderno-lazernykh ustroystv nepreryvnogo deystviya vo VNIIEF). Proceedings of the 2nd International Conference Physics of Nuclear-Excited Plasma and Problems of Nuclear-Pumped Lasers, Arzamas-16, 1995, vol. 1, pp. 16–36
7. T.A. Babicheva, A.M. Voinov, L.Ye. Dovbysh, L.M. Pavlovskaya, A.A. Sinyanskiy, *Nuclear-pumped lasers with liquid active media* (Lazery s yadernoy nakachkoy na zhidkikh aktivnykh sredakh). Proceedings of the Specialist Conference Physics of Nuclear-Excited Plasma and Problems of Nuclear-Pumped Lasers (Fizika yaderno-vozbuzhdayemoy plazmy i problemy lazerov s yadernoy nakachkoy), vol. 3, (Obninsk, 1993), pp. 146–155
8. A.N. Sizov, *Propagation of light through a liquid excited by fission fragments* (Rasprostraneniye sveta cherez zhidkost, vozbuzhdayemuyu oskolkami deleniya). Proceedings of the 2nd International Conference Physics of Nuclear-Excited Plasma and Problems of Nuclear-Pumped Lasers, Arzamas-16, 1995, vol. 1, pp. 397–398.

9. V.M. Andriyakhin, V.V. Vasiltsov, S.S. Krasilnikov, V.D. Pismennyy, V.Y. Khvostionov, On a radiation of the gas mixture Hg-^{3}He irradiated by a neutron beam (Ob izluchennii gazovoy smesi Hg-^{3}He, obluchayemoy potokom neytronov). Pisma v ZhETF **12**(2), 83–85 (1970)
10. G.G. Dolgov-Savelyev, B.A. Knyazev, Y.L. Kozminykh, V.V. Kuznetsov, A.M. Orishich, Pulsed lasing in rare gases at a pressure of up to one atmosphere with fast electron beam pumping (Impulsnaya generatsiya v inertnykh gazakh pri davlenii do odnoy atmosfery s nakachkoy puchkom bystrykh elektronov). Zhurnal prikladnoy spectroscopii **12**(5), 930–933 (1970)
11. A.A. Sinyanskii, S.P. Melnikov, Research on development of continuous nuclear-laser setups in VNIIEF. Proc. SPIE **3686**, 43–55 (1998)
12. P.B. Lyons, J.S. Clarke, D.S. Metzger, Gamma initiated HF laser. IEEE J. Quantum Electron. **QE-10**(9), 736 (1974)
13. D.A. McArthur, Nuclear pumped lasers, in *Encyclopedia of Lasers and Optical Technology*, (San Diego, Academic Press, 1991) pp. 385–393
14. P.J. Ebert, L.J. Ferderber, H.A. Koehler, R.W. Kuckuck, D.L. Redhead, Amplified spontaneous emissions in xenon pumped by gamma rays. IEEE J. Quantum Electron. **QE-10**(9), 736 (1974)
15. L.I. Gudzenko, S.I. Yakovlenko, The atomic reactor-laser (Atomnyy reaktor-lazer). *Kratkiye soobshcheniya po fizike*. (2), 14–15 (1974)
16. S.P. Melnikov, A.A. Sinyanskiy, *On the ultimate efficiency of nuclear-pumped lasers* (O predelnom KPD lazerov s yadernoy nakachkoy). Proceedings of the Specialist Conference "Physics of Nuclear-Excited Plasma and Problems of Nuclear-Pumped Lasers, Obninsk, 1993, vol. 2, pp. 133–148
17. D.A. McArthur, P.B. Tollefsrud, Observations of laser action in CO gas excited only by fission fragments. Appl. Phys. Lett. **26**(4), 187–190 (1975)
18. H.H. Helmick, J.L. Fuller, R.T. Schneider, Direct nuclear pumping of helium-xenon laser. Appl. Phys. Lett. **26**(6), 327–328 (1975)
19. R.J. DeYoung, W.E. Wells, G.H. Miley, J.T. Verdeyen, Direct nuclear pumping of a Ne-N_2 laser. Appl. Phys. Lett. **28**(9), 519–521 (1976)
20. M.A. Prelas, M.A. Akerman, F.P. Boody, G.H. Miley, A direct nuclear pumped 1.45-μ atomic carbon laser in mixtures of He-CO and He-CO_2. Appl. Phys. Lett. **31**(7), 428–430 (1977)
21. M.A. Akerman, G.H. Miley, D.A. McArthur, A helium-mercury direct nuclear pumped laser. Appl. Phys. Lett. **30**(8), 409–412 (1977)
22. A.I. Pavlovskiy, Ye.K. Bonyushkin, V.V. Varaksin, G.S. Vinyarskiy, A.Ye. Lakhtikov, G.M. Mishchenko, A.P. Morovov, B.D. Urlin, *A study of pulsed HF lasers with gamma-radiation pumping* (Issledovaniye impulsnogo khimicheskogo HF-lasera s nakachkoy gamma-izlucheniyem). Doklady AN RF, 1993, vol. 331, No. 3, pp. 299–301
23. B.V. Alekhin, Ye.K. Bonyushkin, V.V. Varaksin, B.V. Lazhintsev, A.Ye. Lakhtnkov, A.P. Morovov, G.F. Nasyrov, V.A. Nor-Arevyan, A.I. Pavlovskiy, R.A. Orlov, V.G. Rogachev, V.B. Shlyakhovoy, *Traveling wave excimer laser with gamma-radiation pumping* (Eksimernyy laser begushchey volny s nakachkoy gamma-izlucheniyem). Proceedings of the Specialist Conference Physics of Nuclear-Excited Plasma and Problems of Nuclear-Pumped Lasers, Obninsk, 1993, vol. 3, pp. 272–273.
24. A.I. Miskevich, Toward a history of the development of the first nuclear-pumped visible range laser (K istoriy sozdaniya pervogo lazera vidimogo diapazona s yadernoy nakachkoy). Inzhenernaya fizika. (2), 6–13 (2007)
25. V.Yu. Matyev, *Gas dynamics of near-central ranges in sealed NPL channels* (Gazodinamika okolotsentralnykh oblastey v germetichnykh kanalakh LYaN). Proceedings of the 2nd International Conference Physics of Nuclear-Excited Plasma and Problems of Nuclear-Pumped Lasers, Arzamas-16, 1995, vol. 1, pp. 410–420
26. V.A. Kryzhanovsky, E.P. Magda, L.V. Semkov, V.F. Tausenev, *About the initial stage of NPL researches in the RFNC-ITP*, Proceedings of the 9th International Conference on Emerging Nuclear Energy Systems, 1998, Tel-Aviv, vol. 2, pp. 874–880

27. V.V. Borovkov, B.V. Lazhintsev, S.P. Melnikov, I.N. Mochkayev, V.A. Nor-Arevyan, A.A. Sinyanskiy, G.I. Fedorov, Investigation of optical non-uniformities in nuclear-pumped lasers (Issledovaniye opticheskikh neodnorodnostey v lazerakh s yadernoy nakachkoy). Izv. AN SSSR, Ser. fizicheskaya **54**(10), 2009–2015 (1990)
28. A.V. Karelin, A.A. Sinyanskiy, S.I. Yakovlenko, Nuclear-pumped lasers and physical problems of reactor-laser development (Lazery s yadernoy nakachkoy i fizicheskiye problemy sozdaniye reaktora-lasera). Kvantovaya elektronika **24**(5), 387–414 (1997)
29. A.M. Voinov, S.P. Melnikov, A.A. Sinyanskiy, A kinetic model of recombination lasers at xenon atom transitions (Kineticheskaya model rekombinatsionnykh lazerov na perekhodakh atoma ksenona). ZhTF **60**(10), 100–113 (1990)
30. P.P. Dyachenko, *On the status of research at the A.I. Leypunskiy state science center (GNTs) RF-FEI on direct conversion of the energy of nuclear fission into the energy of laser radiation* (O sostoyanii issledovaniy v GNTs RF-FEI im. A.I. Leypunskogo po pryamomu preobrazovaniyu energii deleniya yader v energiyu lazernogo izlucheniya). Proceedings of the 3rd International Conference Problems of Nuclear-Pumped Lasers and Pulsed Reactors, Snezhinsk, 2003, pp. 5–30
31. V.V. Borovkov, B.V. Lazhintsev, V.A. Nor-Arevyan, A.N. Sizov, A.A. Sinyanskiy, G.I. Fedorov, Specific features of gas dynamics of flowing-gas lasers excited by uranium fission fragments (Osobennosti gazodinamiki prokachnykh lazerov, vozbuzhdayemykh oskolkami deleniya urana). Kvantovaya elektronika **22**(12), 1187–1191 (1995)
32. A.N. Sizov, V.V. Porkhayev, Influence of the gas flow speed on the lasing power of flowing-gas lasers excited by uranium nuclear fission fragments (Vliyaniye skorosti gazovogo potoka na moshchnost generatsii protochnykh lazerov, vozbuzhdayemykh oskolkami deleniya yader urana). Kvantovaya elektronika **23**(6), 510–514 (1996)
33. A.N. Korzenev, Y.M. Limar, A.N. Sizov, A.A. Sinyanskiy, Distribution of intensity of a nuclear-pumped flow gas laser (Raspredeleniye intensivnosti protochnogo gazovogo lazera s yadernoy nakachkoy). Kvantovaya elektronika **35**(9), 795–798 (2005)
34. J.R. Torczynski, On the motion of a gas experiencing range-dependent volumetric heating. J. Fluid Mech. **201**, 167–188 (1989)
35. J.R. Torczynski, D.R. Neal, Transient gas motion in nuclear-reactor-pumped lasers: computational and experimental results. Nucl. Sci. Eng **113**(3), 189–206 (1993)
36. W.J. Alford, G.N. Hays, Measured laser parameters for reactor-pumped He/Ar/Xe and Ar/Xe lasers. J. Appl. Phys. **65**(10), 3760–3766 (1989)
37. M. Ohwa, T.J. Moratz, M.J. Kushner, Excitation mechanism of the electron-beam-pumped atomic xenon ($5d \rightarrow 6p$) laser in Ar/Xe mixtures. J. Appl. Phys. **66**(11), 5131–5145 (1989)
38. J.W. Shon, R.L. Rhoads, J.T. Verdeyen, M.J. Kushner, Short pulse electron beam excitation of the high-pressure atomic Ne laser. J. Appl. Phys. **73**(12), 8059–8065 (1993)
39. J.W. Shon, M.J. Kushner, Excitation mechanism and gain modeling of the high-pressure atomic Ar laser in He/Ar mixtures. J. Appl. Phys. **75**(4), 1883–1890 (1994)
40. G.H. Miley, *Overview of nuclear pumped lasers*, Proceedings of the Specialist Conference Physics of Nuclear-Excited Plasma and Problems of Nuclear-Pumped Lasers, Obninsk, 1992, vol. 1, pp. 40–53
41. A.V. Bochkov, V.A. Kryzhanovskiy, E.P. Magda, S.L. Mukhin, V.M. Murzin, A.E. Neznakhina, Quasicontinuous lasing at the 7^3S_1–6^3P transition of the mercury atom (Kvazinepreryvnaya generatsiya na 7^3S_1–6^3P perekhode atoma rtuti). Pisma v ZhTF **18**(7), 91–93 (1992)
42. W.J. Alford, G.N. Hays, M. Ohwa, M.J. Kushner, The effect of He addition on the performance of the fission-fragment excited Ar/Xe atomic xenon laser. J. Appl. Phys. **69**(4), 1843–1848 (1991)
43. G.A. Hebner, G.N. Hays, Fission-fragment excited xenon/rare gas mixtures. J. Appl. Phys. **73** (8), 3614–3636 (1993)
44. G.A. Hebner, G.N. Hays, Laser efficiency and gain of the 1.73 μm atomic xenon laser at high He/Ar buffer gas ratios. J. Appl. Phys. **74**(6), 3673–3679 (1993)

45. A.M. Voinov, A.A. Sinyanskiy, A.G. Vasilenko, O.A. Golubyeva, L.Ye. Dovbysh, V.V. Ivanov, M.I. Kuvshinov, B.V. Lazhintsev, A.Ye. Lakhtikov, S.P. Melnikov, A.B. Modenov, V.A. Nor-Arevyan, A.M. Pichugin, A.N. Pokalo, V.V. Porkhayev, Ya.A. Pospelov, A.V. Sinitsyn, I.G. Smirnov, S.L. Turutin, M.V. Khlestkov, *The experimental setup LM4/BIGR. Achieving lasing in an Ar-Xe mixture with a duration of more than one second* (Eksperimentalnyy kompleks LM-4/BIGR. Polucheniye generatsii v smesi Ar-Xe s dlitelnostyu bolee 1s). Proceedings of the 2nd International Conference Physics of Nuclear-Excited Plasma and Problems of Nuclear-Pumped Lasers, Arzamas-16, 1995, vol. 2, pp. 179–185
46. N.M. Barysheva, A.V. Bochkov, N.V. Bochkova, K.F. Grebenkin, E.V. Zaytsev, A.V. Lutsenko, V.A. Kryzhanovskiy, E.P. Magda, V.M. Murzin, A.E. Neznakhina, The first ultraviolet nuclear-pumped laser (Pervyy ultrafioletovyy laser s yadernoy nakachkoy). Pisma v ZhTF **22**(15), 83–86 (1996)
47. A.M. Voinov, S.V. Vorontsov, V.N. Krivonosov, S.P. Melnikov, A.A. Pikulev, V.T. Punin, A.N. Sizov, A.A. Sinyanskiy, S.L. Turutin, Studies Performed at VNIIEF to Investigate the Possibility of Creating a Reactor-Laser (Raboty, vypolnennyye vo VNIIEF, po issledovaniyu vozmozhnosti sozdaniya reaktora-lasera). Voprosy atomnoy nauki i tekhniki. Ser. Fizika yadernykh reaktorov. (2), 19–28 (2007)
48. A.N. Sizov, Yu.N. Deryugin, *Computation of spatial non-uniformities in cylindrical nuclear-pumped gas lasers* (Raschet prostranstvennykh neodnorodnostey v tsilindricheskikh gazovykh lazerakh s yadernoy nakachkoy). Theses of Reports of the Sixth All-Union Conference Laser Optics, Leningrad, 1990, p. 132.
49. V.Yu. Matyev, *Methods of computation of the energy deposition of ions in non-uniform media* (Metody rascheta energovklada ionov v neodnorodnykh sredakh). Proceedings of the Specialist Conference Physics of Nuclear-Excited Plasma and Problems of Nuclear-Pumped Lasers, Obninsk, 1993, vol. 2, pp. 79–88
50. V.Yu. Matyev, A.N. Sizov, The *influence of non-uniformities of the laser medium on gas-dynamic computations of nuclear-pumped lasers* (Vliyaniye neodnorodnostey lazernoy sredy na gazodinamicheskiy raschety lazerov s yadernoy nakachkoy). Proceedings of the Specialist Conference Physics of Nuclear-Excited Plasma and Problems of Nuclear-Pumped Lasers, Obninsk, 1993, vol. 2, pp. 209–218
51. Ya.V. Pupko, A digest of experimental studies on the non-uniformities of nuclear pumping of gas laser-active media (Obzor eksperimentalnykh rabot po neposredstvennoy yadernoy nakachke gazovykh lazerono-aktivnykh sred), FEI Preprint, No. 1245, Obninsk (1981)
52. R.T. Schneider, F. Hohl, Nuclear-pumped lasers. Adv. Nucl. Sci. Technol. **16**, 123–287 (1984)
53. A.I. Mis'kevich, Visible and near-infrared direct nuclear pumped lasers. Laser Physics **1**(5), 445–481 (1991)
54. E.G. Batyrbekov, Direct and combined nuclear pumped lasers, Preprint of IAENNC, Republic of Kazakhstan, Almaty (1994)
55. E.P. Magda, Powerful nuclear pumped lasers (review). Proc. SPIE **3574**, 93–103 (1998)
56. G.A. Mesyats, V.V. Osipov, V.F. Tarasenko, *Pulsed gas lasers (Impulsnyye gazovyye lazery)* (Nauka, Moscow, 1991)
57. J.R. Rusk, R.D. Cook, J.W. Eerkins, J.A. DeJuren, B.T. Davis, Research on direct nuclear pumping of gas lasers (DNPGL), Techical Report AFAL-TR-68-256, Northrop Laboratories (1968)
58. V.M. Andriyakhin, Y.P. Velikhov, S.A. Golubyev, S.S. Krasilnikov, A.M. Prokhorov, V.D. Pismennyy, A.T. Rakhimov, On the increase in lasing power of CO_2 lasers under the action of a beam of fast protons (O vozrastanii moshchnosti generatsii lazera na CO_2 pod vozdeystviyem puchka bystrykh protonov). Pisma v ZhETF **8**(7), 346–349 (1968)
59. F. Allario, H.S. Rhoads, R.T. Schneider, *CO_2 laser experiments using nuclear reactions as the ionization source*, Proceedings of the 2nd Symposium on Uranium Plasmas: Research and Applications, Atlanta, 1971, pp. 53–57

60. T. Ganley, J.T. Verdeyen, G.H. Miley, *Nuclear radiation enhancement of CO_2 laser performance*, Proceedings of the 2nd Symposium on Uranium Plasmas: Research and Applications, Atlanta, 1971, pp. 39–44
61. N.G. Basov, V.A. Danilychev, Condensed and compressed-gas lasers (Lazery na kondensirovannykh i szhatykh gazakh). UFN (Succ. Phys. Sci.) **148**(1), 55–100 (1986)
62. V.Y. Lavrentyuk, I.V. Podmoshenskiy, P.N. Rogovtsev, CO_2 laser with radioisotope pre-ionization (CO_2 lazer s radioizotopnoy predionizatsiyey). Pisma v ZhTF **9**(5), 284–288 (1983)
63. G.A. Batyrbekov, E.G. Batyrbekov, A.B. Tleuzhanov, M.U. Khasenov, Electrodischarge laser with radioisotope pre-ionization (Elektrorazryadnyy lazer s radioizotopnoy predionizatsiyey). ZhTF **57**(4), 783–785 (1987)
64. G.A. Batyrbekov, E.G. Batyrbekov, V.A. Danilychev, A.M. Nazarov, A.B. Tleuzhanov, M.U. Khasenov, An investigation of lasing at neon 3p-3s transitions pumped by independent discharge with radioisotope pre-ionization (Issledovaniye generatsii na 3p-3s perekhodakh neona pri nakachke samostoyatelnym razryadom s radioizotopnoy predionizatsiyey). Kvantovaya elektronika **16**(10), 2060–2062 (1989)
65. A.M. Voinov, A.I. Konak, S.P. Melnikov, A.A. Sinyanskiy, On the possibility of creating a continuous laser with a radioisotope pumping source (O vozmozhnosti sozdaniya nepreryvnogo lazera s radioizotopnym istochnikom nakachki). Kvantovaya elektronika **18** (11), 1297–1299 (1991)
66. Y.R. Shaban, T.P.R. Campos, A proposed continuous wave 585.4-nm $^4He/Ne/H_2$ gas laser mixture pumped by α-emitter radioisotope. Brazilian J. Phys. **27**(2), 129–134 (1997)
67. J.W. Wilson, R.J. DeYoung, Power density in direct nuclear-pumped ^3He lasers. J. Appl. Phys. **49**(3), 980–988 (1978)
68. A.A. Pikulev, The energy deposition in nuclear-pumped gas lasers based on helium-3 (Energovklad v gazovykh lazerakh s yadernoy nakachkoy na osnove geliya-3). ZhTF **76** (10), 91–96 (2006)
69. J.W. Wilson, R.J. DeYoung, Power deposition in volumetric $^{235}UF_6$-He fission-pumped nuclear lasers. J. Appl. Phys. **49**(3), 989–993 (1978)
70. V.F. Kolesov. *Aperiodic pulsed reactors (Aperiodicheskiye impulsnyye reaktory)*. Sarov: RFYaTs-VNIIEF (1999)
71. A.M. Voinov, *Application of pulsed nuclear reactors for studying nuclear-pumped lasers* (Primeneniye impulsnykh yadernykh reaktorov dlya issledovaniya lazerov s yadernoy nakachkoy). Proceedings of the Specialist Conference Physics of Nuclear-Excited Plasma and Problems of Nuclear-Pumped Lasers, Obninsk, 1992, vol. 1, pp. 101–121
72. Y.P. Shabalin, *Pulsed Fast Neutron Reactors (Impulsnyye reaktory na bystrykh neytronakh)* (Atomizdat, Moscow, 1976)
73. S.M. Feynberg, S.T. Konobeyevskiy, N.A. Dollezhal, I.Y. Yemelyanov, V.A. Tsykanov, Y.M. Bulkin, A.D. Zhirnov, A.G. Filippov, O.L. Shchipakin, V.P. Perfilyev, A.G. Samoylov, V.I. Ageyenkov, The SM 50 megawatt research reactor (Issledovatelskiy reaktor SM s moshchnostyu 50 MVt). Atomnaya energiya **8**(6), 493–504 (1960)
74. D.L. Jassby, Feasibility of laser pumping with neutron fluxes from present-day tokamaks. J. Appl. Phys. **61**(4), 1659–1661 (1987)
75. M. Ganciu, I. Tiseanu, V. Zoran, G. Musa, On the possibility of neutron pumping of penning plasma lasers using a plasma focus device. Rev. Roum. Phys. **37**(1), 41–42 (1992)
76. A.M. Voinov, A.S. Koshelev, S.P. Melnikov, A.A. Sinyanskiy, A quasicontinuous gas laser excited by fast neutrons (Kvazinepreryvnyy gazovyy lazer, vozbuzhdayemyy bystrymi neytronami). Pisma v ZhTF **16**(13), 86–89 (1990)
77. A.N. Didenko, B.P. Grigoryev, Y.P. Usov, *Powerful electron beams and their application (Moshchnyye elektronnyye puchki i ikh primeneniye)* (Atomizdat, Moscow, 1977)
78. V.M. Bystritskiy, A.N. Didenko, *Powerful ion beams (Moshchnyye ionnyye puchki)* (Energoatomizdat, Moscow, 1984)

79. S.P. Bugayev, Y.Y. Kreyndel, P.M. Shchanin, *Large-cross-section electron beams (Elektronnyye puchki bolshogo secheniya)* (Energoatomizdat, Moscow, 1984)
80. A.A. Kuznetsov, S.S. Sulakshin, On the excitation of gas lasers with heavy accelerated charged particles (O vozbuzhdenii gazovykh lazerov tyazhelymi uskorennymi zaryazhennymi chastitsami). ZhTF **55**(5), 864–868 (1985)
81. A. Ulrich, J.W. Hammer, W. Biermayer, Lasers pumped by ion beams. J. Appl. Phys. **63**(7), 2206–2211 (1988)
82. A.Y. Dudin, D.A. Zayarnyy, L.V. Semenova, N.N. Ustinovskiy, I.V. Kholin, A. Yu, Chugunov, electron-beam pumped lasers using He Xe, Kr, Ar Mixtures (Lazery s elektronno-puchkovoy nakachkoy na smesyakh He-Xe, Kr, Ar). Kvantovaya elektronika **18**(8), 921–925 (1991)
83. A.Y. Dudin, D.A. Zayarnyy, L.V. Semenova, N.N. Ustinovskiy, I.V. Kholin, A. Yu, Chugunov, Electron-beam pumped lasers using Xe, Kr, and Ar mixtures with two-component buffer gases (Lazery s elektronnopuchkovoy nakachkoy na smesyakh Xe, Kr i Ar s dvukhkomponentnymi bufernymi gazami). Kvantovaya elektroniki **18**(11), 1290–1294 (1991)
84. I.V. Kholin, Powerful infrared Ar-Xe high-pressure lasers (Moshchnyye infrakrasnyye Ar-Xe-lazery vysokogo davleniya). Kvantovaya elektroniki **33**(2), 129–141 (2003)
85. N.N. Koval, Y.Y. Kreyndel, G.A. Mesyats, V.S. Skakun, V.F. Tarasenko, V.S. Tolkachev, A.V. Fedenev), A.A. Chagin, P.M. Shchanin, Lasing in rare gases with large cross-section electron beam pumping with a current pulse duration of up to 2.5 ms (Generatsiya v inertnykh gazakh pri nakachke elektronnym puchkom bolshogo secheniya s dlitelnostyu impulsa toka do 2.5 ms). Pisma v ZhTF **12**(1), 37–42 (1986)
86. F.G. Goryunov, V.I. Derzhiyev, A.G. Zhidkov, A.V. Karelin, D.Y. Nagornyy, V.S. Skakun, V.F. Tarasenko, A.V. Fedenev, S.I. Yakovlenko, He-Cd high-pressure laser pumped by a nanosecond electron beam (He-Cd-lazer vysokogo davleniya, nakachivayemyy nanosekundnym elektronnym puchkom). Kvantovaya elektronica **16**(10), 2039–2046 (1989)
87. A.V. Fedenev, V.F. Tarasenko, Simulation of NPL in experiments with E-beam pumping. Laser and Particle Beams **16**(2), 327–380 (1998)
88. B.M. Berkeliyev, V.A. Dolgikh, I.G. Rudoy, A.M. Soroka, Simultaneous effective lasing in the near IR range of the spectrum at Ar and Xe *d-p* transitions (Odnovremennaya effektivnaya generatsiya v blizhney IK oblasti spektra na *d-p* perekhodakh Ar i Xe). Pisma v ZhTF **17**(21), 80–82 (1991)
89. B.M. Berkeliyev, V.A. Dolgikh, I.G. Rudoy, A.M. Soroka, On the maximal efficiency of lasers based on the *d-p* transitions of heavy rare atoms (O predelnoy effektivnosti lazerov na *d-p*-perekhodakh tyazhelykh inertnykh atomov). Kvantovaya elektronica **17**(12), 1537–1538 (1990)
90. V.A. Makeyev, Y.N. Novoselov, M.Y. Starovoytov, V.V. Uvarin, Cadmium Ion UV lasing excited by a microsecond electronic beam. Pisma v ZhTF **19**(9), 10–13 (1993)
91. Y.N. Novoselov, V.V. Uvarin, Quasistationary ultraviolet lasing in a high-pressure helium-cadmium laser (Kvazistatsionarnaya ultrafioletovaya generatsiya v geliykadmiyevom lazere vysokogo davleniya). Pisma v ZhTF **21**(23), 15–18 (1995)
92. R.L. Watterson, J.H. Jacob, Measurements of intrinsic efficiency and parameters of an electron beam pumped ArXe laser. IEEE J. Quantum Electron. **26**(3), 417–422 (1990)
93. A. Suda, B.L. Wexler, K.J. Riley, B.J. Feldman, Characteristics of the high-pressure Ar-Xe laser pumped by an electron beam and an electron-beam sustained discharge. IEEE J. Quantum Electron. **26**(5), 911–921 (1990)
94. T.T. Perkins, Steady-state gain and saturation flux measurements in a high efficiency, electron-beam-pumped, Ar-Xe laser. J. Appl. Phys. **74**(8), 4860–4866 (1993)
95. E.L. Patterson, G.E. Samlin, Long-pulse, electron-beam-pumped, atomic xenon laser. J. Appl. Phys. **76**(5), 2582–2587 (1994)
96. A. Ulrich, B. Busch, W. Krötz, G. Ribitzki, J. Wieser, D.E. Murnick, *Heavy ion beam pumping as a model for nuclear pumped lasers*, Proceedings of the Specialist Conference Physics of

Nuclear-Excited Plasma and Problems of Nuclear-Pumped Lasers, Obninsk, 1992, vol. 1, pp. 54–64
97. P.J.M. Peters, M. Qi-Chu, W.J. Witteman, Near infrared lasing transitions in Ar, Kr, and Xe atoms pumped by a Coaxial E-beam. Appl. Phys. B **47**(2), 187–190 (1988)
98. S.W.A. Gielkens, W.J. Witteman, V.N. Tskhai, P.J.M. Peters, The optimization of the multi-atmospheric Ar-Xe laser. IEEE J. Quantum Electron. **34**(2), 250–259 (1998)

Chapter 2
Organization of Experiments on Pulsed Reactors to Seek and Study Nuclear-Pumped Lasers

2.1 Specifics of Experiments on Pulsed Reactors

Virtually all of the experimental investigations to seek and study the characteristics of NPLs excited by the products of nuclear reactions were carried out on pulsed nuclear reactors. Pulsed nuclear reactors are distinguished for the composition and structure of their core, the duration and fluence of the neutron pulse, the volume and configuration of the space for the irradiation, and the pulse repetition frequency. Some aspects of the use of pulsed reactors for laser pumping were examined in the survey [1]. This section uses some of the data from study [1], supplemented by information from other sources.

By now—in Russia and elsewhere—more than 10 pulse reactors and a number of modifications have been developed and are in operation. More information about pulse reactors and their characteristics is contained in monograph [2]. The fuel for such reactors includes: metallic highly enriched ^{235}U, uranium-molybdenum alloys, a triple alloy of uranium-zirconium-hydrogen, uranium-aluminum alloys, aqueous solutions of uranium salts, etc. Starting in the 1950s, pulsed reactors were intensively utilized in various fields of science and technology as powerful sources of neutron and γ radiation. The duration of neutron pulses varied from 30 μs to ~0.1 s at neutron flux densities in experimental channels of 10^{17}–10^{19} $cm^{-2} \times s^{-1}$ and neutron fluences of 10^{13}–10^{17} cm^{-2}.

The most complete information about pulsed reactors that operate (or operated) in various countries, the operating principles of these devices, and their characteristics and application, is contained in the monograph [2]. In Russia, pulsed reactors are developed and operated primarily at VNIIEF and VNIITF [2, 3]. At various times in Russia, the pulsed reactors VIR-1, VIR-2, TIBR, BR-1, BIGR (VNIIEF), IIN-3 (IAE), EBR-L (VNIITF), BARS-6 (FEI) have been used, and outside Russia—pulsed reactors such as the Godiva, SPR, TRIGA, ACRR, APRF, etc., which have similar characteristics. In 2003 reports were issued about experimental NPL investigations in China on the CFBR-II reactor [4]. Table 2.1 shows the

S.P. Melnikov et al., *Lasers with Nuclear Pumping*,
DOI 10.1007/978-3-319-08882-2_2

Table 2.1 Pulsed reactors [2, 3] used to investigate NPLs

	Reactor	$\tau_{1/2}$, ms	Q, MJ	F_{max}, 10^{14} cm^{-2}	Field of application
Fast neutron reactors	TIBR, Russia	0.5	8	10	NPLs operating on transitions of Ar, Kr, Xe atoms [5]
	BR-1, Russia	0.05–0.08	12	10	Gas NPLs pumped with fast neutrons [6, 7]; neodymium laser pumped with radioluminescent radiation of scintillators [7, 8]
	BARS-1, Russia	0.05	2	5	NPLs operating on transitions of Cd^+ ion [9]
	EBR-L, Russia	0.2	4	3.5	NPLs operating on transitions of Ne, Ar, Xe, Cd, Hg atoms, Cd^+ and Zn^+ ions, and molecular ion N_2^+ [3, 10, 11]
	BARS-5, Russia (two reactor cores)	0.04	7(×2)	10	NPLs operating on transitions of Xe atom [12]
	BARS-6, Russia (two reactor cores)	~0.2	3(×2)	~3	NPLs operating on transitions of Xe atom [13, 14]
	BIGR, Russia	2.5–10^5	300	120	CW NPLs operating on transitions of Xe atom [15, 16]
	Godiva-IV, United States	0.03	2	3	NPLs operating on transitions of Xe atoms [17]
	SPR-II (SPR-III), United States	0.04 (0.08)	4 (10)	8 (6)	NPLs operating on transitions of Ne, Ar, Xe atoms, Hg^+ ion, and CO molecule [18, 19]; measurement of gain at UV transition of XeF molecule [20]
	APRFR, United States	0.07	5	5	NPLs operating on transitions of Ar, Kr, Xe, and Cl atoms and CO molecule [21, 22]
	HPRR, United States	0.05	3	~4	Search for lasing at initiation of chemical HF laser with nuclear radiation [23]
	CALIBAN, France	0.06	1.8	3.4	NPLs at C atom transitions [24]
	CFBR-II, China	0.19	0.4	0.9	NPLs at Xe atom transitions [4]

(continued)

Table 2.1 (continued)

	Reactor	$\tau_{1/2}$, ms	Q, MJ	F_{max}, 10^{14} cm^{-2}	Field of application
Thermal and intermediate neutron reactors	IIN-3, Russia	~1	~10	5	Search for lasing at transitions of Hg^+ ions [25]
	VIR-2 and VIR-2 M, Russia	3–30	60	7	NPLs operating on transitions of Ne, Ar, Kr, Xe, C, N, O, and Cl atoms and Cd^+ and Zn^+ ions [26–28]
	IGRIK, Russia	2.5	60	15	NPLs operating on transitions of Xe atom [11]
	TRIGA, United States	~10	~30	5–10	NPLs operating on transitions of Ne, Xe, N, C atoms; measurement of gain at UV transition of XeF molecule etc. [29, 30]
	ACRR, United States	7.0	300	60	NPLs operating on transitions of Xe atom [19]

Note: $\tau_{½}$ is the half-height pulse duration; Q is the maximal energy output in the reactor core; F_{max} is the maximal fluence of neutrons inside the reactor core

characteristics of certain of these reactors and experiments with NPLs conducted on them (the results of experiments are discussed in Chap. 3).

Initially the pulsed reactors were developed for testing various materials and electronic devices inside or close to the reactor core, so by no means were all of them adapted for conducting complex laser experiments. The reactor EBR-L [2, 3, 31], the dual-core reactor BARS-6 [3, 32], as well as the water reactor VIR-2 M [2, 26, 33], originally designed for general research uses and later adapted for experiments with NPLs, were perhaps the exceptions.

In experiments to find laser media for NPLs, fast neutron reactors were used as a general rule. Examples are the BR-1, EBR, Godiva, and SPR using small reactor cores (~30 cm) made of metallic ^{235}U or its alloys, with a duration of reactor pulse of 50–100 μs. This selection may be explained by the fact that such reactors assure a maximal neutron flux density, and accordingly, maximal specific power depositions in laser media, thus facilitating attainment of the lasing threshold in NPLs. The reactor core usually is set at a height of 1.5–2 m from the floor of the casemate, the thick walls of which serve as biological shielding. The fluxes of neutrons and γ quanta are maximum at the center of the reactor core and decrease roughly by an order of magnitude at its outside surface. Among the drawbacks of using such reactors include the large spatial non-uniformity of the neutron flux, which limits the length of the laser cells to ~50 cm. The exception is the two-core BARS-6 reactor, which can be used to uniformly irradiate laser cells up to 150 cm in length.

Different versions of the placement of NPLs in experiments using fast neutron-pulsed reactors are shown in Fig. 2.1. The version of Fig. 2.1a is the most widespread, because in this case the influence of NPLs on the parameters of the reactor is not significant. The maximal specific power deposition of the gas media

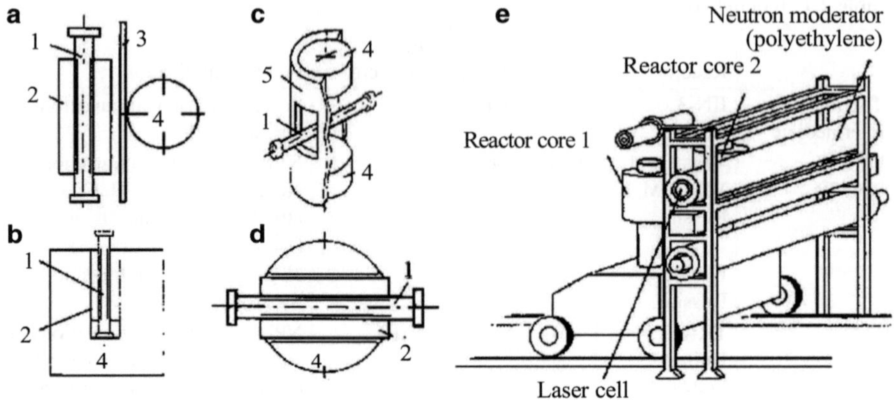

Fig. 2.1 Variations of the placement of laser cells in experiments in fast pulsed reactors: (1) NPL, (2) neutron moderator, (3) cadmium screen, (4) reactor core, (5) neutron reflector; (**a**) placement of NPL near the reactor core (TIBR, Godiva-IV, SPR-II, III, APRF); (**b**) NPL with neutron moderator inside the reactor core (SPR-II, experiments to measure the gain at UV transitions of the XeF molecule); (**c**) placement of NPL without moderator inside the reactor core in experiments to pump NPLs with fast neutrons (BR-1); (**d**) specialized EBR-L reactor; (**e**) specialized facility based on BARS-6 dual-core reactor

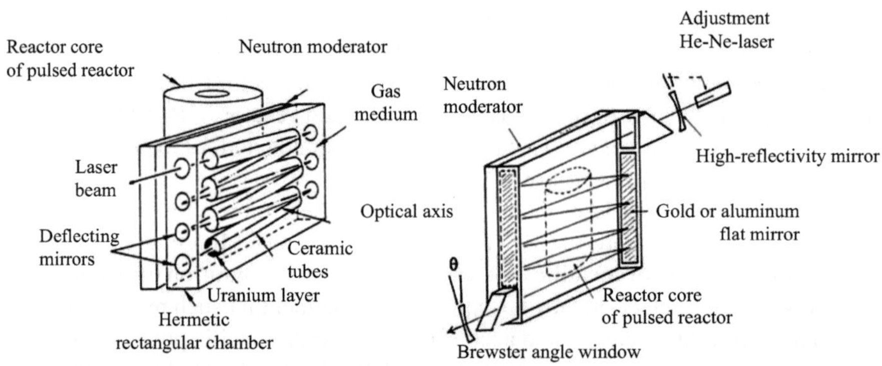

Fig. 2.2 Design layouts of two NPL versions with multipass cells: (a) laser with uranium layers [18, 34], (**b**) laser excited by products of the reaction of $^3He(n, p)^3H$ [35]

of up to 5×10^3 W/cm^3 is implemented with the arrangement of the laser cells inside the reactor core (Fig. 2.1b, d).

The active length and volume of the NPLs can be increased using a multipass laser cell. Two versions of such devices are shown in Fig. 2.2. In the first case [18, 34], up to six ceramic tubes were placed inside a rectangular stainless-steel chamber; layers of $^{235}U_3O_8$ were deposited to the interior surfaces of these tubes having a diameter of 2.54 cm. The full active length of such a laser reached 240 cm [34]. Flat gold-coated mirrors were used as the deflecting mirrors. Another version

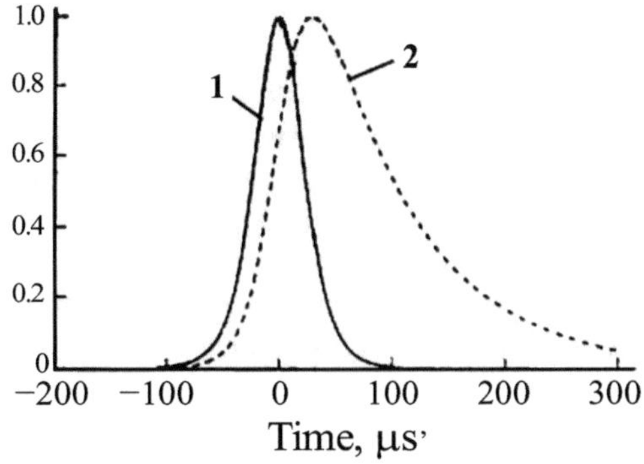

Fig. 2.3 Time dependencies of reactor power (1) and pumping power (2) when a fast neutron-pulsed reactor NPL is used for NPL pumping [36]

of the design [35], intended for excitation of gas media with nuclear reaction products $^3He(n,p)^3H$, differs in that there are no tubes with uranium layers, and rectangular plates with a gold or silver coating were used as the deflecting mirrors. In this case, when the angle between the cavity mirrors is changed, it is possible to vary the number of passes in the cell and consequently the active length of the laser.

Laser cells usually are surrounded by a layer of moderator (polyethylene, Plexiglas, water) 3–5-cm thick for softening the neutron spectrum, making it possible to increase the laser power deposition by a factor of 10–100 when using the nuclear reactions given in Table 1.4 (Chap. 1, Sect. 1.2). When pulsed reactors with a short pulse duration are used (50–100 μs), the time dependencies of the flux of fast neutrons (reactor power) and the pumping power of the laser medium do not coincide, which is related to the process of thermalization of the neutrons inside the moderator. Figure 2.3 shows the results of calculations [36] of the time dependence of the power deposition (flux density of the slowed neutrons) inside the moderator under the conditions of experiments [21] (thickness of the cylindrical polyethylene moderator 5 cm, duration of reactor pulse $\tau_{1/2} = 50$ μs). The use of a moderator spreads the pumping pulse duration to ~150 μs, and shifts the pulse maximum position on ~30 μs. The results of calculations agree with experimental data.

It is more convenient to optimize the design of the NPLs (composition and parameters of laser medium, design of cavity, methods of removing surplus heat) for the purpose of selecting the optimal variation of the cell of the stationary or quasi-stationary nuclear-laser facility using pulsed nuclear reactors based on thermal or intermediate neutrons, with the moderator in the reactor core: the VIR-2 M reactor with reactor core from a solution of ^{235}U salts in water, and the TRIGA pool-type reactor and its modification, the ACRR, the fuel elements of which are made from a uranium-zirconium-hydrogen alloy. These reactors are characterized by a long-duration neutron pulse (1–50 ms), with a maximal neutron fluence of ~5×10^{14} cm^{-2}, as well as the possibility of experimentation with laser cells up to 200-cm long.

The above experimental layouts were used primarily to search for active NPL media and study their characteristics (laser spectrum, energy characteristics, and laser thresholds). To study multichannel NPLs, VNIIEF created the experimental complex LM-4/BIGR [15, 16] and the IKAR nuclear-laser facility, which constitutes a model of the reactor-laser [37]. VNIITF and FEI are developing multichannel LIRA [38] and Stand B [39] facilities. Among the studies outside of Russia with which we are familiar, one can note the experiments on the ACRR reactor (Sandia Laboratories, United States) using the large-scale laser setup ALEC (Advanced Laser Excitation Cavity) [19]. The designs of such complex multichannel facilities and the basic results obtained from studying them are cited in Chap. 6.

2.2 Problems of NPL Radiation Resistance

In experiments with pulsed nuclear reactors, the laser cell and the measurement devices are in the zone of intensive radiation. This imposes a constraint on the choice of structural and optical materials and photodetectors, and also makes it necessary to take special steps to test the functionality of individual unit assemblies at the time they are affected by reactor radiation, and afterwards. It is precisely the inadequate radiation resistance that is the basic reason for the absence of lasing when solid and liquid lasers were pumped by nuclear radiation (see Chap. 11, Sect. 11.1).

All materials and devices used in experiments with NPLs must satisfy the requirements of radiation resistance. Special attention must be paid to the cavity mirrors and windows for output of laser radiation, which in contrast to photodetectors and other measurement devices, cannot be removed from the zone of intensive irradiation. Owing to this, in parallel with the studies of the NPLs themselves, the radiation resistance of their elements and auxiliary devices were also studied [18, 40].

Optical Materials

In the optical materials that are used as substrates for mirrors and laser output windows, under the influence of the reactor radiation, additional losses of light can occur, associated with the reduction in their transparency. As was demonstrated by the results of studies [40–42], the absorption coefficient depends on the type of optical material and the concentration of impurities in it, the temperature of the sample, the wavelength of the light radiation, the absorbed dose, and absorbed dose rate. Absorption of light radiation in optical materials occurs as a result of formation of color centers, which arise with the capture of charge carriers (electrons and holes) on structural defects of the material (for example, see [43, 44]). At the same time, the color centers are formed both as a result of the change of state of already existing defects and as a result of the onset of new defects.

In the majority of studies, materials were irradiated for a long time using stationary reactors, high power isotopic γ-irradiating sources, and their coefficients of absorption were measured before and after irradiation. The data obtained from stationary irradiation cannot be used to predict the value of the induced coefficient of absorption in the process of pulsed irradiation [41, 42], since frequently a large contribution to absorption is made by the color centers with short lifetimes. The coefficients of induced absorption for these two variants of irradiation can differ by a factor of 10 [42].

At VNIIEF, studies into the radiation resistance of optical materials to pulsed radiation have been carried out since the early 1970s. The methods of measuring induced absorption coefficients and radioluminescence of optical materials under the effects of reactor radiations are cited in study [40]. Basic attention was given to the change in optical properties of materials as a function of time during the reactor pulse.

The methods [40–42] used to measure the induced absorption coefficient in the process of pulsed irradiation (and for any time interval after it) are based on a very simple principle of measurement of the intensity of light radiation passed through the specimen prior to irradiation and any subsequent moment of time. CW lasers were used as the sources of light radiation; for example, a helium-neon or helium-cadmium laser, radiating at individual lines in the visible and IR regions of the spectrum, or a lamp with a continuous spectrum. In the latter case, light filters or a monochromator were used to isolate the probe light radiation in the narrow spectral range. To extract the useful signal against the background of various types of noise (including radiation noise in the photodetectors), the probe light signal was modulated.

Such methods make it possible to perform measurements of absorption coefficients simultaneously in several wavelengths. For example, Figure 2.4 shows an oscillogram of one of the experiments with the VIR-2 reactor to measure induced absorption coefficients simultaneously at three wavelengths [40]. The absorbed dose rate of γ radiation at the maximum of the reactor pulse was around 1×10^6 Gy/s. The contribution of the neutron radiation to the absorbed dose did not exceed 10 %. Induced absorption coefficients at wavelengths 633, 1,150, and

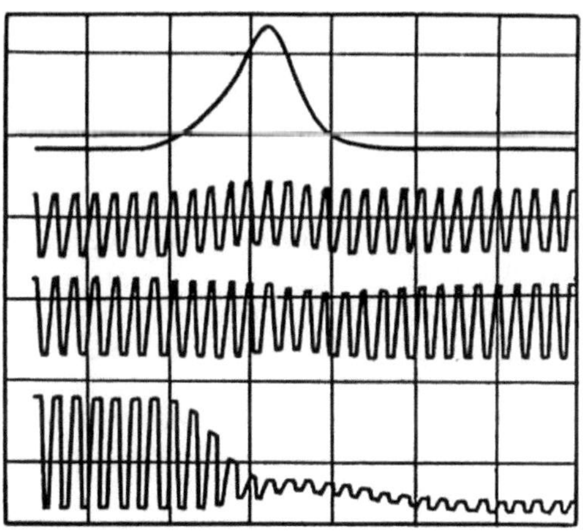

Fig. 2.4 Change in the transmission coefficient of a BaF_2 crystal 10 mm thick when irradiated by a pulse of n,γ radiation of a VIR-2 reactor [40]: (1) reactor pulse, (2–4) signals of modulated probe light at wavelengths of 3,390, 1,150 and 633 nm, respectively. Scale division is 5 ms

3,390 nm in the pulse maximum were 0.65, 0.27, and <0.01 cm^{-1} respectively. It is clear from the data of Fig. 2.4 that the induced absorption coefficient decreases with an increase in light wavelength. This principle is observed for all optical materials.

The results of measuring induced absorption coefficients for certain optical materials obtained in experiments [42] with the pulsed TRIGA reactor are shown in Table 2.2.

Experimental investigations of various optical materials showed that the greatest radiation resistance is possessed by silica glass. Detailed research on the common types of silica glass KU-1, KV, and KI were carried out in experiments with the VIR-2 reactor [41]. In the absorption spectra measured 1 h after pulse irradiation, all the types of quartz showed the band with a maximum at the wavelength of 215 nm, which is characteristic for SiO_2. Apart from the indicated band, the KI quartz (which lacks the hydroxyl group OH) had broad bands with maximums at wavelengths of 300 and 550 nm, while the KV quartz had a band with a maximum at a wavelength of 300 nm. The induced absorption coefficients measured in study [41] are shown in Table 2.3.

The value of the induced absorption coefficient depends greatly on the concentration of coloring impurities and hydroxyl OH. For the purest commercial glass, KU-1 (concentration of coloring impurities ~10 ppm, hydroxyl concentration ≤2,000 ppm), the induced absorption coefficient is minimal, which may be explained not only by the low concentration of coloring impurities, but also by the protective properties of the OH^+ ion.

An additional effect that can influence the recording instruments and the accuracy of measurement of the light intensity is the radioluminescence of optical

Table 2.2 Maximum induced absorption coefficients (cm^{-1}) at an absorbed dose rate of 7×10^4 Gy/s [42]

Material	Wavelength of probe light, nm		
	325	633	3,390
Fused quartz (Corning 7940)	0.0073	–	–
Pyrex (Corning 7740)	–	0.58	–
CsI (monocrystal)	–	0.067	<0.009
Sapphire	–	0.014	<0.008
Spinel	–	0.58	0.011
ALON	–	4.9	0.05

Table 2.3 Maximum induced absorption coefficients (cm^{-1}) at absorbed dose rate of around 1×10^6 Gy/s [41]

Type of silica glass	Wavelength of probe light, nm				
	400	500	600–700	900	1,150
KI	0.6	–	0.28	0.022	0.015
KV	0.36	–	0.18	<0.01	<0.01
KU-1	0.07	0.045	<0.003	<0.005	<0.005

Table 2.4 Maximum specific powers of radioluminescent radiation (mW/cm^3) at absorbed dose rates of 7×10^4 Gy/s [42]

Material	Visible range	IR range
CsI (monocrystal)	10,000	100
Sapphire	200	8–45
Spinel	–	<4
ALON	<50	<3

materials [40, 42]. The intensity of radioluminescent radiation depends on the type of optical material, its temperature, and the spectral range. Table 2.4 shows the specific powers of radioluminescent radiation for certain materials [42].

Optical Fibers

In experiments with pulsed and stationary reactors, optical fibers are used to withdraw the laser or luminescent radiation from the region close to the reactor core to the registering instruments located behind the biological shielding. The influence of reactor radiation on the properties of the optical fibers was examined in a number of studies (see for example [40, 45]).

Resolution of the problem of the fibers' radiation resistance is closely associated with development of radiation-resistant optical materials. For example, development of the method for manufacturing quartz with an extremely low content of impurities made it possible to produce fibers that can be used to withdraw light in the visible range at doses of ~10^7 Gy [46, 47], which corresponds to the annual radiation exposure inside the core of a stationary reactor. To increase radiation resistance of fibers, small controlled impurities can be added, for example, molecular hydrogen [47, 48].

Laser Receivers

Under the influence of reactor n,γ-radiation on the photoreceivers, additional radiation currents arise, integral and spectral sensitivities are altered, the level of intrinsic noise increases, and the shape of the volt–amp characteristic is changed [40]. The degree of influence of reactor radiation depends first of all on the type of receiver and the absorbed dose rate. Some results for the radiation resistance of semiconductor photoreceivers are shown in study [44], while study [40] shows the results of measurements of radiation currents for certain types of semiconductor photoresistors and photodiodes, as well as photomultipliers and photocells.

In experiments with NPLs with pulsed reactors, various types of receivers are used, which make it possible to measure the energy, power, and the spectral and time characteristics of laser radiation. Some of them (thermal gauges of laser pulse energy, photocells with an external photo effect) can be located at distances of 2–5 m from the core of the pulsed reactor with absorbed doses of 1–10 Gy (10^3–10^4 Gy/s), while others (photomultipliers and semiconductor photodiodes) are outside the boundaries of the reactor hall, where absorbed doses do not exceed ~0.01 Gy (~10 Gy/s). The possibility of using specific receivers to measure the characteristics of the NPLs must be tested in additional experiments.

2.3 Methods of First Experiments on VIR-2 and TIBR-1M (VNIIEF) Reactors

The first successful experiment to pump lasers with nuclear radiation occurred on May 12, 1972 at VNIIEF using a He-Xe mixture, but VNIIEF researchers were not able to publish their first article about NPLs until 1979 [5]. This article contained the results of later experiments, while information about first experiments was briefly presented later in a survey paper [7].

In the first series of experiments, the pulsed water reactor VIR-2 [2, 33], which was used for research from 1971 through 1978, was used as the neutron source. In 1979, the modified reactor VIR-2 M [2, 33], with a strengthened reactor core and minor changes in the parameters of the reactor pulse, was placed in operation. A solution of uranium salt (UO_2SO_4) in ordinary water served as the fuel in these reactors.

The arrangement of the VIR-2 and VIR-2 M reactors in the building is shown in Fig. 2.5. The reactor is located in the two-hall building having a concrete wall thickness of 2–6 m. The reactor core vessel (height 2 m; diameter roughly 0.7 m; wall thickness 65 mm) was enclosed in a concrete block measuring $4 \times 4 \times 3.5$ m, which was the biological shielding. The bottom of the vessel is at the level of the lower hall's ceiling, and can be closed by a protective shutter. The following experimental channels are used to locate the following exposed objects:

(a) A central channel with an internal diameter of 142 mm.
(b) A hemispherical cavity with an internal diameter of 300 mm.
(c) Side channels with a diameter of 100 mm abutting the side surface of the reactor core vessel.
(d) A cavity close to the surface of the reactor core with a cross-section of 560×620 mm^2.
(e) A lower reactor hall with a height of 2.5 m.

Figure 2.6 shows a diagram of the first experiments. The vertically arranged laser cell was irradiated in the side channel of the VIR-2 reactor. The average fluence of the thermal neutrons along the length of the cell was 1.3×10^{13} cm^{-2} with a reactor pulse duration of around 4 ms.

An aluminum tube was placed inside the cylindrical body of the cell with an internal diameter of 27 mm and a length of 100 cm. A layer of $^{235}U_3O_8$ was deposited on the tube's surface with a thickness equivalent to around 2–3 mg/cm^2

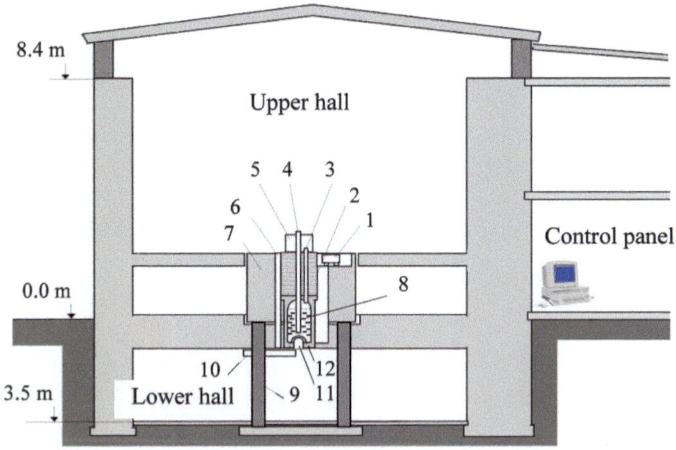

Fig. 2.5 Diagram of building for reactors VIR-2 and VIR-2 M: (1) protective shutter, (2) cavity near reactor core surface, (3) channel for pouring with solution, (4) central channel, (5) drives of control element, (6) side channel, (7) biological shielding, (8) fuel solution, (9) support columns, (10) protective shutter, (11) hemispherical cavity, (12) reactor core vessel

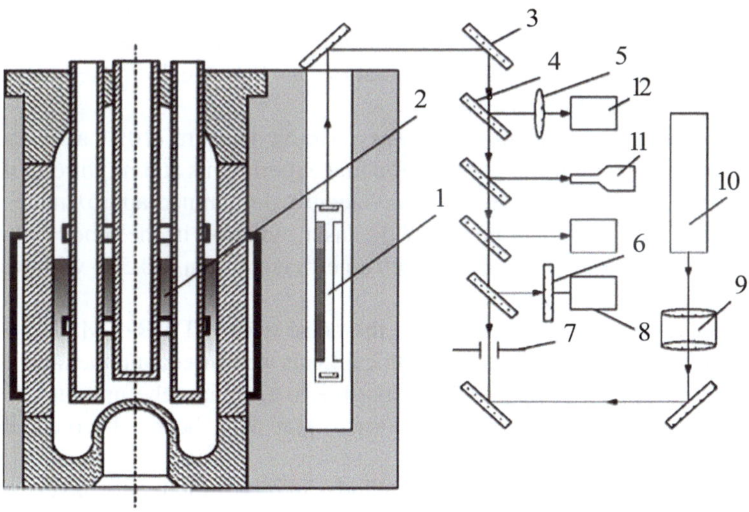

Fig. 2.6 Diagram of first laser experiments with VIR-2 reactor [7]: (1) laser cell, (2) reactor core, (3) deflecting mirrors, (4) semitransparent plates, (5) CaF_2, lens, (6) light filter, (7) diaphragm, (8) Ge:Au photoreceiver, (9) condenser lens, (10) LG-126 helium-neon laser, (11) FEU-28 photomultiplier, (12) IEK calorimeter

of metallic ^{235}U. To increase the flux of the thermal neutrons, the cell was enclosed by a polyethylene moderator with a wall thickness of 30 mm. The laser cavity was formed by mirrors with a silver coating: a spherical 100 % reflectivity mirror with a radius of curvature of 10 m and a flat output coupler on quartz or BaF_2 substrate.

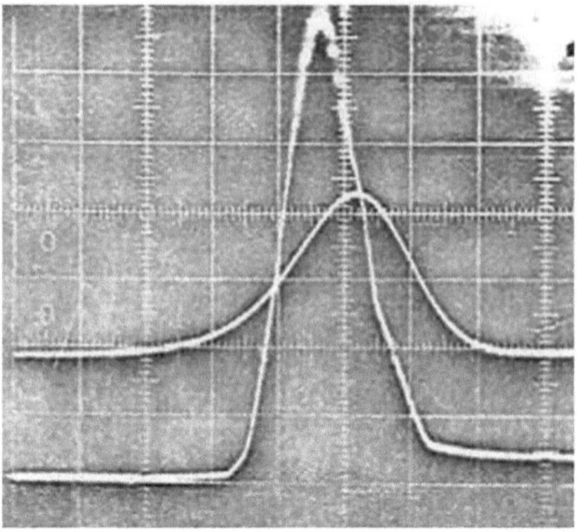

Fig. 2.7 Oscillogram of a reactor pulse (*upper trace*) and laser radiation pulse (*lower trace*) for He-Xe mixture at a pressure of 1 atm. Division value is 2 ms

The laser radiation is extracted from the cell through a coupling hole with a diameter of 1–2 mm in the flat mirror. The distance between the mirrors is 120 cm.

The measurement devices were kept away from the irradiation zone at a distance of around 10 m from the laser cell. In the searching experiments, a FEU-28 photomultiplier was used to register radiation in the range of 400–1,100 nm, while a liquid nitrogen-cooled Ge:Au photoresistor was used to register IR radiation (2–11 μm). After detection of laser radiation, an IEK-1 calorimeter was used to measure the energy of the laser pulse.

Initially, the active medium used was a He-Xe mixture (10:1) at pressures of 0.08–1 atm. An oscillogram of one of the first experiments at a mixture pressure of 1 atm is shown in Fig. 2.7. The laser wavelength, determined approximately using light filters was ~3 μm. The power of the laser radiation in the mode optimal for pressure and composition was 25 W with an efficiency of $\eta_l \sim 0.5$ % with respect to the energy absorbed in the mixture.

In the next series of experiments on the pulse reactor TIBR-1 M (1974–1976), uranium layers more resistant to mechanical loads were used, and the laser cell was arranged horizontally, which made it possible to eliminate the uranium-dust contamination of the lower spherical mirror that had been observed in prior experiments.

The core of the TIBR-1 M reactor includes a $ZrH_{1.9}$ moderator layer, which in comparison with other fast neutron reactors, leads to a reduction in dynamic loads on the fuel elements of the reactor core and to an increase in the duration of the reactor pulse to ~500 μs [2]. The diameter of the uranium-molybdenum alloy reactor core is around 30 cm.

The cylindrical laser cell, enclosed by a polyethylene moderator with a layer of $^{235}U_3O_8$ around 9 mg/cm^2 thick deposited to the internal surface, was arranged close to the surface of the reactor core (see Fig. 2.8). In order to reduce the effect of thermal neutrons emerging from the moderator, on the reactor core the laser cell was surrounded by screens made of cadmium and boron carbide.

Fig. 2.8 Laser cell near TIBR reactor core [7]

A diagram of the experiment is shown in Fig. 2.9. The measuring equipment (except for the IEK-1 calorimeter) was placed on the other side of the biological shielding in a neighboring room. To register the laser radiation, along with the IEK-1 calorimeter, Ge:Au photoresistors and DKPs surface-barrier silicon diodes were used [49]. The length of the uranium layer in the cell was 57 cm, with a diameter of 2.7 cm. The cell was irradiated with a pulsed flux of thermal neutrons with a half-height pulse duration of around 0.8 ms and an average fluence along the length of the uranium layer of 4.2×10^{13} cm^{-2}, which made it possible at a helium pressure of 2 atm to obtain a specific power deposition at the pulse maximum of $q = 600$ W/cm^3. In the experiments, binary mixtures of rare gases were studied: He-Ne (Ar,Kr,Xe), Ne-Ar(Kr,Xe), Ar-Kr(Xe), and Kr-Xe. The total pressure of the mixtures was equal to one atmosphere, and the partial ratio of components ranged from 200:1 to 200:30. Lasing was obtained with use of the mixtures He-Ar(Kr,Xe) and Ar-Kr(Xe) in a range of 2–10 μm, and the mixtures He-Ar and Kr-Xe in a range of 0.2–1.2 μm (see Fig. 2.10).

In this series of experiments [5, 50], the most intensive laser transitions were found for ArI ($\lambda = 1{,}15$; 2.40 μm), KrI ($\lambda = 2.52$ μm), XeI ($\lambda = 2.6$ μm), and thus the existence of the family of NPLs operating on IR transitions of atoms of rare gases was obtained. For the NPLs that were studied in greatest detail, based on mixtures He-Xe ($\lambda = 2.6$ μm) and He-Ar ($\lambda = 1.15$ μm), laser powers of 2,000 W and 250 W were found for $\eta_l = 0.8$ and 0.1 %, respectively.

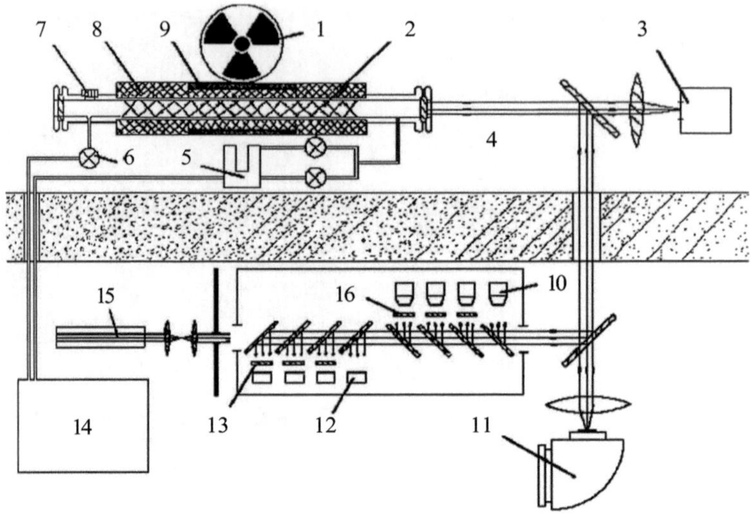

Fig. 2.9 Diagram of experiment with TIBR-1 M reactor: (1) reactor core, (2) aluminum tube with a layer of $^{235}U_3O_8$, (3) IEK-1 calorimeter, (4) biological reactor shielding, (5) diffusion pump, (6) electromagnetic valves, (7) pressure sensor, (8) polyethylene neutron moderator, (9) cadmium and boron-carbide screens, (10) Ge:Au photoresistors, (11) ISP-51 spectrograph, (12) DKPs silicon diodes, (13 and 16) light filters, (14) gas vacuum system, (15) adjustment laser

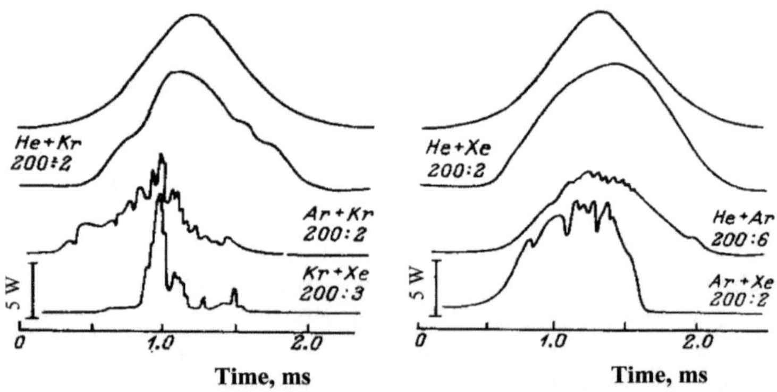

Fig. 2.10 Oscillograms of laser radiation pulses in a wavelength range of 2–10 μm with a total mixture pressure of 1 atm [5]. The upper beams are neutron pulses

2.4 Experimental Complex Based on the VIR-2M Reactor

Studies of nuclear-pumped lasers at VNIIEF were continued after the 1975 development of the laser setup LUNA-2 M, developed to perform experiments in the lower hall of the VIR-2 reactor (see Fig. 2.5), which provides convenient conditions to carry out laser experiments: a sufficient spatial volume to irradiate large-scale

(100–200 cm) laser setups and a circular rail track with remote-controlled cars that deliver prepared laser setups to the irradiation zone. The presence of a hemispherical cavity makes it possible to irradiate laser setups in the lower hall with a spatially localized beam of intense neutron radiation; at the same time, the equipment outside of the beam is in a significantly smaller flux of the reactor radiation.

The LUNA-2 M setup (Fig. 2.11a) is one of the three main laser setups that make up the experimental complex near the VIR-2 M reactor [26, 33]. When this setup was used, VNIIEF performed the bulk of the experiments to study the characteristics of NPL active media in the visible and IR range of the spectrum [27, 28], as well as the problem of extracting the radiation from laser channels [51–53].

The experimental complex also includes the LUNA-2P (Fig. 2.11b) and LYaN-2T setups [26, 28]. The basic characteristics of these setups are given in Table 2.5. Each setup consists of two laser cells (channels) that are identical in design and in irradiating conditions, making it possible to accelerate the rate of the investigations. The layout of the laser setup under the reactor core is shown in Fig. 2.12. The cars can be moved on the track and clamped under the reactor core remotely (from the control panel) or manually (directly in the lower reactor hall).

The LUNA-2P unit (Fig. 2.13) was developed in 1987 and was used to study laser characteristics and the optical quality of the gas medium in the gas flowing mode. The gas was circulated in the transverse direction relative to the optical axis at a speed of 5–10 m/s. Heat exchangers (radiators) were used to cool the gas—packets 100 cm long made of aluminum or beryllium plates 0.5-mm thick, with 0.5-mm gaps between them.

The LYaN-2T setup, developed in 1991 to study lasers operating on metal vapors and to study the influence of temperature on laser parameters, used quartz laser cells 3.2 cm in diameter with electrical heaters filled with a mixture of the isotope ^{3}He and laser gas. In this case, the active media were excited by the nuclear

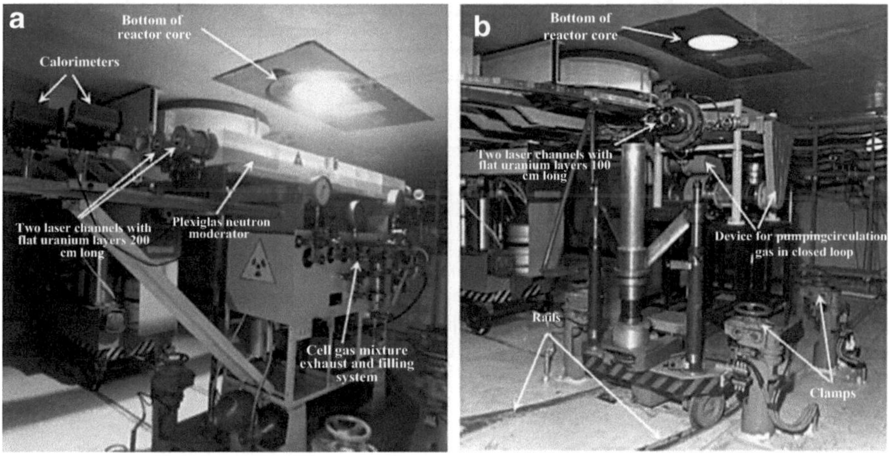

Fig. 2.11 LUNA-2 M (**a**) and LUNA-2P (**b**) unit setups under VIR-2 M reactor core [7]

Table 2.5 Basic characteristics of laser setups making up the research complex based on the VIR-2 M reactor [28]

Setup	LUNA-2M	LUNA-2P	LYaN-2T
Number of laser cells	2	2	2
Active length of cell	200 cm	100 cm	100 cm
Neutron moderator	Plexiglas	Plexiglas	Water
Average flux density of thermal neutrons at pulse maximum along cell length	2.5×10^{15} cm$^{-2} \times$ s^{-1}	2.2×10^{15} cm$^{-2} \times$ s^{-1}	3×10^{15} cm$^{-2} \times$ s^{-1}
Pumping method	Two parallel uranium layers of 200×6 cm^2 (2 cm apart). Thickness of $^{235}U_3O_8$ layer: 3 mg/cm^2	Two parallel uranium layers of 100×6 cm^2 (2 cm apart). Thickness of $^{235}U_3O_8$ layer: 3 mg/cm^2	Nuclear reaction products: $^3He(n,p)^3H$
Initial temperature of gas medium	300 K	300 K	300–1,000 K
Gas flowing	None	Transverse gas circulation relative to cell axis	None

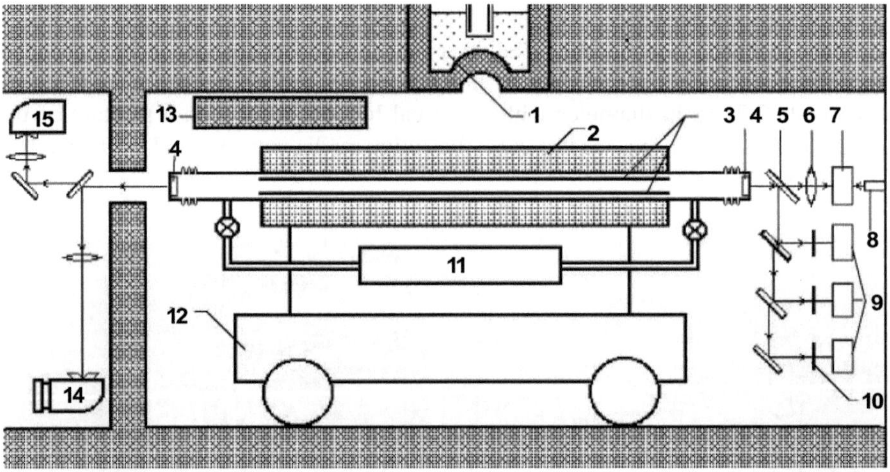

Fig. 2.12 Diagram of laser setup in lower hall of reactor VIR-2 M [28]: (1) reactor core, (2) neutron moderator, (3) plates with uranium layers, (4) cavity mirrors, (5) beam splitters, (6) focusing lens, (7) laser energy meter, (8) adjustment laser, (9) photoreceiver, (10) light filters, (11) cell gas mixture exhaust and filling system, (12) car, (13) protective shutter, (14 and 15) spectral devices

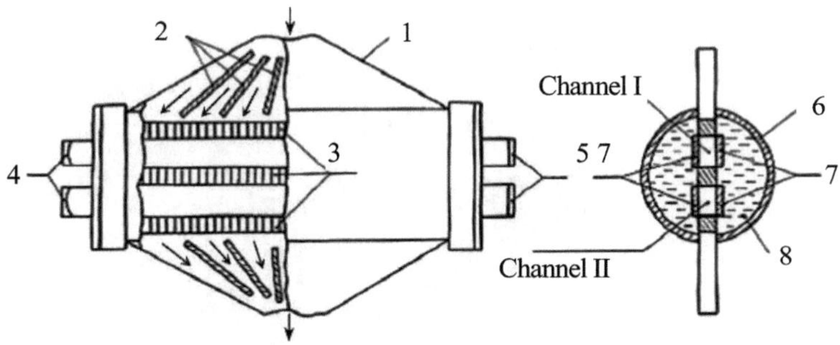

Fig. 2.13 Design of laser setup LUNA-2P: (1) aluminum funnel, (2) plates for distribution of gas flow, (3) heat exchangers (radiators), (4) 100 % reflectivity mirrors, (5) output couplers, (6) hull of water neutron moderator, (7) plates with uranium layers, (8) water. The *arrows* show the direction of gas flow

reaction products $^3He(n,p)^3H$. A detailed description of the LYaN-2T setup is considered in study [54].

The energy parameters of the laser radiation were determined using the laser pulse energy meters IEK-1 and IMO-2 N, various photocells (F-28, FEK-11, FEK-14, and SDF-7) and photoresistors (PbS, PbSe). For measuring the spectra of laser radiation in UV, visible, and near-IR ranges of the spectrum, the spectrographs DFS-452 and MS257 were used, and in the mid-IR range, the monochromator MDR-2 with a PbSe photoresistor array. The absorption (reflection) spectra of optical materials and cavity mirrors were determined using various spectrophotometers (IKS-14A, SF-9, etc.).

In the first stages of the investigations, electrical signals were registered using various multibeam oscillographs. At present an automated system is used for recording and processing measurement results using computer technologies.

From 1975 to 2010, around 3,000 pulses were produced on the VIR-2 and VIR-2 M reactors to study the characteristics of gas NPLs. As a result of these investigations, lasing was achieved and studied in the visible and IR ranges of the spectrum at roughly 40 transitions of the atoms Xe, Kr, Ar, Ne, C, O, N, and Cl and the ions Cd^+ and Zn^+ with excitation of gas mixtures at atmospheric pressure with uranium fission fragments and nuclear reaction products $^3He(n,p)^3H$ [1, 7, 27, 28]. In addition, a number of important tasks related to the study of optical and gas-dynamic characteristics of gaseous media were completed, and the information needed to develop CW multichannel NPLs was obtained [15, 16, 37].

2.5 EBR-L Experimental Setup

VNIITF began studying NPL problems in 1979, and developed the EBR-L experimental complex for these purposes [2, 3, 31]. At present, apart from the EBR-L pulsed reactor, the complex includes a set of laser cells, a system of evacuation and

Fig. 2.14 EBR-L pulsed reactor [3]

Fig. 2.15 Diagram of EBR-L reactor core: [3, 31]: (1) mechanism for fine adjustment of reactivity, (2) upper copper reflector, (3) hemispherical parts of reactor core, (4) cylindrical copper reflector, (5) polyethylene moderator, (6) lower reflector, (7) safety unit, (8) pulse rod. (9) stop unit

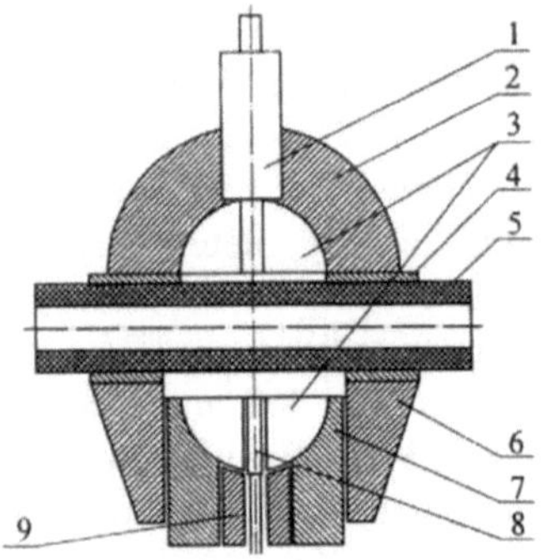

filling of laser cells with gas mixtures, an arrangement to withdraw optical radiation from the reactor hall and register its parameters. Ten types of NPLs were studied with this complex, and lasing was achieved at 21 laser transitions [10, 55], including metal-vapor lasers (Cd, Zn, Hg) and laser at UV transitions of the molecular ion N_2^+.

The EBR-L reactor (Figs. 2.14 and 2.15), which is a modified version of the pulsed reactor EBR-200, was put in operation in 1981. The reactor core consists of two hemispherical parts made of ^{235}U-Mo alloy (3 % molybdenum by mass), enclosed by a copper reflector. Between the two parts of the reactor core there is a cylindrical copper reflector with a through channel 12.5 cm in diameter. A polyethylene moderator 60 cm in long is placed in this channel, and it accommodates the laser cells. Some characteristics of the EBR-L reactor are shown in Table 2.6.

The average flux density of the thermal neutrons along the length of the uranium layer at the pulse maximum is 1.3×10^{17} cm^{-2} s^{-1}.

A scheme of one of the laser cells for metal-vapor NPLs is shown in Fig. 2.16. The central part of the cell is in the form of a stainless steel tube 80-cm long, with a wall thickness of 3 mm. A three-section wire heater is wound around the tube and is used to heat the tube up to 850°K. Water-cooled extension pipes and adjustment units are connected to the ends of the tube. The total length of the cell is 120–

Table 2.6 Parameters of maximal pulse in EBR-L reactor [3]

Filling of channel in copper reflector	None	Polyethylene moderator in cadmium casing
Total number of reactor core fissions	1.5×10^{17}	1.5×10^{17}
Maximal heating of reactor core per pulse, K	730	730
Pulse half-width, μs	60	140
Neutron fluence, cm^{-2}	3.5×10^{14}	5.8×10^{13}
Density of neutron flux, $cm^{-2} \cdot s^{-1}$	3.8×10^{18}	2.7×10^{17}

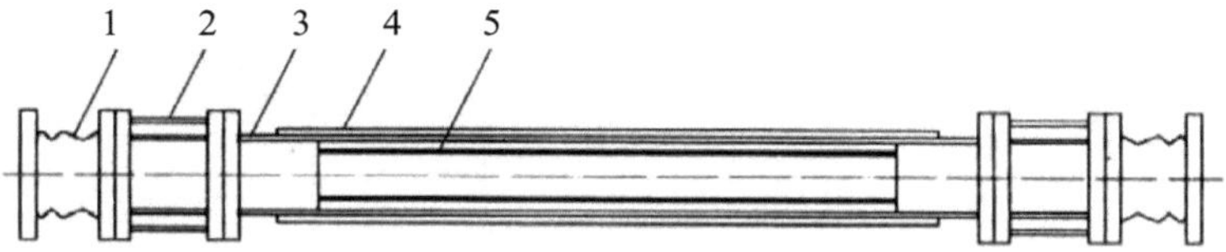

Fig. 2.16 Scheme of laser cell with heater [3, 31]: (1) adjustment unit, (2) connector, (3) extension pipe, (4) electric heater, (5) uranium layer

Table 2.7 Characteristics of laser cells for experiments with NPLs on the EBR-L setup [3]

Diameter of uranium layer, cm	1.2	2.8	4.8
Material of tube with uranium layer	Aluminum	Stainless steel	Aluminum
Active laser volume, cm^3	50	290	500
Thickness of $^{235}U_3O_8$ layer, mg/cm^2	2–3	2–3	2–3
Maximal specific energy deposition, J/cm^3	0.9	0.4	0.2
Maximal specific power deposition, kW/cm^3	4	2	1

140 cm. Inside the cell there are tubes with an internal diameter of 2.8 cm, on whose surface a layer of U_3O_8 is deposited (90 % ^{235}U enrichment) at a thickness of 2–3 mg/cm^2 and a length of 60 cm. For experiments with NPLs, laser cells are also used without heaters, with uranium layer diameters of 1.2 and 4.8 cm. The basic characteristics of the cells are shown in Table 2.7.

To measure the characteristics of laser or luminescent radiation emerging from the cell, optical equipment (Fig. 2.17) is used, which can determine the laser spectra, energy and threshold characteristics of the laser pulse. In experiments to study the luminescent characteristics of gas and vapor media, the spectral composition of radiation, the temporal pulse shape, and the intensity of luminescence were measured for several lines. The spectral range for registration of optical signals was 0.2–5 μm.

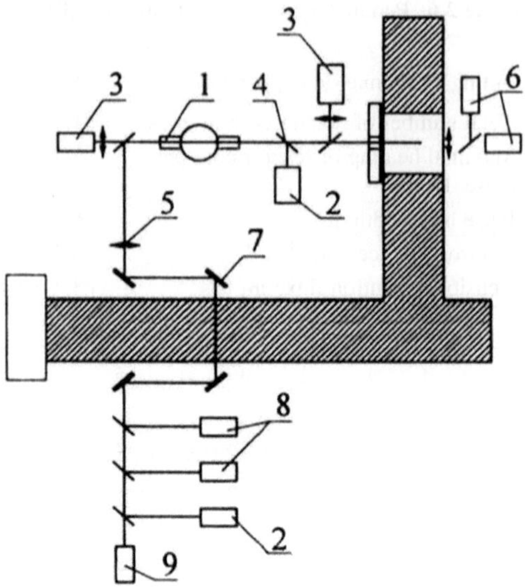

Fig. 2.17 Diagram of the withdrawal and registration of optical radiation [3, 31]: (1) laser cell, (2) adjustment laser, (3) calorimeter, (4) semitransparent plate, (5) focusing lens, (6) spectrograph, (7) deflecting mirror, (8) photoreceivers, (9) monochromator

2.6 "Stand B" for Studying NPL Characteristics

The RF State Science Center FEI (in Obninsk) began studying problems of NPLs in 1981 at the initiative of V. Ya Pupko [39]. Within the scope of the program to develop a powerful laser amplifier with nuclear pumping, at present an evaluation unit of the laser system has been developed which was called "Stand B" [13, 39, 56]. Here the characteristics of one of the two parts of "Stand B" are examined, the so-called "first workplace," which was put in operation in 1996, and was intended to study the characteristics of individual laser cells. Information about the "second workplace" of Stand B is provided in Sect. 6.3 of Chap. 6.

The basic element of the laser stand is the BARS-6 dual-core pulsed reactor, made on the basis of the BARS-5 reactor [3, 32] in accordance with the design documentation and technical assistance of VNIITF. The two reactor cores are arranged on a platform that can be moved through the reactor hall depending on the tasks of the experiment. The distance between the axes of the reactor cores is varied between 340 and 1,500 mm, making it possible to irradiate laser cells more uniformly than in the case of a pulsed reactor with a single reactor core. The half-height duration of the reactor pulse is 150–200 μs.

In one of the positions ("first workplace") of the reactor core, experiments are carried out to study the characteristics of individual NPLs (see Fig. 2.1e). By way of example, Fig. 2.18 shows a diagram of one of the experiments studying the characteristics of an Ar-Xe laser [13]: the distance between the reactor cores is 70 cm; the cylindrical stainless steel laser cell has a length of 250 cm and internal diameter of 49 mm (wall thickness 0.5 mm). A 5-μm thick layer of $^{235}UO_2$ was deposited to the internal surface. The distribution of the specific power deposition

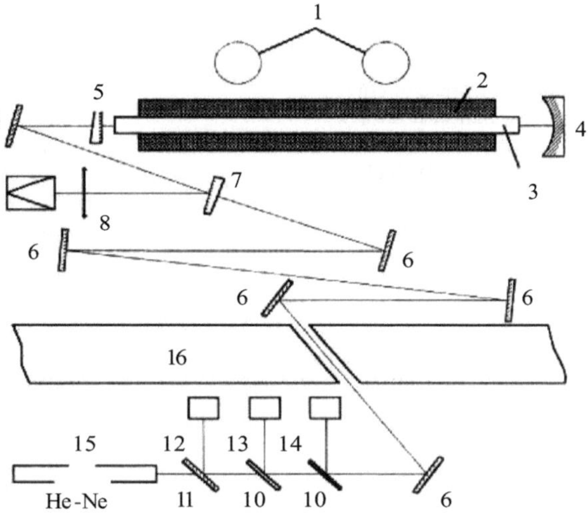

Fig. 2.18 Diagram of experiments with NPLs with the use of the BARS-6 dual-core reactor [13]: (1) reactor core, (2) polyethylene neutron moderators, (3) laser cell, (4 and 5) cavity mirrors, (6) deflecting mirrors, (7, 10, and 11) beam splitters, (8) lenses, (9) calorimeter, (12–14) photoreceivers, (15) adjustment laser, (16) biological shielding

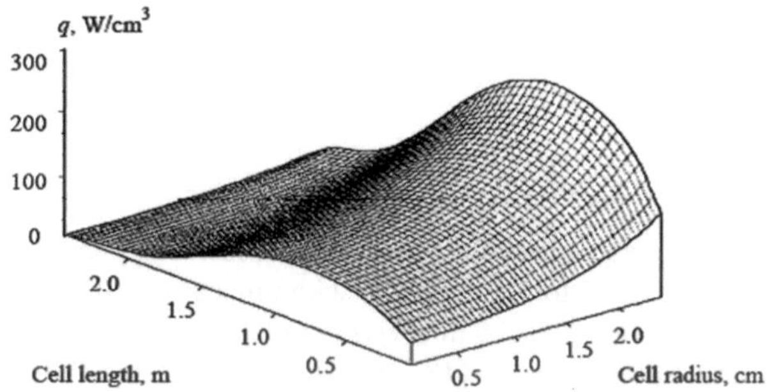

Fig. 2.19 Distribution of specific power deposition along the length and radius of the laser cell for an Ar-Xe mixture (100:1) at a pressure of 0.5 atm in experiments on the BARS-6 reactor [13]

along the length and radius of the laser cell at the reactor pulse maximum for the Ar-Xe mixture (100:1) at a pressure of 0.5 atm is shown in Fig. 2.19.

In experiments in the "first workplace" of Stand B, studies were also carried out on the gain measurements of some laser media [57, 58] and the dynamics of development of optical non-uniformities [59].

2.7 Experiments with the SPR and TRIGA Reactors

The pulsed nuclear reactors listed in Table 2.1 were used outside of the USSR to search for and to study the characteristics of NPL active media. The bulk of the studies were done in the United States, at the Sandia Laboratory (reactors SPR-II, SPR-III) and the University of Illinois (TRIGA reactor).

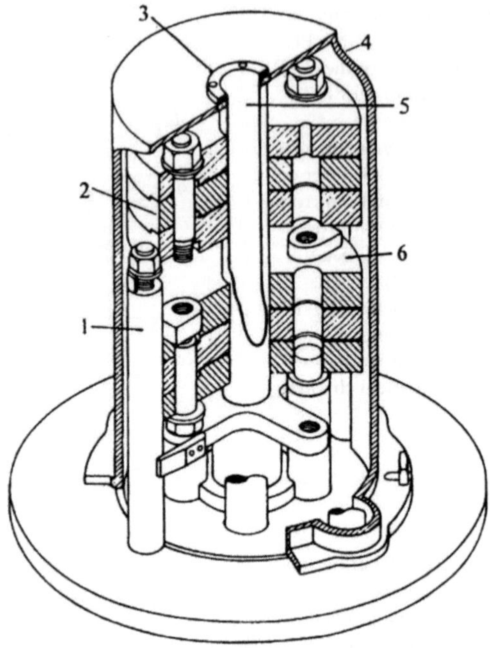

Fig. 2.20 Schematic view of the SPR-II reactor [2]: (1) support column for upper half of reactor core, (2) upper half of reactor core, (3) flange of protective tube in cavity, (4) shielding enclosure, (5) axial cavity for irradiation, (6) lower half of reactor core

A report concerning the first successful experiment outside of the USSR in a pulsed reactor to pump a laser with nuclear radiation appeared in 1975: a molecular CO-laser ($\lambda = 5.1$–5.6 μm) was pumped by uranium fission fragments in experiments using SPR-II reactor [18, 60]. Figures 2.20 and 2.21 show a diagram of the fast neutron-pulsed reactor SPR-II [2] and the scheme of experiments on this reactor with a CO-laser [60]. The cylindrical reactor core consists of six annular discs and four rods made of uranium alloy (93 % enriched ^{235}U) with molybdenum (10 %). The external diameter and height of the reactor core are 20.3 cm and 20.8 cm, respectively, and the total mass of the uranium-molybdenum alloy is 105 kg. Detailed characteristics of the SPR-II reactor can be found in monograph [2].

The laser cell was placed at a distance of around 20 cm from the reactor core. The neutron flux density at the reactor pulse maximum was 1×10^{17} cm$^{-2} \times$ s^{-1}. Excitation of the CO at a pressure of 0.13 atm was carried out with uranium fission fragments emerging from the cylindrical $^{235}U_3O_8$ layer. The specific power deposition was around 200 W/cm^3. A specific feature of these experiments was the necessity of cooling the carbon monoxide to a temperature of 77 K. The laser radiation receivers (calorimeter and Ge:Au photoresistor) were disposed beyond the biological shielding at a distance of around 15 m from the laser cell. In subsequent experiments with the SPR-II reactor with CO laser, a laser cell with a multipass cavity was used (see Fig. 2.2a).

Significantly more experiments with various NPLs were carried out at the Sandia Laboratories with the SPR-III pulsed reactor, which was put into operation in 1975

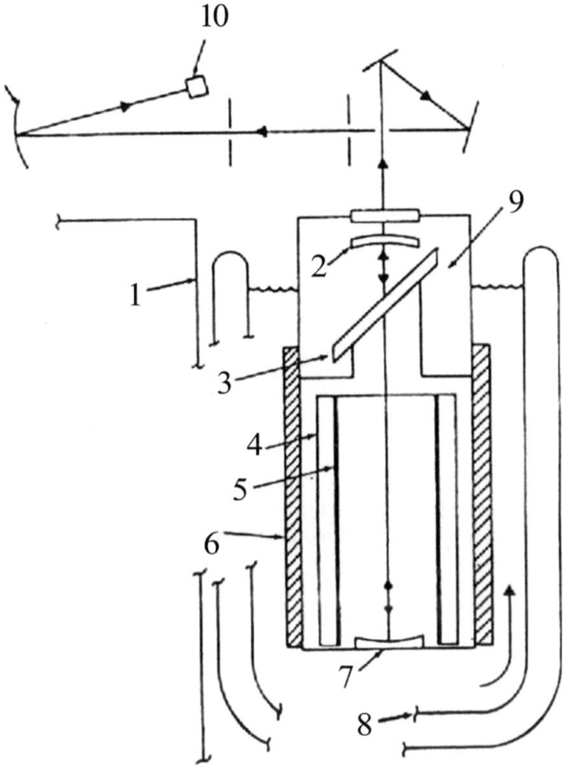

Fig. 2.21 Diagram of experiments to pump a CO-laser with uranium fission fragments [60]: (1) SPR-II reactor core, (2) output coupler, (3) CaF_2 window at Brewster angle, (4) aluminum tube, (5) $^{235}U_3O_8$ layer, (6) polyethylene neutron moderator, (7) 100 % reflectivity mirror, (8) Dewar flask with liquid nitrogen, (9) region with vacuum, (10) laser radiation receivers

[2]. The reactor has a cavity 17 cm in diameter and 35-cm long that passes through the reactor core; it can be used to irradiate quite large objects. For example, this cavity accommodated a cell enclosed with polyethylene, which could be used to experimentally measure the gain in the mixture ^{3}He-Xe-NF_3 ($\lambda = 351$ nm) [20]. Under such conditions, it was possible to obtain a specific power deposition of around 5 kW/cm^3, the maximum possible using pulsed reactors.

In most of the experiments with the SPR-III reactor, the laser cell was positioned close to the surface of the reactor core (see studies [19, 61]). Most often a rectangular laser cell of $60 \times 7 \times 1$ cm^3 was used, placed inside a polyethylene neutron moderator (Fig. 2.22). The cell walls were coated with a layer of $^{235}UO_2$ that was 3-μm thick. To register the shape of the thermal neutron pulse, a 1-m long cobalt-based neutron detector was placed inside the moderator. The distribution of the fluence of the thermal neutrons (specific energy deposition) along the 60-cm length of the cavity is non-uniform. At the ends of the moderator (30 cm from the center), the fluence of the thermal neutrons is around 25 % of the maximal value in the central part.

To study NPLs operating on the transitions of rare gas atoms, which can operate at lower specific power depositions of ≤ 100 W/cm^3, the pulse duration of the SPR-III reactor was increased to several milliseconds. For this purpose, a "pulse

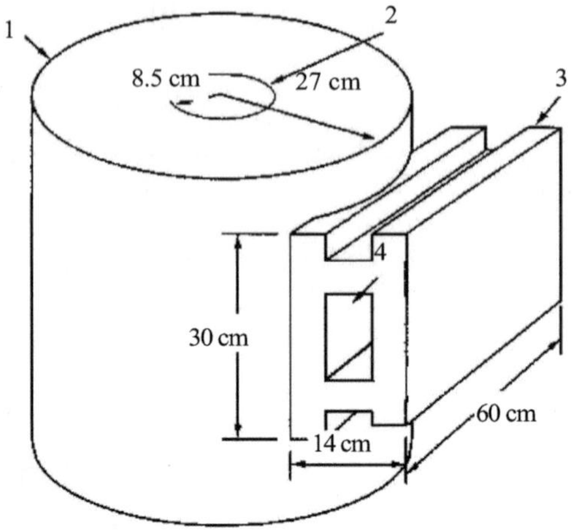

Fig. 2.22 The scheme of experiments with rectangular laser cell placed close to the SPR-III reactor core [61]: (1) reactor core, (2) cavity for irradiating of specimens, (3) polyethylene neutron moderator, (4) cavity in moderator to accommodate the laser cell

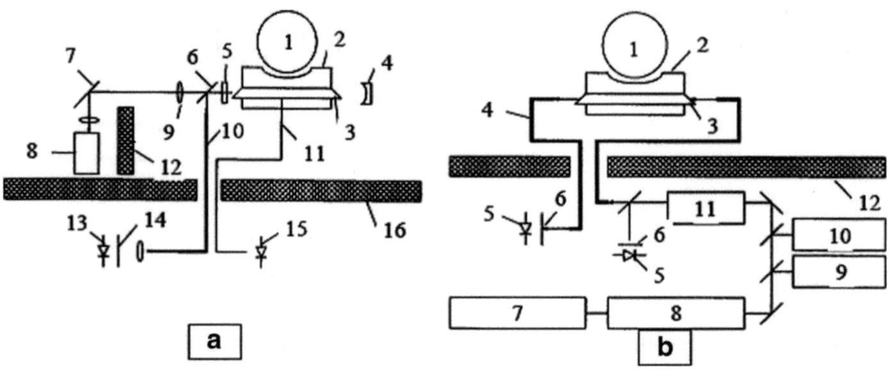

Fig. 2.23 Configurations of SPR-III reactor experiments [62]. Figure (**a**) is a study of NPL characteristics in the lasing mode. (1) reactor core, (2) polyethylene moderator, (3) laser cell, (4 and 5) cavity mirrors, (6) beam splitter, (7) deflecting mirror, (8) calorimeter, (9) lens, (10) lightguide, (11) quartz optical fiber, (12) lead shielding, (13) InAs photodiode, (14) light filter, (15) photoreceiver for registration of luminescent radiation, (16) biological shielding. Figure (**b**) displays gain measurement. (1) reactor core, (2) polyethylene moderator, (3) laser cell, (4) lightguide, (5) InAs photodiode, (6) light filter, (7) CW YAG:Nd^{3+} laser, (8) tunable F-center laser, (9) wavelength meter, (10) spectrum analyzer, (11) modulator, (12) biological shielding

extender"—several grams of ^{235}U—was placed in a cavity inside the reactor core in a polyethylene moderator with a wall thickness of a few centimeters. This device increased the effective lifetime of the neutrons in the reactor core, resulting in a threefold increase in the pulse duration [19].

To study the NPL characteristics, as a rule the experimental configurations shown in Fig. 2.23 were used [62]. Using these configurations, NPLs were studied

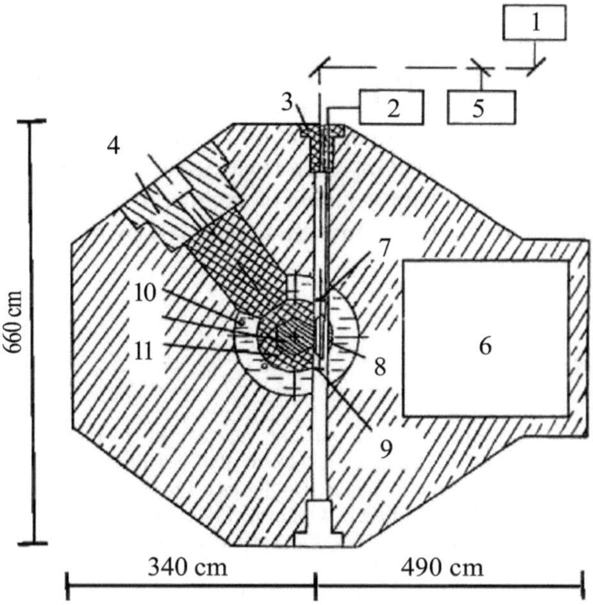

Fig. 2.24 Configuration of laser experiments on the TRIGA reactor [2, 30, 45]: (1 and 5) instruments registering the characteristics of the laser radiation, (2) system for supplying laser cells with the gases being studied, (3) protective plug, (4) thermal column (arrangement for withdrawal the beam of thermal neutrons), (6) experimental chamber, (7 and 9) cavity mirrors, (8) laser cell, (10) neutron detector, (11) reactor core, (12) graphite neutron reflector, (13) concrete shielding

at the transitions of atoms Ne ($\lambda = 585.3$, 703.2, 724.5 nm), Ar ($\lambda = 1.27$, 1.79 μm), and Xe ($\lambda = 1.73$, 2.03 μm).

The laser pulse energies were measured with calorimeters, while the shape was measured using silicon or InAs photodiodes. The energy deposition to the gas medium was determined from the jump in pressure, measured using a piezoelectric pressure sensor. To transport the laser radiation to and from the cell over a distance of around 20 m, lightguides—tubes with a diameter of 2.54 cm with a gold reflective coating—were used. Luminescent radiation emerging through the side surface of the cell was withdrawn beyond the biological shielding using a quartz optical fiber and was registered using a photodiode or a multichannel spectrometer with a registration range of 450–750 nm. Auxiliary lasers—a tunable F-center laser excited by radiation from a solid-state YAG:Nd^{3+} laser or a gas discharge laser—were used to measure the gain of the NPL.

TRIGA reactors are thermal neutron pool-type reactors, whose cores are made of the triple alloy $UZrH_x$. These reactors are the most common of the research pulsed reactors—the number of their modification in the world in 1987 was 53 [2].

One of the reactors of the TRIGA family, namely the TRIGA Mark II, was used for many years to study NPL problems at the University of Illinois [29, 30]. The characteristic parameters of TRIGA pulsed reactors are: total number of fissions in the reactor core ~1×10^{18} fissions (~30 MJ), half-height pulse width ~10 ms, neutron fluence at the reactor center $(0.5–1.0) \times 10^{15}$ cm^{-2} [2]. The reactor can operate in stationary mode at a power of up to 1.5 MW with a thermal neutron flux density of 3×10^{12} $cm^{-2} \times s^{-1}$ [42].

The configuration of experiments on the TRIGA reactor is shown in Fig. 2.24. The typical length of the laser cell is 70–100 cm. The gas mixtures He-CO(CO_2)

(λ = 1.45 μm), Ne-N_2 (λ = 862.9; 939.3 nm), Ar-Xe (λ = 1.73 μm), and ^3He-Ne-H_2 (λ = 585.3 nm) were pumped by the nuclear reaction products $^{10}B(n,\alpha)^7Li$ or $^3He(n,p)^3H$. Various types of instruments were used to register the laser radiation (photomultipliers, photoresistors, monochromators, and power meters).

Apart from the studies of NPLs with direct pumping of active media by nuclear radiation, the TRIGA reactor was used to pump a photodissociation iodine laser (λ = 1.31 μm) with radiation of excimer molecules XeBr*, which are formed upon irradiation of the mixture Xe-$CHBr_3$-^3He by the neutron flux [29].

References

1. A.M. Voinov, *Pulsed nuclear reactors in nuclear-pumped laser research* (Primeneniye impulsnykh yadernykh reaktorov dlya issledovaniya lazerov s yadernoy nakachkoy), Proceedings of the Specialist Conference Physics of Nuclear-Excited Plasma and Problems of Nuclear-Pumped Lasers, Obninsk, 1992, vol. 1, pp. 101–121.
2. V.F. Kolesov, *Aperiodic Pulse Reactors (Aperiodicheskiye impulsnyye reaktory)* (RFYaTa-VNIIEF, Sarov, 1999)
3. B.G. Levakov, A.V. Lukin, E.P. Magda, I.S. Pogrebov, A.A. Snopkov, V.A. Terekhin, *Pulsed Nuclear Reactors of RFYaTS-VNIITF (Impulsnyye yadernyye reaktory RFYaTSVNIITF)* (RFYaTs-VNIITF, Snezhinsk, 2002)
4. C.D. Yang, H.D. Chen, C. Zheng, X.L. Zhao, *The progress of nuclear pumped laser in chinese pulse reactors*, Proceedings of the 3rd International Conference "Problems of Nuclear-Pumped Lasers and Pulsed Reactors, Snezhinsk, 2003, pp. 31–33.
5. A.M. Voinov, L.Ye. Dovbysh, A.T. Kazakevich, V.N. Krivonosov, S.P. Melnikov, I.V. Podmoshenskiy, A.A. Sinyanskiy, Nuclear-pumped infrared lasers operating on ArI, KrI, and XeI transitions (Infrakrasnyye lazery s yadernoy nakachkoy na perekhodakh ArI, KrI and XeI). Pisma v ZhTF. **5**(7), 422–424 (1979)
6. A.M. Voinov, A.S. Koshelev, S.P. Melnikov, A.A. Sinyanskiy, Quasi-CW gas laser excited by fast neutrons (Kvazinepreryvnyy gazovyy lazer, vozbuzhdayemyy bystrymi neytronami). Pisma v ZhTF **16**(13), 86–89 (1990)
7. A.A. Sinyanskiy, *Research into development of stationary nuclear laser devices at VNIIEF* (Issledovaniye po sozdaniyu yaderno-lazernykh ustroystv nepreryvnogo deystviya vo VNIIEF), Proceedings of the 2nd International Conference Physics of Nuclear-Excited Plasma and Problems of Nuclear-Pumped Lasers, Arzamas-16, 1995, vol. 1, pp. 16–36
8. A.M. Voinov, L.Ye. Dovbysh, V.N. Krivonosov, S.P. Melnikov, A.A. Sinyanskiy, *Lasers pumped by optical radiation of a converter based on radioluminescent and plasma light sources* (Lazery s nakachkoy opticheskim izlucheniyem konvertora na osnove radiolyuminestsentnogo i plazmennogo istochnikov sveta), Proceedings of the Specialist Conference Physics of Nuclear-Excited Plasma and Problems of Nuclear-Pumped Lasers," Obninsk, 1993, vol. 3, pp. 224–225
9. A.I. Miskevich, Toward a history of development of the first nuclear-pumped visible-range laser (K istorii sozdaniya pervogo lazera vidimogo diapazona s yadernoy nakachkoy). Inzhenernaya fizika. (2), 6–13 (2007)
10. E.P. Magda, *Analysis of experimental and computational-theoretical work on nuclear-pumped lasers performed at VNIITF* (Analiz eksperimentalnykh i raschetnoteoreticheskikh rabot po lazeram s yadernoy nakachkoy, vypolnennykh vo VNIITF), Proceedings of the Specialist Conference Physics of Nuclear-Excited Plasma and Problems of Nuclear-Pumped Lasers, Obninsk, 1992, vol. 1, pp. 65–74

11. A.V. Bochkov, E.P. Magda, V.V. Mironenko, V.M. Murzin, G.S. Sofiyenko, *Research on laser parameters at the transitions of atoms of rare gases under the conditions of a high energy load* (Issledovaniya parametrov lazerov na perekhodakh atomov inertnykh gazov v usloviyakh vysokoy energeticheskoy nagruzki). Proceedings of the 3rd International Conference "Problems of Nuclear-Pumped Lasers and Pulsed Reactors, Snezhinsk, 2003, pp. 134–139
12. E.P. Magda, A.V. Bochkov, V.A. Kryzhanovskiy, S.L. Mukhin, *Measurement of gain at transition* $5d[3/2]_1$-$6p[5/2]_2$ *of the xenon atom with excitation of the active medium with fission products* (Izmereniye koeffitsienta usileniya na perekhode atoma ksenona 5*d*[3/2]1-6*p* [5/2]2 pri vozbuzhdenii aktivoy sredy oskolkami deleniya). Proceedings of the 2nd International Conference Physics of Nuclear-Excited Plasma and Problems of Nuclear-Pumped Lasers, Arzamas-16, 1995, vol. 1, pp. 265–269
13. E.D. Poletaev, A.F. Gamaly, P.P. Dyachenko, V.N. Smolsky, M.Yu. Zaitsev, V.R. Agafonov, Yu.A. Dyuzhov, A.V. Gulevich, B.V. Kachanov, O.F. Kukharchuk, *Experimental study of the laser-active element characteristics for high-power nuclear reactor-pumped laser system.* Proceedings of the 9th International Conference on Emerging Nuclear Energy Systems, Tel-Aviv, 1998, pp. 832–839
14. M.V. Bokhovko, A.P. Budnik, I.V. Dobrovolskaya, V.N. Kononov, O.Y. Kononov, The influence of the specific energy deposition on the characteristics of the nuclear-pumped Ar/Xe laser (Vliyaniye udelnogo energovklada na kharakteristiki Ar/Xe lazera s yaderno y nakachkoy). Pisma v ZhTF **24**(14), 16–20 (1998)
15. V.N. Bogdanov, A.G. Vasilenko, V.V. Zherbtsov, V.V. Ivanov, M.I. Kuvshinov, A.Ye. Lakhtikov, V.S. Mayornikov, A.B. Modenov, A.V. Nikulin, V.A . Nor-Arevyan, A.N. Pokalo, A.A. Sinyanskiy, I.G. Smirnov, S.L. Turutin, M.V. Khlestov, V.V. Shirokov, *LM4/BIGR experimental complex. design and operating principle* (Eksperimentalnyy kompleks LM-4/BIGR. Ustroystvo i printsip raboty). Proceedings of the 2nd International Conference Physics of Nuclear-Excited Plasma and Problems of Nuclear-Pumped Lasers, Arzamas-16, 1995, vol. 2, pp. 172–178
16. A.A. Sinyanskiy, *Multichannel quasi-stationary nuclear laser setups based on the BIGR reactor* (Mnogokanalnyye yaderno-lazernyye ustanovki kvazinepreryvnogo deystviya na reaktore BIGR). Proceedings of the 3rd International Conference Problems of Nuclear-Pumped Lasers and Pulsed Reactors, Snezhinsk, 2003, pp. 377–387
17. H.H. Helmick, J.L. Fuller, R.T. Schneider, Direct nuclear pumping of a He-Xe laser. Appl. Phys. Lett. **26**(6), 327–328 (1975)
18. D.A. McArthur, Nuclear Pumped Lasers. *Encyclopedia of Lasers and Optical Technology.* (Academic Press, San Diego, 1991), pp. 385–393
19. D.E. Bodette, D.R. Neal, D.A. McArthur, P.S. Pickard, G.A. Hebner, R. Michie, *Reactor Pumped Laser Research at the Sandia National Laboratories Pulsed Reactor Facilities.* International Embedded Topical Meeting on Physics, Safety, and Applications of Pulse Reactors, Washington, 1994, pp. 163–171
20. G.N. Hays, D.A. McArthur, D.R. Neal, J.K. Rice, Gain Measurements near 351 nm in $^3He/Xe/NF_3$ mixtures excited by fragments from the $^3He(n,p)^3H$ reaction. Appl. Phys. Lett. **49**(7), 363–365 (1986)
21. R.J. DeYoung, N.W. Jalufka, F. Hohl, Direct nuclear-pumped lasers using the 3He(n, p)3H reaction. AIAA J. **16**(9), 991–998 (1978)
22. N.W. Jalufka, Direct Nuclear-Pumped Lasers. NASA Technical Paper 2091, 1983
23. T.G. Miller, G.J. Dezenberg, G.W. Morrison, Nuclear-initiated chemical laser experiments. Trans. Am. Nucl. Soc. **21**, 22–23 (1975)
24. D. Auphelle, F. Euve, M. Fitaire, A.M. Pointu, M. Vialle, Recherche d'un effet Laser dans le Melange He-CO_2 Irradie par le Reacteur Caliban. Phys. Lett **72A**(6), 439–441 (1979)
25. V.M. Andriyakhin, V.V. Vasiltsov, S.S. Krasilnikov, V.D. Pismennyy, V.Y. Khvostionov, On radiation of the gas mixture Hg-He irradiated by neutron flux. Pisma v ZhETF **12**(2), 83–85 (1970)

26. A.M. Voinov, L.Ye. Dovbysh, V.N. Krivonosov, S.P. Melnikov, S.F. Melnikov, A.A. Sinyanskiy, The experimental complex based on the VIR-2M reactor for studying nuclear-pumped lasers (Eksperimentalnyy kompleks na osnove reaktora VIR-2M dlya issledovaniya lazerov s yadernoy nakachkoy). Voprosy atomnoy nauki i tekhniki. Ser. Fizika yadernykh reaktorov. (2/3), pp. 63–68 (2000)
27. S.P. Melnikov, A.A. Sinyanskiy, *On the maximal efficiency of nuclear pumped gas lasers* (O predelnom KPD gazovykh lazerov s yadernoy nakachkoy). Proceedings of the Specialist Conference Physics of Nuclear-Excited Plasma and Problems of Nuclear-Pumped Lasers, Obninsk, 1993, vol. 2, pp. 133–148
28. A.A. Sinyanskiy, S.P. Melnikov, Research on development of continuous nuclear-laser setups in VNIIEF. Proc. SPIE **3686**, 43–55 (1998)
29. G.H. Miley, *Overview of nuclear pumped lasers*. Proceedings of the Specialist Conference Physics of Nuclear-Excited Plasma and Problems of Nuclear-Pumped Lasers. Obninsk, 1992, vol. 1, pp. 40–53
30. G.H. Miley, *Dual-media nuclear-pumped lasers*. International Symposium on Short-Wavelength Lasers and Their Applications, Osaka, 1987, pp. 14–25
31. V.A. Kryzhanovskiy, E.P. Magda, *The EBR-L experimental setup for studying nuclear-pumped lasers* (EBR-L-eksperimentalnaya ustanovka dlya issledovaniya lazerov s yadernoy nakachkoy). Proceedings of the Specialist Conference Physics of Nuclear-Excited Plasma and Problems of Nuclear-Pumped Lasers, Obninsk, 1993, vol. 3, pp. 136–145
32. A.A. Snopkov, V.I. Cherashev, V.I. Litvin, V.P. Koshmyakov, E.P. Magda, *An extended neutron source based on the BARS-5 dual-core pulsed reactor* (Protyazhennyy neytronnyy istochnik na baze dvukhzonnogo impulsnogo reaktora BARS-5). Proceedings of the Specialist Conference Physics of Nuclear-Excited Plasma and Problems of Nuclear-Pumped Lasers, Obninsk, 1992, vol. 1, pp. 144–155
33. A.M. Voinov, V.F. Kolesov, A.S. Matveyenko, S.F. Melnikov, I.V. Levitskiy, G.N. Maslov, L.M. Pavlovskaya, L.A. Samodurov, V.S. Gladkov, The water pulsed reactor VIR-2M and its predecessors (Vodnyy impulsnyy reaktor VIR-2M i yego predshestvenniki). Voprosy atomnoy nauki i tekhniki. Ser. Fizika yadernykh reaktorov. (3), 3–15, (1990)
34. R.J. DeYoung, Multiple-path fission-foil nuclear lasing of Ar-Xe. Appl. Phys. Lett. **39**(8), 585–587 (1981)
35. R.J. DeYoung, Kilowatt multiple-path ^{3}He-Ar nuclear-pumped laser. Appl. Phys. Lett. **38**(5), 297–298 (1981)
36. M.G. Anuchin, K.F. Grebenkin, Y.Z. Kandiyev, Y.I. Cherepanova, A computational investigation of pumping a gas medium with charged particles produced by nuclear reactions (Raschetnoye issledovaniye nakachki gazovoy sredy zaryazhennymi chastitsami–produktami yadernykh reaktsii). ZhTF **61**(1), 3–8 (1991)
37. A.M. Voinov, S.V. Vorontsov, V.N. Krivonosov, S.P. Melnikov, A.A. Pikulev, V.T. Punin, A.N. Sizov, A.A. Sinyanskiy, S.L. Turutin, Studies carried out at VNIIEF to investigate the possibility of creating a reactor-laser (Raboty, vypolnennyye vo VNIIEF, po issledovaniyu vozmozhnosti sozdaniya reaktora-lazera). Voprosy atomnoy nauki i tekhniki. Ser. Fizika yadernykh reaktorov. (2), 19–28, (2007)
38. E.P. Magda, I.S. Pogrebov, I.S. Putnikov, A.V. Bochkov, A.I. Volkov, A.N. Maloyaroslavtsev, Yu.G. Gornitsyn, V.A. Vdovenko, Yu.M. Kononenko, Ya.Z. Kandiyev, Laboratory laser reactor facility LIRA (Laboratornaya lazerno-reaktornaya ustanovka LIRA). Voprosy atomnoy nauki i tekhniki. Ser. Fizika yadernykh reaktorov. (1/2), 39–43, (2001)
39. P.P. Dyachenko, *On the status of research at the A.I. Leypunskiy GNTs RF-FEI on the direct transformation of nuclear fission into the energy of laser radiation* (O sostoyanii issledovaniy v GNTs RF-FEI im. A.I. Leypunskogo po pryamomy preobrazovaniyu energii deleniya yader v energiyu lazernogo izlucheniya). Proceedings of the 3rd International Conference Problems of Nuclear-Pumped Lasers and Pulsed Reactors, Snezhinsk, 2003, pp. 5–30
40. A.M. Voinov, L.Ye. Dovbysh, V.N. Krivonosov, S.P. Melnikov, A.A. Sinyanskiy, *Methodological support of research into the radiation resistance of optical elements of nuclear pumped*

lasers (Metodicheskoye obespecheniye issledovaniy po radiatsionnoy stoykosti opticheskikh elementov lazerov s yadernoy nakachkoy). Proceedings of the 2nd International Conference Physics of Nuclear-Excited Plasma and Problems of Nuclear-Pumped Lasers, Arzamas-16, 1995, vol. 1, pp. 52–69

41. A.M. Voinov, L.Ye. Dovbysh, S.N. Lebedev, A.A. Sinyanskiy, *The transparency of silica glass under conditions of reactor irradiation* (Prozrachnost stekloobraznogo kvartsa v usloviyakh reaktornogo oblucheniya). Proceedings of the Specialist Conference Physics of Nuclear-Excited Plasma and Problems of Nuclear-Pumped Lasers, Obninsk, 1993, vol. 3, pp. 21–31
42. G.H. Miley, *Influence of high-intensity radiation on optical materials for nuclear diagnostics/ control systems*. International Embedded Topical Meeting on Physics, Safety, and Applications of Pulse Reactors, Washington, 1994, pp. 172–178
43. S.M. Brekhovskikh, V.A. Tyulnin, *Radiation Centers in Inorganic Glass (Radiatsionnyye tsentry v neorganicheskikh steklakh)* (Energoizdat, Moscow, 1988)
44. F.A. Zaitov, N.N. Litvinova, V.G. Savitskiy, V.G. Sredin, *Radiation Resistance in Optronics (Radiatsionnaya stoykost v optoelektronike)* (Voyenizdat, Moscow, 1987)
45. F.P. Boody, M.A. Prelas, *Transient radiation-induced absorption in fused-silica optical fibers, 450–950 nm*. Proceedings of the Specialist Conference Physics of Nuclear-Excited Plasma and Problems of Nuclear-Pumped Lasers, Obninsk, 1993, vol. 3, pp. 32–40
46. D.L. Griscom, K.M. Golant, A.L. Tomashuk, D.V. Pavlov, Y.A. Tarabrin, γ-Radiation resistance of aluminum-coated all-silica optical fibers fabricated using different types of silica in the core. Appl. Phys. Lett. **69**(3), 322–324 (1996)
47. A.L. Tomashuk, K.M. Golant, Radiation resistant and radiation-sensitive silica optical fibers. Proc. SPIE **4083**, 188–201 (2000)
48. A.L. Tomashuk, E.M. Dianov, K.M. Golant, A.O. Rybaltovskii, γ-Radiation-induced absorption in pure-silica-core fibers in the visible spectral region: the effect of H_2-loading. IEEE Trans. Nucl. Sci. **45**(3), 1576–1579 (1998)
49. S.N. Afonin, L.Y. Dovbysh, A.T. Kazakevich, S.P. Melnikov, A.A. Sinyanskiy, A laser radiation receiver in a wavelength range of 0.27-1.2 μm with high time resolution (Priyemnik lazernogo izlucheniya v diapazone dlin voln 0.27-1.2 mkm s vysokim vremennym razresheniyem). Kvantovaya elektronika **3**(2), 459–461 (1976)
50. S.P. Melnikov, A.A. Sinyanskiy, Quasi-CW lasing at the transition 4p-4s of an argon atom when pumping a He-Ar mixture with uranium fission fragments (Kvazinepreryvnaya generatsiya na perekhode 4*p*-4*s* atoma argona pri nakachke lazera na smesi He-Ar oskolkami deleniya urana). Kvantovaya elektronika **18**(12), 1444–1446 (1991)
51. A.A. Pikulev, V.M. Tsvetkov, P.V. Sosnin, A.A. Sinyanskiy, *Experimental investigation of serial addition o f two laser channels* (Eksperimentalnoye issledovaniye posledovatelnogo slozheniya dvukh lazernykh kanalov). Proceedings of the 4th International Conference Physics of Nuclear-Pumped Lasers and Pulsed Reactors, Obninsk, 2009, vol. 1, pp. 290–297
52. A.A. Pikulev, V.M. Tsvetkov, P.V. Sosnin, A.A. Sinyanskiy, Serial addition of two laser channels during excitation of He-Ar-Xe (2.03 μm) and Ar-Xe (1.73 μm) mixtures with uranium fission fragments (Posledovatelnoye slozheniye dvukh lazernykh kanalov pri vozbuzhdenii smesey He-Ar-Xe (2.03 mkm) i Ar-Xe (1.73 mkm) oskolkami deleniya urana). Kvantovaya elektronika **39**(6), 494–500 (2009)
53. A.A. Pikulev, V.M. Tsvetkov, P.V. Sosnin, A.A. Sinyanskiy, *Operating of nuclear-pumped laser at the parallel addition of two laser channels*, (Rabota lazera s yadernoy nakachkoy v rezhime parallelnogo slozheniya dvukh lazernykh kanalov). Proceedings of the 4th International Conference Physics of Nuclear-Pumped Lasers and Pulsed Reactors, Obninsk, 2009, vol. 1 pp. 213–218
54. A.I. Konak, S.P. Melnikov, V.V. Porkhayev, A.A. Sinyanskiy, *Nuclear-pumped gas laser at temperatures up to 800 °C* (Gazovyy lazer s yadernoy nakachkoy pri temperaturakh do 800°C). Proceedings of the Specialist Conference Physics of Nuclear-Excited Plasma and Problems of Nuclear-Pumped Lasers, Obninsk, 1993, vol. 3, pp. 98–106

55. E.P. Magda, Nuclear pumped lasers, problems and perspectives: review. Proc. SPIE **3343**, 158–170 (1998)
56. P.P. Dyachenko, O.A. Yelovskiy, Y.A. Prokhorov, G.N. Fokin, P.A. Yakubov, A.A. Snopkov, The stand B—reactor-laser complex (Reaktorno-lazerniy kompleks Stend B). Atomnaya Energiya **88**(5), 337–342 (2000)
57. V.N. Kononov, M.V. Bokhovko, P.P. Dyachenko, G.N. Kazantsev, N.Ya. Maksimov, V.I. Regushevskiy, Experimental Model of Master Generator-Amplifier Pulsed Laser System With Nuclear Pumping Using Uranium Fission Fragments (Eksperimentalnaya model impulsnoy lazernoy sistemy 'zadayushchiy generator-usilitel' s yadernoy nakachkoy oskolkami deleniya urana), FEI Preprint, No. 2477, Obninsk (1995)
58. E.A. Seregina, A.F. Dobrovolsky, P.P. Daychenko, A.A. Seregin, G.V. Tikhonov, *Liquid laser with nuclear pumping*. Proceedings on 9th International Conference on Emerging Nuclear Energy Systems, Tel-Aviv, 1998, pp. 824–831
59. P.P. Dyachenko, Yu.A. Dyuzhov, A.A. Orlov, Ye.D. Poletayev, V.N. Smolskiy, *Experimental study of the dynamics of formation of optical non-uniformities in Ar and He-Ar media with excitation by fission fragments in the BARS-6 reactor* (Eksperimentalnoye izucheniye dinamiki obrazovaniya opticheskikh neodnorodnostey v Ar i He-Ar sredakh pri vozbuzhdeniy oskolkami deleniya na reaktore BARS-6), Proceedings of the 3rd International Conference "Problems of Nuclear-Pumped Lasers and Pulsed Reactors, Snezhinsk, 2003, pp. 233–237
60. D.A. McArthur, P.B. Tollefsrud, Observations of laser action in CO gas excited only by fission fragments. Appl. Phys. Lett. **26**(4), 187–190 (1975)
61. D.A. McArthur, G.N. Hays, W.J. Alford, D.R. Neal, D.E. Bodette, J.K. Rice, Recent results on reactor-pumped laser studies at Sandia national laboratories. Laser Interaction **8**, 75–86 (1988)
62. G.A. Hebner, G.N. Hays, Fission-fragment excited xenon/rare gas mixtures. I. Laser parameters of the 1.73 μm Xenon Transition. J. Appl. Phys. **73**(8), 3614–3626 (1993)

Chapter 3
Investigations of Reactor-Pumped Gas NPLs

Presently, known gas NPLs radiate in a wide band of the optical spectrum of 390–5,600 nm using approximately 50 transitions of atoms Xe, Kr, Ar, Ne, C, N, Cl, O, Hg, Cd, and I; the ions Cd^+, Zn^+, and Hg^+; the molecule CO; and the molecular ion N_2^+.

3.1 IR Lasers Operating on Transitions of the Xe, Kr, and Ar Atoms

Nuclear-pumped lasers operating on transitions of Xe, Kr, and Ar atoms have been studied in greatest detail and have the maximal energy parameters of NPLs studied to date. Primarily, uranium fission products and nuclear reaction products $^3He(n, p)^3H$ were used to excite these lasers. Diagrams of the energy levels of atoms Xe, Kr, and Ar with laser transitions are shown in Fig. 3.1. Lasing occurred in the IR range of the spectrum, and the most intensive laser lines belong to the transitions nd–$(n+1)p$ of Xe, Kr, and Ar atoms ($n = 5, 4, 3$ for Xe, Kr, and Ar respectively).

Experimental investigations of NPLs operating on IR transitions of Xe, Kr, and Ar atoms in Russia were done chiefly at VNIIEF and VNIITF, and elsewhere at the Sandia Laboratories (United States).

Research by VNIIEF

Already in the first experiments performed with the VIR-2 reactor in 1972, the output power of the xenon laser with an optimal pressure and composition of the He-Xe mixture was 25 W with $\eta_l \sim 0.5$ % [1]. In subsequent experiments using the TIBR-1M, VIR-2, and VIR-2M reactors, lasing was obtained in a spectral range of 1.15–3.65 μm at 24 transitions of the Xe, Kr, and Ar atoms, with excitation of active

S.P. Melnikov et al., *Lasers with Nuclear Pumping*,
DOI 10.1007/978-3-319-08882-2_3

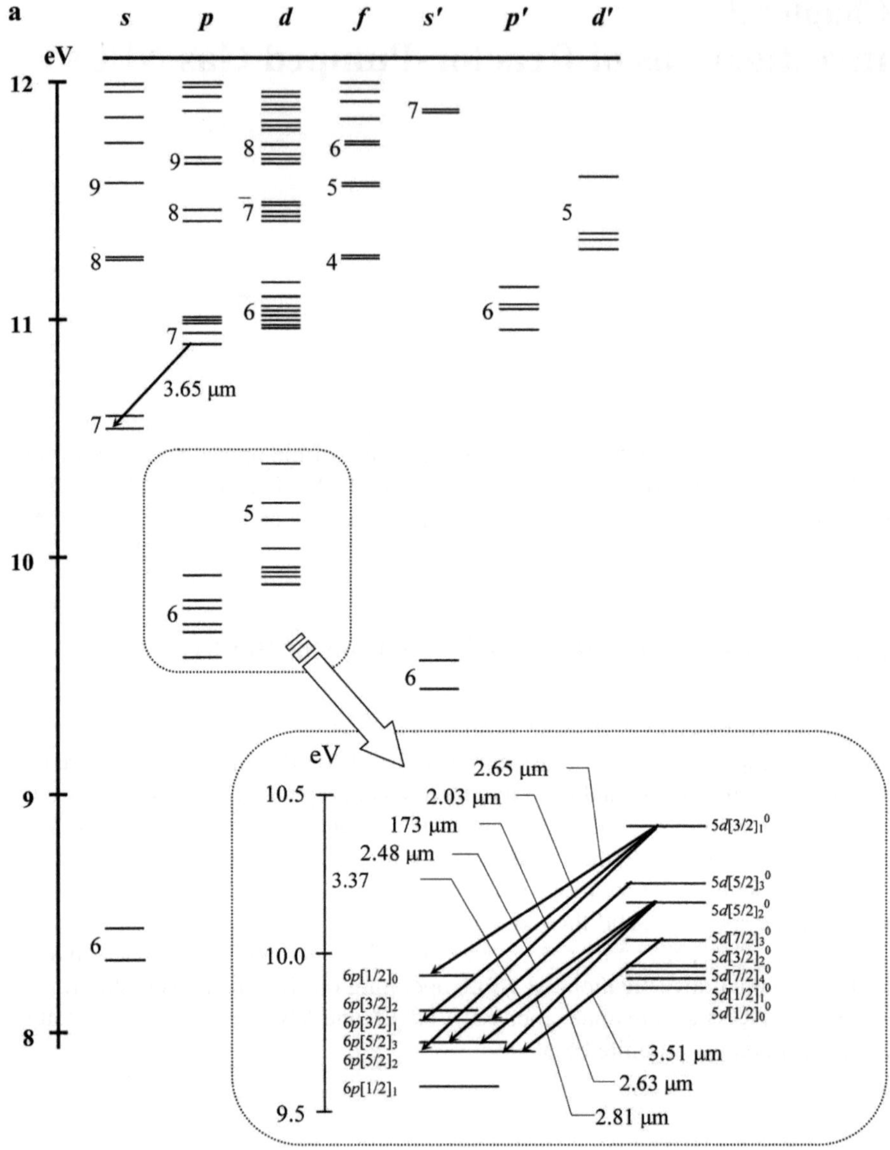

Fig. 3.1 (continued)

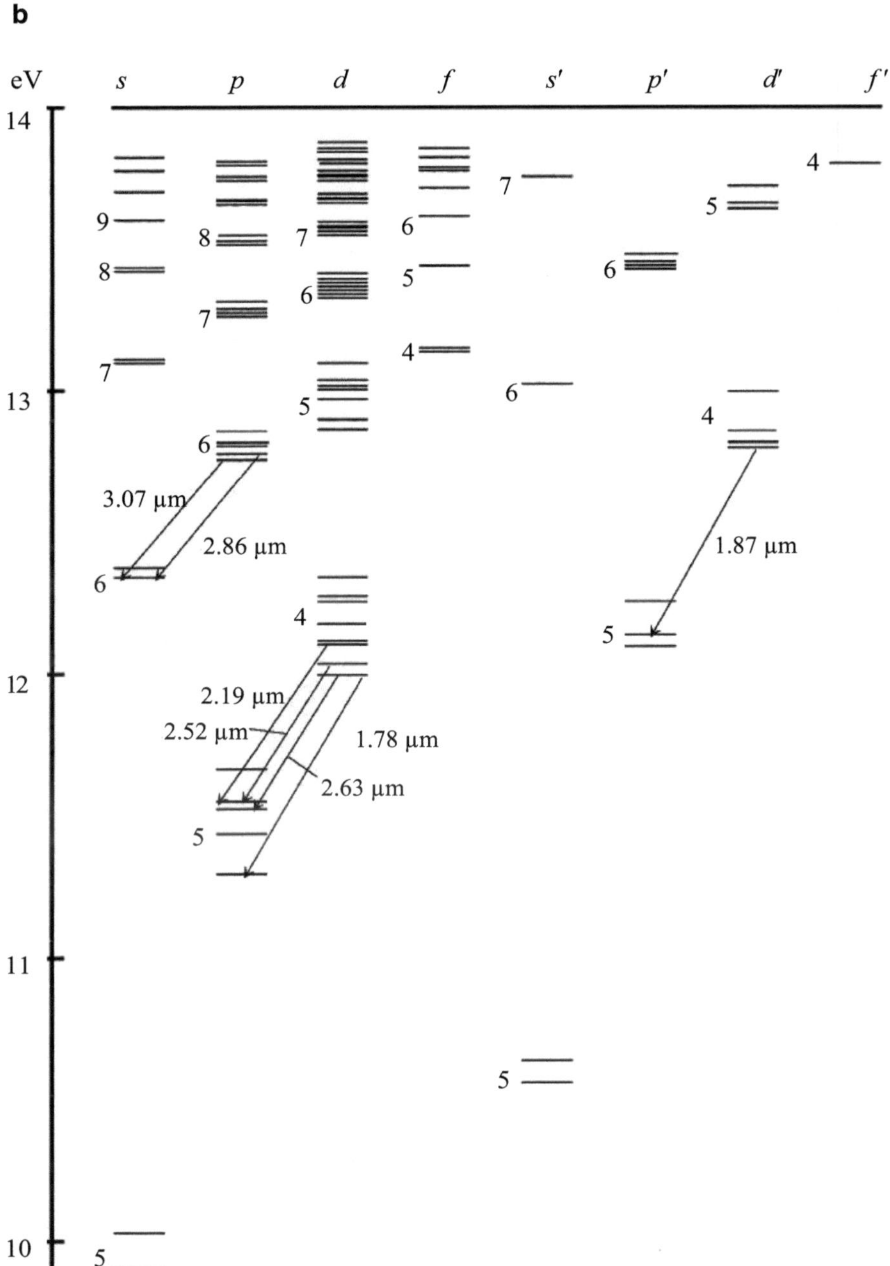

Fig. 3.1 (continued)

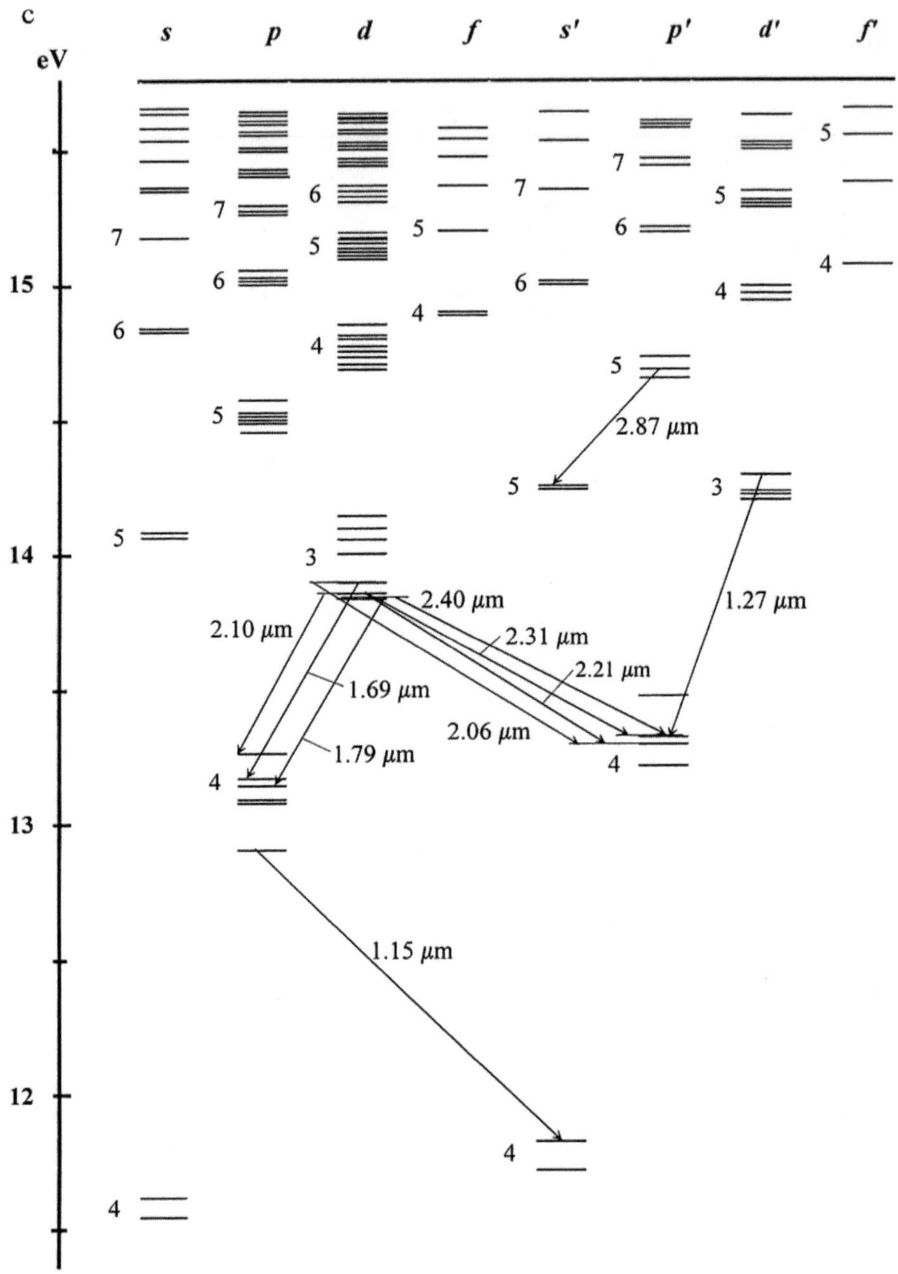

Fig. 3.1 Diagrams of excited levels of atoms Xe (**a**), Kr (**b**), and Ar (**c**) with laser transitions

media by uranium fission fragments. Their spectral, energy, and threshold characteristics were studied in detail. In part these results are cited in the reviews [1–4]. Table 3.1 shows the basic data for all the lasers studied at VNIIEF operating on transitions of Xe, Kr, and Ar atoms with references to the original studies. Experimental units and methods are examined in Chap. 2, Sects. 2.3 and 2.4.

The majority of laser lines belong to the transitions *nd*–(*n* + 1)*p* of Xe, Kr, and Ar atoms (see Fig. 3.1). Apart from that, lasing was observed at 3.65 μm (transition 6*p*-6*s* XeI); 3.07 and 2.86 μm (transitions 5*p*-5*s* KrI); 1.87 μm (transition 4*d*′-5*p*′ KrI); 2.87 μm (transition 5*p*′-5*s*′ ArI); 1.27 μm (transition 3*d*′-4*p*′ ArI) and 1.15 μm (transition 4*p*-4*s* ArI). We note that nearly all of the laser transitions shown in Table 3.1 were observed early in low-pressure gas discharge lasers [18] except for the 2.81 μm line of the Xe atom and the 1.87 μm line of the Kr atom. The mechanism of creating the population inversion on transitions of Xe, Kr, and Ar atoms is reviewed in Sect. 5.2 of Chap. 5.

The laser operating on transitions of the Xe atom has the maximal energy parameters. The active medium of this laser is mixtures of He-Xe, Ar-Xe, Kr-Xe, He-Ar-Xe, Ne-Ar-Xe and pure Xe at pressures ≤6 atm. Helium, neon, argon, and krypton are buffer gases, while the concentration of xenon is 1–10 %. Depending on the conditions of the experiment (pressure and composition of mixtures, power deposition, properties of the cavity mirrors), lasing was observed at different lines in a spectral range of 1.7–3.5 μm.

The maximal energy parameters (output power $W_{out} \leq 2$ kW, $\eta_l \leq 2.5$ %) were registered at the 1.73; 2.03 and 2.65 μm lines of the Xe atom, which originate from the level 5*d*[3/2]1^0. For these same lines, the minimal laser thresholds $\Phi_{th} \leq 10^{13}$ cm^{-2} s^{-1} were observed. The lowest laser threshold, at $\Phi_{th} = 1.5 \times 10^{12}$ cm^{-2} s^{-1} (specific power deposition at laser threshold $q_{th} \sim 0.1$ W/cm^3) was registered for mixtures of Ar-Xe ($\lambda = 2.03$ μm) [6, 7]. Such low thresholds make it possible to use not only neutron radiation of stationary nuclear reactors to pump an Ar-Xe laser, but also radioisotope sources [7].

NPLs operating on transitions of Kr and Ar atoms have lower energy parameters—in pumping mixtures of He(Ne)-Ar and He-(Ne)-Kr at atmospheric pressure with uranium fission fragments, $\eta_l \leq 1$ % was obtained. Among NPLs using Ar atom transitions, one should note the He-Ar laser ($\lambda = 1.15$ μm) [8, 16], in which quasi-CW lasing occurs at transition 4*p*[1l/2]1-4*s*′[1/2]1^0 as a result of collision “quenching” of the lower metastable level 4*s*′[1/2]1^0 by atoms of the buffer gas, helium.

In experiments using an Ar-Xe mixture carried out on the LUNA-2M setup in 1985, a competition effect was observed among the laser lines 1.73; 2.03 and 2.65 μm of the Xe atom, which have a common upper laser level. Figure 3.2 shows oscillograms of pulses of neutron and laser radiation for the mixtures He-Xe, Ar-Xe, and Ar-Xe-He [6]. In the He-Xe mixture, lasing occurs only at $\lambda = 2.03$ μm (Fig. 3.2a). In the mixture Ar-Xe, lasing first occurs at $\lambda = 2.03$ μm. With an increase in the pumping power, the line 1.73 μm appears, and there is stopping of lasing at $\lambda = 2.03$ μm. The 2.03-μm line arises again at the end of the pumping pulse after cessation of lasing at $\lambda = 1.73$ μm. Small additions of helium

Table 3.1 Results of investigations of NPLs operating on transitions of Xe, Kr, and Ar atoms (VNIIEF)

Atom	λ, μm	Mixture	P, atm	W_{out}, W	η_l, %	$\Phi_{th} \times 10^{-14}$, cm^{-2} s^{-1}	$\Phi_{max} \times 10^{-15}$, cm^{-2} s^{-1}	Reactor	Work cited
Xe	1.73	He-Xe	2	80	0.2	4.0	1.1	VIR-2	[5]
	1.73	Ar-Xe	0.5	500	2.2	0.1	1.1	VIR-2	[5]
	1.73	He-Ar-Xe	1	410	2.5	0.1	1.1	VIR-2	[3]
	2.03	He-Ar-Xe	1	700	1.0	0.6	2.5	VIR-2M	[3]
	2.03	Ar-Xe	0.25	–	–	0.015	2.5	VIR-2M	[6, 7]
	2.48	He-Xe	1	–	–	–	26	TIBR-1M	[8]
	2.65	He-Xe	5	2,000	0.8	10	26	TIBR-1M	[8]
	2.63 2.81	Kr-Xe	0.35	150	0.2	0.03	2.5	VIR-2M	[9]
	2.63 2.65	Ar-Xe	0.5	230	2.0	0.1	1.1	VIR-2	[10]
	2.65	He-Xe	3	250	1.2	0.6	1.1	VIR-2	[10]
	2.65	He-Ne-Xe	1.5	520	0.6	1.2	2.5	VIR-2M	[11]
	2.65	He-Ar-Xe	1	320	0.5	0.1	2.2	VIR-2M	[12, 13]
	2.65	Ne-Ar-Xe	1.1	570	0.6	0.4	2.5	VIR-2M	[11]
	3.11 (possibly 3.37)	Ar-Xe	0.5	–	–	–	1.1	VIR-2	[10]
	3.51	Xe	0.6	90	0.1	0.8	2.5	VIR-2M	[9]

Kr	1.78	He-Kr	2	50	0.3	0.6	1.1	VIR-2	[14]
	2.52	He-Kr	2	110	0.6	2.6	1.1	VIR-2	[14]
	3.07	He-Kr	2	40	0.2	1.7	1.1	VIR-2	[14]
	1.78 1.87 2.19	He-Kr	2	120	0.2	0.29	2.5	VIR-2M	[15]
	1.78 1.87 2.19	He-Ne-Kr	1	140	0.2	0.19	2.5	VIR-2M	[15]
	2.52 3.07	He-Kr	2	420	0.6	3.2	2.5	VIR-2M	[15]
	2.19 2.86 3.07	Ne-Kr	1	100	0.1	0.68	2.5	VIR-2M	[15]
	2.19 2.52 2.86 3.07	He-Ne-Kr	1	460	0.6	1.2	2.5	VIR-2M	[15]
	2.63	Kr	0.25	2	–	0.2	2.5	VIR-2M	[15]

(continued)

Table 3.1 (continued)

Atom	λ, μm	Mixture	P, atm	W_{out}, W	η_l, %	$\Phi_{th} \times 10^{-14}$, $cm^{-2}\,s^{-1}$	$\Phi_{max} \times 10^{-15}$, $cm^{-2}\,s^{-1}$	Reactor	Work cited
Ar	1.15 1.19	He-Ar	2	250	0.1	100	26	TIBR-1M	[8, 16]
	1.27	He-Ar	1	10	–	5.0	2.1	VIR-2M	[17]
	1.69 1.79	He-Ar	2	390	0.6	3.2	2.1	VIR-2M	[17]
	1.69 1.79	Ne-Ar	1	140	0.2	0.9	2.1	VIR-2M	[17]
	2.10	He-Ar	2	220	0.3	6.3	2.5	VIR-2M	[15]
	2.06 2.10 2.21	Ne-Ar	1	110	0.15	1.2	2.5	VIR-2M	[15]
	2.40	He-Ar	1	–	–	–	26	TIBR-1M	[8]
	2.40 2.87	He-Ar	2	190	0.25	8.4	2.5	VIR-2M	[15]
	2.21 2.31 2.40	Ne-Ar	1	60	0.1	1.3	2.5	VIR-2M	[15]
	2.31 2.40 2.87	He-Ne-Ar	1.3	140	0.15	2.0	2.5	VIR-2M	[15]

Note: P pressure of active medium, W power of laser radiation (output power), η_l power efficiency (ratio of output power to power deposition), Φ_{th} thermal-neutron flux density averaged over active medium length, at which the laser threshold is achieved (threshold thermal-neutron flux density), Φ_{max} maximum thermal-neutron flux density averaged over active medium length, at the reactor pulse maximum

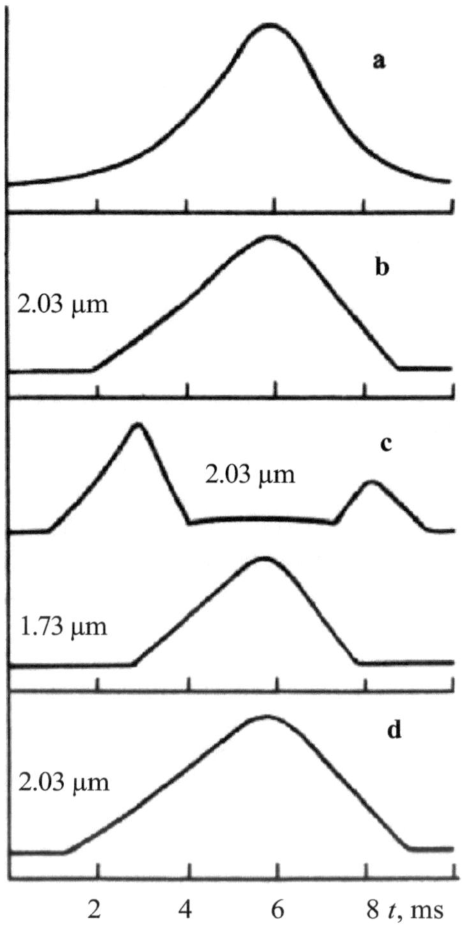

Fig. 3.2 Oscillograms of thermal neutron pulse (**a**) and of laser pulses: (**b**) mixture of He-Xe (1,000:1), $P = 2$ atm; (**c**) mixture of Ar-Xe (100:1), $P = 0.5$ atm; (**d**) mixture of Ar-Xe-He (100:1:100), $P = 1$ atm [6]

(0.25–0.5 atm), which insignificantly affect the power deposition, lead to elimination of the 1.73-μm line (Fig. 3.2d).

Competition of laser lines having the common upper laser level $5d[3/2]_1{}^0$ may be explained by the differences in probabilities of radiative decay and the rate constants of the processes of collision "quenching" of the lower laser levels by atoms of the buffer gas. An analogous effect was observed in the Ar-Xe mixture for two other lines of the Xe atom (2.65 and 1.73 μm; 2.03 and 2.65 μm) [6, 7], in the Kr-Xe mixture for the 2.63 and 2.81 μm lines of the Xe atom [9], as well as in the mixture Ne-Ar for the 2.31 and 2.21 μm lines of the Ar atom [15]. Competition of lines was also observed in excitation of the mixture Ar-Xe by an electron beam [19, 20]. The most complete analysis of the different variants of competition of laser lines having a common upper or lower laser level was done in studies [21–23].

For a more detailed study of xenon NPLs using mixtures of He-Xe (1.73; 2.03 and 2.65 μm), Ar-Xe (1.73 μm), and He-Ar-Xe (2.03 and 2.65 μm), a series of

Table 3.2 Small-signal gains (α_0) and saturation intensities (I_s) for NPLs transitions of the Xe atom at optimal xenon pressure

Mixture Parameters	He-Xe (1.73 μm)	He-Xe (2.03 μm)	He-Xe (2.65 μm)	Ar-Xe (1.73 μm)	He-Ar-Xe (2.03 μm)	He-Ar-Xe (2.65 μm)
Buffer gas	He	He	He	Ar	He-Ar(1:1)	He-Ar(1:1)
Pressure, atm	2	2	2	0.5	1	1
P_{Xe}, Torr	3.8	0.95	0.95	0.95	0.95	0.95
α_0, cm^{-1}	6.8×10^{-4}	1.7×10^{-3}	3.6×10^{-3}	3.6×10^{-3}	1.6×10^{-2}	2.4×10^{-2}
I_s, W/cm^2	120	80	72	110	20	7.3
Works cited	[24]	[24]	[12]	[12, 13]	[12]	[12, 13]

experiments was carried out on the LUNA-2M setup to measure the energy and threshold characteristics as a function of the partial pressure of the xenon with various transmission coefficients of the cavity mirrors [12, 13, 24]. Narrow-band dielectric mirrors were used in the experiments, which made it possible to preclude simultaneous lasing at several lines. The basic goal of the experiments was to obtain information about the parameters of the active media of xenon NPLs (small-signal gain α_0, saturation intensity I_s, and coefficient of distributed losses β) as a function of the partial pressure of the xenon. This information, obtained as a result of processing of experimental data using an approximation by Rigrod's formula [25], is shown in Table 3.2 for $q = 40$ W/cm^3 (maximum of pumping pulse) and optimal partial pressures of xenon.

To study the parameters of the active medium of a xenon NPL, a direct method of measurement by the "oscillator-amplifier" scheme was also used [26]. Experiments were carried out on the LUNA-2M setup for the mixture He-Ar-Xe (380:380:1) at a pressure of 1 atm at the 2.03 μm line. The small-signal gain increases approximately linearly with an increase in the specific power deposition, and at a maximal value of $q = 40$ W/cm^3 reaches 1.2×10^{-2}, while the saturation intensity is virtually independent of q and is 70–90 W/cm^2. The output laser power at the pumping pulse maximum at the outlet from the amplification channel (790 W) proved to be virtually twice as high as at the input (400 W).

In the experiments considered above, simple, stable two-mirror cavities were used. The presence of two identical laser channels in the LUNA-2M setup makes it possible to study more complex schemes for combining the NPL channels and the radiation withdrawal, which is of interest for forming output radiation of multichannel laser facilities, particularly to reduce the number of light beams (see Chap. 10). VNIIEF proposed two methods of combining the NPL channels, which by analogy with the parallel and serial connection of elements in electrical circuits, can be called "serial" and "parallel" combining of laser channels (see Fig. 10.7). The results of computational and experimental investigations in this area are shown in studies [27–30].

The basic results of investigations on combining the radiation of laser channels on the LUNA-2M setup are shown in Table 3.3. The active laser media used were mixtures of He-Ar-Xe (380:380:1) at a pressure of 1 atm ($\lambda = 2.03$ μm) and Ar-Xe

Table 3.3 Maximal energy parameters with serial and parallel combining of two laser channels of the LUNA-2M setup [28–30]

λ, μm	Mixture	Pressure	Type of laser channel	r_{opt}, %	E, J	W_{out}, kW
2.03	He-Ar-Xe (380:380:1)	1 atm	*A*	71	1.5	0.54
			B	46	1.8	1.0
			C	No output coupler	1.9	0.87
1.73	Ar-Xe (380:1)	0.5 atm	*A*	68	0.96	0.52
			B	47	1.1	0.62
			C	No output coupler	1.7	0.98

Note: *A* is one standard channel (two-mirror cavity with radiation output in the longitudinal direction relative to the optical axis); *B* is two serial combined laser channels (see Fig. 10.7a); *C* is two parallel combined laser channels (see Fig. 10.7b); r_{opt} is the optimal reflectivity of the output mirror; *E* is the output energy of laser radiation per pulse; W_{out} is the maximal output power of laser radiation in the process of the pumping pulse

(380:1) at a pressure of 0.5 atm ($\lambda = 1.73$ μm). The active volumes of the laser channels were identical and equal to 1.9 l. It is clear from Table 3.3 that the output power of two serially combined channels ($\lambda = 2.03$ μm) and parallel combined channels ($\lambda = 1.73$ μm) is virtually twice the output power of a single channel. It should be noted that optical non-uniformities arising in the laser medium in the process of the pumping pulse have a significant effect on the laser parameters of the double laser channel with serial combining. This is particularly distinctly demonstrated for the mixture Ar-Xe. In the case of the parallel combining, the shape of the laser pulse of several combined channels does not depend on the number of channels, but the output power and energy are proportional to the number of combined channels. This is explained by the fact that with parallel combining (in contrast to serial), there is no accumulation of optical non-uniformities. The parallel scheme is therefore very promising for use in multichannel NPLs.

Research by VNIITF

Investigations of NPL problems were begun at VNIITF in 1979, and in 1981 the first experiments were carried out on pumping a He-Xe mixture with uranium fission fragments [31]. In contrast to the majority of experimental studies, in which the laser cells were placed close to the cores of the pulsed reactors, the VNIITF experiments used the specialized EBR-L setup, which included a fast neutron pulsed reactor and laser cell (see Chap. 2, Sect. 2.5).

The EBR-L setup was used to research IR lasers operating on transitions of the Xe and Ar atoms, the basic results of which are shown in Table 3.4.

Table 3.4 Results of studies of NPLs operating on IR transitions of rare-gas atoms (*VNIITF*)

Atom	Mixture	λ, μm	$\Phi_{th} \times 10^{-14}$, $cm^{-2}\ s^{-1}$	W_{out}, W	η_l, %	Works cited
Xe	He-Xe	1.73	50	300	0.2	[32, 33]
	Ar-Xe	1.73	0.6	430	2.1	[33]
	He-Ar-Xe	1.73	1.0	2,600	1.7	[32, 33]
	Ne-Ar-Xe	1.73	1.0	700	1.7	[32, 33]
	He-Xe	2.03	19	430	0.3	[32, 33]
	He-Ar-Xe	2.03	0.5	2,000	1.5	[32, 33]
	He-Ar-Xe	2.03	–	1.3×10^6	3	[34]
	He-Xe	2.65	25	170	0.15	[32, 33]
	He-Ar-Xe	2.63; 2.65	40	1,000	0.9	[32, 33]
	He-Ar-Xe	2.48	30	15	~0.01	[33, 35]
Ar	He-Ar	1.79	3.2	–	1.2	[36, 37]

The high efficiency of the Xe-laser, confirmation of the results of experiments at VNIIEF, as well as the $\eta_l \geq 1$ % obtained for the He-Ar laser ($\lambda = 1.79$ μm), must be considered among the most important results. One should also note the study [37], in which experiments using the EBR-L and IGRIK pulsed reactors determined the energy characteristics for lasers using the mixtures He-Ar ($\lambda = 1.79$ μm) and He-Ar-Xe ($\lambda = 2.03$ μm), at high specific energy depositions of up to 2.5 J/cm^3. As a result of the investigations, quite high specific output energies were obtained—7.5 J/l at $\lambda = 1.79$ μm and 9 J/l at $\lambda = 2.03$ μm.

Some important experiments were recently carried out at VNIITF using the He-Ar-Xe mixture ($\lambda = 2.03$ μm). These experiments (discussed in reference [34]) demonstrated the maximal energy parameters for NPLs excited using pulsed reactors. The laser output energy was 520 J (output power of 1.3 MW) with an efficiency $\eta_l = 3$ %. More detailed information is presented later in Chap. 6, Sect. 6.4.

Research Outside of Russia

Outside of Russia, lasers operating on transitions of rare gas atoms were first pumped by nuclear radiation in 1975 at LANL in the United States, with excitation of the mixture He-Xe ($\lambda = 3.51$ μm) by uranium fission fragments [38]. Further investigations were aimed at finding and studying parameters of NPLs operating on various transitions of Xe, Kr, and Ar atoms. The basic results of the experiments are shown in Table 3.5.

It is shown that in these studies, carried out mainly in the United States, lasing was obtained on the same transitions *nd*–$(n+1)p$ of the Xe, Kr, and Ar atoms (except for the 3.65 μm line of the Xe atom and 1.27 μm line of the Ar atom). Maximal output powers up to 1 kW were registered in the experiments [51] with the

Table 3.5 Results of foreign studies of NPLs operating on IR transitions of Xe, Kr, and Ar atoms

Atom	λ, µm	Mixture	P, atm	W_{out}, W	η_l, %	$\Phi_{th} \times 10^{-14}$, $cm^{-2}\ s^{-1}$	$\Phi_{max} \times 10^{-15}$, $cm^{-2}\ s^{-1}$	Pumping method	Reactor (laboratory)	Works cited
Xe	3.51	He-Xe	0.26	0.01	0.0001	0.5	–	Fission fragments	Godiva-IV (LANL)	[38]
	2.03	^{3}He-Xe	0.53	0.08	–	40	150	$^3He(n,p)^3H$	APRF (NASA)	[39]
	2.03	^{3}He-Xe	3.9	0.35	–	40	150	$^3He(n,p)^3H$	APRF (NASA)	[40]
	2.03	^{3}He-Xe	0.76	0.1	0.005	–	33	$^3He(n,p)^3H$	Godiva-IV (LANL)	[41]
	3.51			0.4						
	3.65			0.5						
	2.6	Ar-Xe	0.8	4.6	0.1	40	80	Fission fragments	APRF (NASA)	[42]
	2.6	Ar-Xe	1.0	50	–	–	40	Fission fragments	APRF (NASA)	[43]
	2.63*	^{3}He-Xe	3.0	215	0.05	300	60	$^3He(n,p)^3H$	APRF (NASA)	[44]
	1.73	Ar-Xe	0.68	250	5.6	–	~10	Fission fragments	SPR-III (Sandia)	[45]
	2.03	He-Ar-Xe	1.35	250	3.3	–	~10	Fission fragments	SPR-III (Sandia)	[45]
	1.73	He-Ar-Xe	1.0	150	2.5–3.0	–	~1	Fission fragments	SPR-III (Sandia)	[46, 47]
	1.73	Ar-Xe	0.44	–	–	~0.2	~1	Fission fragments	ACRR (Sandia)	[48]
	1.73	Ar-Xe	0.4	–	–	1	2.5	$^{10}B(n,p)^7Li$	TRIGA (UI)	[49]
Kr	2.52	^{3}He-Kr	–	–	–	1,100	150	$^3He(n,p)^3H$	APRF (NASA)	[39]
	2.52	^{3}He-Kr	0.53	–	–	1,000	150	$^3He(n,p)^3H$	APRF (NASA)	[40]

(continued)

Table 3.5 (continued)

Atom	λ, μm	Mixture	P, atm	W_{out}, W	η_l, %	$\Phi_{th} \times 10^{-14}$, $cm^{-2}\ s^{-1}$	$\Phi_{max} \times 10^{-15}$, $cm^{-2}\ s^{-1}$	Pumping method	Reactor (laboratory)	Works cited
Ar	1.79	^{3}He-Ar	0.92	0.05	0.0001	140	150	^{3}He(n,p)^{3}H	APRF (NASA)	[50]
	1.79	^{3}He-Ar	2.0	3.4	0.001	250	150	^{3}He(n,p)^{3}H	APRF (NASA)	[40]
	1.27	^{3}He-Ar	–	–	–	–	150	^{3}He(n,p)^{3}H	APRF (NASA)	[40]
	1.79	^{3}He-Ar	2.9	550	0.01	20	30	^{3}He(n,p)^{3}H	APRF (NASA)	[51]
	1.79	^{3}He-Ar	3.0	1,012	0.01	47	43	^{3}He(n,p)^{3}H	APRF (NASA)	[52]
	1.79	He-Ar	1.0	270	1.4	~1	~10	Fission fragments	SPR-III (Sandia)	[47, 53]
	1.27	He-(Ne)-Ar	1.0	30	1.1	~1	~10	Fission fragments	SPR-III (Sandia)	[47, 54]

Note: *In the He-Xe mixture, lasing occurs most likely at the 2.65 μm line, rather than at 2.63 μm line (see Chap. 5, Sect. 5.2)

APRF fast pulsed reactor ($\Phi_{max} = 4.3 \times 10^{16}$ cm^{-2} s^{-1}) for a laser with the mixture ^{3}He-Ar ($\lambda = 1.79$ μm) when a multiple-path laser cell with a volume of 3.6 l was used (see Chap. 2, Sect. 2.1). However, the laser efficiency was not great, and amounted to ~0.01 %.

In studies of American researchers published prior to 1989, the efficiency for IR lasers operating on the transitions of Xe, Kr, and Ar atoms did not exceed 0.1 %, which may be explained by the design features of the laser cells, which were sealed at the ends by windows arranged at the Brewster angle. In these windows, owing to the reduction of light transmission under the effects of the reactor n,γ-radiation, additional losses occur inside the resonator (see Chap. 2, Sect. 2.2). In addition, the presence of Brewster windows prevented lasing at the intensive 2.63 and 2.65 μm lines of the Xe atom, because of absorption of radiation in atmospheric water vapor in the cavity areas located between the windows and the mirrors.

Significantly higher values of $\eta_l = 3.3$ and 5.6 % were obtained at the Sandia Laboratories for lasers using the mixtures He-Ar-Xe ($\lambda = 2.03$ μm) and Ar-Xe ($\lambda = 1.73$ μm) when cells with windows made from radiation-resistant optical materials were used, and with use of comparatively low specific power depositions $q < 10$ W/cm^3 [45]. At higher specific power depositions and specific absorbed energies, there was a reduction in efficiency, and lasing was observed at the leading edge of the pumping pulse, which could be caused by the significant deteriorations of optical uniformity of the medium (especially in the Ar-Xe mixture) or by plasma effects. Of the research done on the xenon NPL at Sandia, one should note the studies [46, 55], which examined the influence of helium additions on the characteristics of an Ar-Xe laser. The introduction of helium to the Ar-Xe mixture leads to a change in the laser spectrum (instead of the 1.73 μm line, the 2.03 μm line appears), and to an increase in the laser pulse duration.

Sandia carried out research into the amplification properties of NPL active media operating on transitions of the atoms Xe [45, 46, 56, 57] and Ar [53, 54]. It should be noted that the first experiments [58] to determine the gain of NPLs operating on transitions of the Xe atom were carried out in 1981 for a ^{3}He-Xe-laser at the 2.65 μm line (in study [58], the 2.63 μm line was mistakenly given as the lasing line). In the studies [45, 53, 54], the Rigrod theory [25] was used to determine the small-signal gain, while in the other studies, direct measurements were made by the "oscillator-amplifier" scheme using cw tunable color-center lasers, or low-pressure gas-discharge lasers using an He-Xe mixture.

Of the studies in the United States, one should mention [42], where the influence of additives of ^{235}UF$_6$ on the characteristics of the Ar-Xe laser ($\lambda = 2.6$ μm) was investigated. These experiments were aimed at determining the possibility of using the gaseous compound ^{235}UF$_6$ as a volume source of fission fragments. It was shown that reduction in the energy parameters is observed at a low concentration of uranium hexafluoride of ~0.5 %, which testifies to the effective "quenching" of the excited atoms Xe* by the molecules of UF$_6$. An analogous conclusion follows from calculations of characteristics of a laser based on the mixture ^{235}UF$_6$-Xe [59] and luminescence investigations [60], which detected a more intense "quenching" of the ArI lines in the ^{235}UF$_6$-Ar mixture than in ^{3}He-Ar. Thus at present there are no data that would allow us to hope for the possibility of creating lasers based on $^{235}UF_6$ as the buffer gas.

Efficiencies of NPLs Operating on Transitions of Xe, Kr, and Ar Atoms

The efficiency of transformation of deposited power (energy) into laser radiation is the most important parameter for all lasers. In the case of NPLs, the maximal efficiency of transformation of deposited nuclear energy into laser radiation belongs to the lasers operating on transitions of rare gas atoms, and the highest $\eta_l \geq 1$ % were registered for lasers operating on transitions of the Xe atom (1.73, 2.03, and 2.65 μm) and the Ar atom (1.27 and 1.79 μm). Table 3.6 compares the results obtained by various laboratories, and also provides the maximal efficiencies (η_{max}) constituting the ratio of the laser photon energy to the energy consumption for forming one primary active particle of plasma (ion or excited atom). This question is discussed in greater detail in the first section of Chap. 5. For comparison, Table 3.6 also includes information on efficiencies obtained from excitation of gas media by electron beams.

It follows from Table 3.6 that in some cases the real efficiency (η_l) approaches maximal, which testifies to minor energy losses in plasmochemical reactions of energy transfer and high selectivity of the final process of populating of the upper laser levels. Possibly the high efficiency of NPLs is explained by the noticeable

Table 3.6 Maximal efficiencies (η_l) of Xe and Ar atoms

Atom	Medium	λ, μm	Experimental values of η_l, %				Maximal efficiency (η_{max}), %
			Uranium fission fragments			Electron beam	
			VNIIEF	VNIITF	Sandia		
Xe	He-Xe	2.65	1.2 [10]	0.15 [33]	–	1.6 [61]	1.6
	Ar-Xe	1.73	2.2 [5]	2.1 [33]	5.6 [45]	1.5 [62] 2.6 [63]	3.8
	Ar-Xe	2.63 2.65	2.0 [10]	–	–	–	2.5
	He-Ar-Xe	1.73	2.5 [3]	1.7 [33]	3.0 [55]	–	4.0
	He-Ar-Xe	2.03	1.0 [3]	1.5 [33] 3 [34]	3.3 [45]	1.1 [64]	3.4
	He-Ar-Xe	2.65	0.5 [12, 13]	0.9 [33]	2.0 [47]	–	2.6
Ar	He-Ar	1.79	0.6 [17]	1.2 [36] [37]	1.4 [47, 53]	1.4 [61]	2.3
	He-Ar	1.27	–	–	1.1 [47, 54]	–	3.3

contribution to populating the upper laser levels from the processes of associative ionization with the participation of metastable states 6*s* of the Xe atom or 4*s* of the Ar atom [65].

For quasi-CW NPLs with a laser duration of $\geq$100 μs, which is substantially greater than the characteristic times of plasma processes, the efficiency usually is defined as the ratio of the output power to the instantaneous power absorbed in the laser's active volume. As a rule the active volume is understood to be the entire excited volume of the laser cell. In particular, this method was used to determine the efficiency in the studies by VNIIEF and VNIITF. However, sometimes η_l was determined with respect to the deposition of power to the mode (radiating) volume, which was less than the total cell volume. Naturally, in this case the value of η_l will be greater. For example, in study [40], the efficiencies for NPLs using the mixture ^{3}He-Ar ($\lambda = 1.79$ μm), calculated by two methods, differ by roughly a factor of 100.

Another reason for the discrepancy in the efficiency data obtained in the various laboratories is the use of two different methods to determine the energy deposition to the laser medium: from the results of measurement of the pressure jump in the gas medium, or by calculation. The results of investigations in different laboratories showed that the energy deposition measured in experiments is less than the computed value by roughly a factor of 1.5–2. The reasons for the discrepancy between the computed and experimental data are considered in Sect. 7.4 of Chap. 7. It cannot be excluded that the values of η_l obtained in the Sandia experiments, which exceed maximal values, may be explained by errors in determining the energy deposition.

Influence of Temperature and Gas Impurities on Laser Characteristics

In experiments with NPLs based on pulsed reactors, the specific power deposition $q < 5$ kW/cm^3 (specific energy deposition up to 1 J/cm^3), which can lead to a marked increase in the temperature of the gas medium, of up to ~1,000 °K at pressures of ~1 atm. An increase in the temperature in the process of pulse pumping can influence the output power and possibly be the reason for a cessation of lasing until the moment of attainment of the pumping pulse maximum for some mixtures (for example, see [8, 45, 66]). Marked heating of the active medium can occur in powerful stationary nuclear-laser devices, so that a study of the influence of heating temperature of the NPL active medium on their characteristics is of great interest.

The first experiments in this field were carried out at VNIITF for NPLs operating on transitions of the Xe atom. By now the influence of temperature on NPL characteristics has been studied for mixtures of He-Xe, He-Ar-Xe, Ar-Xe [33, 67, 68]; ^{3}He-Ar-Xe [69]; Ar-Xe, He-Ar-Xe, Ne-Ar-Xe [70] and Ar-Xe, He-Ar-Xe, He-Ar, He-Ne-Ar [71]. In all the experiments, an electric heater was used to change the temperature of the gas media. In the experiments, a marked reduction (by a factor of ~2) in the energy parameters was recorded at comparatively low temperatures (see Table 3.7). Substantially different data were obtained in study [72], in

Table 3.7 Influence on temperature of the gas medium on NPL energy parameters

Medium	He-Xe (400:1)	He-Xe (400:1)	He-Xe (400:1)	Ar-Xe (200:1)	Ar-Xe (330:1)	He-Ar-Xe (200:200:1)	He-Ar-Xe (2,000:200:1)	He-Ar-Xe (200:200:1)
Initial pressure, atm	–	–	–	0.8	0.34	1	2	–
λ, μm	1.73	2.03	2.65	1.73	1.73	1.73	1.73	2.03
$T_{1/2}$, °K	350	360	390	400	440	430	350	520
Works cited	[67]	[67]	[67]	[67, 68]	[71]	[67]	[67]	[33]
Medium	^{3}He-Ar-Xe (100:100:1)	^{3}He-Ar-Xe (100:100:1)	He-Ar-Xe (107:107:1)	He-Ar-Xe (107:107:1)	He-Ar (333:1)	He-Ne-Ar (220:110:1)	He-Ne-Ar (220:110:1)	
Initial pressure, atm	1	1	0.57	0.57	0.68	0.68	0.57	
λ, μm	2.03	2.65	1.73	2.03	1.79	1.79	1.27	
$T_{1/2}$, °K	530	420	500	550	400	450	440	
Works cited	[69]	[69]	[71]	[71]	[71]	[71]	[71]	

Note: The initial temperature is 300 K, $T_{1/2}$ is the temperature at which the energy parameters decrease by a factor of 2

which no changes were noted in the laser parameters up to a temperature of 900 K for an electroionization laser using a mixture of ^{3}He-Ar-Xe at a pressure of 1.5 atm with preliminary ionization of the gas medium by nuclear reaction products ^{3}He(n, p)^{3}H.

Analysis of the data of Table 3.7 shows that the influence of temperature on the NPL parameters depends on the pressure (gas density), composition of the gas medium, and laser wavelength. The reasons for this influence have not been finally ascertained to this point, and are the subject of discussion.

The most likely reasons considered for the reduction in the energy laser parameters with the growth in temperature were the following: the influence of temperature on the processes of formation of the population inversion, for example, collisional "quenching" and "mixing" of laser levels by the atoms of the buffer gas [69]; the reduction in the rate of formation of heteronuclear $ArXe^+$ ions and "mixing" of laser levels by electrons [70]; and the destruction of ions $ArXe^+$ during collisions with atoms of the buffer gas [73].

Another possible cause is contamination of the laser medium by outside gas impurities (in particular water vapor) as a result of their desorption from the cell walls as the temperature increases [68, 74, 75]. The electron attachment to molecules of H_2O can lead to a decrease in the concentration of electrons, and consequently to a reduction in the populating rate of the upper laser level as a result of reduction in the rate of recombination processes.

Interesting information was obtained in studies [76, 77], in which, given the excitation of lasers using mixtures of Ar-Xe ($\lambda = 1.73$; 2.03 μm) and He-Ar-Xe ($\lambda = 1.73$ μm) by a beam of $^{32}S^{9+}$ ions with an energy of 100 MeV (duration of rectangular pulses 20–50 μs, pulse repetition frequency 30–45 Hz), it was possible to study the separate influence of the temperature of the medium and H_2O impurities on the laser parameters. The energy parameters of a laser using an Ar-Xe mixture decrease both with an increase in the active medium without H_2O impurities, and with an increase in the content of H_2O at constant room temperature. Thus for the mixture Ar-Xe ($P = 0.16$ atm; 0.3 % Xe), the output power decreases by a factor of 2 when the temperature increases to 400 K. When there is an increase in the water vapor content in the Ar-Xe mixture ($P = 0.5$ atm; 0.5 % Xe), a similar reduction is observed when the concentration of H_2O is about 1×10^{15} cm^{-3}. For mixtures of He-Ar-Xe ($P = 0.2$ atm; 23 % He; 0.3 % Xe), an output power reduction by a factor of 2 occurs at a temperature of 520 K. In the opinion of the authors of [76, 77], the influence of water vapors lies not only in the reduction of electron concentration, but also in the collisional "quenching" of the upper laser level $5d[3/2]_1{}^0$ of the Xe atom, because the rate constant of the "quenching" process is very high and amounts to 4×10^{-9} cm^3 s^{-1}. The output power reduction observed in the experiments [77] with growth in temperature for mixtures without impurities of H_2O testifies that the gas temperature influences the processes of formation of the inverse population of laser levels, which were noted previously.

Apart from the water vapors, the NPL active media also contain the impurities N_2, CO_2, O_2, H_2, etc., which can appear both because of an inadequate degree of evacuation of the laser cells and the desorption processes of these impurities from the cell walls. At high power depositions ($q \geq 10$ kW/cm^3), which are achieved with the use of electron beams, the addition of molecule impurities sometimes leads to an increase in output power, which is due primarily to the reduction in the electron temperature [78]. In the case of NPLs operating at $q \leq 5$ kW/cm^3, the presence of molecular impurities leads to a reduction in the energy parameters and an increase in the laser thresholds [79].

3.2 Visible-Range Lasers Operating on Ne Atom Transitions

The spectra of rare-gas atoms have a number of intensive lines belonging to the transitions $(n+1)p$–$(n+1)s$ (where $n = 5, 4, 3, 2$ for Xe, Kr, Ar, and Ne, respectively), which end in metastable states $(n+1)s$. The efficiency of populating the upper levels $(n+1)p$ in a recombination-nonequilibrium plasma is very high [80, 81], so these transitions are promising in the search for lasing. However, in order to obtain lasing on the transitions $(n+1)p$–$(n+1)s$ in cw mode, a high depopulating rate of $(n+1)s$-levels is necessary. This can be ensured by two methods: through collisional "quenching" by atoms of the buffer gas (as, for example, in a nuclear-pumped laser at the line of 1.15 μm of the Ar atom [8, 16]), and using additional "quenching" impurities.

The difficulty in choosing the impurity is related first of all to the fact that it must not substantially influence the population of the upper level. This scheme was implemented in a neon NPL operating on transitions $3p$–$3s$ of the Ne atom in the lines 585.3-; 703.2- and 724.5-nm [2, 32, 33, 82–91]. Apart from nuclear radiation, electron [92–96] and ion [97] beams were used to pump the neon laser; when electron beams were used, lasing was also observed at the 626.7-; 633.4-; 659.9- and 743.9-nm lines [95]. The quenching impurities in the case of NPLs were $M =$ Ar, Kr, Xe, H_2, and with electron beam pumping, NF_3 also. Quenching of the lower $3p$-levels occurs through the Penning reaction: $Ne^*(3p) + M \rightarrow M^+ + Ne + e$. A diagram of the Ne atom levels with laser transitions is provided in Fig. 3.3.

A neon laser was first pumped with nuclear radiation at VNIIEF in 1985, immediately after the first reports [92, 93] about successful pumping of this laser with an electron beam. These results were published in the public press in 1995 [2, 82]. The basic results of researching NPLs operating on $3p$–$3s$ transitions of the Ne atom, obtained in various laboratories, are shown in Table 3.8.

In lasers using binary mixtures Ne-M ($M =$ Ar, Kr, Xe), lasing was observed at two lines—703.2 and 724.5 nm—while with a change in the concentration of the quenching impurity of M, there was restructuring of the lasing spectrum [82–84]. By way of illustration, Fig. 3.4 shows data for the mixture Ne-Kr. The ranges of impurity M pressure at which simultaneous lasing was observed at two lines are diverse. Thus, in the mixture Ne-Ar, simultaneous lasing occurs at $P_{Ar} \approx 28$–55 Torr, in the mixture Ne-Kr—$P_{Kr} \approx 30$–50 Torr, while for the mixture Ne-Xe

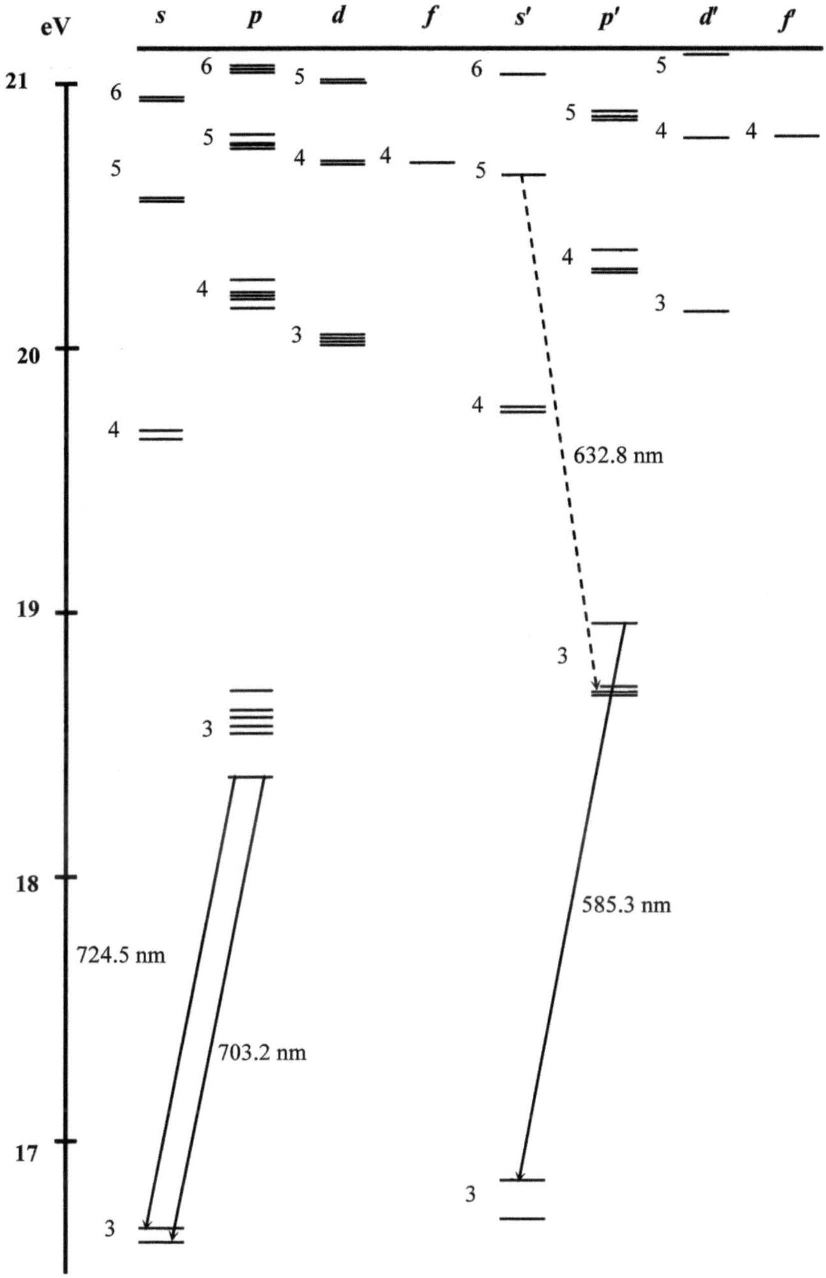

Fig. 3.3 Diagram of Ne atom levels with laser transitions

Table 3.8 Results of research on NPLs operating on $3p$-$3s$ transitions of the Ne atom

λ, nm	Mixture	P, atm	W_{out}, W	η_l, %	$\Phi_{th} \times 10^{-14}$, $cm^{-2}\ s^{-1}$	$\Phi_{max} \times 10^{-15}$, $cm^{-2}\ s^{-1}$	Pumping method	Reactor (laboratory)	Works cited
585.3	He-Ne-Ar (100:4:1)	1	10	~0.1	4.3	1.3	Fission fragments	VIR-2M (VNIIEF)	[82]
	He-Ne-Kr	1	–	–	–				
	He-Ne-Xe	1	–	–	–				
	^{3}He-Ne-Ar (100:2:1)	1	0.35	–	17	2.2	$^3He(n,p)^3H$	VIR-2M (VNIIEF, MIFI)	[86]
	He-Ne-Ar (190:10:3)	–	140	0.1	4	130	Fission fragments	EBR-L (VNIITF)	[32, 33]
	He-Ne-Ar (50:2:1)	1.5	–	>0.01	~100	~100	Fission fragments	SPR-III (Sandia)	[47, 87]
	^{3}He-Ne-H_2 (21:10:8)	2.8	–	–	~1	2.5	$^3He(n,p)^3H$, $^{10}B(n,\alpha)^7Li$	TRIGA (University of Illinois)	[88]
	He-Ne-Ar (250:5:2)	2	35	~0.1	6	2.5	Fission fragments	VIR-2M (VNIIEF)	[83, 84]
	He-Ne-Kr	2	25	~0.1	6				
	He-Ne-Xe	2	12	~0.05	7				
	Ne-H_2 (2:1)	0.6	~10	0.05	~100	130	Fission fragments	EBR-L (VNIITF)	[90, 91]
703.2 724.5	Ne-Kr (66:1)	1.5	2.5	–	2.4	1.3	Fission fragments	VIR-2M (VNIIEF)	[82]
703.2 724.5	Ne-Kr	–	30	0.02	120	130	Fission fragments	EBR-L (VNIITF)	[32, 33]
724.5	He-Ne-Kr	–	40	0.02	70				
703.2	He-Ne-Kr	2	40	0.06	~100	~100	Fission fragments	SPR-III (Sandia)	[47, 89]
724.5	He-Ne-Ar	1.7	90	0.1					

703.2	Ne-Kr (30:1)	1	7	~0.03	5.6	2.5	Fission fragments	VIR-2M (VNIIEF)	[83, 84]
724.5	Ne-Ar	1	2	–	14				
	Ne-Xe	1	1	–	18				
724.5	He-Ne-Kr (90:300:1)	1.3	40	~0.1	4				

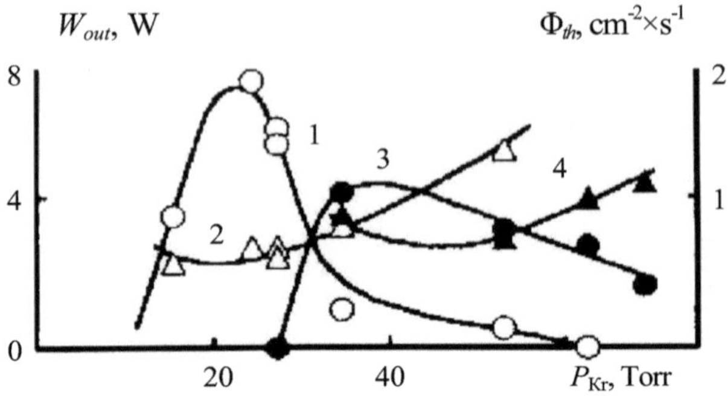

Fig. 3.4 Dependencies of output power W_{out} (1, 3) and threshold thermal-neutron flux density Φ_{th} (2, 4) at lines 724.5 (1, 2) and 703.2 nm (2, 4) for a Ne-Kr mixture on the Kr partial pressure at a Ne pressure of 1 atm [83, 84]

this pressure range is very narrow (close to $P_{Xe} \approx 28$ Torr). Similar results were also obtained in the mixtures He-Ne-Kr (Ar) [89].

The use of hydrogen as the quenching impurity is expedient only when there are sufficiently high specific power depositions ($q \geq 1$ kW/cm^3) because of competition of the processes of depopulation of 3*s*-levels of the Ne atom and charge-transfer of the molecular ions Ne_2^+ on molecules of H_2 [83, 85]. Therefore, the report [88] about obtaining lasing at the 585.3-nm line in NPLs using the mixture He-Ne-H_2 for $q \sim 100$ W/cm^3 is doubtful.

The $\eta_l \sim 0.1$ % obtained in experiments for neon NPLs is much lower than in the case of NPLs operating on IR transitions of rare gas atoms, although in experiments by FIAN associates with electron beams for an He-Ne-Kr laser ($\lambda = 585.3$ nm), $\eta_l = 1.6$ % [94] was noted. As was shown by the results of theoretical analysis [98], the maximal efficiency of a neon laser at the 585.3-nm line cannot exceed 0.5 %, therefore the high efficiency cited in study [94] is evidently explained by an error in determining the energy deposition. In later studies by FIAN associates, values of $\eta_l = 0.1$–0.2 % are cited for the neon laser at the 585.3-, 703.2-, and 724.5-nm lines.

By 1980, a report [99] was published stating that lasing had been obtained at transition $5s'[1/2]_1^0$-$3p'[3/2]_2$ of the Ne atom Ne ($\lambda = 632.8$ nm) while the mixture ^{3}He-Ne was excited by nuclear reaction products ^{3}He$(n,p)^3$H. According to the authors of [99], the laser efficiency was around 0.03 %, and the laser threshold was achieved at a very low thermal-neutron flux density of $\Phi_{th} = 2 \times 10^{11}$ cm^{-2} s^{-1}. The results of these experiments are doubtful, and are the subject of discussion in references [100, 101]. The simple evaluations cited in study [100] show that under the conditions of [99], it is not possible to obtain lasing even in the maximal case, when all power deposited to the active medium is transferred without losses to populate the $5s'[1/2]_1^0$-level and this level is not "quenched" in collisions with atoms. In later experimental investigations, it is noted that attempts at obtaining lasing at the 632.8-nm line with excitation of the mixture He-Ne using an ion beam [102] and radiation of a stationary nuclear reactor [103] yielded negative results. Numerous investigations [60, 103–107] of the luminescence spectra of the mixture He-Ne, excited by various types of nuclear radiation, showed the absence of a

632.8-nm line. The data cited allow us to conclude that the information [99] about creating a low-threshold NPL at the 632.8-nm line was erroneous.

In conclusion, let us examine the possibility of creating NPLs operating on the transitions $(n+1)p$–$(n+1)s$ of other rare gas atoms: Xe, Kr, and Ar. Such exploratory experiments were carried out at VNIIEF on the LUNA-2M setup in a range of 700–1,000 nm for those transitions $(n+1)p$–$(n+1)s$ of Ar, Kr, and Xe atoms at which maximal values of luminescence efficiency were obtained [81]. Studies were performed on Ar, Kr, and Xe, and the mixtures He-Ar (Kr,Xe) and Ar-Xe at pressures of up to 2 atm. For quenching of the lower $(n+1)s$ states, impurities of the molecular gases CO, H_2, N_2, and NF_3 were used, as well as Kr and Xe at partial pressures of 8–60 Torr. The numerous experiments did not yield a positive result, which possibly may be explained by the insufficient rate of quenching of the lower laser levels and/or the essential reduction in the populations of upper laser levels in the presence of impurities. The absence of lasing can be provoked by the fact that the experiments were conducted at low specific power depositions ($q < 50$ W/cm^3), so that the small-signal gain could be deficient for achieving the lasing threshold. It is helpful to conduct such research experiments at higher specific power depositions, for example using beams of fast electrons. This is confirmed by the successful experiments [108] in pumping of a laser using the mixture Ar-NF_3 (transition $4p$-$4s$ of an Ar atom, $\lambda = 750.4$ nm) by an electron beam.

3.3 Metal Vapor Lasers

The use of mixtures of rare gases and metal vapors as active media for NPLs is of interest first of all in connection with the possibility of obtaining lasing in the visible and UV ranges of the spectrum. We note that lasing at a large number of transitions of metal ions and atoms was obtained in recombination mode with the excitation of active media using an electron beam and in a gas discharge afterglow (for example, see [18, 109], and the literature cited there). This information helps in the search for NPL active media.

Virtually all of the experimental and theoretical research known to us concerning metal vapor NPLs was carried out in Russia. At VNIIEF, the possibility of using mixtures of rare gases with vapors of metals (Li, Na, Mg, Hg, Cd, etc.) as the NPL active media was considered from 1975 to 1976 [3].

Mercury-Vapor Lasers

The first experiments to study the possibility of developing metal-vapor NPLs were carried out in 1970 using the IIN water-pulsed reactor [110]. In these experiments, using the mixture ^{3}He-Hg at the transition $7p^2P_{3/2}$-$7s^2S_{1/2}$ of the ion Hg^+ ($\lambda = 615.0$ nm), light signals that exceeded the level of spontaneous radiation were registered. The pressure of the mixture ^{3}He-Hg was equal to 0.46 atm. These investigations were developed further in the study [111], in which, during experiments with the SPR-II pulsed reactor, with excitation of the mixture He-Hg ($P = 0.8$ atm) using nuclear reaction products $^{10}B(n,\alpha)^7Li$, lasing was achieved and studied at the 615.0-nm line. The output laser power was ~1 mW ($\eta_l \sim 10^{-6}$ %), and the laser threshold was reached at $\Phi_{th} \sim 10^{16}$ cm^{-2} s^{-1}. The authors of [111] note that attempts to obtain lasing for the conditions of [110], where the mercury-vapor pressure was three orders of magnitude greater, did not yield a positive result. Study [112], based on spectroscopic investigations of the mixture ^{3}He-Hg, concluded that it is not possible to obtain a high efficiency at the 615.0-nm line. This conclusion is confirmed by the calculations of [113], which show that the maximal efficiency at this line is no more than 0.03–0.04 %.

Investigations [112, 114, 115] of luminescence spectra of ^{3}He-Hg showed a high populating efficiency of level 7^3S_1 of the Hg atom, which is the upper of the triplet transitions 7^3S_1-$6^3P^0{}_{2,1,0}$ ($\lambda = 546.1$; 435.8, and 404.7 nm). The levels $6^3P^0{}_{2,1,0}$ of the Hg atom are metastable, so to obtain lasing in quasi-cw or cw modes, quenching of these levels is necessary using an additional impurity, for which in [112, 114] nitrogen was proposed. However, attempts to achieve lasing in the mixture ^{3}He-Hg-N_2 at $\lambda = 564.1$ nm were unsuccessful. The authors of [114] attributed the absence of lasing to the insufficient quenching influence of the N_2 impurities in the presence of large quantities of helium.

The proposals to use the molecules H_2 and D_2 to quench the levels $6^3P^0{}_{2,1,0}$ were more successful. Studies [103, 115, 116] concluded that the most promising medium for NPLs is a mixture of (He)-Xe-Hg-H_2, in which Xe is a buffer gas, and He is used to cool the plasma electrons. Even before the appearance of the first lasers, Fabrikant and Butayeva in 1959 conducted research [117] into selective "quenching" of low levels of the mercury triplet with molecules of H_2 in a gas discharge.

The spectroscopic investigations cited above made it possible to pump laser on the mixture He-Xe-Hg-H_2 ($\lambda = 546.1$ nm) using uranium fission fragments in experiments on the EBR-L setup [33, 118, 119]. For the mixture He-Xe-Hg-H_2 (35:35:2:10), at an initial pressure of 0.4 atm and optimal temperature of 480 K, output laser power of 20 W was obtained ($\eta_l = 0.4$ %). The laser threshold was reached at $\Phi_{th} \approx 2 \times 10^{15}$ cm^{-2} s^{-1}. We note that pumping of the Hg laser ($\lambda = 546.1$ nm) was also carried out using excitation of the active medium by an electron beam [120].

Cadmium- and Zinc-Vapor Lasers

The majority of investigations of metal-vapor NPLs were dedicated to the cadmium-vapor laser. A scheme of energy levels of the cadmium atom and ion with laser transitions is shown in Fig. 3.5. The first successful pumping of a Cd-vapor laser by nuclear radiation was carried out by MIFI associates in 1979, in experiments with the BARS-1-pulsed reactor [121, 122]. When the mixture

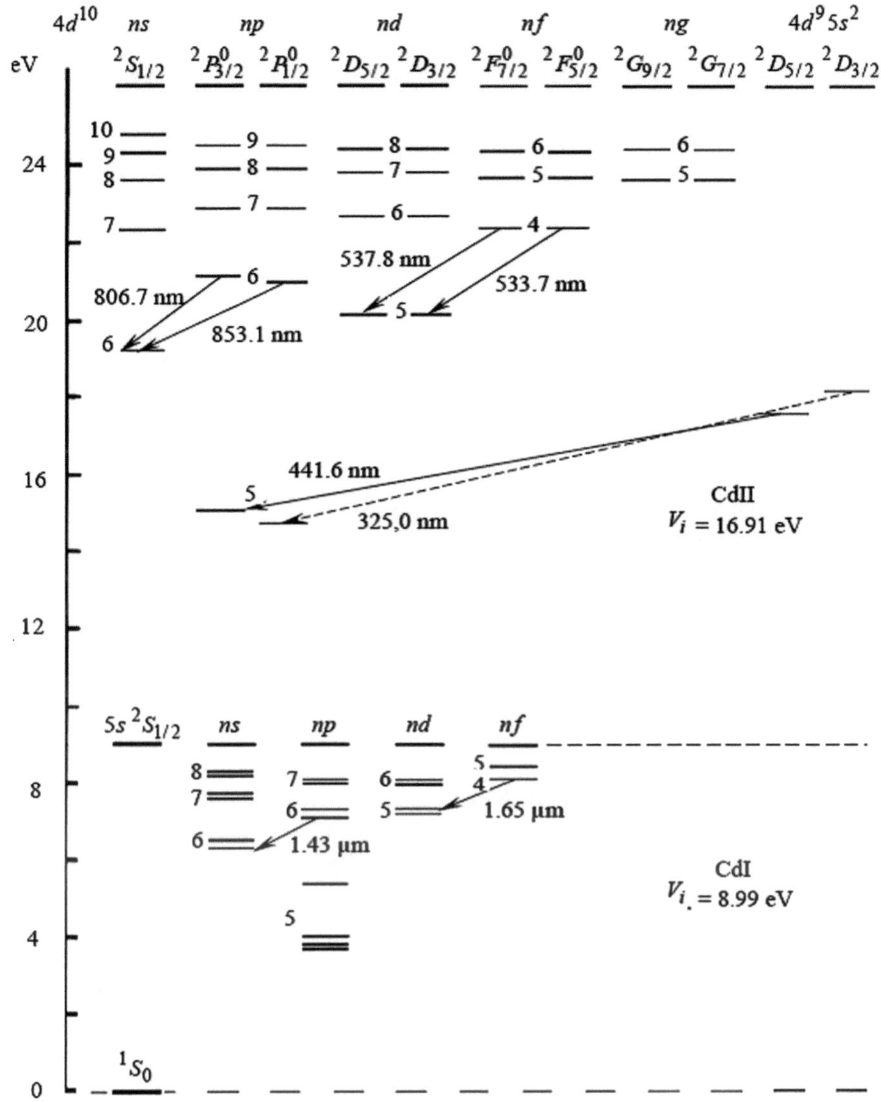

Fig. 3.5 Diagram of energy levels of the cadmium atom and ion with laser transitions

^{3}He-^{116}Cd was used, lasing was achieved at transitions $4f^2F^0_{5/2,7/2}$-$5d^2D_{3/2,5/2}$ of the Cd$^+$ ion ($\lambda = 533.7$ and 537.8 nm). The first successful experiments to pump NPLs based on metal vapors (the mixture He-^{116}Cd, $\lambda = 441.6$; 533.7 and 537.8 nm) with uranium fission fragments were carried out in 1982 by associates of VNIITF and VNIIEF on the EBR-L reactor [31].

The basic characteristics of cadmium- and zinc-vapor NPLs, which have similar lasing mechanisms, are shown in Table 3.9. There is no information about the

Table 3.9 Results of experimental investigations of Cd- and Zn-vapor NPLs

Mixture	λ, nm	P_{opt}, atm	W_{out}, W	η_l %	$\Phi_{th} \times 10^{-14}$, $cm^{-2}\ s^{-1}$	T_{opt}, °K	Reactor (laboratory)	Works cited
^3He-^{116}Cd	533.7; 537.8 (Cd^+)	0.53	0.1	–	3	600	BARS-1 (MIFI)	[122]
^3He-^{116}Cd	441.6 (Cd^+)	0.53	0.05	–	2.5	680	BARS-1 (MIFI)	[123]
^3He-^{116}Cd	441.6 (Cd^+)	1	–	–	3	660	VIR-2M (VNIIEF, MIFI)	[124]
^3He-Zn	747.9 (Zn^+)	1.1	2	–	17	740	VIR-2M (VNIIEF, MIFI)	[125]
He-^{116}Cd	441.6 (Cd^+)	1.8	1,000	0.4	7	660	EBR-L (VNIITF)	[32, 33]
He-^{116}Cd	533.7; 537.8 (Cd^+)	–	470	0.3	6	740	EBR-L (VNIITF)	[32, 33]
He-^{116}Cd	806.7; 853.1 (Cd^+)	–	35	0.02	320	740	EBR-L (VNIITF)	[32, 33]
He-^{116}Cd	1,430; 1,650 (Cd)	–	1–2	–	100	650	EBR-L (VNIITF)	[32, 33]
He-Zn	747.9 (Zn^+)	–	60	0.05	100	770	EBR-L (VNIITF)	[32, 33]

Note: T_{opt} is the optimal heating temperature of the active medium; P_{opt} is the optimal pressure of the medium at an initial temperature of 300 K

investigations of Cd and Zn vapor NPLs outside of Russia. Only the survey study [126], published in 1983, notes that such experiments were carried out on a pulsed reactor (probably the APRF) using the mixture ^3He-Cd ($P = 0.8$ atm). Intensive radiation was observed at the 533.7- and 537.8-nm lines, but reliable proofs of the presence of lasing were not obtained.

Information about high-pressure NPLs using Cd and Zn vapors may be supplemented with the results of investigations of these lasers when pumped by electron beams (see review [95] and the studies cited there). For Cd-vapor lasers, apart from the laser lines 441.6, 533.7, and 537.8 nm, registered during nuclear pumping, lasing was also obtained in the UV range of the spectrum at transitions of the Cd^+ ion ($\lambda = 325.0$ nm) and Cd atom ($\lambda = 361.0$ nm). In the mixture He-Zn, lasing was obtained at the transition of the Zn^+ ion ($\lambda = 610.2$ nm), which was not observed under nuclear pumping conditions. The efficiency of Cd-vapor lasers was ~0.1 %. For electron beam pumping, lasing was also observed in the mixture He-Sr ($P = 3.5$ atm) at transitions of the Sr^+ ion ($\lambda = 416.5$ and 430.5 nm) [127]. We note that the possibility of achieving lasing at the transition of the ion Sr^+ ($\lambda = 430.5$ nm) for the mixture He-Sr pumped by nuclear radiation was considered previously in the study [128].

3.4 Lasers Operating on Transitions of C, N, and Cl Atoms

One area of NPL development is the search for active media containing molecular gases as the laser component. In the active media of these lasers, consisting of mixtures of rare gases He, Ne, and Ar with molecular gases CO, CO_2, N_2, O_2, Cl_2, CCl_4, F_2, NF_3, etc., lasing can occur at the transitions of the C, N, O, Cl, and F atoms. Attempts to achieve lasing at the transitions of these atoms under nuclear pumping began immediately after development of the first NPLs. In study [129], on excitation of the mixture ^{3}He-Ne-O_2 (2,000:30:1) at a pressure of 0.8 atm with nuclear reaction products ^{3}He$(n,p)^3$H, the observed small-signal gain was ~10^{-4} cm^{-1} at the transition of the O atom ($\lambda = 844.6$ nm). These experiments were carried out on a TRIGA reactor. Somewhat later, this reactor was used to conduct experiments [130] in which, when the mixture Ne-N_2 ($P \leq 0.5$ atm) was pumped with uranium fission fragments and nuclear reaction products ^{10}B$(n,\alpha)^7$Li, lasing occurred at transitions of the N atom with $\lambda = 862.9$ and 939.3 nm. The output power was 1.5 mW, $\eta_l \sim 10^{-4}$ % and $\Phi_{th} = 1.5 \times 10^{15}$ cm^{-2} s^{-1}.

Pumping of a laser at the transition $3p^1P_1$-$3s^1P_1{}^0$ of the C atom ($\lambda = 1.45$ μm) with nuclear radiation (in this case with nuclear reaction products ^{10}B$(n,\alpha)^7$Li) was carried out for the first time in experiments [131, 132] on a TRIGA reactor at pressures of the mixtures He(Ne)-CO and He(Ne)-CO_2 ranging from 0.03 to 1 atm and partial pressures of CO (CO_2) of up to 0.02 Torr. The output power was ~1 mW ($\eta_l \sim 10^{-3}$ %). The laser threshold was reached at $\Phi_{th} = 3 \times 10^{14}$ cm^{-2} s^{-1}. Later the analogous experiments were performed in France on the Caliban-pulsed reactor [133].

There is no detailed information on research of NPLs outside Russia using the transition of the Cl atom. The reviews [40, 126] reported that lasing was observed at the 1.59 μm line at the mixture ^{3}He-Cl_2 ($P = 0.8$ atm). The laser threshold was reached at $\Phi_{th} = 7 \times 10^{15}$ cm^{-2} s^{-1}, and the output power was ~0.1 W.

In Russia, experimental research of NPLs operating on transitions of the atoms C, N, O, and Cl were conducted at VNIIEF. In the first experiments with the VIR-2 reactor, carried out in 1978, lasing occurred at transitions of the C atom ($\lambda = 833.5$; 940.6 and 1,454 nm) while the mixtures He(Ne,Ar)-CO were excited by uranium fission fragments, and the maximal output power was 7 W ($\lambda = 1.45$ μm) [134].

In subsequent experiments [135–139] on the VIR-2M reactor, a cycle of investigations were carried out to find lasing at transitions of the atoms C, N, O, and Cl and to optimize laser characteristics with regard to the pressure and composition of mixtures and the cavity parameters. The LUNA-2M setup was used in these experiments. The basic results of the investigations are shown in Table 3.10. We note that earlier, lasing in nearly all of the wavelengths cited in Table 3.10 was obtained in lower-pressure gas discharge lasers [18].

The energy parameters obtained at VNIIEF for lasers operating on transitions of C, N, O, and Cl atoms are two or three orders of magnitude greater than the parameters of analogous lasers studied in other laboratories, while the laser using

Table 3.10 Basic results of research of NPLs operating on transitions of C, N, O, and Cl atoms

Atom	Mixture	λ, nm	P, atm	W_{out}, W	η_l, %	$\Phi_{th} \times 10^{-14}$, $cm^{-2}\ s^{-1}$	Works cited
O	He-Ne-O_2 (10,000:10,000:1)	2,653; 2,760	2	30	0.05	3.0	[135, 136]
	He-Ne-NO (10,000:10,000:1)	2,653; 2,760	2	8	<0.01	9.9	
	He-Ne-CO (10,000:10,000:1)	2,653; 2,760	2	6	<0.01	21	
Cl	He-Ne-CCl_4 (25,000:25,000:1)	1,587	2	130	0.2	0.2	[136, 137]
	He-Ne-$CHCl_3$ (25,000:25,000:1)	1,587	2	40	0.08	0.56	
	He-Ne-CF_2Cl_2 (25,000:25,000:1)	1,587	2	8	0.01	0.46	
	He-Ne-$C_2F_3Cl_3$ (25,000:25,000:1)	1,587	2	10	0.01	2.2	
	Ar-CCl_4 (10,000:1)	1,587	0.4	2.5	<0.01	18	
	He-Ne-CCl_4 (25,000:25,000:1)	2,447	2	5	<0.01	10	
N	Ne-N_2 (25,000:1)	859.4, 862.9	0.25	5	<0.01	12	[138, 139]
		904.6		20	0.05	15	
	Ne-N_2 ($[N_2] \leq 10^{-3}$ %)	938.7	0.4	0.4	–	13	
C	Ne-CO (40,000:1)	833.5	0.75	6	<0.01	3.6	[139]
		940.6		36	0.05	7.6	
	Ne-CO_2 (40,000:1)	940.6	0.4	4	<0.01	3.4	
	Ne-Ar-CO (36,000:1,000:1)	940.6	0.75	120	0.3	3.7	

the transition of the Cl atom ($\lambda = 1.59$ μm) is one of the most low-threshold of the NPLs. A specific feature of C, N, O, and Cl lasers is the low concentration (~10^{-2} Torr) of laser component (molecular gas), compared to the concentration of molecular impurities in high-purity rare gases. Therefore, to increase the energy characteristics it is necessary to have a higher degree of exhaustion of the laser cells, and additional purification of the initial gases.

The LUNA-2M setup was also used to perform experiments to find lasing at oxygen atom lines 777.4 and 844.6 nm (mixture of He-O_2 and He-Ne-O_2) [139] and fluorine atom lines in a range of 620–910 nm (mixture He-NF_3) [140]. The experiments yielded no positive results, although these oxygen lines and a number of lines of the fluorine atom showed high intensity in luminescence research [81, 140, 141]. The absence of lasing at fluorine atom lines 703.8 and 712.8 nm in experiments with the mixture ^{3}He-NF_3 using the APRF reactor is also reported in the review [126]. We note that with electron beam pumping, lasing was obtained in the mixture He-NF_3 at the fluorine atom lines 634.8, 641.4, 712.8, 731.1, 739.8, and 755.2 nm [142].

3.5 Lasers Operating on Transitions of Molecules CO, N_2^+, and CO_2

CO Laser

A molecular CO laser was the first laser to be pumped with nuclear radiation in experiments outside Russia with a pulsed reactor [143]. As with gas-discharge CO lasers, lasing was observed on vibrational–rotational transitions of the CO molecule with $\lambda = 5.1$–5.6 μm. The pumping of carbon monoxide at a pressure of 0.13 atm and temperature of a 77 K was carried out using uranium fission fragments. The neutron source was an SPR-II pulsed reactor (see Chap. 2, Sect. 2.7). The output laser power was 2–6 W; the efficiency, determined in [143] with respect to the energy deposition to the volume of the principal mode, was 0.1–0.3 %. The small-signal gain at transitions 10–9, 9–8, and 7–6 of the CO molecule reached 5×10^{-3} cm^{-1} [144]. Study [51] reports that later on, the authors of [143, 144] obtained output power of around 100 W for CO NPLs using a multiple-path cavity with a 120-cm active length (see Fig. 2.2a).

Lasing at $\lambda \approx 5$ μm of the CO molecule was also obtained in study [145] on exciting the mixture ^{3}He-CO with nuclear reaction products ^{3}He$(n,p)^3$H. The output power at a mixture pressure of 3 atm was ≥200 W from an active volume of 300 cm^3. The laser threshold was reached at $\Phi_{th} = 3 \times 10^{16}$ cm^{-2} s^{-1}.

N_2^+ Laser

Pulsed lasing at electron-vibration transitions of the first negative system of the molecular ion N_2^+ ($\lambda = 427.8$ nm) with pumping of a high-pressure He-N_2 mixture with an electron beam with a duration of 20 ns was obtained in 1974 [146]. Later on, with pumping of the mixture He-N_2-H_2 ($\lambda = 391.4$; 427.8 ns) with an electron beam with a duration of 0.7 μs, it was possible to realize a quasi-CW operating mode using hydrogen to quench low laser levels: $N_2^+(X) + H_2 \rightarrow N_2H^+ + H$ [147]. The efficiency of the laser was ~1 % at a specific power deposition $q = 5$–10 kW/cm^3.

The possibility of pumping this laser with nuclear radiation was examined in study [148]. As a result of calculating the kinetics of plasma processes in the mixture He-N_2-H_2, it was concluded that it was possible to achieve lasing at $\lambda = 391.4$ and 427.8 nm for $q = 1$–3 kW/cm^3, which is quite attainable in experiments on fast reactors EBR-L and SPR III.

Soon afterward, the same authors used an EBR-L setup to obtain lasing at the 391.4-nm line when the He-N_2-H_2 mixture was excited by uranium fission fragments with a output power of ~10 W and η_l ~ 0.01 % [149]. This laser became the first NPL radiating in the UV range of the spectrum. Later on, in experiments with the EBR-L reactor, lasing was also registered at $\lambda = 427.8$ nm of the ion N_2^+ [150]. For an optimal mixture of He-N_2-H_2 at pressures of 3.5–5 atm (partial pressures of nitrogen and hydrogen 3–10 Torr), at lines 391.4 nm and 427.8 nm, laser efficiencies of 0.3 % and 0.2 %, respectively, were obtained. Confirmation of these results was obtained in experiments [151] with a BARS-6 pulsed reactor, in which a He-N_2-H_2 mixture was excited by uranium fission fragments. Lasing was observed at the lines 391.4 and 427.8 nm for substantially lower specific threshold power depositions of 50–60 W/cm^3 (η_l ~ 0.1–0.2 %).

CO_2 Laser

The possibility of pumping molecular CO_2 lasers ($\lambda = 10.6$ μm) with nuclear radiation was considered in many studies in the early stages of NPL research (see Chap. 1, Sect. 1.1). Despite the numerous experiments, attempts to pump CO_2 lasers with nuclear radiation yielded negative results. Experiments [152] to excite a ^{3}He-N_2-CO_2 mixture with nuclear reaction products ^{3}He$(n,p)^3$H showed the absence of gain at the 10.6 μm line in a wide range of changes in the total pressure (0.26–0.8 atm) and composition of the mixture. In these experiments, absorption of radiation of the probe laser with $\lambda = 10.6$ μm was observed, which testifies to the predominant populating of the lower laser level when a CO_2 laser is pumped with high-energy charged particles. Calculations [153] of kinetic processes in the He-N_2-CO_2 plasma also confirm the ineffectiveness of direct pumping of CO_2 lasers with nuclear radiation.

Another variation of pumping the CO_2 laser with preliminary excitation of a nitrogen molecule and subsequent transfer of energy from these molecules to CO_2 molecules was implemented in experiments reported in reviews [126, 154]. A diagram of the experiments is shown in Fig. 3.6. Nitrogen passing through a tube with a layer of ^{10}B at a velocity of 7.6×10^{-4} m^3 s^{-1} was excited by the nuclear reaction products ^{10}B$(n,\alpha)^7$Li in the neutron flux of a pulse reactor with a pulse duration of about 200 μs ($\Phi_{max} = 5 \times 10^{16}$ cm^{-2} s^{-1}). Then the excited nitrogen was mixed in the laser cavity with CO_2 and He. The laser radiation arose within 30 ms after the reactor pulse and had a power of about 100 W with a laser pulse duration of about 1 ms.

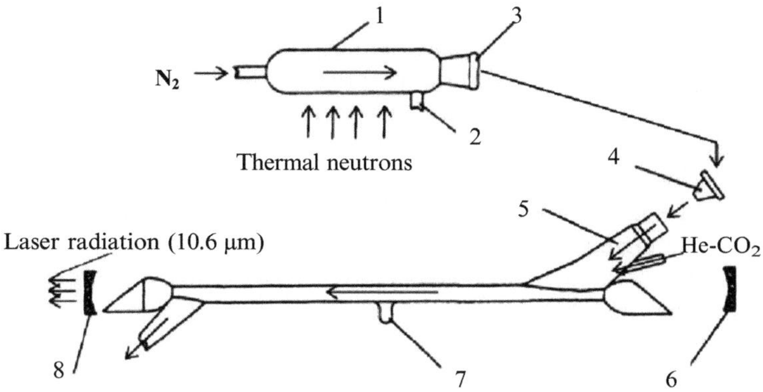

Fig. 3.6 Diagram of experiments on pumping a CO_2 laser with nuclear radiation [126]: (*1*) tube with ^{10}B layer; (*2*) channel to measure nitrogen pressure; (*3*) connector socket; (*4*) glass nozzle; (*5*) inlet tube; (*6*) high-reflectivity mirror; (*7*) pin electrode; (*8*) exit mirror

3.6 Excimer, Iodine, and Chemical Lasers

Excimer Laser Media

Excimer lasers using molecular halogenides of rare gases *RX* ($R = $ Ar, Kr, Xe; $X = $ F, Cl, Br), excited by electron beams and in a gas discharge, are powerful sources of laser UV radiation [155, 156]. To pump these lasers, the radiation of nuclear explosions was also used (see Chap. 12). Excimer lasers operate most effectively in pulsed mode at high specific pumping powers of $q \geq 0.1$ MW/cm^3, which is explained by the great radiation line width.

Excimer lasers can operate in quasi-CW mode, because after photon radiating, excimer molecules fall into the lower repulsive or weakly bound state. In this connection, some studies were carried out on the possibility of pumping excimer laser media with nuclear radiation using pulsed reactors as the neutron sources.

The largest number of investigations were devoted to the XeF-laser ($\lambda = 351$ and 353 nm), which has the lowest laser threshold [60, 157–162]. In experiments [157] with the TRIGA reactor, a small-signal gain was registered of ~10^{-4} cm^{-1} at line 351 nm of the XeF molecule at a low specific power deposition of $q \approx 40$ W/cm^3. Experiments performed with the SPR-III reactor showed that for the mixture ^{3}He-Xe-NF_3, the gain at the 351-nm line is around 7×10^{-3} cm^{-1} ($q \approx 5$ kW/cm^3) [158], while for pumping the mixture Ne(Ar)-Xe-NF_3 with uranium fission fragments, a gain of ~2×10^{-3} cm^{-1} ($q \approx 2$ kW/cm^3) was registered [48, 159]. Experiments using pulsed reactors with active media based on the molecule XeF [48, 160] and KrF ($\lambda = 248$ nm) [60], aimed at obtaining lasing, yielded no positive results. These experiments used the mixtures ^{3}He-Xe-NF_3, Ne-Xe-NF_3 [48], Ne(Ar)-Xe-NF_3(SF_6) [160], and ^{3}He-Kr-NF_3 [60].

Calculations showed that for the mixture Ne-Xe-NF_3, it is possible to obtain $\eta_l \approx 1$ % for $q \sim 100$ W/cm^3 [161], and 500 W/cm^3 [162]. The possibility of creating a nuclear-pumped XeF laser is determined primarily by the absorption coefficient in the active medium, the uncertainty of which greatly influences the results of calculations [161].

In conclusion, we note that study [163] reports achieving laser action at the transition of the XeCl molecule ($\lambda = 308$ nm) when pumping the mixture Ar-Xe-CCl_4(HCl) with uranium fission fragments in experiments with the EBR-L reactor ($q \sim 1$ kW/cm^3). Although the authors assert that lasing was observed in the experiments, the cited data were insufficient for this conclusion.

Iodine Lasers

Lasers operating on transitions of the I atom ($\lambda = 1.315$ μm) excited as a result of photodissociation of the molecules CF_3I and C_3F_7I, or by transfer of energy to iodine atoms from metastable molecules of oxygen $O_2^*(^1\Delta) + I(^2P_{3/2}) \rightarrow O_2^*(^3\Sigma) + I^*(^2P_{1/2})$, are well known as some of the most powerful sources of laser IR radiation [164, 165]. The literature reviewed several variations of excitation of the iodine laser by nuclear radiation: (a) formation of metastable molecules $O_2^*(^1\Delta)$ with subsequent energy transfer to iodine atoms [166, 167]; (b) formation of vibrationally excited molecules $H_2^*(\nu \geq 2)$ in the mixture Ar-H_2-I_2 with subsequent transfer of energy to iodine atoms [168]; use of radioluminescent radiation of excimer molecules (KrF, XeBr, etc.) for photodissociation of molecules CF_3I or C_3F_7I [169, 170].

Studies [166, 167] provide the results of experiments and theoretical investigations into excitation by nuclear radiation of mixtures of He, Ne, and Ar with O_2, aimed at determining the concentrations of $O_2^*(^1\Delta)$ necessary to achieve the laser threshold. Formation of $O_2^*(^1\Delta)$ can occur both during direct excitation of the oxygen-containing mixtures, and during photolytic decomposition of O_3 by the radiation of excimer molecules, for example KrF. Pumping of an oxygen-iodine laser, or an Ar-H_2-I_2 laser [168], does not exist as yet.

A method of pumping an iodine laser using radioluminescent radiation of excimers is based on high conversion efficiencies of luminescence, which for many excimers is 20–30 % [170]. When the ^{3}He-Xe-$CHBr_3$ mixture is irradiated with the pulsed neutron flux of the TRIGA reactor, intensive radiation of the XeBr* molecules occurred ($\lambda = 282$ nm), which then was used to photodissociate C_3F_7I [171, 172]. The conversion efficiency of luminescence for the XeBr* molecules ($\lambda = 282$ nm) was around 1 %. The design of the laser cell is shown in Fig. 3.7. The power of the laser radiation at the line 1.315 μm was not great: ~20 mW.

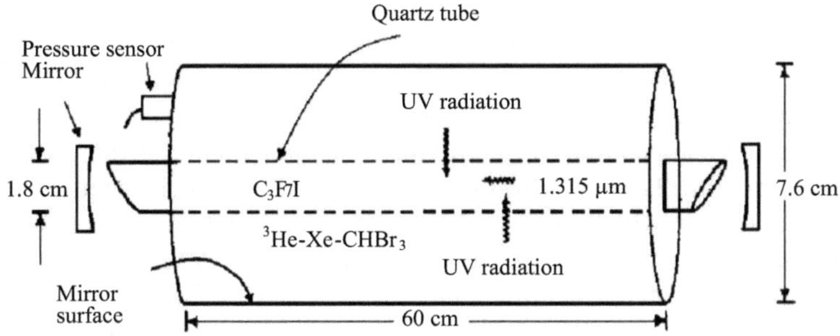

Fig. 3.7 Design of cell for pumping an iodine laser with radioluminescent radiation of XeBr* molecules [171, 172]

Chemical Lasers

The interest in chemical lasers [173] may be explained by the fact that for them, in CW mode, high output powers of up to several megawatts have been obtained [174]. Various methods of acting on the active medium are used to initiate chemical lasers: gas discharge, photoinitiation, fast electron beams, γ radiation from a nuclear explosion, etc.

Initiation of chemical lasers using HF(DF) molecules with ionizing radiation is one of the methods of creating powerful IR lasers in a spectral range of 2.7–4.4 μm. Marked successes in this area were achieved in pulsed mode using electron beams and the γ radiation from nuclear explosions as the source of initiation. In the first case, the output energy of the HF laser using the mixture H_2-F_2-O_2-NF_3 with a chain reaction reached 4.5 kJ ($\eta_l = 226$ %) at the pulse duration of 50 ns [175], and in the second case, for the mixture SF_6-H_2 with a non-chain reaction, it reached 70 kJ ($\eta_l \sim 5$ %) in ~10 ns pulse (see Chap. 12).

When chemical lasers are initiated by nuclear radiation with a pulse duration ≥100 μs, which is typical of pulsed reactors, the specific power deposition is not great, so it is hard to ensure high rates of dissociation of the fluorine-containing substances, and accordingly, the rate of formation of the excited molecules HF*. As a result, HF molecules accumulate in their ground state, and effectively quench molecules of HF* in collisions. This circumstance leads to an increase in the laser threshold and a reduction in efficiency.

The possibility of creating chemical NPLs using pulsed reactors was examined in several studies. The first of them [176], carried out in 1970 and published in 1989, examined the mixtures $^{235}UF_6$-H_2 and $^{235}UF_6$-H_2-F_2. It was concluded that for the first mixture, $\Phi_{th} = 5 \times 10^{14}$ cm^{-2} s^{-1}, while for the second, $\Phi_{th} \approx 10^{17}$ cm^{-2} s^{-1}. Study [176] briefly discussed one version of chemical NPL at the mixture $^{235}UF_6$-D_2-F_2-CO_2 ($\lambda = 10.6$ μm), with transfer of energy from the excited DF* molecules to the CO_2 molecules. Study [177] examined several problems associated with development of the chemical laser DF-CO_2, intended for experiments with the reactors Godiva-IV ($\tau_{1/2} \sim 30$ μs) and SKUA ($\tau_{1/2} \leq 400$ μs).

The $^{235}UF_6$-H_2 mixture is interesting in that at present it is perhaps the most realistic NPL active medium in which the "fissionable" material $^{235}UF_6$ can be a component part of the laser medium. Study [178] calculated the gain for this mixture depending on the duration of the neutron pulse, the neutron flux density, the pressure and composition of the mixture, the temperature of the medium, and the dimensions of the laser cell. In contrast to other NPLs, which usually operate for an interval of time in which the thermal neutron flux density exceeds the threshold value, the characteristics of NPLs using the mixture $^{235}UF_6$-H_2 depend on the duration and shape of the neutron pulse. It was shown that for typical reactor pulse parameters ($\Phi_{max} \le 10^{17}$ cm^{-2} s^{-1}, $\tau_{1/2} = 30$–200 μs), the laser threshold is achieved at $\Phi_{th} < 10^{14}$ cm^{-2} s^{-1}, and the small-signal gain can be $\ge 5 \times 10^{-3}$ cm^{-1}.

When mixtures based on H_2-F_2 with a chain reaction are used, it is possible to obtain a laser efficiency of greater than 100 % owing to the chemical energy stored in the active medium [173]. The kinetics of one such laser based on the mixture He-F_2-H_2-O_2, initiated by nuclear radiation, was examined in study [179] for experimental conditions of the EBR-L reactor. Calculations showed the possibility of achieving lasing at a number of vibrational transitions of the HF molecule. Lasing can occur at the leading edge of the reactor pulse, and for the optimal mixture He-F_2-H_2-O_2 (12:2:1:2) at a pressure of 2 atm, the duration of the laser radiation is ~10 μs with an efficiency of up to 100 %.

Unfortunately, the few experiments attempting to create chemical NPLs did not yield positive results. Such experiments were carried, for example, with the HPRR fast pulsed reactor ($\tau_{1/2} \sim 50$ μs) using the mixtures $^{235}UF_6$-H_2 ($P = 0.07$–0.8 atm) and ^{3}He-F_2-H_2 ($P = 0.4$ atm) [180]. The thermal-neutron flux density inside a laser cell 46 cm in length was $\ge 10^{17}$ cm^{-2} s^{-1}. Measurements of the chemical composition of mixtures showed that the concentration of HF molecules formed in the pulse ([HF]/[UF_6] > 70 %) is entirely sufficient to achieve the laser threshold. In the opinion of the authors of [180], the lack of lasing may be explained by contamination of the active medium with impurities due to the increased temperature of the laser cell or an insufficiently high appearance rate of atomic fluorine. Analogous experiments performed at VNIIEF in 1973 with the BIR-2-pulsed reactor ($\tau_{1/2} \sim 100$ μs) with mixtures ^{3}He-SF_6-D_2 and ^{3}He-SF_6-D_2-CO_2 also did not yield positive results.

3.7 Lasers Excited by Fast Neutrons

NPLs excited by products of nuclear reactions occurring at the interaction of the nuclei of several isotopes with thermal neutrons were considered earlier in this chapter. There is yet another method of pumping, using fast neutrons. It was proposed at VNIIEF and was used to excite NPLs operating on transitions of the Xe atom [181]. This method is based on elastic scattering of fast neutrons on atoms (nuclei) of the medium, and makes it possible to homogeneously excite gas media at pressures of tens and hundreds of atmospheres. For the efficient transfer of energy from fast neutrons to recoil nuclei, it is advisable to use buffer gases with a low atomic weight, for example helium or hydrogen. Direct ionization and excitation of the gas medium is carried out by recoil nuclei. For helium at a pressure of 100 atm, the flux of fast neutrons with an energy of 1 MeV is weakened by a factor of e over a length of approximately 60 cm, so that it is possible to realize a quite homogeneous pumping of gas media at pressures of ~100 atm with characteristic transverse dimensions of ~1 m.

Gas lasers were first pumped with fast neutrons at VNIIEF in 1981 in experiments with the BR-1-pulsed reactor; the results were published in 1990 [181]. Successful pumping of the lasers became possible thanks to the BR-1 reactor, in which, in one of the operating modes, the reactor core can be separated to create an internal gap (slot) up to 10 cm in height.

The experiments used a cylindrical laser cell with an internal diameter of 3.5 cm and an active length of 40 cm, placed in the gap between the halves of the reactor core (see Fig. 2.1c). The fluence of the fast neutrons with an energy of more than 0.5 MeV was equal to 5.6×10^{14} cm^{-2}, while the neutron flux density at the pulse maximum was $\Phi_{max} = 6.4 \times 10^{18}$ cm^{-2} s^{-1}.

The specific power deposition near the axis of the laser cell at the pulse maximum may be evaluated from the expression $q = [\mathrm{He}]\sigma_s\Phi_{max}E_\alpha$ ([He] is the concentration of helium atoms, $\sigma_s = 6.5 \times 10^{-24}$ cm^2 is the cross-section of elastic scattering of neutrons with an average energy of $E_n = 1.3$ MeV in helium nuclei, E_α is the average energy of a recoil nucleus). With isotropic scattering, $E_\alpha = 2A_sE_n/(A_s + 1)^2$, where A_s is the mass number of the atoms of the medium. The specific power deposition is proportional to the pressure of the helium and for the conditions cited above, $q \approx 80$ W/cm^3 atm. The energy deposition owing to the absorption of γ radiation was approximately two orders of magnitude lower.

The mixture He-Xe was used to prove the possibility of pumping high-pressure gas lasers with fast neutrons; it was quite well studied from pumping with nuclear reaction products (see Sect. 3.1) and involves a buffer gas of low atomic weight. In experiments, the changes in laser power and threshold characteristics of a He-Xe laser ($\lambda = 2.65$ μm) as a function of the pressure and composition of the mixture were studied. An oscillogram from one of the experiments is shown in Fig. 3.8. The dependence of the laser parameters on the helium pressure at optimal xenon pressure of around 4 Torr is shown in Table 3.11. A maximal output power of 290 W ($\eta_l \sim 0.2$ %) and minimal value of $\Phi_{th} = 1.9 \times 10^{18}$ cm^{-2} s^{-1} were registered for the mixture He-Xe (950:1) at a pressure of 5 atm. These energy parameters evidently are not maximal, since the laser cavity parameters were not optimized.

At present, among NPL applications being considered is the possibility of creating powerful reactor-laser facility (see Chap. 10), in the reactor core of which the kinetic energy of fission fragments is converted into laser radiation. Investigations in the field of neutron pumping of lasers can be of interest for future power plants based on controlled thermonuclear fusion, in which the bulk of the energy is released in the form of kinetic energy of fast neutrons.

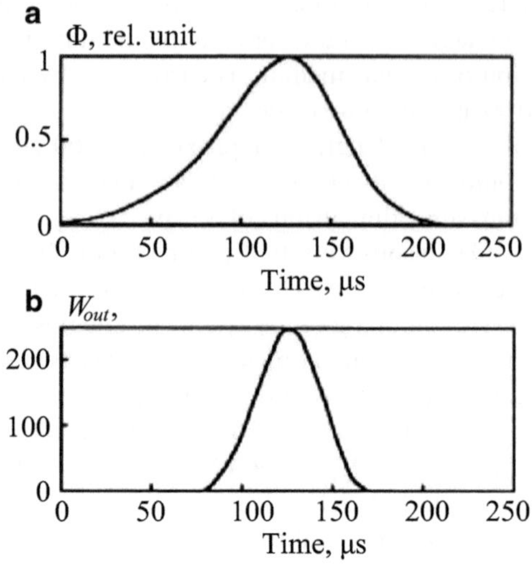

Fig. 3.8 Oscillograms of fast neutron pulse of reactor BR-1 (**a**) and laser pulse (**b**) for the He-Xe mixture (2,000:1) at a pressure of 5 atm [181]

Table 3.11 Parameters of a He-Xe laser ($P_{Xe} = 4$ Torr) at different helium pressures [181]

Helium pressure, atm	3	5	10
Output power, W_{out}	70	290	220
Threshold flux density of fast neutrons, 10^{18} cm^{-2} s^{-1}	1.8	2.0	6.1

References

1. A.A. Sinyanskiy, *Research into the development of stationary nuclear laser devices at VNIIEF* (Issledovaniye po sozdaniyu yaderno-lazernykh ustroystv neprerynogo deystviya vo VNIIEF). Proceedings of the 2nd International Conference "Physics of Nuclear-Excited Plasma and Problems of Nuclear-Pumped Lasers," Arzamas-16, 1995, vol. 1, pp. 16–36.
2. A.M. Voinov, L.E. Dovbysh, V.N. Krivonosov, S.P. Melnikov, A.A. Sinyanskii, *A study of nuclear-pumped lasers operating on rare gas atomic transitions*. Proc. 17th Int. Quantum Electronics Conf., Anaheim, USA, 1990, pp. 348–350.
3. S.P. Melnikov, A.A. Sinyanskiy, *On the ultimate efficiency of nuclear-pumped lasers* (O predelnom KPD lazerov s yadernoy nakachkoy). Proceedings of the Specialist Conference "The Physics of Nuclear-Excited Plasma and the Problems of Nuclear-Pumped Lasers," Obninsk, 1993, vol. 2, pp. 133–148
4. A.A. Sinyanskii, S.P. Melnikov, Research on development of continuous nuclear-laser setups in VNIIEF. Proc. SPIE **3686**, 43–55 (1998)
5. A.M. Voinov, L.Y. Dovbysh, V.N. Krivonosov, S.P. Melnikov, I.V. Podmoshenskiy, A.A. Sinyanskiy, Infrared nuclear-pumped laser using mixtures He + Xe and Ar + Xe (Infrakrasnyy lazer s yadernoy nakachkoy na smesyakh He + Xe i Ar + Xe). PIsma v ZhTF **7**(16), 1016–1020 (1981)
6. A.M. Voinov, V.G. Zobnin, A.I. Konak, S.P. Melnikov, I.N. Mochkayev, A.A. Sinyanskiy, Quasi-CW low-threshold lasing and line competition in nuclear-pumped lasers operating on transitions of the xenon atom (Kvazinepreryvnaya nizkoporogovaya generatsiya i

konkurentsiya liniy v lazerakh s yadernoy nakachkoy na perekhodakh atoma ksenona). Pisma v ZhTF **16**(8), 34–38 (1990)

7. A.M. Voinov, A.I. Konak, S.P. Melnikov, A.A. Sinyanskiy, On the possibility of creating a CW laser with a radioisotope pumping source (O vozmozhnosti sozdaniya nepreryvnogo lazera s radioizotopnym istochnikom nakachki). Kvantovaya elektronika **18**(11), 1297–1299 (1991)
8. A.M. Voinov, L.Y. Dovbysh, A.T. Kazakevich, V.N. Krivonosov, S.P. Melnikov, I.V. Podmoshenskiy, A.A. Sinyanskiy, Infrared nuclear-pumped lasers operating on transitions of ArI, KrI, and XeI (Infrakrasnyye lazery s yadernoy nakachkoy na perekhodakh ArI, KrI i XeI). Pisma v ZhTF **5**(7), 422–424 (1979)
9. A.I. Konak, S.P. Melnikov, V.V. Porkhayev, A.A. Sinyanskiy, Lasing at Xe atom transitions when a Kr-Xe mixture and pure xenon are excited by uranium fission fragments (Generatsiya na perekhodakh atoma Xe pri vozbuzhdenii smesi Kr-Xe i chistogo ksenona oskolkami deleniya urana). Kvantovaya elektronika **22**(12), 1184–1186 (1995)
10. A.M. Voinov, L.Y. Dovbysh, V.N. Krivonosov, S.P. Melnikov, I.V. Podmoshenskiy, A.A. Sinyanskiy, Low-threshold nuclear-pumped lasers operating on transitions of atomic xenon (Nizkoporogovyye lazery s yadernoy nakachkoy na perekhodakh atomarnogo ksenona). Doklady AN SSSR **245**(1), 80–83 (1979)
11. A.I. Konak, S.P. Melnikov, A.A. Sinyanskiy, Quasi-CW nuclear-pumped lasers using the mixture Ne-Xe-(He, Ar) (Kvazinepreryvnyy lazer s yadernoy nakachkoy na smesi Ne-Xe-(He, Ar)). ZhTF **62**(3), 24–29 (1992)
12. A.A. Abramov, S.P. Melnikov, A.K. Mukhamatullin, A.A. Pikulev, A.A. Sinyanskii, V.M. Tsvetkov, Study of fission-fragment excited xenon laser in He-Xe, Ar-Xe, and He-Ar-Xe mixtures. Proc. SPIE **5483**, 1–13 (2004)
13. V.D. Grigoryev, A.V. Lisenkov, S.P. Melnikov, C.V. Patyanin, A.A. Pikulev, A.A. Sinyanskiy, P.V. Sosnin, V.M. Tsvetkov, *Parametric investigations of a nuclear-pumped xenon laser using mixtures of Ar-Xe ($\lambda = 1.73$ μm) and He-Ar-Xe ($\lambda = 2.65$ μm)* (Parametricheskiye issledovaniya ksenonvogo lazera s yadernoy nakachkoy na smesyakh Ar-Xe ($\lambda = 1.73$ μm) i He-Ar-Xe ($\lambda = 2.65$ μm)), Proceedings of the 3rd international conference "Problems of Nuclear-Pumped Lasers and Pulsed Reactors," Snezhinsk, 2003, pp. 178–184.
14. A.M. Voinov, L.Y. Dovbysh, V.N. Krivonosov, S.P. Melnikov, I.V. Podmoshenskiy, A.A. Sinyanskiy, High-pressure helium-krypton laser excited by uranium fission fragments (Geliy-kriptonovyy lazer vyskogo davleniya, vozbuzhdayemyy oskolkami deleniya urana). ZhTF **52** (7), 1346–1350 (1982)
15. A.I. Konak, S.P. Melnikov, V.V. Porkhayev, A.A. Sinyanskiy, Lasing at IR transitions of krypton and argon atoms during excitation of active media by uranium fission fragments (Generatsiya na IK-perekhodakh atomov kriptona i argona pri vozbuzhdenii aktivnykh sred oskolkami deleniya urana). Kvantovaya elektronika **22**(6), 537–541 (1995)
16. S.P. Melnikov, A.A. Sinyanskiy, Quasi-CW lasing at the transition *4p-4s* of the argon atom during pumping of a He-Ar mixture by uranium fission fragments (Kvazinepreryvnaya generatsiya na perekhode 4*p*-4*s* atoma argona pri nakachke lazera na smesi He-Ar oskolkami deleniya urana). Kvantovaya elektronika **18**(12), 1444–1446 (1991)
17. A.M. Voinov, V.N. Krivonosov, S.P. Melnikov, I.N. Mochkayev, A.A. Sinyanskiy, Quasicontinuous nuclear-pumped lasers operating on transitions of the argon atom (Kvazinepreryvnyy lazer s yadernoy nakachkoy na perekhodakh atoma argona). Kvantovaya elektronika **18**(2), 177–179 (1991)
18. R. Beck, W. Englisch, K. Gürs, *Table of Laser Lines in Gases and Vapors* (Springer, New York, 1980)
19. A. Suda, B.L. Wexler, B.J. Feldman, K.J. Riley, Measurements of gain, saturation, and line competition in an electron beam pumped high-pressure Ar/Xe laser. Appl. Phys. Lett. **54**(14), 1305–1307 (1989)
20. D.A. Zayarnyy, A.G. Korolev, N.N. Sazhina, N.N. Ustinovskiy, I.V. Kholin, On the influence of pumping power on the spectral-temporal characteristics of the Ar-Xe laser (O vliyanii

moshchnosti nakachki na spektralno-vremennyye kharakteristiki Ar-Xe lazera). Kvantovaya elektronika **18**(5), 538–544 (1991)
21. A.A. Pikulev, Lasing line competition in gas lasers (Konkurentsiya generatsionnykh liniy v gazovykh lazerakh). Kvantovaya elektronika **30**(8), 675–680 (2000)
22. A.A. Pikulev, Investigation of the parameters of simultaneous lasing at two lines having a common upper level (Issledovaniye parametrov odnovremnnoy generatsii na dvukh liniyakh, imeyushchikh obshchiy verkhniy uroven). Kvantovaya elektronika **32**(8), 689–691 (2002)
23. A.A. Pikulev, Competition of laser lines having upper or lower common level. Proc. SPIE **6263**, 120–130 (2006)
24. A.A. Abramov, S.P. Melnikov, A.A. Pikulev, V.M. Tsvetkov, *Investigation of the parameters of an active laser medium using the mixture He-Xe ($\lambda = 1.73$ and 2.03 μm) during excitation by uranium fission fragments.* (Issledovaniye parametrov aktivnoy sredy lazera na smesi He-Xe ($\lambda = 1.73$ i 2.03 μm) pri vozuzhdenii oskolkami deleniya urana), Proceedings of the 4th International Conference "Problems of Nuclear-Pumped Lasers and Impulse Reactors," Obninsk, 2009, vol. 1, pp. 298–304.
25. W.W. Rigrod, Homogeneously broadened cw lasers with uniform distributed loss. IEEE J. Quant. Electron. **QE-14**(5), 377–381 (1978)
26. A.A. Pikulev, V.M. Tsvetkov, P.V. Sosnin, A.A. Sinyanskiy, Operation of nuclear-pumped lasers in the master oscillator-amplifier mode (Rabota lazera s yadernoy nakachkoy v rezhime zadayushchiy generator-usilitel). Kvantovaya elektronika **38**(7), 655–660 (2008)
27. A.A. Pikulev, A.A. Abramov, The numerical investigation efficiency of parallel and successive laser cells composition. Proc. SPIE **6263**, 186–196 (2006)
28. A.A. Pikulev, V.M. Tsvetkov, P.V. Sosnin, A.A. Sinyanskiy, *Experimental investigation of serial addition of laser channels* (Eksperimentalnoye issledovaniye posledovatelnogo slozheniya lazernykh kanalov). Proceedings of the 4th International Conference "Physics of Nuclear-Pumped Lasers and Pulsed Reactors," Obninsk, 2009, vol. 1, pp. 290–297.
29. A.A. Pikulev, V.M. Tsvetkov, P.V. Sosnin, A.A. Sinyanskiy, Serial addition of two laser channels during excitation of He-Ar-Xe mixtures (2.03 μm) and Ar-Xe (1.73 μm) by uranium fission fragments (Posledovatelnoye slozheniye dvukh lazernykh kanalov pri vozbuzhdenii smesey He-Ar-Xe (2.03 μm) i Ar-Xe (1.73 μm) oskolkami deleniya urana). Kvantovaya elektronika **39**(6), 494–500 (2009)
30. V.M. Tsvetkov, A.A.Pikulev, P.V. Sosnin, Investigation of two laser channels parallel composition at Ar:Xe (1.73 μm) and He:Ar:Xe (2.03 μm) mixtures excited by uranium fission fragments (Isslidovaniye parallelnogo slozheniya izlycheniya dvukh lazernykh kanalov pri vozbuzhdenii smesey Ar-Xe (1.73 μm) i He-Ar-Xe (2.03 μm) oskolkami deleniya urana). Trudy RFYaTS-VNIIEF (17), 218–227 (2012).
31. V.A. Kryzhanovsky, E.P. Magda, L.V. Semkov, V.F. Tausenev, *About the initial stage of NPL researches in the RFNC-ITP*. Proc. 9th Int. Conf. on Emerging Nuclear Energy Systems, Tel-Aviv, 1998, vol. 2, pp. 874–880.
32. E.P. Magda, K.F. Grebyonkin, V.A. Kryzhanovsky, *Nuclear pumped lasers at the Institute of Technical Physics*. Proc. Int. Conf. LASERS '90, McLean, 1991, pp. 827–834.
33. E.P. Magda, *Analysis of experimental and computational-theoretical studies on nuclear-pumped lasers peformed at VNIITF* (Analiz eksperimentalnykh i raschetnoteoreticheskikh rabot po lazeram s yadernoy nakachkoy, vypolnennych vo VNIITF). Proceedings of the Specialist Conference "The Physics of Nuclear-Excited Plasma and the Problems of Nuclear-Pumped Lasers," Obninsk, 1992, vol. 1, pp. 65–74.
34. A.V. Zagidulin, A.V. Bochkov, V.V. Mironenko, G.S. Sofienko, The 500-J nuclear-pumped laser (500-dzoulnyy lazer s yadernoy nakachkoy). Pisma v ZhTF **38**(23), 31–39 (2012)
35. A.V. Bochkov, V.A. Kryzhanovskiy, O.A. Lyubimov, E.P. Magda, S.L. Mukhin, *Xenon nuclear-pumped laser at the 2.48 μm wavelength* (Ksenonovyy lazer s yadernoy nakachoy na dline volny 2.48 μm). Proceedings of the Specialist Conference "The Physics of Nuclear-Excited Plasma and the Problems of Nuclear-Pumped Lasers," Obninsk, 1993, vol. 2, pp. 130–132.

36. A.V. Bochkov, V.A. Kryzhanovsky, E.P. Magda, *Research of characteristics of nuclear pumped laser on* $3d[1/2]$*-*$4p[3/2]_2$ *(*$\lambda = 1.79$ *μm) transition of atomic argon*. Proc. 9th Int. Conf. on Emerging Nuclear Energy Systems, Tel-Aviv, 1998, vol. 2, pp. 868–873.
37. A.V. Bochkov, E.P. Magda, V.V. Mironenko, V.M. Murzin, G.S. Sofiyenko, *Research on parameters of lasers operating on transitions of rare gas atoms under a high energy load* (Issledovaniya parametrov lazerov na perekhodakh atomov inertnykh gazov v usloviyakh vysokoy energeticheskoy nagruzki). Proceedings of the 3rd International Conference "Problems of Nuclear-Pumped Lasers and Pulsed Reactors," Snezhinsk, 2003, pp. 134–139.
38. H.H. Helmick, J.L. Fuller, R.T. Schneider, Direct nuclear pumping of a helium-xenon laser. Appl. Phys. Lett. **26**(6), 327–328 (1975)
39. R.J. DeYoung, N.W. Jalufka, F. Hohl, Nuclear-pumped lasing of ^{3}He-Xe and ^{3}He-Kr. Appl. Phys. Lett. **30**(1), 19–21 (1977)
40. R.J. DeYoung, N.W. Jalufka, F. Hohl, Direct Nuclear-Pumped Lasers using the He^3(n, p)H^3 reaction. AIAA J. **16**(9), 991–998 (1978)
41. C.R. Mansfield, P.F. Bird, J.F. Davis, T.F. Wimett, H.H. Helmick, Direct nuclear pumping of a ^{3}He-Xe laser. Appl. Phys. Lett. **30**(12), 640–641 (1977)
42. R.J. DeYoung, Y.J. Shiu, M.D. Williams, Fission-fragment nuclear lasing of Ar(He)-Xe. Appl. Phys. Lett. **37**(8), 679–681 (1980)
43. R.J. DeYoung, Multiple-path fission-foil nuclear lasing of Ar-Xe. Appl. Phys. Lett. **39**(8), 585–587 (1981)
44. N.W. Jalufka, Nuclear-pumped lasing of ^{3}He-Xe at 2.63 μm. Appl. Phys. Lett. **39**(7), 535–536 (1981)
45. W.J. Alford, G.N. Hays, Measured laser parameters for reactor-pumped He/Ar/Xe and Ar/Xe lasers. J. Appl. Phys. **65**(10), 3760–3766 (1989)
46. G.A. Hebner, G.N. Hays, Laser efficiency and gain of the 1.73 μm atomic xenon laser at high He/Ar buffer gas ratio. J. Appl. Phys. **74**(6), 3673–3679 (1993)
47. G.A. Hebner, G.N. Hays, Reactor-pumped laser experimental results. Proc. S. P. I. E. **2121**, 10–20 (1994)
48. D.A. McArthur, G.N. Hays, W.J. Alford, D.R. Neal, D.E. Bodette, J.K. Rice, Recent results on reactor-pumped laser studies at Sandia National Laboratories. Las. Interact. **8**, 75–86 (1988)
49. E.G. Batyrbekov, G.H. Miley, Y.D. Poletayev, E. Suzuki, *Ar-Xe laser With 1.73 μm wavelength pumped by nuclear reaction products* $^{10}B(n,\alpha)^7Li$ (Ar-Xe lazer s dlinoy volny 1.73 μm, nakachivayemyy produktami yadernoy reaktsii ^{10}B(n,α)^{7}Li). Proceedings of the 2nd International Conference "Physics of Nuclear-Excited Plasma and Problems of Nuclear-Pumped Lasers," Arzamas-16, 1995, vol. 1, pp. 329–337.
50. N.W. Jalufka, R.J. DeYoung, F. Hohl, M.D. Williams, Nuclear-pumped ^{3}He-Ar laser excited by the ^{3}He(n, p)^{3}H reaction. Appl. Phys. Lett. **29**(3), 188–190 (1976)
51. R.J. DeYoung, F. Hohl, Large volume multiple path nuclear lasing of ^{3}He-Ar. I. E. E. E. J. Quant. Electron. **QE-16**(10), 1114–1117 (1980)
52. R.J. DeYoung, Kilowatt multiple-path ^{3}He-Ar nuclear-pumped laser. Appl. Phys. Lett. **38**(5), 297–298 (1981)
53. G.A. Hebner, G.N. Hays, Parametric investigation of the fission-fragment excited helium/argon laser at 1.79 μm. J. Appl. Phys. **71**(4), 1610–1616 (1992)
54. G.A. Hebner, G.N. Hays, 1.27-μm atomic argon laser parameters in fission-fragment excited He/Ar and He/Ne/Ar gas mixtures. I. E. E. E. J. Quant. Electron. **29**(8), 2356–2363 (1993)
55. W.J. Alford, G.N. Hays, M. Ohwa, M.J. Kushner, The effect of He addition on the performance of the fission-fragment excited Ar/Xe atomic xenon laser. J. Appl. Phys. **69**(4), 1843–1848 (1991)
56. G.A. Hebner, G.N. Hays, Fission-fragment excited xenon/rare gas mixtures. I. Laser parameters of the 1.73 μm xenon transition. J. Appl. Phys. **73**(8), 3614–3626 (1993)
57. G.A. Hebner, G.N. Hays, Fission-fragment excited xenon/rare gas mixtures. II. Small signal gain of the 2.03 μm xenon transition. J. Appl. Phys. **73**(8), 3627–3636 (1993)

58. N.W. Jalufka, Direct nuclear-pumped laser amplifier. Appl. Phys. Lett. **39**(9), 690–692 (1981)
59. H.A. Hassan, Analysis of the UF_6-Xe Direct Nuclear-Pumped Lasers. AIAA J. **18**(12), 1490–1494 (1980)
60. R.J. DeYoung, W.R. Weaver, Spectra from nuclear-excited plasmas. J. Opt. Soc. Am. **70**(5), 500–506 (1980)
61. A.Y. Dudin, D.A. Zayarnyy, L.V. Semenova, N.N. Ustinovskiy, I.V. Kholin, A.Y. Chugunov, Electron beam-pumped lasers using He-Xe, Kr, and Ar mixtures (Lazery s elektronnopuchkovoy nakachoy na smesyakh He-Xe, Kr, Ar). Kvantovaya elektronika **18**(8), 921–925 (1991)
62. N.G. Basov, V.A. Danilychev, A.Y. Dudin, D.A. Zayarnyy, N.N. Ustinovskiy, I.V. Kholin, A.Y. Chugunov, Electroionization IR Xe laser (Elektroionizatsionnyy IK lazer na atomakh Xe). Kvantovaya elektronika **11**(9), 1722–1736 (1984)
63. A. Suda, B.L. Wexler, K.J. Riley, B.J. Feldman, Characteristics of the high-pressure Ar-Xe laser pumped by an electron beam and electron-beam sustained discharge. IEEE J. Quant. Electron. **26**(5), 911–921 (1990)
64. A.Y. Dudin, D.A. Zayarnyy, L.V. Semenova, N.N. Ustinovskiy, I.V. Kholin, A.Y. Chugunov, Electron beam pumped lasers using mixtures Xe, Kr, and Ar with two-component buffer gases (Lazery s elektronnopuchkovoy nakachkoy na smesyakh Xe, Kr i Ar s dvukhkomponentnymi bufernymi gazami). Kvantovaya elektronika **18**(11), 1290–1294 (1991)
65. B.M. Berkeliyev, V.A. Dolgikh, I.G. Rudoy, A.M. Soroka, On the ultimate efficiency of lasers operating on the *d-p* transitions of heavy rare atoms (O predelnoy effektivnosti lazerov na *d* - *p*-perekhodakh tyazhelykh inertnykh atomov). Kvantovaya elektronika **17**(12), 1537–1538 (1990)
66. M.V. Bokhovko, A.P. Budnik, I.V. Dobrovolskaya, V.N. Kononov, O.Y. Kononov, The influence of the specific energy deposition on the characteristics of the nuclear-pumped Ar/Xe laser (Vliyaniye udelnogo energovklada na kharakteristiki Ar/Xe lazera s yadernoy nakachkoy). Pisma v ZhTF **24**(14), 16–20 (1998)
67. E.P. Magda, K.F. Grebenkin, V.A. Kryzhanovskiy, A.V. Bochkov, S.L. Mukhin, Experimental investigation of the influence of temperature on the characteristics of the nuclear-pumped xenon laser (Eksperimentalnoye issledovaniye vliyaniya temperatury na kharakteristiki ksenonovogo lazera s yadernoy nakachkoy). Pisma v ZhTF **18**(5), 96–98 (1992)
68. V.A. Kryzhanovskiy, A.A. Mavlyutov, A.I. Miskevich, Characteristics of lasing of the nuclear-pumped Ar-Xe laser at increased temperature (Kharakteristiki generatsii Ar-Xe lazera s yadernoy nakachkoy pri povyshennykh temperaturakh). Pisma v ZhTF **21**(13), 90–94 (1995)
69. A.I. Konak, S.P. Melnikov, V.V. Porkhayev, A.A. Sinyanskiy, *A nuclear-pumped gas laser at temperatures of 800°C* (Gazovyy lazer s yadernoy nakachkoy pri temperaturakh do 800°C). Proceedings of the Specialist Conference "The Physics of Nuclear-Excited Plasma and the Problems of Nuclear-Pumped Lasers," Obninsk, 1993, vol. 3, pp. 98–106.
70. G.A. Hebner, J.W. Shon, M.J. Kushner, Temperature-dependent gain of the atomic xenon laser. Appl. Phys. Lett. **63**(21), 2872–2874 (1993)
71. G.A. Hebner, Gas temperature dependent output of the atomic argon and xenon lasers. I. E. E. E. J. Quant. Electron. **31**(9), 1626–1631 (1995)
72. G.A. Batyrbekov, E.G. Batyrbekov, V.A. Danilychev, A.B. Tleuzhanov, M.U. Khasenov, Gas-discharge xenon laser with weak ionization by an outside source (Elektrorazryadnyy ksenonovyy lazer so slaboy ionizatsiey vneshnim istochnikom). Kvantovaya elektronika **16** (11), 2165–2169 (1989)
73. N.M. Barysheva, A.V. Bochkov, N.V. Bochkova, K.F. Grebenkin, V.A. Kryzhanovskiy, E.P. Magda, *On the possible mechanism of overheating of the active medium of an NPL operating on ir transitions of the xenon atom* (O Vozmozhnom mekhanizme nepegreva aktivnoy sredy LYaN na IK perekhodakh atoma ksenona). Proceedings of the Specialist

Conference "The Physics of Nuclear-Excited Plasma and the Problems of Nuclear-Pumped Lasers," Obninsk, 1992, vol. 1, pp. 374–380.

74. A.A. Mavlyutov, A.I. Miskevich, B.S. Salamakha, *Rehabilitation of the concept of the powerful Ar-Xe nuclear-pumped laser* (Reabilitatsiya kontseptsii moshchnogo Ar-Xe lazera s yadernoy nakachkoy). Proceedings of the 2nd International Conference "Physics of Nuclear-Excited Plasma and Problems of Nuclear-Pumped Lasers," Arzamas-16, 1995, vol. 1, pp. 318–328.
75. A.A. Mavlyutov, A.I. Miskevich, B.S. Salamakha, Excitation of the A-X bands of OH in dense rare gases by high-energy nuclear particles (Vozbuzhdeniye A-X-polos OH v plotnykh inertnykh gazakh yadernymi chastitsami vysokikh energy). Optika i spektroskopiya **76**(6), 949–954 (1994)
76. H. Tomizawa, M. Salvermoser, J. Wieser, A. Ulrich, Influence of water vapor impurities on the atomic Xe laser. J. Phys. B. At. Mol. Opt. Phys. **33**, 181–190 (2000)
77. H. Tomizawa, J. Wieser, A. Ulrich, *Aspects of gas purity and gas temperature in nuclear pumped laser systems*. Proceedings of the 3rd International Conference "Problems of Nuclear-Pumped Lasers and Pulsed Reactors," Snezhinsk, 2003, pp. 146–161.
78. V.S. Skakun, V.F. Tarasenko, A.V. Fedenev, The influence of additions of N_2, CO_2, and He on the output power of the xenon laser with electron beam pumping (Vliyaniye dobavok N_2, CO_2 i He na moshchnost izlucheniya ksenonovogo lazera pri nakachke elektronnym puchkom). Optika i spektroskopiya **71**(4), 669–674 (1991)
79. V.V. Porkhayev, Influence of the impurities N_2, O_2, H_2 and H_2O on the lasing of nuclear-pumped lasers operating on transitions of xenon and chlorine atoms (Vliyaniye primesey N_2, O_2, H_2 i H_2O na generatsiyu lazerov s yadernoy nakachkoy na perekhodakh atomov ksenona i khlora). Kvantovaya elektronika **25**(12), 1095–1099 (1998)
80. V.A. Ivanov, Dissociative recombination of molecular ions in a plasma of rare gases (Dissotsiativnaya rekombinatsiya molekulyarnykh ionov v plazme inertnykh gazov). Uspekhi fizicheskikh nauk **162**(1), 35–70 (1992)
81. A.A. Abramov, V.V. Gorbunov, S.P. Melnikov, A.K. Mukhamatullin, A.A. Pikulev, A.V. Sinitsyn, A.A. Sinyanskiy, V.M. Tsvetkov, Luminescence of nuclear-induced rare-gas plasmas in near infrared spectral range. Proc. SPIE **6263**, 279–296 (2006)
82. A.M. Voinov, V.N. Krivonosov, S.P. Melnikov, A.I. Pavlovskiy, A.A. Sinyanskiy, Quasi-CW lasing at *3p-3s* transitions of the neon atom with excitation of mixtures of rare gases by uranium fission fragments (Kvazinepreryvnaya generatsiya na perekhodakh 3*p*-3*s* atoma neona pri vozbuzhdenii smesey inertnykh gazov oskolkami deleniya urana). Doklady AN SSSR **312**(4), 864–867 (1990)
83. A.I. Konak, S.P. Melnikov, V.V. Porkhayev, A.A. Sinyanskiy, Characteristics of nuclear-pumped lasers operating on the *3p-3s* transitions of the neon atom (Kharakteristiki lazera s yadernoy nakachoy na perekhodakh 3*p*-3*s* atoma neona). Kvantovaya elektronika **22**(3), 225–230 (1995)
84. A.I. Konak, S.P. Melnikov, V.V. Porkhayev, A.A. Sinyanskiy, Research on *Lasing in the mixtures He-Ne-M and Ne-M (M = Ar, Kr, Xe) at the 585.3, 703.2 and 724.5 nm lines during excitation by uranium fission fragments* (Issledovaniye generatsii v smesyakh He-Ne-M i Ne-M (M = Ar, Kr, Xe) na liniyakh 585.3, 703.2 i 724.5 nm pri vozbuzhdenii oskolkami deleniya urana). Proceedings of the 2nd International Conference "Physics of Nuclear-Excited Plasma and Problems of Nuclear-Pumped Lasers," Arzamas-16, 1995, vol. 1, pp. 293–306.
85. A.I. Konak, S.P. Melnikov, V.V. Porkhayev, A.A. Sinyanskiy, *The influence of hydrogen on the characteristics of the neon nuclear-pumped laser* (Vliyaniye vodoroda na kharakteristiki neonovogo lazera s yadernoy nakachkoy). Proceedings of the 2nd International Conference "Physics of Nuclear-Excited Plasma and Problems of Nuclear-Pumped Lasers," Arzamas-16, 1995, vol. 1, pp. 256–263.

86. A.P. Kopay-Gora, A.I. Miskevich, B.S. Salamakha, Generating laser radiation with a wavelength of 585.2 nm in a dense ^{3}He-Ne-Ar plasma (Generatsiya lazernogo izlucheniya s dlinoy volny 585.2 nm v plotnoy ^{3}He-Ne-Ar plazme). Pisma v ZhTF **16**(11), 23–26 (1990)
87. G.A. Hebner, G.N. Hays, Fission-fragment excited lasing at 585.3 nm in He/Ne/Ar gas mixtures. Appl. Phys. Lett. **57**(21), 2175–2177 (1990)
88. Y.R. Shaban, G.H. Miley, *A practical visible wavelength nuclear-pumped laser*. Proceedings of the Specialist Conference "The Physics of Nuclear-Excited Plasma and the Problems of Nuclear-Pumped Lasers," Obninsk, 1993, vol. 2, pp. 240–250.
89. G.A. Hebner, Fission-fragment excitation of the high-pressure atomic neon laser at 703.2 and 724.5 nm. J. Appl. Phys. **74**(4), 2203–2207 (1993)
90. A.V. Bochkov, V.A. Kryzhanovskiy, E.P. Magda, S.L. Mukhin, Quasi-CW lasing with $\lambda = 585.2$ nm in a Ne-H_2 mixture (Kvazinepreryvnaya generatsiya s $\lambda = 585.2$ nm v smesi Ne-H_2). Pisma v ZhTF **19**(23), 30–33 (1993)
91. A.V. Bochkov, V.A. Kryzhanovskiy, E.P. Magda, S.L. Mukhin, *Quasi-CW lasing of a nuclear-pumped laser at the wavelength* $\lambda = 585.2$ *nm in a Ne-*H_2 *mixture* (Kvazinepreryvnaya generatsiya lazera s yadernoy nakachkoy na dline volny $\lambda = 585.2$ nm v smesi Ne-H_2). Proceedings of the 2d International Conference "Physics of Nuclear-Excited Plasma and Problems of Nuclear-Pumped Lasers," Arzamas-16, 1995, vol. 1, pp. 270–275.
92. F.V. Bunkin, V.I. Derzhiev, G.A. Mesyats, V.S. Skakun, V.F. Tarasenko, S.I. Yakovlenko, A plasma laser at 585.3 nm with penning quenching in dense mixtures with neon excited by an electron beam (Plazmennyy lazer na dline volny 585.3 nm s penningovskoy ochistkoy na plotnykh smesyakh s neonom, vozbuzhdayemikh elektronnym puchkom). Kvantovaya elektronika **12**(2), 245–246 (1985)
93. N.G. Basov, A.Y. Aleksandrov, V.A. Danilychev, V.A. Dolgikh, O.M. Kerimov, Y.F. Myznikov, I.G. Rudoy, A.M. Soroka, Powerful quasi-CW high-pressure laser in the visible range of the spectrum operating on *p-s* transitions of the Ne atom (Moshchnyy kvazinepreryvnyy lazer vysokogo davleniya v vidimoy oblasti spectra na *p-s* perekhodakh atoma Ne). Kvantovaya elektronika **12**(1), 8 (1985)
94. A.Y. Aleksandrov, V.Y. Ananyev, N.G. Basov, V.A. Danilychev, V.A. Dolgikh, A.A. Ionin, O.M. Kerimov, A.P. Lytkin, Y.F. Myznikov, I.G. Rudoy, A.M. Soroka, An efficient visible range laser operating on the *3p-3s* transitions of neon (Effektivnyy lazer vidimogo diapazona na 3*p*-3*s*-perekhodakh neona). Doklady AN SSSR **284**(4), 851–854 (1985)
95. F.V. Fedenev, V.F. Tarasenko, Simulation of NPL in experiments with e-beam pumping. Laser and Particle Beams **16**(2), 327–380 (1998)
96. D.A. Zayarnyy, I.V. Kholin, Penning high-pressure lasers operating on *3p-3s* transitions of neon at 703 and 725 nm (Penningovskiye lazery vysokogo davleniya na 3*p*-3*s* perekhodakh neona s dlinami voln 703 i 725 nm). Kvantovaya elektronika **33**(6), 474–484 (2003)
97. A. Ulrich, J. Wieser, A. Brunnhuber, W. Krötz, Heavy ion beam pumped visible laser. Appl. Phys. Lett. **64**(15), 1902–1904 (1994)
98. A.V. Karelin, V.F. Tarasenko, A.V. Fedenev, S.I. Yakovlenko, On the ultimate efficiency of the penning neon plasma laser (O predelnom KPD penningovskogo plazmennogo lazera na neone). Kvantovaya elektronika **23**(4), 299–302 (1996)
99. B.D. Carter, M.J. Rowe, R.T. Schneider, Nuclear-pumped cw lasing of the ^{3}He-Ne system. Appl. Phys. Lett. **36**(2), 115–117 (1980)
100. M.A. Prelas, G.A. Schlapper, Comments on nuclear-pumped cw lasing of the ^{3}He-Ne system. J. Appl. Phys. **52**(1), 496–497 (1981)
101. R.T. Schneider, B.D. Carter, M.J. Rowe, Reply to 'Comments on nuclear-pumped cw lasing of the 3He-Ne system'. J. Appl. Phys. **52**(11), 6980 (1981)
102. A. Ulrich, J.W. Hammer, W. Biermayer, Lasers pumped by ion beams. J. Appl. Phys. **63**(7), 2206–2211 (1988)
103. E.G. Batyrbekov, Direct and combined nuclear pumped lasers, (1994) Preprint of IAENNC, Republic of Kazakhstan, Almaty

104. V.V. Gorbunov, V.D. Grigoryev, L.Y. Dovbysh, S.P. Melnikov, A.V. Sinitsyn, A.A. Sinyanskiy, V.M. Tsvetkov, Luminescence spectra of dense gases in the region of 350–875 nm with excitation by uranium fission fragments. I. Rare gases and their mixtures (Spektry lyuminestsentsii plotnykh gazov v oblasti 350–875 nm pri vozbuzhdenii oskolkami deleniya urana. I. Inertnyye gazy i ikh smesi). *Trudy RFYaTSVNIIEF*, (6), 148–173 (2004).
105. Y.D. Poletayev, Y.B. Dorofeyev, P.P. Dyachenko, A.P. Kopay-Gora, A.A. Mavlyutov, A.I. Miskevich, B.S. Salamakha, Radiating characteristics of pure neon and high-pressure He-Ne mixtures during excitation by nuclear particles (Izluchatelnyye kharakteristiki chistogo neona i He-Ne smesey vysokogo davleniya pri vozbuzhdemii yadernymi chastitsami). ZhTF **62**(2), 1–8 (1992)
106. A.B. Dmitriyev, V.C. Ilyashenko, A.I. Miskevich, B.S. Salamakha, Luminescence of neon and certain mixtures thereof at high pressures (Lyuminestsentsiya neona i nekotorykh yego smesey pri vysokikh davleniyakh). Optiika i spektroskopiya **43**(6), 1165–1168 (1977)
107. G.A. Batyrbekov, E.G. Batyrbekov, V.A. Danilychev, M.U. Khasenov, The influence of helium on the populating efficiency of 3p-levels of the neon atom (Vliyaniye geliya na effektivnost zaseleniya 3p-urovney atomov neona). Kvantovaya elektronika **17**(9), 1175–1180 (1990)
108. M.I. Lomayev, D.Y. Nagornyy, V.F. Tarasenko, A.V. Fedenev, G.V. Kirillin, Lasing at atomic transitions of rare gases in mixtures with NF_3 (Generatsiya na atomarnykh perekhodakh inertnykh gazov v smesyakh s NF_3). Kvantovaya elektronika **16**(10), 2053–2056 (1989)
109. I.G. Ivanov, Y.L. Latush, M.F. Sem, *Metal Vapor Ion Lasers (Ionnye lazery na parakh metallov)* (Energoatomizdat, Moscow, 1990)
110. V.M. Andriyakhin, V.V. Vasiltsov, S.S. Krasilnikov, V.D. Pismennyy, V.Y. Khvostionov, On radiation of the $Hg\text{-}He^3$ gas mixture irradiated by a neutron flux (Ob izluchenii gazovoy smesi $Hg\text{-}He^3$, obluchayemoy potokom neytronov). Pisma v ZhTF **12**(2), 83–85 (1970)
111. M.A. Akerman, G.H. Miley, D.A. McArthur, A helium-mercury direct nuclear pumped laser. Appl. Phys. Lett. **30**(8), 409–412 (1977)
112. A.B. Dmitriyev, V.S. Ilyashenko, A.I. Miskevich, B.S. Salamakha, Spectroscopy of ^3He-Hg and ^3He-Kr-Hg high-pressure plasma excited by reaction products ^3He(n, p)T, (Spektroskopiya ^3He-Hg i ^3He-Kr-Hg plazmy vysokogo davleniya, vozbuzhdayemoy produktami reaktsii ^3He(n, p)T). Optika i spektroskopiya **47**(1), 63–66 (1979)
113. A.M. Boychenko, V.I. Derzhiev, A.G. Zhidkov, A.V. Karelin, A.V. Koval, O.V. Sereda, S.I. Yakovlenko, Kinetic models of certain plasma lasers pumped by a hard ionizer (Kineticheskiye modeli nekotorykh plazmennykh lazerov, nakachivayemykh zhestkim ionizatorom). Trudy IOFAN **21**, 44–115 (1989)
114. A.B. Dmitriyev, V.S. Ilyashenko, A.I. Miskevich, B.S. Salamakha, Excitation by products of neutron nuclear reactions of laser transitions in metal-vapor gas mixtures (Vozbuzhdeniye produktami neytronnykh yadernykh reaktsiy lazernykh perekhodov v parometallicheskikh gazovykh smesyakh). ZhTF **52**(11), 2235–2237 (1982)
115. G.A. Batyrbekov, E.G. Batyrbekov, V.A. Dolgikh, I.G. Rudoy, A.M. Soroka, A.B. Tleuzhanov, M.U. Khasenov, Luminescence of mixtures of mercury and rare gases with molecular additives during excitation by ionizing radiation (Lyuminestsentsiya smesey rtuti i inertnykh gazov s molekulyarnymi dobavkami pri vozbuzhdenii ioniziruyushchimi izlucheniyami). Zhurnal prikladnoy spektroskopii **49**(5), 770–774 (1988)
116. G.A. Batyrbekov, E.G. Batyrbekov, V.A. Dolgikh, I.G. Rudoy, A.M. Soroka, A.B. Tleuzhanov, M.U. Khasenov, On the possibility of creating a quasi-CW laser operating on *7s-6p* transitions of mercury with pumping by ionizing radiation (O vozmozhnosti sozdaniya kvazinepreryvnogo lazera na 7*s*-6*p*-perekhodakh rtuti pri nakachke ioniziruyushchimi izlucheniyami). Kvantovaya elektronika **14**(6), 1216–1218 (1987)
117. F.A. Butayeva, V.A. Fabrikant, On a medium with a negative absorption coefficient (O srede s otritsatelnym koeffitsiyentom pogloshcheniya). In the digest *Investigations in Experimental*

and Theoretical Physics (*Issledovaniya po eksperimentalnoy i teoreticheskoy fizike*) (AN SSSR Publishing, Moscow, 1959), pp. 62–70.

118. A.V. Bochkov, V.A. Kryzhanovskiy, E.P. Magda, S.L. Mukhin, V.M. Murzin, A.E. Neznakhina, Quasi-CW lasing at the 7^3S_1-6^3P_2 transition of the mercury atom (Kvazinepreryvnaya generatsiya na 7^3S_1-6^3P_2, perekhode atoma rtuti). Pisma v ZhTF **18** (7), 91–93 (1992)
119. A.V. Bochkov, E.P. Magda, V.V. Mironenko, A.E. Neznakhina, G.S. Sofienko, *Investigation of NPLs at the* 7^3S_1-6^3P_2 *transition* ($\lambda = 546.1$ *nm*) *of the mercury atom* (Issledovaniye LYaN na 7^3S_1-6^3P_2 ($\lambda = 546.1$ nm) perekhode atoma rtuti). Proceedings of the 3rd International Conference "Problems of Nuclear-Pumped Lasers and Pulsed Reactors," Snezhinsk, 2003, pp. 212–217.
120. R.L. Rhoades, J.T. Verdeyen, Electron-beam pumping of the 546.1 nm mercury laser. Appl. Phys. Lett. **60**(24), 2951–2953 (1992)
121. A.I. Miskevich, Toward a history of the development of a first visible-range nuclear-pumped laser (K istorii sozdaniya pervogo lazera vidimogo diapazona s yadernoy nakachkoy). Inzhenernaya fizika, (2), 6–13 (2007).
122. A.I. Miskevich, A.B. Dmitriyev, V.S. Ilyashenko, B.S. Salamakha, V.A. Stepanov, Y.M. Gorodkov, Obtaining of laser radiation using Cd vapors with excitation by nuclear reaction products $^3He(n, p)T$ (Generatsiya lazernogo izlucheniya na parakh Cd pri vozbuzhdenii produktami yadernoy reaktsii $^3He(n, p)T$). Pisma v ZhTF **6**(13), 818–821 (1980)
123. A.I. Miskevich, V.S. Ilyashenko, B.S. Salamakha, A.A. Sipaylo, V.A. Stepanov, Y.M. Gorodkov, Obtaining laser radiation at a wavelength of 446.6 nm in a high-pressure ^3He-^{116}Cd mixture (Generatsiya lazernogo izlucheniya na dline volny 441.6 nm v ^3He-^{116}Cd smesi vysokogo davleniya). ZhTF **52**(2), 402–404 (1982)
124. A.P. Kopay-Gora, A.I. Miskevich, B.S. Salamakha, Purification of helium by forced circulation through cooled sorbers in nuclear-pumped lasers (Ochistka geliya prinuditelnoy tsirkulyatsiyey cherez okhlazhdayemyye sorbery v lazerakh s yadernoy nakachkoy). Pribory i tekhnika experimenta, (5), 171–174 (1986).
125. A.I. Miskevich, A.P. Kopay-Gora, B.S. Salamakha, Quasi-CW lasing at the beutler transition ZnII in a dense ^3He-Zn plasma (Kvazinepreryvnaya generatsiya na beytlerovskom perekhode ZnII v plotnoy ^3He-Zn plazme). Pisma v ZhTF **16**(9), 62–64 (1990)
126. N.W. Jalufka, Direct nuclear-pumped lasers, NASA Technical Paper 2091, 1983.
127. A.M. Rybalov, A.N. Soldatov, V.I. Solomonov, Y.V. Sharabarin, V.M. Muratov, N.K. Kapishnikov, A unit for exciting gases and vapors with an electron beam (Ustanovka dlya vozbuzhdeniya gazov i parov elektronnym puchkom), Pribory i tekhnika experimenta, (4), 127–129 (1986).
128. S.M. Babenko, S.I. Yakovlenko, Possibility of nuclear pumping of a He-Sr laser. Phys. Lett. **76A**(3, 4), 237–239 (1980)
129. R.J. DeYoung, W.E. Wells, G.H. Miley, Optical gain in a neutron-induced ^3He-Ne-O_2 plasma. Appl. Phys. Lett. **28**(4), 194–197 (1976)
130. R.J. DeYoung, W.E. Wells, G.H. Miley, J.T. Verdeyen, Direct nuclear pumping of a Ne-N_2 laser. Appl. Phys. Lett. **28**(9), 519–521 (1976)
131. M.A. Prelas, M.A. Akerman, F.P. Boody, G.H. Miley, A direct nuclear pumped 1.45-μm atomic carbon laser in mixtures of He-CO and He-CO_2. Appl. Phys. Lett. **31**(7), 428–430 (1977)
132. M.A. Prelas, J.H. Anderson, F.P. Boody, S.J.S. Nagalingam, G.H. Miley, A nuclear pumped laser using Ne-CO and Ne-CO_2 mixtures. Bull. Am. Phys. Soc. **23**(2), 152 (1978)
133. D. Auphelle, F. Euve, M. Fitaire, A.M. Pointu, M. Vialle, Recherche d'un effet laser dans le melange He-CO_2 irradie par le reacteur Caliban. Phys. Lett. **72A**(6), 439–441 (1979)
134. A.M. Voinov, V.N. Krivonosov, S.P. Melnikov, A.A. Sinyanskiy, *Quasi-CW nuclear-pumped laser operating on carbon atom transitions* (Kvazinepreryvnyy lazer s yadernoy nakachkoy na perekhodakh atoma ugleroda). Proceedings of the Specialist Conference "The

Physics of Nuclear-Excited Plasma and the Problems of Nuclear-Pumped Lasers," Obninsk, 1993, vol. 2, pp. 197–202.

135. S.P. Melnikov, V.V. Porkhayev, Lasing at oxygen atom transitions during pumping of a He-(Ne)-O_2 mixture with uranium fission fragments (Generatsiya na perekhodakh atoma kisloroda pri nakachke smesi He-(Ne)-O_2 oskolkami deleniya urana). Pisma v ZhTF **21**(1), 86–89 (1995)
136. S.P. Melnikov, V.V. Porkhayev, Laser action on infrared chlorine and oxygen atomic transitions using uranium fission fragment pumping. Proc. SPIE **2619**, 22–28 (1995)
137. S.P. Melnikov, V.V. Porkhayev, Lasing at IR transitions of the chlorine atom with pumping of gas mixtures by uranium fission fragments (Generatsiya na IK perekhodakh atoma khlora pri nakachke gazovykh smesey oskolkami deleniya urana). Kvantovaya elektronika **22**(9), 891–894 (1995)
138. S.P. Melnikov, V.V. Porkhayev, Nuclear-pumped laser operating on nitrogen atom transitions (Lazer s yadernoy nakachkoy na perekhodakh atoma azota). Pisma v ZhTF **22**(1), 52–55 (1996)
139. S.P. Melnikov, V.V. Porkhayev, Quasi-CW near-IR nuclear-pumped laser operating on transitions of nitrogen and carbon atoms (Kvazinepreryvnyy lazer blizhnego IK diapazona na perekhodakh atomov azota i ugleroda s yadernoy nakachkoy). Kvantovaya elektronika **23** (10), 956–960 (1996)
140. V.V. Gorbunov, V.D. Grigoryev, S.P. Melnikov, A.V. Sinitsyn, A.A. Sinyanskiy, V.M. Tsvetkov, *The search for laser transitions with excitation of mixtures of helium with halogen-containing gases by uranium fission fragments* (Poisk lazernykh perekhodov pri vozbuzhdenii oskolakami deleniya urana smesey geliya s galogenosoderzhashchimi gazami). Proceedings of the 3d International Conference "Problems of Nuclear-Pumped Lasers and Pulsed Reactors," Snezhinsk, 2003, pp. 191–198.
141. A.A. Mavlyutov, A.I. Miskevich, B.S. Salamakha, *Kinetics of populating of levels in high-pressure Ne-O_2 and Ar-I_2 mixtures* (Kinetika zaseleniya urovney v Ne-O_2 i Ar-I_2 smesyakh visokogo davleniya). Proceedings of the Specialist Conference "The Physics of Nuclear-Excited Plasma and the Problems of Nuclear-Pumped Lasers," Obninsk, 1992, vol. 1, pp. 308–317.
142. F.G. Goryunov, D.Y. Nagornyy, V.F. Tarasenko, Lasing at transitions of atomic fluorine in a He-NF_3 mixture during pumping by an electron beam with a duration of ~10 ns (Generatsiya na perekhodakh atomarnogo ftora v smesi He-NF_3 pri nakachke elektronnym puchkom dlitelnostyyu ~ 10 ns). Kvantovaya elektronika **16**(7), 1354–1357 (1989)
143. D.A. McArthur, P.B. Tollefsrud, Observations of laser action in CO gas excited only by fission fragments. Appl. Phys. Lett. **26**(4), 187–190 (1975)
144. D.A. McArthur, P.B. Tollefsrud, Measurements of optical gain in CO gas excited only by fission fragments. I. E. E. E. J. Quant. Electron. **QE-12**(4), 244–253 (1976)
145. N.W. Jalufka, F. Hohl, A direct nuclear-pumped ^{3}He-CO laser. Appl. Phys. Lett. **39**(2), 139–142 (1981)
146. C.B. Collins, A.J. Cunningham, M. Stockton, A nitrogen ion laser pumped by charge transfer. Appl. Phys. Lett. **25**(6), 344–345 (1974)
147. N.G. Basov, A.Y. Aleksandrov, V.A. Danilychev, V.A. Dolgikh, O.M. Kerimov, Y.F. Myznikov, I.G. Rudoy, A.Y. Samarin, A.M. Soroka, An effective quasi-CW high-pressure laser in the first negative system of nitrogen (Effektivnyy kvazinepreryvnyy lazer vysokogo davleniya na pervoy otritsatelnoy sisteme azota). Pisma v ZhTF **42**(1), 39–42 (1985)
148. N.M. Barysheva, N.V. Bochkova, K.F. Grebenkin, E.P. Magda, O.V. Skvortsova, *The kinetics of a nuclear-pumped nitrogen laser* (Kinetika azotnogo lazera s yadernoy nakachkoy). Proceedings of the Specialist Conference "The Physics of Nuclear-Excited Plasma and the Problems of Nuclear-Pumped Lasers," Obninsk, 1992, vol. 1, pp. 337–347.
149. N.M. Barysheva, A.V. Bochkov, N.V. Bochkova, K.F. Grebenkin, E.V. Zaytsev, A.V. Lutsenko, V.A. Kryzhanovskiy, E.P. Magda, V.M. Murzin, A.E. Neznakhina, The

first ultraviolet nuclear-pumped laser (Pervyy ultrafioletovyy lazer s yadernoy nakachkoy). Pisma v ZhTF **22**(15), 83–86 (1996)
150. A.V. Bochkov, E.P. Magda, V.V. Mironenko, V.M. Murzin, G.S. Sofienko, *Investigation of a laser using the first negative band of nitrogen* (Issledovaniye lazera na 1-oy otritsatelnoy polose azota). Proceedings of the 3rd International Conference "Problems of Nuclear-Pumped Lasers and Pulsed Reactors," Snezhinsk, 2003, pp. 167–172.
151. Y.A. Dyuzhov, Y.D. Poletayev, V.N. Smolskiy, *An investigation of lasing at transitions of the first negative system of nitrogen ($\lambda = 391.4$; 428.1 nm) in He-N_2-H_2 mixtures during pumping by fission fragments from the BARS-6 pulsed reactor* (Issledovaniye generatsii na perekhodakh pervoy otritsatelnoy sistemy azota ($\lambda = 391.4$; 428.1 nm) v He-N_2-H_2 smesyakh pri nakachke oskolkami deleniya ot impulsnogo reaktora BARS-6). Proceedings of the 4th International Conference "Physics of Nuclear-Pumped Lasers and Pulsed Reactors," Obninsk, 2009, vol. 1, pp. 151–155.
152. N.W. Jalufka, Direct nuclear excitation of a ^{3}He-CO_2 gas mixture. Appl. Phys. Lett. **39**(3), 190–192 (1981)
153. H.A. Hassan, Kinetics of a CO_2 Nuclear Pumped Laser. AIAA J. **18**(10), 1221–1222 (1980)
154. R.T. Schneider, F. Hohl, Nuclear-pumped lasers. Adv. Nucl. Sci. Technol. **16**, 123–287 (1984)
155. C. Rhodes (ed.), *Excimer Lasers (Eksimernyy lazery)* (Mir, Moscow, 1981)
156. G.A. Mesyats, V.V. Osipov, V.F. Tarasenko, *Pulsed Gas Lasers (Impulsnyye gazovyye lazery)* (Nauka, Moscow, 1991)
157. G.H. Miley, S.J.S. Nagalingam, F.P. Boody, M.A. Prelas, Nuclear pumping of XeF(*B*), a candidate laser fusion driver. Trans. Am. Nucl. Soc. **30**(1), 26–27 (1978)
158. G.N. Hays, D.A. McArthur, D.R. Neal, J.K. Rice, Gain measurements near 351 nm in ^{3}He/Xe/NF_3 mixtures excited by fragments from the ^{3}He(n, p)^{3}H reaction. Appl. Phys. Lett. **49**(7), 363–365 (1986)
159. J.K. Rice, G.N. Hays, D.R. Neal, D.A. McArthur, W.J. Alford, *Nuclear reactor excitation of XeF laser gas mixtures*. Proc. Int. Conf. on LASERS'86, 1986, pp. 571–578.
160. A.V. Bochkov, V.A. Kryzhanovskiy, E.P. Magda, V.M. Murzin, S.L. Mukhin, A.E. Neznakhina, *Investigation of the characteristics of excimer laser media* (Issledovaniye kharakteristik eksimernykh lazernykh sred). Proceedings of the 2nd International Conference, "Physics of Nuclear-Excited Plasma and Problems of Nuclear-Pumped Lasers," Arzamas-16, 1995, vol. 1, pp. 154–161.
161. A.V. Demyanov, N.A. Dyatko, I.V. Kochetov, A.P. Napartovich, *Modeling of nuclear-pumped excimer lasers* (Modelirovaniye eksimernykh lazerov s yadernoy nakachkoy). Proceedings of the Specialist Conference "The Physics of Nuclear-Excited Plasma and the Problems of Nuclear-Pumped Lasers," Obninsk, 1992, vol. 1, pp. 252–261.
162. A.M. Boychenko, A.V. Karelin, S.I. Yakovlenko, Calculation of threshold characteristics of a nuclear-pumped Ne-Xe-NF_3-laser (Raschet porogovykh kharakteristik Ne-Xe-NF_3-lazera s yadernoy nakachkoy). Kvantovaya elektronika **22**(6), 547–550 (1995)
163. A.A. Mavlyutov, A.I. Miskevich, A nuclear-pumped excimer laser with a lasing wavelength of 308 nm (Eksimernyy lazer s yadernoy nakachkoy s dlinoy volny generatsii 308 nm). Pisma v ZhTF **22**(8), 48–52 (1996)
164. G.B. Brederlow, E. Fill, K. Witte, *High-Power Iodine Laser (Moshchnyy yodnyy lazer)* (Energoatomizdat, Moscow, 1985)
165. N.N. Yuryshev, A chemically pumped oxygen-iodine laser (Kislorodno-yodnyy lazer s khimicheskoy nakachkoy). Kvantovaya elektronika **23**(7), 583–600 (1995)
166. F.P. Boody, C.A. Ottinger, M.S. Zediker, G.H. Miley, Nuclear production of singlet-delta O_2. Trans. Am. Nucl. Soc. **34**(1), 820–821 (1980)
167. G.H. Miley, *Dual-media nuclear-pumped lasers*. Int. Symp. on Short-Wavelength Lasers and Their Applications, Osaka, 1987, pp. 14–25.
168. N.M. Barysheva, N.V. Bochkova, A.A. Kosorukova, E.P. Magda, *Numerical modeling of a nuclear-pumped Ar-H_2-I_2 Laser* (Chislennoye modelirovaniye Ar-H_2-I_2 lazera s yadernoy

nakachkoy). Proceedings of the 3rd International Conference "Problems of Nuclear-Pumped Lasers and Pulsed Reactors," Snezhinsk, 2003, pp. 69–75.
169. J.W. Wilson, Nuclear-induced XeBr* photolytic laser model. Appl. Phys. Lett. **37**(8), 695–697 (1980)
170. J.W. Wilson, A. Shapiro, Nuclear-induced excimer fluorescence. J. Appl. Phys. **51**(5), 2387–2393 (1980)
171. W.H. Williams, G.H. Miley, *Nuclear-induced XeBr* fluorescence pumping of the atomic iodine laser*. Proceedings of the Specialist Conference "The Physics of Nuclear-Excited Plasma and the Problems of Nuclear-Pumped Lasers," Obninsk, 1993, vol. 2, pp. 223–232.
172. W.H. Williams, G.H. Miley, Lasing in atomic iodine with a nuclear-pumped XeBr* flash lamp. Appl. Phys. Lett. **62**(15), 1724–1726 (1993)
173. N.G. Basov (ed.), *Chemical Lasers (Khimicheskiye lazery)* (Nauka, Moscow, 1982)
174. A.S. Boreysho, Powerful mobile chemical lasers (Moshchnyye mobilnyye khimicheskiye lazery). Kvantovaya elektronika **35**(5), 393–406 (2005)
175. F. Kannari, H. Inagaki, M. Obara, High efficiency multikilojoule HF chemical lasers using an electron beam initiated low-pressure mixtures of $H_2/F_2/NF_3$ or $H_2/F_2/SF_6$. Appl. Phys. Lett. **48**(4), 266–268 (1986)
176. N.G. Basov, A.G. Molchanov, A.N. Orayevskiy, Initiation of lasing of chemical gas lasers by neutron fluxes (Initsiirovaniye generatsii khimicheskikh gazovykh lazerov neytronnymi potokami). Trudy FIAN **194**, 71–86 (1989)
177. R.E. Anderson, E.M. Leonard, R.F. Shea, R.R. Berggren, *Nuclear-pumped lasers for large-scale applications*. Proc. Int. Conf. LASERS'88, 1989, pp. 236–244.
178. M.J. Kushner, Characteristics of a UF_6-H_2/HF nuclear-pumped laser. J. Appl. Phys. **51**(5), 2421–2428 (1980)
179. N.M. Barysheva, A.V. Bochkov, N.V. Bochkova, A.A. Kosorukova, E.P. Magda, *A nuclear-pumped HF laser* (HF-lazer s yadernoy nakachkoy). Proceedings of the 3rd International Conference "Problems of Nuclear-Pumped Lasers and Pulsed Reactors," Snezhinsk, 2003, pp. 82–86.
180. T.G. Miller, G.J. Dezenberg, G.W. Morrison, Nuclear-initiated chemical laser experiments. Trans. Am. Nucl. Soc. **21**, 22–23 (1975)
181. A.M. Voinov, A.S. Koshelev, S.P. Melnikov, A.A. Sinyanskiy, A quasi-cw gas laser excited by fast neutrons (Kvazinepreryvnyy gazovyy lazer, vozbuzhdayemyy bystrymi neytronami). Pisma v ZhTF **16**(13), 86–89 (1990)

Chapter 4
Basic Parameters of Nuclear-Excited Plasma

At present, direct nuclear pumping is implemented for gas laser media in which the populating of lasing levels occurs as a result of processes occurring in a high-pressure low-temperature plasma formed by ionizing radiation. Sometimes this plasma is called nuclear-excited plasma.

A plasma formed during the stopping of high-energy charged particles in a gas at atmospheric pressure is a thermodynamically non-equilibrium plasma. A plasma of this type, in which the concentration of electrons substantially exceeds its equilibrium value, is called a supercooled or recombination plasma [1, 2]. In a nuclear-excited plasma (in contrast to a gas-discharge plasma with an ionization type of non-equilibrium), the majority of electrons have energies close to thermal, and recombination processes play a key role in the kinetics of the charged particles.

4.1 Initial Stage of Ionization Processes in Gas Media

Dependence of Ionization Processes on the Type of Charged Particles

Various types of ionizing radiation are used to act on the gas medium—uranium and transuranium fission fragments, α particles, γ quanta, fast electrons and protons. In the case of γ quanta, the gas ionization is provoked by fast electrons arising as a result of Compton scattering, the photoeffect, and effect of electron-positron pair formation. In the initial stage of ionization processes, primary ionization during the immediate interaction of the charged particle with the atoms is distinct from secondary ionization due to the interaction of atoms with electrons formed as a result of primary ionization.

The process of ionization of an atom may be viewed as a binary collision of an oncoming charged particle with one of the electrons of the atom's shell [3]. Because

S.P. Melnikov et al., *Lasers with Nuclear Pumping*,
DOI 10.1007/978-3-319-08882-2_4

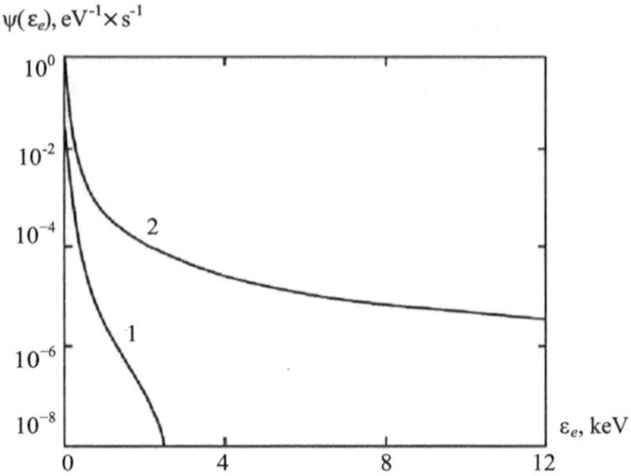

Fig. 4.1 Normalized spectra of primary electrons in neon during ionization by uranium fission fragments (*1*) and by fast electrons with an initial energy of 1 MeV (*2*) [4]

of the great difference in the masses of heavy charged particles (for example, a fission fragment) and the electron, only a comparatively small percentage of the fragment energy can be transferred to the orbital electron. Thus the electron spectrum arising due to primary ionization by heavy particles is softer than the analogous spectrum formed during ionization of a gas by fast electrons [4, 5]. For example, Fig. 4.1 shows the spectra of primary electrons during the ionization of neon by uranium fission fragments and fast electrons with an initial energy of 1 MeV [4]. Although both spectra have maxima in the low-energy region, the spectrum formed by fast electrons has a significant percentage of electrons with energies up to ~10 keV. The average energy of the primary electrons formed in neon as a result of ionization by fast electrons and fission fragments is equal to 150 eV and 40 eV, respectively. In the case of fission fragments, these electrons may provide an additional one or two acts of ionization (on average), while in the case of the fast electrons, it is from five to ten [5]. Information about the average energy of primary electrons in ionization by fission fragments for other rare gases agrees with the above data for neon. It follows from the calculations in [6] that the average energies of primary electrons in the ionization of He, Ar, Kr, and Xe by fission fragments are 60 eV, 44 eV, 41 eV, and 35 eV, respectively.

Despite the specifics of primary ionization by heavy and light charged particles discussed previously, differences in the radiation effects on the gas media of different types of ionizing particles are not substantial, because the ultimate result is a combined effect of primary and secondary ionization. This is confirmed by the data cited below, from which it follows that the final electron energy distribution and the energy cost of forming an ion–electron pair in different gases are virtually independent of the type of charged particles. The same conclusion can be drawn from analysis of the luminescence spectra of gas media and the characteristics of gas NPLs, which depend not on the type of the charged particles, but on the specific power deposition. In this regard, the same kinetic models are used to calculate the characteristics of NPLs excited by heavy charged particles and those excited by fast electrons.

Plasma Track Structure

One of the specific features of the nuclear-excited plasma is related to the formation of tracks (or columns) of ionization with passage of heavy charged particles (fission fragments and other heavy ions) through high-pressure gases, which in turn leads to non-uniform ionization of the gas media, and as a consequence, can influence populating of laser levels in NPLs [7, 8]. Depending on the parameters of the gas media, the transverse dimensions of the tracks are 1–10 μm, and the track lifetime, or the time to establish uniform ionization through diffusion, is 0.1–1 μs [7–9]. The degree of influence of the track structure on the characteristics of the nuclear-excited plasma depends on the type of ionizing particles, the specific power deposition, and the type and pressure of the gas. The ionization non-uniformity, provoked by the formation of tracks, will be the most significant in the following cases: (a) when the gas is ionized by fission fragments and by other heavy ions; (b) for high-pressure gases with high atomic mass; (c) at low specific power deposition, when there is no overlapping of tracks.

The influence of the track nature of gas ionization on recombination processes was examined in study [9], where the recombination coefficients were measured for ^{3}He, ^{4}He and the mixtures ^{3}He-CO_2 and Ar-N_2 on ionization by γ quanta, products of the nuclear reaction ^{3}He(n,p)^{3}H and ^{235}U fission fragments. In the first case, the ionization chamber was irradiated with ^{60}Co γ quanta, in the two other cases—with neutron radiation of a stationary nuclear reactor, with a neutron flux density of $\leq 10^{10}$ cm^{-2} s^{-1}. The results of these measurements for several pressures of the gas media are shown in Table 4.1.

It follows from a comparison of the data in Table 4.1 that with the ionization of ^{4}He (10 atm) and the mixture Ar-N_2 (1 atm) by uranium fission fragments, the recombination coefficients are 3–4 times greater than with ionization by γ quanta. It is in these two cases that the track structure of the plasma became apparent, and the so-called columnar recombination occurs, which is in accord with the calculations

Table 4.1 Comparison of recombination coefficients of gas media with different methods of ionization [9]

Ionization method	Gas medium	Pressure, atm	Recombination coefficient, cm^3/s: Neutron reaction products	Recombination coefficient, cm^3/s: γ quanta
Gas ^{3}He; reaction ^{3}He(n,p)^{3}H (kinetic energy of reaction products 0.76 MeV)	^{3}He	1	5.4×10^{-7}	5.4×10^{-7}
		6	2.0×10^{-6}	1.8×10^{-6}
		10	3.0×10^{-6}	2.9×10^{-6}
	^{3}He-CO_2 (2 %)	1	1.8×10^{-5}	1.8×10^{-5}
		10	1.7×10^{-5}	1.8×10^{-5}
Thin layer of $^{235}UO_8$; reaction ^{235}U(n,f) (kinetic energy of fission fragments 176 MeV)	^{4}He	1	2.4×10^{-6}	–
		10	5.1×10^{-6}	1.6×10^{-6}
	Ar-N_2 (5 %)	1	3.6×10^{-5}	9.2×10^{-6}

performed in study [9] based on the theory of columnar recombination [10]. Also important in study [9] is the conclusion that columnar recombination becomes noticeable when there are ionization losses of $\theta_i \geq 10^7$ ion pairs/cm. The optimal buffer-gas pressures at which gas NPLs operate are $P_{He} \leq 3$ atm, $P_{Ne} \leq 1$ atm, and $P_{Ar} \leq 0.5$ atm (see Chap. 3, Sect. 3.1, and Chap. 7, Sect. 7.5). At such pressures, $\theta_i \leq 10^6$ ion pairs/cm and the influence of the track structure of the plasma on the recombination processes will be insignificant.

Fluctuation of plasma component concentrations induced by the track structure can exert some influence on the characteristics of NPLs excited by fission fragments, if the upper laser levels in these lasers are populated owing to fast charge-transfer processes, for example in the case of a He-Cd laser [7, 8]. When lasers operating on transitions of rare gas atoms are pumped by fission fragments, the influence of the plasma track structure can evidently be ignored, because the track lifetime is considerably less than the characteristic times of the recombination processes [8].

At high specific power deposition, the tracks are closed and the track structure of the plasma disappears. Assessments performed in study [8] show that for helium at a pressure of 1 atm, the track overlapping occurs at specific power deposition of $q \geq 2$ W/cm^3.

Kinetics of Gas Ionization

Gas ionization in the initial stage occurs directly by charged particles and secondary electrons. A picture of the electron energy distribution in the gas is shown in Fig. 4.2 [11], where f_e is the energy distribution function of the electrons, normalized by the constraint $\int_0^\infty f_e(\varepsilon_e) d\varepsilon_e = 1$; and ε_e is the electron energy. The entire electron energy range can be divided into three regions:

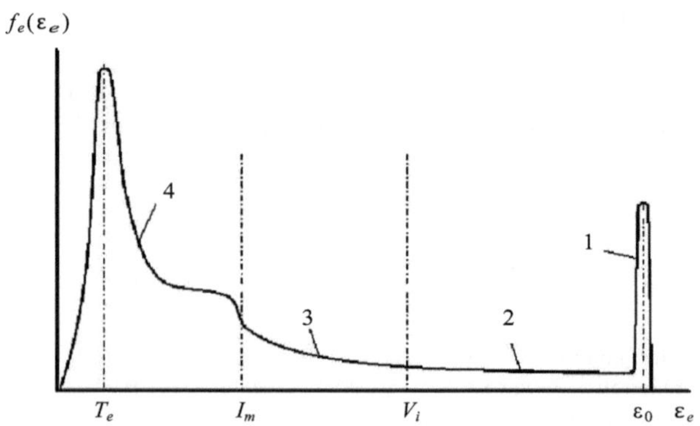

Fig. 4.2 Electron energy distribution in a gas [11]: (*1*) primary electrons of source; (*2*) electrons of ionization cascade; (*3*) electrons in the inelastic excitation region; (*4*) thermal and subthreshold electrons

1. The region of the ionization cascade $V_i < \varepsilon_e < \varepsilon_0$ (ε_0 is the particle initial energy, V_i is the ionization potential of the gas atoms and molecules), in which the energy of the electrons is sufficient for ionization of gas particles, and is expended for ionization and excitation of particles.
2. The region $I_m < \varepsilon_e < V_i$ (I_m is the minimal threshold of electron, or vibrational excitations) in which the energy of the electrons is reduced, primarily due to excitation of electron and vibrational states of the gas particles. In this region, as in the previous one, electrons lose energy in comparatively large "portions," equal to or greater than I_m (for helium, $I_m = 19.8$ eV is the energy of the lowest 2^3S state; for molecular gases I_m is substantially lower and can be ~1 eV).
3. In the subthreshold region ($\varepsilon_e < I_m$), electrons lose energy in small "portions" due to elastic collisions with gas particles, and the share of energy lost by an electron in one collision, $\delta_{ea} = 2m_e/M_a$ (where m_e and M_a are the masses of the electron and atom, respectively). This process is sometimes called electron thermalization.

An estimate of the characteristic time of energy loss by electrons in the high-energy region ($\varepsilon_e > I_m$) for gases at atmospheric pressure gives $\tau_{nel} \sim (u_e \sigma_{ex}[A])^{-1} \sim 0.1$ ns [2], where $u_e \sim 10^8$ cm/s is the electron velocity, $\sigma_{ex} \sim 10^{-17}$ cm^2 is the characteristic cross section for inelastic processes, and $[A]$ is the concentration of gas particles. In study [12], the Monte Carlo method was used to calculate the time required to slow primary electrons jointly with secondary electrons of an ionization cascade from an initial energy of ε_0 to I_m. This time τ_{ex} was calculated for He, Ne, Ar, Kr, and Xe on the assumption that an electron beam pulse constitutes a δ function at the moment of time $t = 0$ (Fig. 4.3). Calculations

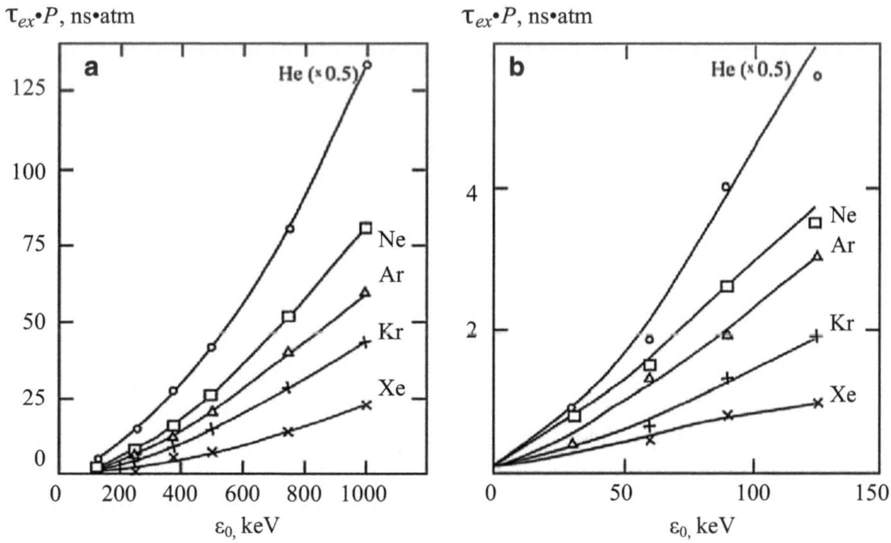

Fig. 4.3 Dependence of the product $\tau_{ex} \cdot P$ (P is the gas pressure) on the initial electron beam energy ε_0 for He, Ne, Ar, Kr, Xe: (**a**) 150 keV $\leq \varepsilon_0 \leq$ 1 MeV; (**b**) $\varepsilon_0 \leq$ 125 keV [12]. For helium, the values of $\tau_{ex} \cdot P$ should be doubled

show that τ_{ex} increases with growth in ε_0 and for $\varepsilon_0 = 1$ MeV, at atmospheric pressure τ_{ex} varies from 25 ns (xenon) to 250 ns (helium). It should be noted that the time to slow the secondary electrons (τ_s), which have lower energies and accordingly higher cross sections of inelastic interaction with the gas atoms, is much lower than τ_{ex}, and amounts to $\tau_s \geq 0.01$ ns atm [12]. These are the times that are realized when fast electrons excite a gas volume of small size (on the order of a few centimeters).

Calculations of the high-energy part of f_e for argon ($\varepsilon_e > I_m = 11.6$ eV) showed that stationary distribution is established rather quickly in times of ≤ 0.1 ns after switching on the electron beam [13]. The time to establish stationary distribution increases with the growth in the initial energy of the electrons: from ~0.001 ns at $\varepsilon_0 = 1$ keV to ~0.1 ns at $\varepsilon_0 = 1$ MeV. These results agree with calculations conducted in study [14] on temporal evolution $f_e(\varepsilon_e > I_m)$ when $\varepsilon_0 = 10$ keV for neon (4 atm) and xenon (1 atm). The time required to reach stationary mode for xenon is roughly an order of magnitude less than for neon, and amounts to ~0.001 ns, which is explained by the higher cross sections of inelastic processes.

For electrons with energies $\varepsilon_e < I_m$, the mechanism of energy loss is related to the processes of elastic scattering with gas particles. The characteristic time of this energy-loss process $\tau_{el} \sim (u_e \sigma_{ea} \delta_{ea}[A])^{-1}$ where $u_e \sigma_{ea} \sim 10^{-9}$ cm^3/s is the rate constant of the elastic process for rare gases [2]. With an average energy of plasma electrons $\bar{\varepsilon}_e \sim 1$ eV and at atmospheric pressure, $\tau_{el} \geq 100$ ns, which greatly exceeds τ_{nel}. Measurements of the characteristic times of the electron thermalization processes at an initial energy of 0.15 eV and a gas temperature of $T_g = 300$ K, carried out using microwave methods in study [15] for Ar (0.4 atm), Kr (0.15 atm) and Xe (0.14 atm), showed that the products $\tau_{el} \cdot P = 1{,}050$ ns · atm, 240 ns · atm, and 260 ns · atm, respectively.

Experimental data for He and Ne obtained in study [16] with an initial electron energy of around 1 eV and $T_g = 300$ K are 4.8 ns · atm and 120 ns · atm, respectively.

The time to establish stationary distribution f_e in the subthreshold region $\varepsilon_e < I_m$ considerably exceeds the analogous time for the high-energy region. From the calculations of $f_e(\varepsilon_e < I_m)$, obtained as a result of a numerical solution of the Boltzmann non-stationary kinetic equation for He [6], Ne [17], Ar [18, 19], and Xe [19] in the case of ionization by electron beams, it follows that the time required to reach stationary mode at atmospheric pressure is 5–10 ns.

To a significant extent, the electron energy distribution function in the subthreshold range is determined by the frequency of electron–electron collisions, which depends on the degree of plasma ionization $\zeta = n_e/[A]$ (n_e is the electron concentration). From the data shown in Fig. 4.4 [17] for the mixture Ne-Xe-HCl, it follows that the influence of electron–electron collisions on the formation of the electron energy distribution function is great even for a comparatively low specific power deposition $q = 2$ kW/cm^3 ($\zeta \leq 10^{-6}$). The complicated structure of the distribution shown in Fig. 4.4, calculated without taking electron–electron collisions into account, is explained by the influence of vibrational excitation of HCl molecules.

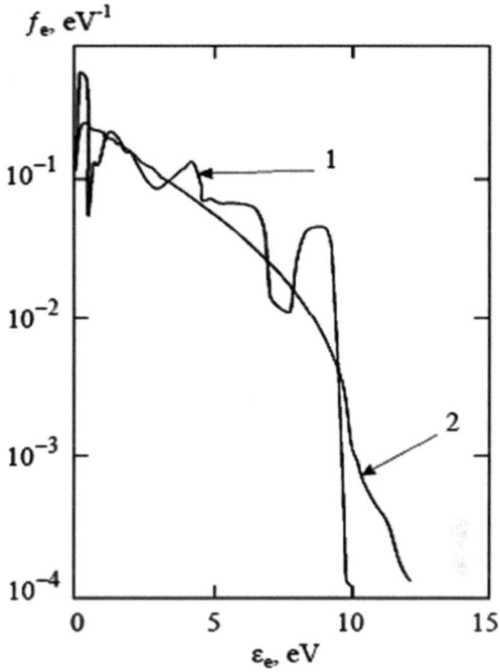

Fig. 4.4 Calculation of stationary electron energy distribution in the low-energy region for a Ne-Xe-HCl mixture (596:3:1) at a pressure of 4.9 atm, excited by an electron beam, without taking into account (*1*) and with taking into account (*2*) for electron–electron collisions [17]

For a low degree of plasma ionization, the electron energy distribution in the subthreshold region ($\varepsilon_e < I_m$) can differ from the Maxwell distribution. To establish the Maxwell energy distribution for slow electrons at temperature T_e, it is necessary for the frequency of the electron–electron collisions to greatly exceed the frequency of energy loss in elastic electron–atom collisions. This constraint can be written as follows [2]:

$$\zeta = \frac{n_e}{[A]} \gg \frac{m_e}{M_a} \frac{\sigma_{ea} \bar{\varepsilon}_e^2}{e^4 \Lambda}, \tag{4.1}$$

where σ_{ea} is the cross section of elastic electron scattering on the atoms; e is the electron charge; and Λ is the Coulomb logarithm. For helium, which is frequently used in NPL gas media as the buffer gas, $\sigma_{ea} \sim 5 \times 10^{-15}$ cm^{-2}; $\bar{\varepsilon}_e \sim 0.15$ eV; $\Lambda \sim 10$; $m_e/M_a = 1.4 \times 10^{-4}$; and consequently, $\zeta \gg 10^{-6}$. We note that in experiments with pulsed reactors for atmospheric pressure NPLs, $q \leq 5$ kW/cm^3 and the degree of ionization $\zeta < 10^{-5}$, so that the condition (4.1) is fulfilled only in the case of maximal possible power deposition for NPLs.

Methods of calculating electron energy distributions are considered, for example, in studies [11, 20]: the use of the continuous slowing down model, in which it is presumed that a fast electron loses energy in the degradation process continuously; modeling of the process of electron energy exchange by the Monte Carlo method, as well as numerical solution of the Boltzmann kinetic equation. The continuous slowing down model is applicable primarily to calculate the electron energy distribution with energies $\varepsilon_e > I_m$. For precise determination of electron energy distribution in the region of subthreshold and thermal energies, numerical solution of the Boltzmann kinetic equation is necessary.

For analysis of the kinetics of plasma processes in NPLs, information about the electron energy distribution for He, Ne, or Ar is the most interesting, because these are the gases that are used as the basic component (buffer gases) in nearly all known gas NPLs. It should be noted that all the data relating to electron energy distributions in rare gases were obtained by computation except for study [21], where such distributions were found using spectroscopic and probing measurements for He and Ar during excitation by ^{235}U fission fragments (Figs. 4.5 and 4.6) in a pressure range of 0.03–1 atm ($q \sim 0.01$ W/cm^3 at a pressure of 1 atm, $n_e \sim 10^{11}$ cm^{-3}, $\zeta \sim 10^{-8}$). For comparison, Fig. 4.5 also shows the results of the calculation [22] for helium excited by α particles.

Information about the electron energy distributions for helium with excitation by fission fragments at an energy of 85 MeV and protons with an energy of 0.6 MeV ($P_{He} = 0.4$ atm; $q = 1$ kW/cm^3) is contained in study [6], where these data were obtained as a result of solving the non-stationary kinetic equation. It follows from Fig. 4.7 that the electron energy distributions for two methods of excitation in the region $\varepsilon_e < I_m$ virtually coincide and are close to the Maxwell distribution.

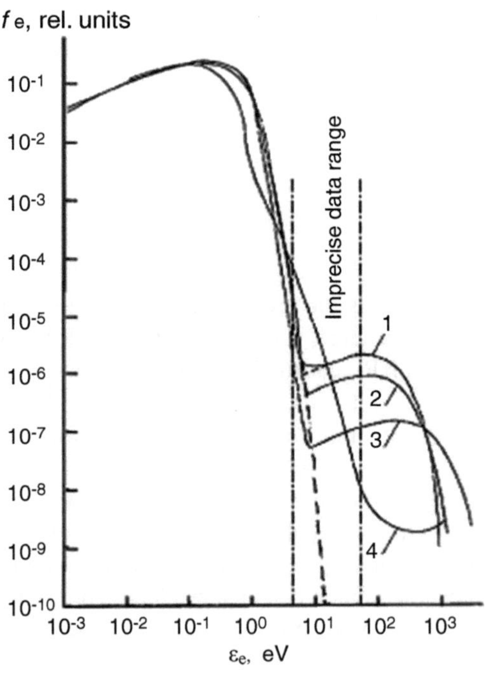

Fig. 4.5 The electron energy distribution in helium excited by fission fragments [21]: $P_{He} = 0.07$ atm (*1*); 0.13 atm (*2*) and 1 atm (*3*); (*4*) calculation results [22] with excitation of helium by α particles; *dotted line* is the Maxwell distribution

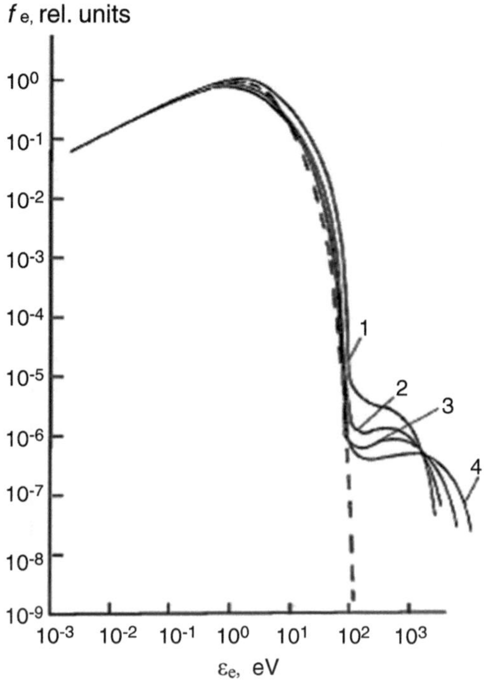

Fig. 4.6 Electron energy distribution in argon excited by fission fragments [21]: $P_{Ar} = 0.4$ atm (*1*); 0.6 atm (*2*); 0.8 atm (*3*) and 1 atm (*4*); *dotted line* is the Maxwell distribution

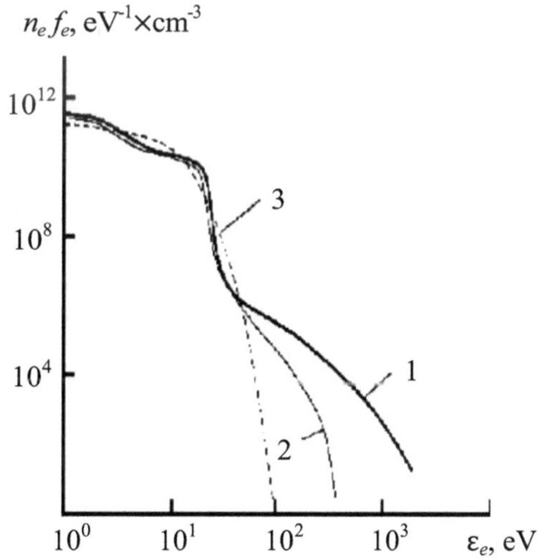

Fig. 4.7 Electron energy distribution in helium ($P_{He} = 0.4$ atm) excited by fission fragments (*1*) with an energy of 85 MeV and by protons (*2*) with an energy of 0.6 MeV [6]. The *dotted line* (*3*) is the Maxwell distribution with $T_e = 4$ eV

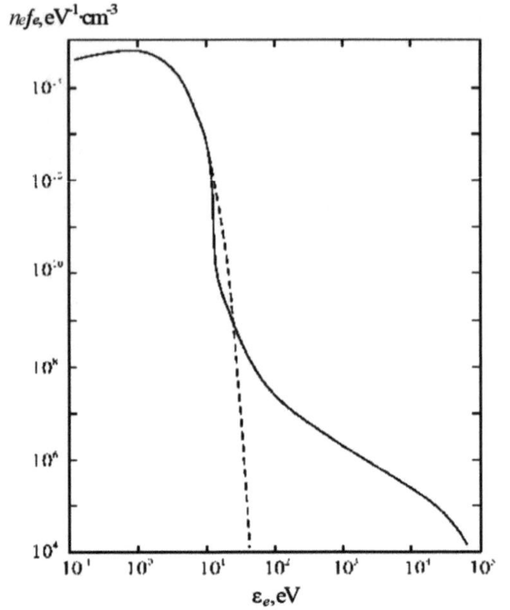

Fig. 4.8 Electron energy distribution in argon ($P_{Ar} = 3$ atm) excited by an electron beam [18]. The *dotted line* is the Maxwell distribution when $T_e = 1.5$ eV

The authors are not aware of any computational or experimental data on the electron energy distributions in neon during excitation by nuclear particles.

Information on electron energy distributions for plasma excited by electron beams is more extensive. Calculations by the Monte Carlo method for He, Ar and Xe in a range of initial electron energies $\varepsilon_0 = 20$–10,000 eV showed that the electron energy distribution in the subthreshold region ($\varepsilon_e < 20$ eV) does not depend on ε_0 for $\varepsilon_0 > 300$ eV [23].

The electron energy distribution for Ar ($P_{Ar} = 3$ atm) during excitation by an electron beam with a duration of 5 ns and $\varepsilon_0 = 200$ keV ($q = 1.5$ MW/cm^3; $n_e = 2 \times 10^{15}$ cm^{-3}, $\zeta = 2.5 \times 10^{-5}$) is shown in Fig. 4.8 [18]. From the data cited, it follows that for electrons with $\varepsilon_e < 10$ eV, the electron distribution roughly corresponds with the Maxwell distribution. This circumstance is more clearly illustrated by the data in Fig. 4.9, which shows the ratio of the electron distribution to the Maxwell distribution for three different pressures of Ar [18]. With higher ε_e, these functions differ: the number of electrons in the region $\varepsilon_e = 10$–30 eV is less, while for $\varepsilon_e > 30$ eV it is greater than the Maxwell distribution. From a comparison of the data of Figs. 4.6 and 4.8, one can conclude that the shape of the distribution for argon under significantly differing conditions of excitation is roughly equal.

Calculation of the electron energy distribution for Ne was carried out in the studies [14, 17] for the following conditions: an electron beam with $\varepsilon_0 = 200$ keV, $P_{Ne} = 4$ atm, $q = 200$ kW/cm^3, $n_e \approx 5 \times 10^{14}$ cm^{-3}, $\zeta \approx 5 \times 10^{-6}$. For neon (as for argon), the electron distribution function in the low-energy region ($\varepsilon_e < I_m$) also

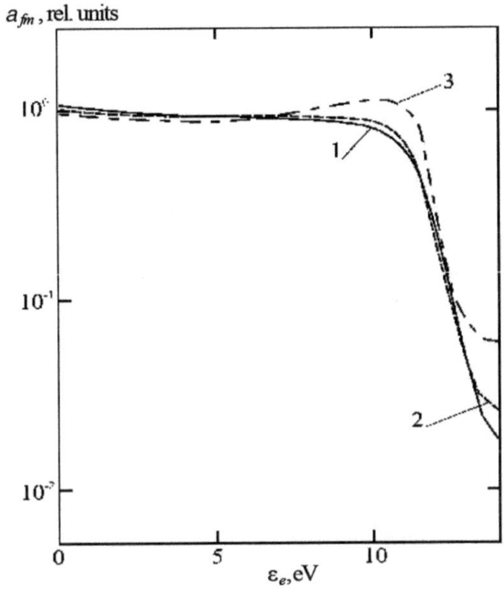

Fig. 4.9 Ratio a_{fm} of the electron energy distribution to the Maxwell distribution at argon pressures of 1 (*1*), 3 (*2*), and 10 (*3*) atm [18]. The excitation conditions are the same as in Fig. 4.8

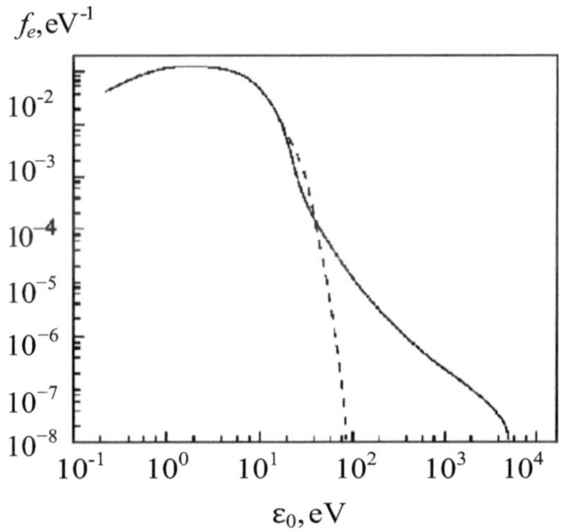

Fig. 4.10 Stationary electron energy distribution in neon ($P_{Ne} = 4$ atm) excited by a beam of fast electrons with $\varepsilon_0 = 10$ keV [14, 17]. The *dotted line* is the Maxwell distribution when $T_e = 4.6$ eV

coincides with the Maxwell distribution, while for $\varepsilon_e > 40$ eV the number of electrons significantly exceeds the equilibrium value (Fig. 4.10).

From the data cited it follows that in the range of conditions in which atmospheric-pressure gas NPLs were studied in experiments using pulsed reactors with excitation by nuclear reaction products ($q \leq 5$ kW/cm^3, $\zeta \leq 10^{-5}$), the electron energy distribution consists of two parts. The chief quantity of plasma electrons are concentrated in the subthreshold region, $\varepsilon_e < I_m$, and their energy distribution is similar to the Maxwell distribution, with a certain characteristic temperature $T_e = 2\int \varepsilon_e f_e(\varepsilon_e)d\varepsilon/3k = 2\,\bar{\varepsilon}_e/3k$. The electron energy distribution function in the region $\varepsilon_e < I_m$ is virtually independent of the type of charged particles. The low-energy electrons of this region participate in processes of great importance in the kinetics of populating and relaxation of lasing levels of NPLs: electron–ion recombination, "quenching" of excited states, the processes of attachment to electronegative atoms and molecules, and excitation and ionization in collisions with gas particles that are in excited states.

For $\varepsilon_e > I_m$, the distribution function decreases as a consequence of the effective inelastic electron–atom collisions and differs from the Maxwell distribution. The electrons in this region scarcely participate in the recombination processes, and this region itself is a source of electrons for the subthreshold region.

4.2 Formation of Ions and Excited Atoms at the Initial Ionization Stage

Ions and excited atoms can occur both directly in the processes of collisions between a primary charged particle and gas atoms, and during the collision of atoms with secondary electrons. As a rule, computational methods are used to determine the actual contribution of the primary and secondary processes to the formation of ions and excited atoms.

Figure 4.11 shows the results of computation in [24] of the energy that is expended for the formation of ions, excited atoms, and subthreshold electrons during excitation of helium ($P_{He} = 0.53$ atm) by a proton beam with an energy of 4 MeV. The absorbed energy on a proton path length of 1 cm was taken as the baseline; it amounted to 9.1 keV. It is evident from Fig. 4.11 that the percentage of

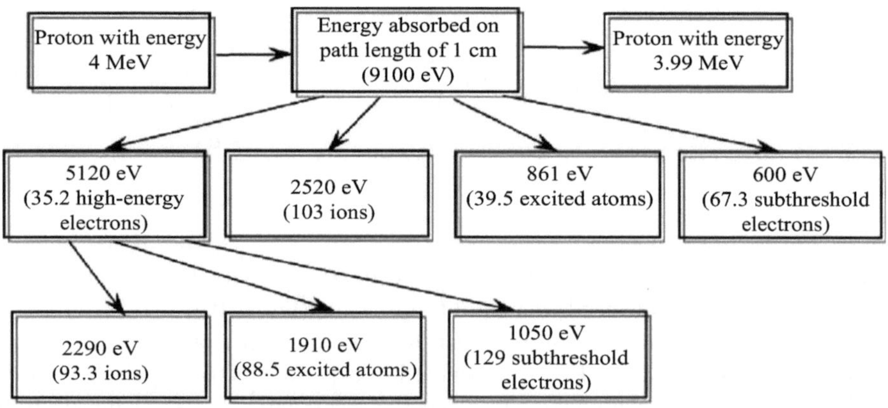

Fig. 4.11 Channels for conversion of absorbed energy during excitation of helium ($P_{He} = 0.53$ atm) by a proton beam with an energy of 4 MeV [24]

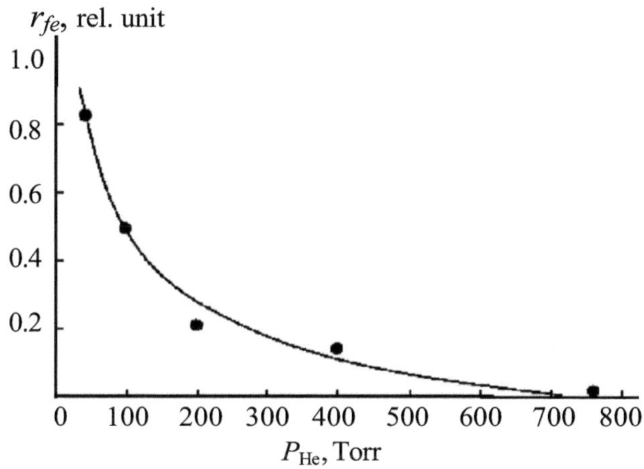

Fig. 4.12 Dependence of r_{fe} on helium pressure [25]

the primary ionization is 52 %, and the percentage of excited atoms formed as a result of primary processes is 31 %. According to the data of study [5], the percentage of the primary ionization for Ne and Xe is 15–20 % when excited by a beam of fast electrons. When excited by fission fragments the percentage is 30–50 %. The authors of [5] attribute the difference to the harder, more severe spectrum of electrons formed as a result of primary ionization by an electron beam.

It appears that the only study in which the ratio of the primary and secondary inelastic processes was determined experimentally was study [25], in which spectroscopic methods were used to study the excitation processes during ionization of He by uranium fission fragments. The ratio r_{fe} of the number of excited atoms formed directly by fission fragments to the number of excited atoms formed by secondary electrons of the ionization cascade is shown in Fig. 4.12. With a helium pressure of 25 Torr, the ratio $r_{fe} \approx 1$, while at a pressure of 100 Torr, the contribution of secondary electrons to the formation of excited helium atoms is twice as great as for fission fragments. At helium pressures of around 1 atm, the contribution of fission fragments to direct formation of excited atoms may be ignored.

Frequently, the energy cost of the elementary process is used to characterize the rates of the inelastic processes of ionization and excitation in gas media. The energy cost of the *j*th inelastic process occurring with a gas particle of the type *s* constitutes the ratio of the specific excitation power to the frequency of the occurring actions of this process [26]:

$$w_{sj} = \frac{q}{[A_s]n_e \int\limits_{E_{sj}}^{\infty} \sigma_{sj}(\varepsilon_e)\sqrt{2\varepsilon_e/m_e} f_e(\varepsilon_e)d\varepsilon_e}, \tag{4.2}$$

where $[A_s]$ the concentration of atoms of the type s; σ_{sj}, E_{sj} are the cross section and energy threshold of the jth inelastic process. The expression (4.2) is applicable to determine the energy costs of ionization and excitation both in the case of single-component gas media, and of gas mixtures.

The simplest method of determining the energy expended on ionization and excitation of atoms is to use the semi-empirical Platzmann formula [27], which relates the energy cost of formation of an ion–electron pair w_i to the ionization potential of an atom, V_i:

$$w_i = E_i + (N_x/N_i)E_x + E_e = 1.71V_i, \tag{4.3}$$

where $E_i = 1.06V_i$ is the average energy expended on ionization; $E_x = 0.85V_i$ is the average energy expended on excitation; $E_e = 0.31V_i$ is the average kinetic energy of a subthreshold electron; and $N_x/N_i = 0.4$ is the ratio of the number of excited atoms to the number of ionized atoms, which is roughly identical for all rare gases.

The expression (4.3) was obtained for rare gases excited by α and β particles. For fission fragments, the value w_i is somewhat higher than for α and β particles. In this case, the Platzmann formula has the same form, but the ratio $N_x/N_i = 0.53$ and $w_i = 1.82V_i$ [28]. Using w_i, it is possible to determine the formation rates of ions f^+ and excited atoms f^* per unit of volume:

$$f^+ = q/w_i \qquad f* = (N_x/N_i)f, \tag{4.4}$$

where q is the specific power deposition of the gas medium.

Table 4.2 shows the results of calculation of w_i by formula (4.3) for light particles and fission fragments, the results of numerical calculations at ionization of rare gases by electron beams, as well as the results of experiments for rare gases excited by α and β particles and uranium fission fragments. From the data of Table 4.2, it is possible to draw the following conclusions: (a) precise numerical calculations and calculations by formula (4.3) of values of w_i agree with one another with an accuracy of no worse than 10 %; (b) α and β particles are virtually equivalent to one another from the standpoint of energy expenditures for the formation of an ion–electron pair (ratio $w_i(\alpha)/w_i(\beta) = 1$ with an accuracy no worse than ±2 % [27]); (c) the value of w_i for fission fragments is 6–7 % greater than for light particles [29]; (d) the maximal differences between the results of different authors is observed for the ratio N_x/N_i, which is evidently associated with differences in the excitation cross sections included in the calculations.

For calculations of the plasma kinetics and characteristics of NPLs, it is necessary to know not only the percentages of energy that are expended on ionization and excitation, but also the distribution of absorbed energy between excitation states.

Table 4.2 Energy cost of formation of an ion–electron pair (in eV) in rare gases and the ratio of the number excited atoms to the number of ionized atoms (N_x/N_i)

Rare gas		He	Ne	Ar	Kr	Xe
Ionization potential, V_i, eV		24.6	21.6	15.8	14.0	12.1
Calculation by formula (4.3)	Light particles $w_i = 1.71\ V_i$, eV	42.1	36.9	27.0	23.9	20.1
	Fission fragment $w_i = 1.82\ V_i$, eV	44.8	39.3	28.8	25.5	22.0
Numerical calculations	Electron beam	46.8 [12]	36.5 [12]	27.3 [12]	23.6 [12]	20.9 [12]
		45.9 [23]	38 [14]	25.4 [13]	21 [14]	22 [14]
		46.2 [30]	38.5 [23]	25 [14]	24.3 [23]	22.2 [23]
		46.0 [31]	36.1 [26]	26.1 [23]	24.0 [31]	22.3 [26]
		46.4 [32]	36.6 [31]	26.0 [26]		21.7 [31]
				26.4 [31]		
	Proton beam	46.4 [24]	–	–	–	–
Experiment	^{14}C and ^{63}Ni β particles	42.3 [33]	36.6 [33]	26.4 [33]	24.2 [33]	22.0 [33]
	^{210}Po and ^{239}Pu α particles	42.7 [33]	36.8 [33]	26.4 [33]	24.1 [33]	21.9 [33]
		46.0 [34]		26.4 [34]		
	^{241}Am α particles	–	–	26.5 [35]	23.9 [35]	21.0 [35]
	^{235}U fission fragments	–	39.2 [29]	28.2 [29]	–	–
N_x/N_i		0.65 [12]	0.45 [12]	0.51 [12]	0.53 [12]	0.60 [12]
		0.64 [23]	0.54 [14]	0.44 [14]	0.32 [14]	0.38 [14]
		0.66 [24]	0.55 [23]	0.45 [23]	0.56 [23]	0.70 [23]
		0.68 [30]	0.51 [26]	0.57 [26]	0.39 [31]	0.51 [31]
		0.54 [31]	0.33 [31]	0.32 [31]	0.55 [35]	0.60 [35]
		0.77 [32]		0.52 [35]		

The initial spectra of excited states in gases excited by nuclear radiation were determined as a rule by calculation, with the reliability and accuracy of the calculations depending primarily on the values of the excitation cross sections used in the calculations.

From expression (4.2) it is clear that calculation of the energy costs of the processes is closely connected with the precision of the determination of $f_e(\varepsilon_e)$ in the region of inelastic electron–atom interactions ($\varepsilon_e > I_m$). Because in this region, the shape of electron energy distribution does not depend on the energy of the primary electrons (see the data of [13] for argon, for example), the absence of energy dependence for w_{sj} in a wide range of electron energies is an important practical consequence. The absence of energy dependence of w_i (the cost of an ion–electron pair formation) was observed in experiments [27] in the early stages of investigation, in which it was shown that such a rule is valid with an accuracy of ±2 % for β particles in an energy range of 5–60 keV.

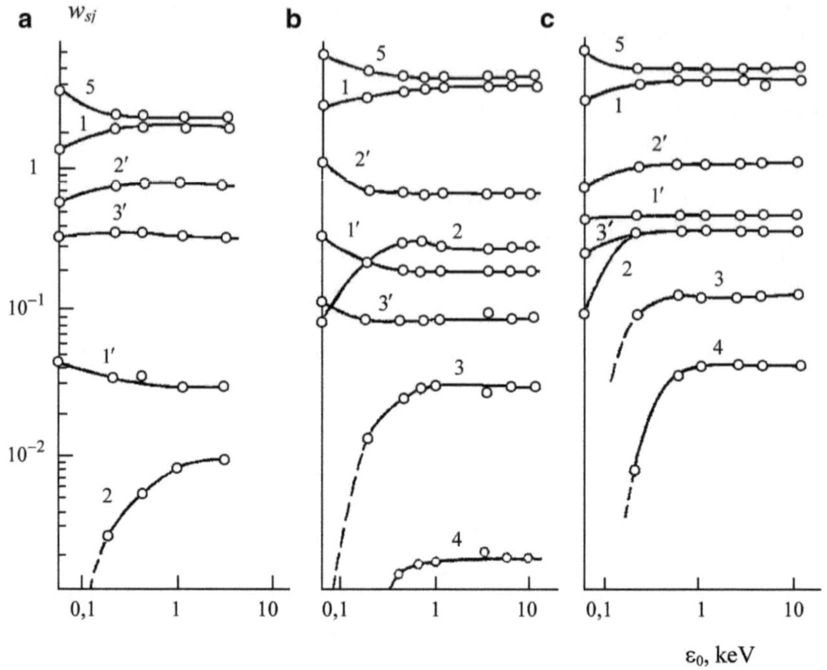

Fig. 4.13 Dependence of the number of ions and excited atoms at 100 eV of absorbed energy ω_{sj} on the initial energy of the electron ε_0 in helium (**a**), argon (**b**), and xenon (**c**) [23]: (*1–4*) ions A^+, A^{2+}, A^{3+}, and A^{4+}, respectively; (*5*) subthreshold electrons; (*1′*, *2′*, *3′*) excited states of atoms of helium (1′–2^1S and 2^3S; 2′–2^1P and 2^3P; 3′–3^1S, 3^1P, 3^1D, 3^3P and 3^3D), argon (1′–4*s*; 2′–4*s*′; 3′–5*s*, 3*d*, 5*s*′ and 3*d*′) and xenon (1′–6*s*; 2′–6*s*′, 6*p*, 5*d*; 3′–higher levels)

As an example, Fig. 4.13 shows the dependence of the number of ions and excited atoms at 100 eV of absorbed energy ω_{sj} ($\omega_{sj} = 100/w_{sj}$) on the initial electron energy, as obtained from calculations [23] by the Monte Carlo method for He, Ar, and Xe. For all three gases, ω_{sj} and, consequently, w_{sj} cease to depend on ε_0 if $\varepsilon_0 \geq 500$ eV. This result agrees with the computation of the ionization cascade in helium obtained in study [30] by the degradation spectrum method.

The most complete information on energy costs of inelastic processes was obtained as a result of calculations for single-component gas media. Tables 4.3, 4.4, 4.5, 4.6, and 4.7 show the calculated data for He, Ne, Ar, Kr, and Xe, respectively. It was presumed in calculations [24] that the gas medium (helium) was excited by a proton beam with an energy of 4 MeV; in all the other studies by electron beams with an initial energy of 1–2 keV.

The most detailed information about the formation of excited states in the initial stage of the processes of interaction of the primary particle and secondary electrons with the gas medium was obtained for helium, because helium has quite detailed information about the cross sections of inelastic processes. From the data of Table 4.3, we see that around 50 % of excited atoms of He are formed in the state 2^1P. This agrees with the results of study [36], in which, on the basis of analysis of the luminescence spectra and the kinetics of plasma processes, it was

Table 4.3 Energy costs w_{sj} of inelastic excitation processes for He

Level	2^3S	2^1S	2^3P	2^1P	3^3S	3^1S	3^3P	3^1P	3^1D	3^3D	$n \geq 4$	Works cited
Threshold energy, eV	19.8	20.6	20.9	21.2	22.7	22.9	23.0	23.1	23.1	23.1	23.7	
w_{sj}, eV	2,560		120		290						670	[23]
	569	890	1,050	172	5,830	4,500	5,140	725	827	4,920	480	[24]
	343	840	1,050	132	410						700	[32]
	547	860	1,020	168	5,600	4,240	6,150	711	6,470	19,500	1,170	[30]

concluded that the majority of initially excited He* atoms are formed in the states n^1P, and half of them in the states 2^1P. The validity of the conclusion is confirmed by calculated data [24], from which it follows that when helium is irradiated with a proton beam, roughly 30 % of the absorbed energy is expended on excitation of helium atoms, and half of them are formed in the states 2^1P. The basic channels of destruction of the atoms $He^*(2^1P)$ are radiative and collisional transitions to metastable states 2^1S [36, 37]. The probability of radiative transition is $1.9 \times 10^6\ s^{-1}$ [38], while for the process $He^*(2^1P) + He \rightarrow He^*(2^1S) + He$, the rate constant $k_q = 1.8 \cdot 10^{-12}\ cm^3/s$ [37]. Therefore, for $P_{He} \sim 1$ atm, as a result of collision processes, the atoms $He^*(2^1P)$ are converted rather quickly to the metastable atoms $He^*(2^1S)$, in a time of $\tau_q = (k_q[He])^{-1} \sim 20$ ns. Then, examining the kinetics of plasma processes in quasicontinuous helium gas NPLs, one can assume that atoms $He^*(2^1S)$ are the primary excited atoms.

It follows from the data of Tables 4.4 through 4.7 that in the case of heavy rare gases, 80–90 % of excited atoms are formed in lower *ns* and *np* states ($n = 3, 4, 5, 6$, respectively, for Ne, Ar, Kr, and Xe) roughly in an equal proportion. Excited atoms are also formed through plasmochemical reactions, such as recombination for example. However, the contribution of the initial ionization stage to the formation of excited atoms in the case of heavy rare gases may be significant. Thus, an investigation of the spectral-kinetic characteristics of Ne luminescence on excitation by single fission fragments ^{252}Cf showed that all ten *3p* levels of the Ne atom are populated very quickly, in the time that the fission fragment passes through the gas (~4 ns) [39, 40]. This circumstance allowed the authors of [39, 40] to conclude that a significant contribution to population of the *3p* levels is provided by the processes of direct excitation of the neon in the initial stage of ionization of the gas medium.

Above we looked at the energy expenditures for the formation of ions and excited atoms in single-component gas media. In the case of gas mixtures, the energy cost of the inelastic process can be represented in the form [41]:

$$\overline{w}_{sj} = \left(\sum_s \frac{\nu_s}{w_{sj}} \right)^{-1}, \qquad \nu_s = \frac{[A_s]}{R_s} \left(\sum_s \frac{[A_s]}{R_s} \right)^{-1}, \tag{4.5}$$

where ν_s is the percentage of energy absorbed in the component s of the gas mixture, R_s is the path length of the charged particle in the component s of the gas mixture at a pressure of 1 atm. We note that the formula (4.5) is equivalent to the expression for w_{sj} proposed in the study [34] for binary mixtures.

Binary mixtures are most often the active media of gas NPLs, and He, Ne, or Ar are used as the main component (buffer gas). Information about the energy costs of formation, $\overline{w}_{sj}$, for mixtures that are of interest for NPLs is very limited and was obtained as a result of calculations for the mixtures He-Xe [12]; Ar-Kr [14]; He-Ar [34]; Ar-Xe and Ne-Xe [26]; Ar-Xe(Kr,N_2) [42], and He-Cd [43]. As an example, Figs. 4.14 and 4.15 show the costs of forming ions and excited Xe* atoms for the mixture Ar-Xe [26].

Table 4.4 Energy costs w_{sj} of inelastic excitation processes for Ne

Level	$3s$	$3p$	$4s$, $5s$, $3d$	Works cited
Threshold energy, eV	16.7	18.6	20.1	
w_{sj}, eV	170	130	1,500	[14]
	110	150	1,200	[26]

Table 4.5 Energy costs w_{sj} of inelastic excitation processes for Ar

Level	$4s[3/2]_2^0$, $4s'[1/2]_0^0$	$4s[3/2]_1^0$	$4s'[1/2]_1^0$	$4p$	$3d$, $5s$	$4d$, $6s$	Works cited
Threshold energy, eV	11.55	11.62	11.83	13.0	14.0	15.0	
w_{sj}, eV	280			640	550	–	[13]
	190			270	290	660	[14]
	130			–	1,000	–	[23]
	110	800	120	500	800	–	[26]

Table 4.6 Energy costs w_{sj} of inelastic excitation processes for Kr

Level	$5s$	$5p$	$6s$, $4d$	Higher levels	Works cited
Threshold energy, eV	10.0	11.5	12.2	12.9	
w_{sj}, eV	110	200	1,500	4,700	[14]

Table 4.7 Energy costs w_{sj} of inelastic processes for Xe

Level	$6s$	$6s'$, $6p$, $5d$	Higher levels	Works cited
Threshold energy, eV	8.4	9.7	11.2	
w_{sj}, eV	130	120	910	[14]
	400	90	270	[23]

For a xenon concentration of ~1 % (optimal for the most powerful IR laser operating on transitions $5d$-$6p$ of the Xe atom), the cost of forming the excited atom Xe* in the upper laser state $5d[3/2]_1^0$ is >10 keV, which is roughly two orders of magnitude greater than the cost of forming Xe* atoms in the lower lasing states $6p$ (Fig. 4.14), so population of the state $5d[3/2]_1^0$ in the initial stage is not significant.

From the data provided in Fig. 4.15 it is clear that when there is an increase in the xenon concentration, the cost of forming an electron in the Ar-Xe mixture smoothly decreases from the cost of ionizing pure Ar (26 eV) to the cost of ionizing pure Xe (22 eV). For a Xe concentration of around 45 %, the costs and rates of ionization of Xe and Ar atoms are roughly identical.

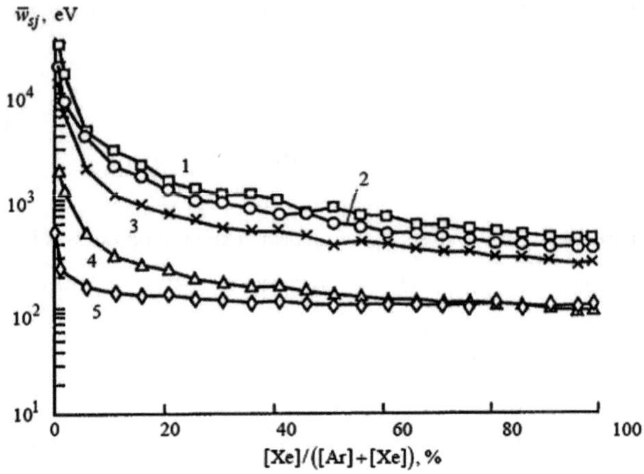

Fig. 4.14 Costs of excitation of some Xe levels for the Ar-Xe mixture [26]: (*1*) $5d[3/2]_1^0$; (*2*) $5d[1/2]_1^0$; (*3*) $6s'[1/2]_1^0$; (*4*) $6s'[1/2]_1^0$; (*5*) $6p$

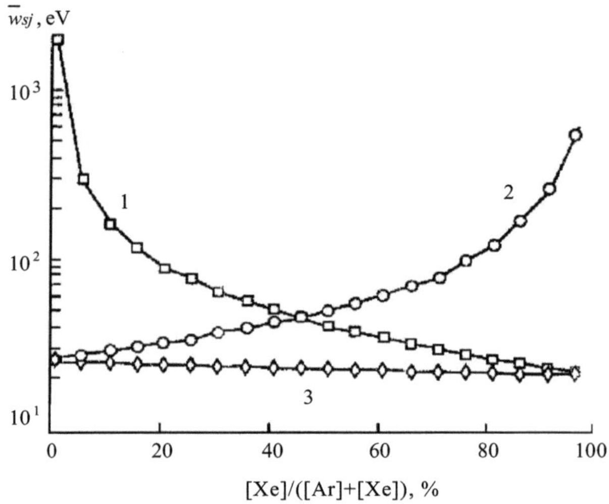

Fig. 4.15 Cost of ionization of atoms Xe (*1*) and Ar (*2*) and cost of forming electrons in the Ar-Xe mixture (*3*) [26]

4.3 Plasmochemical Processes

Following the initial stage, a sequence of plasmochemical processes occur in the plasma, tending to return it to its equilibrium state. An important place among such processes is held by charged particle recombination reactions, as a result of which plasma neutralization and the formation of charged atoms occur.

Kinetics of Plasma Processes for a Single-Component Mixture

For a single-component gas medium *A* at atmospheric pressure, the basic recombination processes are:

1. collisional-radiative recombination

$$A^{+} + e + e \rightarrow A* + e, \tag{4.6}$$

2. three-body recombination

$$A^{+} + e + A \rightarrow A* + A, \tag{4.7}$$

3. dissociative recombination

$$A_2^{+} + e \rightarrow A* + A. \tag{4.8}$$

Molecular ions A_2^{+} are formed as a result of ion conversion:

$$A^{+} + 2A \rightarrow A_2^{+} + A, \tag{4.9}$$

and the rate constants of this process differ insignificantly for all ions of rare gases and amount to $k_{ic} = (0.6\text{–}3.5) \times 10^{-31}$ cm^6/s at a gas temperature of $T_g = 300$ K [44].

As a result of recombination processes (4.6) and (4.7), initially highly excited atom states are formed, which are stabilized through collisions with a third particle (electron or atom). Further relaxation of excited states occurs as a result of collisions with plasma electrons and gas atoms, and through spontaneous decay. A special place among the recombination processes is held by the dissociative recombination reaction (4.8), which at a high pressure often is not only the basic process of neutralization of charged particles, but also one of the chief channels for the formation of excited particles, including populating of the upper lasing levels of NPLs [45]. In the case of dissociative recombination, the binding energy of the recombination electron is converted into the kinetic energy of recession of the atoms.

Values of the rate constants of recombination processes differ greatly and depend first of all on the electron temperature T_e. For recombination processes (4.6) and (4.7) of atomic ions A^+, the rate constants equal:

- for collisional-radiative recombination $k_{cr} = 4 \times 10^{-9} T_e^{-4.5}$ cm^6/s (T_e, K) for any atomic ions of rare gases [46];
- for three-body recombination $k_{tr} \approx (0.5\text{–}30) \times 10^{-22} T_e^{-2.5}$ cm^6/s (T_e, K) for $A =$ Xe, Kr, Ar, Ne, He [47].

Rate constants of the dissociative recombination (4.8) for molecular ions of rare gases are shown in Table 4.8 [48]. The dependence on electron temperature for k_{dr} is significantly weaker than for k_{cr} and k_{tr}, so with an increase in T_e, the influence of dissociative recombination grows sharply.

For clarity, let us examine an argon plasma and estimate the characteristic times of plasma processes for $q \approx 0.01\text{–}5$ kW/cm^3, which occur during operation of gas NPLs. Given an atmospheric pressure of the argon, $T_e \approx 0.5\text{–}2.0$ eV; $n_e \approx 5 \times 10^{12}$–

Table 4.8 Rate constants of dissociative recombination processes for molecular ions of rare gases with electrons [48]

Molecular ion	He_2^+	Ne_2^+	Ar_2^+	Kr_2^+	Xe_2^+
k_{dr}, cm^3/s	$1.5 \times 10^{-7} T_e^{-1}$	$2 \times 10^{-6} T_e^{-0.43}$	$3.9 \times 10^{-5} T_e^{-0.67}$	$3.7 \times 10^{-5} T_e^{-0.55}$	$8.1 \times 10^{-5} T_e^{-0.6}$

Note: T_e, K

2×10^{14} cm^{-3} and $\zeta \approx 2 \times 10^{-7}$–$7 \times 10^{-5}$. A comparison of the characteristic times of the recombination process $\tau_{cr} = (k_{cr} n_e^2)^{-1} \approx 1$ s, $\tau_{tr} = (k_{tr} n_e[\mathrm{Ar}])^{-1} \approx 0.3$ s ([Ar] = 2.7×10^{19} cm^{-3}, the concentration of Ar atoms), the process of ion conversion $\tau_{ic} = (k_{ic}[\mathrm{Ar}]^2)^{-1} \approx 5 \times 10^{-9}$ s and dissociative recombination $\tau_{dr} = (k_{dr} n_e)^{-1}$ $\approx 1 \times 10^{-7}$–2×10^{-6} s shows that in argon plasma, neutralization of charged particles occurs exclusively by the channel of dissociative recombination of molecular ions.

A similar conclusion can be drawn for plasma based on Ne, Kr, and Xe. First of all, for helium plasma, the constant k_{dr} is two or three orders of magnitude lower; secondly, the process of electron thermalization occurs more efficiently, and accordingly, T_e is significantly less. In this case, one cannot entirely neglect the recombination processes (4.6) and (4.7) with the participation of atomic He^+ ions.

Balance equations of charged particles in plasma for the conditions examined above, without taking into account recombination processes (4.6) and (4.7), can be written as follows

$$\frac{d[A^+]}{dt} = f^+ - k_{ic}[A^+][A]^2, \tag{4.10}$$

$$\frac{d[A_2^+]}{dt} = k_{ic}[A^+][A]^2 - k_{dr}[A_2^+]n_e, \tag{4.11}$$

$$n_e = [A^+] + [A_2^+], \tag{4.12}$$

where $[A^+]$ and $[A_2{}^+]$ are the concentrations of atomic and molecular ions, $f^+ = q/w_i$ is the rate of ion formation (w_i is the energy cost of forming an ion–electron pair). For Eq. (4.10), the solution has the simple form [2, 49]:

$$[A^+] = f^+\tau_{ic}[1 - \exp(-t/\tau_{ic})], \tau_{ic} = \left(k_{ic}[A]^2\right)^{-1}, \tag{4.13}$$

which was obtained given the condition that $[A^+] = 0$ for $t = 0$.

In the plasma of gas NPLs, the characteristic times of all the main plasma processes are considerably less than the duration of the excitation pulse (minimal duration of reactor pulse is ~50 μs), so that for $q \approx 0.01$–5 kW/cm^3, a quasi-stationary mode is established in the plasma over a time on the order of $(f^+k_{dr})^{-1/2} \approx 5 \times 10^{-8}$–$2 \times 10^{-6}$ s. The concentration of electrons and ions can be found from the solution to Eqs. (4.10)–(4.12), if we equate the values of the derivatives with zero:

$$n_e = \frac{f^+\tau_{ic}}{2}\left(\sqrt{S+1} + 1\right); [A^+] = f^+\tau_{ic};$$
$$[A_2^+] = \frac{f^+\tau_{ic}}{2}\left(\sqrt{S+1} - 1\right), S = \frac{4}{f^+\tau_{ic}^2 k_{dr}}. \tag{4.14}$$

The characteristics of the plasma being examined depend significantly on the dimensionless parameter S, the insertion of which makes it possible to carry out a convenient classification of the excitation conditions. With so-called "weak" excitation, the condition $S \gg 1$ is fulfilled, and from the expression (4.14) we obtain:

$$n_e \approx \left[A_2^+\right] = \sqrt{\frac{f^+}{k_{dr}}} \gg \left[A^+\right]. \tag{4.15}$$

In the case of "weak" excitation, the non-stationary system of Eqs. (4.10)–(4.12) has an analytical solution [49]:

$$n_e \approx \left[A_2^+\right] = \sqrt{\frac{f^+}{k_{dr}}} \mathrm{th}\left[\left(f^+ k_{dr}\right)^{1/2} t\right]. \tag{4.16}$$

In the case of "strong" excitation ($S \ll 1$) we have:

$$n_e \approx \left[A^+\right] = f^+ \tau_{ic}; \left[A_2^+\right] \approx \frac{1}{\tau_{ic} k_{dr}} \ll n_e, \tag{4.17}$$

If we assume $S = 1$, it is possible to obtain the expression for the degree of plasma ionization, ζ_{ws}, which separates the "weak" and "strong" excitation regions:

$$\zeta_{ws} = \frac{n_e}{[A]} \approx \frac{5 k_{ic} [A]}{k_{dr}}. \tag{4.18}$$

For example, for argon at atmospheric pressure, $\zeta_{ws} \approx 5 \times 10^{-4}$ ($T_e \approx 0.5$–2 eV). In experiments with pulsed reactors, $q \leq 5$ kW/cm^3 and $\zeta < 10^{-5} \ll \zeta_{ws}$, so that the estimate of parameters of the plasma of gas NPLs can be made in an approximation of the "weak" excitation.

For the conditions in which atmospheric pressure gas NPLs were studied with using of pulsed reactors, there is "weak" excitation ($S > 10^3$), so that for evaluative calculations of the plasma characteristics, one can use the formula (4.15). The validity of Eq. (4.15) is confirmed, for example, in experiment [50], where the electron concentration was measured for excitation of neon ($P_{Ne} = 0.24$ atm) by a proton beam with an energy of 20 MeV in a range $q \approx 2 \times 10^{-5}$–$5 \times 10^{-2}$ W/cm^3 ($S > 10^4$).

The balance Eqs. (4.10)–(4.12) are written on the assumption that neutralization of the charged particles occurs exclusively owing to the dissociation recombination of molecular ions. This assumption is based on estimates of the rates of recombination processes, and it was assumed that the values T_e are known, and for gas NPLs are 0.5–2.0 eV.

To determine T_e, it is necessary to consider the energy balance of plasma electrons. Studies [2, 49] analyzed the basic processes leading to establishment of some average energy of plasma electrons $\overline{\varepsilon}$. There are four such processes: (1) formation, in the interval $0 < \varepsilon_e < I_m$, of electrons with an average energy $I_m/2$, exceeding $\overline{\varepsilon}_e$; (2) dissociative recombination resulting in the disappearance of the

Table 4.9 Values of the function $F(Z)$ [2, 49]

Z	1	2	4	6	8	10	20
$F(Z)$	0.53	0.40	0.28	0.23	0.20	0.17	0.11

slowest electrons; (3) inelastic electron–atom processes of atom excitation; (4) elastic electron–atom collisions. As a result of the first two processes, electrons are heated, and as a result of the latter two they are cooled.

For the region of "weak" excitation ($S \gg 1$), the electron energy balance equation has the form [2]:

$$\frac{f^+}{n_e}\left(\frac{I_m}{2} - \frac{2\bar{\varepsilon}_e}{3}\right) = [1 - F(Z)]\frac{4\exp(-1/g)}{g^{3/2}}\Lambda e^4 n_e \sqrt{\frac{2\pi}{m_e I_m}}, \tag{4.19}$$

where $g = T_e/I_m$; $Z = \dfrac{2f_{0n}[A]g^{3/2}}{5\Lambda n_e}$ (f_{0n} is the transition oscillator strength); $F(Z) = \int\limits_0^\infty \exp\left[-\left(t + Zt^{3/2}\right)\right]dt$. Values of the function $F(Z)$ are given in Table 4.9.

As a result of transformations, from expression (4.19) it is possible to obtain a rather simple transcendent equation:

$$\frac{n_e}{[A_2^+]}\frac{\exp(-1/g)}{g(0.5 - g)}[1 - F(Z)] = C, \tag{4.20}$$

where $C = \dfrac{k_{dr}^0}{4\Lambda e^4 I_m}\sqrt{\dfrac{m_e}{2\pi}}$ (k_{dr}^0 is the rate constant of dissociative recombination for $T_e = 300$ K). The dimensionless constant C is equal to 0.015, 0.03, 0.053, and 0.06, respectively, for Ne, Ar, Kr, and Xe. For large values $Z \gg 1$, Eq. (4.20) is simplified, since in this case $F(Z) \ll 1$ (Table 4.9):

$$e^{-1/g} = Cg(0.5 - g). \tag{4.21}$$

From solving of the approximate Eq. (4.21) given the condition that $n_e = [A_2^+]$ ($S \gg 1$), it follows that the values $g = T_e/I_m$ for Ne, Ar, Kr, and Xe differ insignif icantly and are found in the range of 0.14–0.17. In particular, the calculation for Ar ($C = 0.03$; $I_m \approx 11.5$ eV) yields $g \approx 0.16$ and $T_e \approx 2.2 \times 10^4$ K.

From the cited results, it follows that in the case of "weak" excitation for $\zeta \ll 10^{-4}$, the electron energy balance and T_e scarcely depend on ζ. In the opinion of the authors of study [2], this is explained by the fact that with an increase in the gas pressure, there is a sharp drop in the energy distribution function in the region of $\varepsilon_e > I_m$ and consequently a reduction in the percentage of plasma electrons expending their energy on the excitation of atoms.

Calculation of Plasma Parameters for Gas NPL Mixtures

The basic plasma processes occurring in a single-component gas mixture were examined previously in this chapter. Most often binary mixtures *A–B* are used as the active medium, where *A* is a buffer gas with a high ionization and excitation potential, and *B* is a laser additive with lower ionization and excitation potential. A diagram of the main plasma processes in a binary mixture is shown in Fig. 4.16.

The basic channels to transfer energy from ions and atoms of the buffer gas *A* to the atoms of the additive *B* are the charge-transfer process $A^+(A_2^+)+B \to (B^+)*+A$ (2*A*), the Penning reaction $A*+B \to (B^+)*+A+e$ (if the energy of the excited atom $A*$ is greater than the ionization potential of atom *B*), or transfer of excitation $A*+B \to B*+A$.

In high-pressure plasma, the molecular ions and A_2^+, B_2^+ AB^+ are the main ion types. These are formed as a result of three-body processes $A^+(B^+)+2A(A,B) \to A_2^+$ $(B_2^+, AB^+)+A$. Neutralization of plasma occurs as a result of recombination processes, among which, depending on the specific conditions, either the processes of three-body recombination of atomic ions $A^+(B^+)+2e(e,A) \to A*(B*)+e(A)$, or the processes of dissociative recombination $A_2^+(B_2^+, AB^+)+e \to A*(B*)+A(B,A)$ predominate.

Kinetic equations, which represent the balance of rates of formation and decay of individual plasma components, are used to calculate the plasma parameters, and subsequently, the laser characteristics. In certain models, the number of considered plasmochemical reactions reaches several hundred (see [51], for example). For a satisfactory description of plasma phenomena relating to calculation of the characteristics of a specific laser, it is entirely sufficient to use 10–15 main processes for binary mixtures. In this regard, for the calculations it is sometimes helpful to use so-called "small" models, in which only the basic plasma processes are included. The results of calculation of the plasma parameters for mixtures He-Xe and Ar-Xe using "small" models are provided next as an example [52].

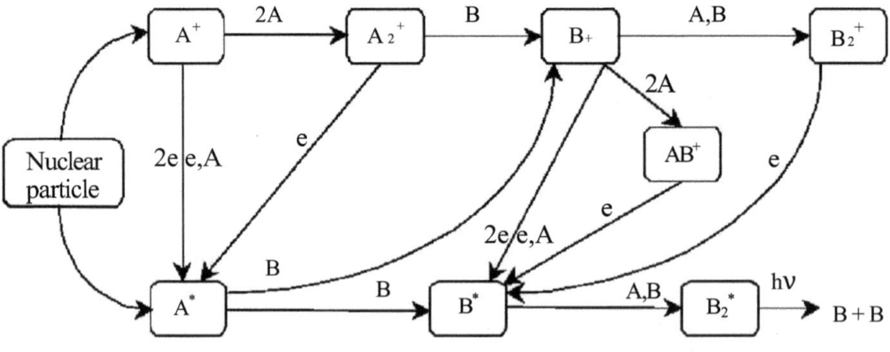

Fig. 4.16 Diagram of main plasma processes in the binary mixture *A–B*

Table 4.10 Basic plasma processes in a He-Xe mixture

Process number	Process	Rate constant	Works cited
1	$He^*(2^1S) + Xe \rightarrow Xe^+ + He + e$	4.4×10^{-10} cm^3/s	[55]
2	$He + 2He \rightarrow He_2^+ + He$	6.4×10^{-32} cm^6/s	[44]
3	$He_2^+ + Xe \rightarrow Xe^+ + 2He$	4.7×10^{-10} cm^3/s	[56]
4	$Xe^+ + Xe + He \rightarrow Xe_2^+ + He$	1.1×10^{-31} cm^6/s	[57]
5	$Xe^+ + e + e \rightarrow Xe^* + e$	$4.0 \times 10^{-9} T_e^{-4.5}$ cm^6/s	[46]
6	$Xe^+ + e + He \rightarrow Xe^* + He$	$1.7 \times 10^{-21} T_e^{-2.5}$ cm^6/s	[47]
7	$Xe_2^+ + e \rightarrow Xe^* + Xe$	$8.1 \times 10^{-5} T_e^{-0.6}$ cm^3/s	[48]

Note: Rate constants of processes are shown for $T_g = 300$ K, electron temperature T_e, K

Table 4.11 Basic plasma processes in Ar-Xe mixture

Process number	Process	Rate constant	Works cited
1	$Ar^+ + 2Ar \rightarrow Ar_2^+ + Ar$	2.5×10^{-31} cm^6/s	[44]
2	$Ar_2^+ + Xe \rightarrow Xe^+ + 2Ar$	1.2×10^{-9} cm^3/s	[58]
3	$Ar_2^+ + e \rightarrow Ar^* + Ar$	$3.9 \times 10^{-5} T_e^{-0.67}$ cm^3/s	[48]
4	$Xe^+ + Xe + Ar \rightarrow Xe_2^+ + Ar$	2.0×10^{-31} cm^6/s	[59]
5	$Ar^* + Xe \rightarrow Xe^* + Ar$	6.8×10^{-10} cm^3/s	[60]
6	$Xe_2^+ + e \rightarrow Xe^* + Xe$	$8.1 \times 10^{-5} T_e^{-0.6}$ cm^3/s	[48]
7	$Xe^+ + 2Ar \rightarrow ArXe^+ + Ar$	1.0×10^{-31} cm^6/s	[59]
8	$Ar_2^+ + Xe \rightarrow ArXe^+ + Ar$	2.0×10^{-10} cm^3/s	[61]
9	$ArXe^+ + Xe \rightarrow Xe_2^+ + Ar$	7.0×10^{-10} cm^3/s	[61]
10	$ArXe^+ + Ar \rightarrow Xe^+ + 2Ar$	5.0×10^{-11} cm^3/s	[61]
11	$ArXe^+ + e \rightarrow Xe^* + Ar$	$1.7 \times 10^{-5} T_e^{-0.5}$ cm^3/s	[61]

Note: Rate constants of processes are shown for $T_g = 300$ K, electron temperature T_e, K

Calculations of the concentrations of electrons and ions and the electron temperature depending on the partial pressure of Xe are performed for the experimental conditions [53, 54], in which He-Xe and Ar-Xe NPLs excited by uranium fission fragments with a neutron pulse duration of 4 ms were studied. The specific power deposition at the neutron pulse maximum for $P_{He} = 2$ atm and $P_{Ar} = 0.5$ atm was around 20 W/cm^3. The plasma processes that were included in the kinetic models are shown in Tables 4.10 and 4.11. The processes were selected as a result of preliminary analysis of the characteristic times of several tens of plasmochemical reactions with the participation of atomic and molecular ions of rare gases, as well as excited atoms and molecules.

The kinetic equations were supplemented with electron energy balance equations. For the He-Xe mixture, this equation included the following processes of electron energy change: (a) ionization of the gas, as a result of which subthreshold electrons with an energy $\varepsilon_e = 0.31 V_i = 7.6$ eV arise at a rate of $f^+ = q/w_i$; (b) the Penning process (1), as a result of which electrons with an energy of 8.5 eV are produced; (c) inelastic collisions of electrons with molecular ions He_2^+, which can transfer a maximal energy of 2.4 eV to the electrons [62]; (d) elastic collisions of

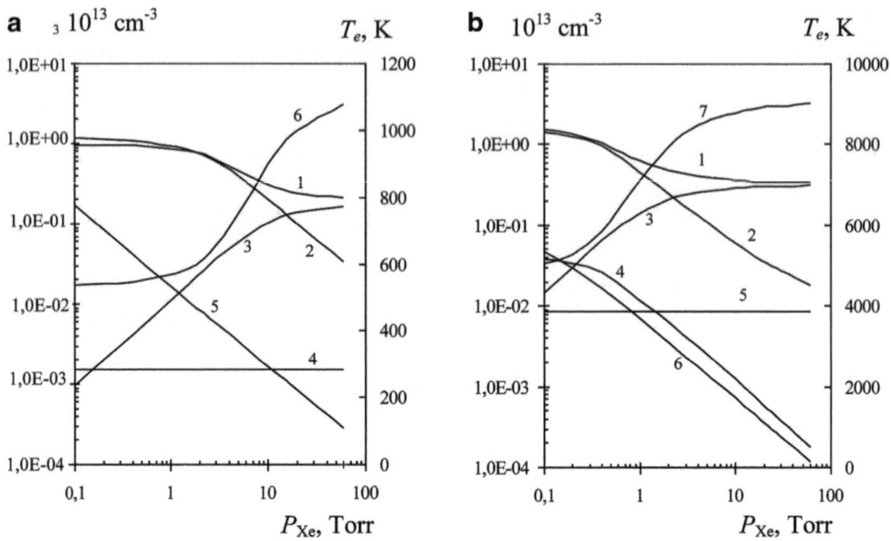

Fig. 4.17 Dependencies of plasma parameters on partial pressure of Xe for He-Xe and Ar-Xe mixtures [52]: (**a**) He-Xe [*(1)* n_e; *(2)* [Xe^+]; *(3)* [Xe_2^+]; *(4)* [He^+]; *(5)* [He_2^+]; *(6)* T_e; (**b**) mixture Ar-Xe [*(1)* n_e; *(2)* [Xe^+]; *(3)* [Xe_2^+]; *(4)* [$ArXe^+$]; *(5)* [Ar^+]; *(6)* [Ar_2^+]; *(7)* T_e

electrons with atoms of helium, as a result of which electron thermalization occurs. For the Ar-Xe mixture, the electron energy balance equation includes the processes similar to those cited in points (a), (c), and (d) for the mixture He-Xe.

Characteristic times of all plasma processes are much shorter than the pumping pulse durations, so a quasistationary mode is established in the plasma. The results of calculations of ion and electron concentrations and the electron temperature for the mixtures He-Xe and Ar-Xe at the neutron pulse maximum are shown in Fig. 4.17. With the increase in Xe partial pressure, there is a change in the ion composition of the plasma: concentrations of atomic ions and molecular ions of the buffer gas (He_2^+, Ar_2^+) decrease, while the concentration of the molecular ions Xe_2^+ increases. The low concentration of heteronuclear ions $ArXe^+$ (Fig. 4.17b) is explained by their effective destruction as a result of collisions with atoms of Ar and Xe. Partial pressures of Xe for the mixtures He-Xe and Ar-Xe, at which the maximal output laser powers are achieved, are 1–2 Torr [53, 54]. In this case, as follows from Fig. 4.17 data, the process of dissociative recombination of molecular ions Xe_2^+ with electrons is the basic channel of formation of excited atoms Xe*.

In calculating plasma characteristics for a He-Xe mixture, the processes of formation and decay of heteronuclear ions $HeXe^+$ were not taken into account. The dissociation energy of heteronuclear ions decreases with an increase in the difference in masses of the atoms making them up. Thus for the ions $ArXe^+$ and $HeAr^+$ the dissociation energy is 0.18 eV and 0.026 eV, respectively [63], while for the ion $HeXe^+$ it is $\leq$0.02 eV. Consequently, $HeXe^+$ ions are effectively destroyed in collisions with atoms, and their equilibrium concentration is insignificant. Mass-spectrometry measurements [64], which registered different heteronuclear ions except for the ion $HeXe^+$, can serve as a confirmation of this.

Analogous calculations of plasma parameters using the "small" model were carried out for He-Ar and He-Kr mixtures [65]. As in the case of the He-Xe mixture, the processes of formation and decay of heteronuclear ions $HeAr^+$ and $HeKr^+$ were not taken into account. One of the significant differences between mixtures based on helium and mixtures in which the buffer gases are Ne, Ar, or Kr, is the appreciable role of the processes of collisional-radiative and three-body recombination of atomic ions in the processes of plasma neutralization and formation of excited atoms (Fig. 4.17a). In mixtures based on Ne, Ar, or Kr, these processes are virtually entirely suppressed because of the high electron temperature.

The main problem that has to be resolved as a result of analysis of plasma processes in NPL gas media is identification of channels for the population of upper and lower laser levels, as well as determination of their populating rates.

Experimental Research

The majority of studies dedicated to examining the characteristics of nuclear-excited plasma are computational-theoretical, and the data presented previously are essentially a brief review of them. The number of experimental studies in this area is very small, especially as regards experiments performed with the help of reactors.

Experiments with stationary reactors were devoted primarily to measuring of electrons and the electron temperature using microwave diagnostics [28, 66, 67] and electrical probes [68–70]. Measurements were carried out for ^4He [66]; ^3He [69]; Ne [67, 70]; Ar and Xe [70]; and for the Ne-Ar mixture [28, 68, 70] at ionization of gas media directly by n,γ radiation of a reactor [66, 67], as well as by uranium fission fragments [28, 68, 70] and products of the nuclear reaction $^3He(n,p)^3H$ [69].

The most interesting are the optical methods of plasma diagnostics, which do not introduce any perturbations in the characteristics to be measured. Information about the properties of the plasma in this case can be obtained from investigation of the characteristics of its radiation or absorption—the intensity and wavelength of individual lines, the width and shape of contours of the lines, etc. [71, 72]. We know of only one study [73] in which, during reactor experiments, the concentrations of metastable atoms in He, Ne, and the He-Ne mixture were measured using the optical method. Excitation of gas media at pressures of up to 1.2 atm was carried out by products of the nuclear reaction $^{10}B(n,\alpha)^7Li$ in the process of irradiation of a gas cell having a thin layer of ^{10}B by a neutron beam of a TRIGA-pulsed reactor. To determine the concentrations of metastable atoms $He^*(2^1S)$, $He^*(2^3S)$, $Ne^*(3s[1/2]^0_0)$, and $Ne^*(3s[3/2]^0_2)$, absorption of radiation by these atoms at transitions with wavelengths of 501.6 nm; 388.9 nm; 626.6 nm and 621.7 nm, respectively, was used. The concentrations of metastable atoms, depending on the pressure of the gas media and the neutron flux density, can bc significant. Thus for helium at $P_{He} = 0.2–0.5$ atm and a neutron flux density $\Phi = 4 \times 10^{13}$ cm^{-2} s^{-1}, the concentration of atoms He*(2^3S) was (2–4) $\times 10^{11}$ cm^{-3}, which is comparable in size to the concentration of electrons (~10^{12} cm^{-3}).

4.4 Luminescence Characteristics

This section provides a brief survey of research on the luminescent spectra of gas media at high pressure in the visible and near IR spectral ranges with excitation by nuclear radiation and fast electron beams. The results of spectral-luminescence investigations in the UV range, aimed at creation of powerful excimer sources of coherent and spontaneous radiation, are not examined here, but are contained, for example, in the surveys [74, 75].

Research on luminescence provides information about the spectroscopic characteristics of transitions, making it possible to ascertain the kinetics of populating and "quenching" of excited states and to estimate the possibility of obtaining laser action at individual transitions. Early stages of investigations of radioluminescent characteristics of condensed and gas media performed before 1964 are considered in the survey [76].

From various methods used in spectral-luminescent investigations, we single out the method of single-photon spectrometry [39, 40, 77, 78], which is based on the counting of single photons and allows analysis of luminescent radiation depending on wavelengths and de-excitation times. Single-photon spectrometry is the basis of the method of subthreshold diagnostics of active NPL media, developed at FEI [79, 80]. Further, we shall briefly examine the most significant studies dedicated to researching the luminescence characteristics of gas media excited by ionizing radiation.

Rare Gases and their Mixtures

Rare gases are a component part of virtually all gas NPLs. Helium, neon, or argon are used as buffer gases, while lasing is achieved at a number of transitions of Xe, Kr, Ar, and Ne atoms, so the luminescence spectra of rare gases are of the greatest interest. Unfortunately, in the scientific literature one can to find information only about the luminescence spectra in the visible and near IR spectral ranges ($\lambda < 1.2$ μm), which may be explained by the absence of sensitive photodetectors for the longer-wave radiation. Table 4.12 shows a list of the studies in which the most significant results were achieved.

Table 4.12 Fundamental experimental research on radioluminescent characteristics of rare gases

Gas medium	Method of excitation	Spectral range, nm	Pressure, atm	Works cited
He, Ne, Ar, Kr, Xe	^{210}Po α particles	240–550	≤1.3	[36]
Ne, He-Ne	^{252}Cf fission fragments; products of the reaction $^{3}He(n,p)^{3}H$	<800	≤1.8	[40]
Ne, He-Ne, Ne-Ar(Xe)	^{239}Pu α particles	300–820	0.1–6	[77]
He, Ne, Ar, Kr, Xe	^{210}Po α particles	240–1,200	1	[81]
^{3}He	Products of the reaction $^{3}He(n,p)^{3}H$	400–800	0.57	[82]
Ne, He-Ne	^{238}Pu α particles	500–830	≤1	[78, 83]
He, Ne, Ar, He-Ne	^{210}Po α particles	330–900	≤1	[84]
He, Ne, Ar, Kr, Xe, Ne-Ar(Kr, Xe), Ar-(Kr, Xe)	Electrons with an energy of 500 keV	350–1,070	0.3–2	[85]
^{3}He-Ne(Ar, Kr, Xe)	Products of the reaction $^{3}He(n,p)^{3}H$	300–900	≤2	[86]
^{3}He-Ar(Kr, Xe)	Products of the reaction $^{3}He(n,p)^{3}H$	400–900	~1	[87, 88]
He, Ne, Ar, Kr, Xe, He(Ne)-Ar, He(Ne, Ar) -Kr; He(Ne, Ar, Kr)-Xe	^{235}U fission fragments	350–1,100	0.2–4.4	[89–91]

Despite the fact that study of the luminescence characteristics of rare gases excited by ionizing radiation (electron and ion beams, nuclear reaction products) began more than 50 years ago, investigations in this area cannot be deemed complete. As a rule such investigations were conducted under different experimental conditions for a limited number of gas media and pressures, which makes it difficult to compare results. Studies [40, 82, 87, 88, 91] carried out measurements of the absolute intensities of spectral lines.

The most complete information is contained in [89–91], in which all the rare gases and their binary mixtures that were of most interest from the standpoint of NPLs were studied under identical experimental conditions. The gas mixtures were excited by ^{235}U fission fragments escaping from a thin uranium layer when the ^{235}U nuclei interacted with neutrons. A VIR-2M-pulsed reactor was used as the neutron source. As a result of the investigations, information was obtained on the intensity of the roughly 300 spectral lines belonging to the transitions of HeI, NeI, ArI(II), KrI(II), and XeI(II). Also observed were lines belonging to transitions of the molecules N_2 and N_2^+, and the atoms O, N, and C, which were there because of the presence of impurities N_2, CO_2, CH_4, O_2, H_2, and H_2O in the rare gases, with a total concentration of no more than 5×10^{-4} %. The intensities of certain of them (for example, the 777.4 and 844.6 nm lines of the O atom) were comparable with the intensities of the basic lines of rare gas atoms.

The calibration of spectral apparatus performed in study [91] by using a reference source made it possible to determine the absolute values of radiating powers on individual lines in the range of 740–1,100 nm, while the conversion efficiencies obtained for the mixtures He-(Ar, Kr, Xe) conform well with the results of studies [87, 88]. Virtually all of the most intensive lines belong to the transitions $(n+1)p$–$(n+1)s$ of the atoms Ar, Kr, Xe ($n = 3, 4, 5$ for Ar, Kr, Xe respectively). The levels $(n+1)p$ are the lower levels for numerous laser IR transitions in the 1–4 μm region, arising on excitation of rare gases by nuclear radiation (see Chap. 3, Sect. 3.1). The maximal conversion efficiencies (~0.1 %) were registered for the lines 912.3; 965.8 nm (ArI, mixture of He-Ar); 892.9 nm (KrI, mixture of He-Kr) and 980.0 nm (XeI, mixture of Ar-Xe).

Mixtures of Rare Gases with Molecular Gases

Interest in the study of luminescence characteristics of mixtures of rare gases with molecular gases is related primarily to the search for new lasing transmissions for NPLs and to the study of the influence of molecular impurities on the parameters of known NPLs. The most significant studies in this area are shown in Table 4.13.

Table 4.13 Experimental research on radioluminescent characteristics of mixtures of rare gases with molecular gases

Gas media (excitation method)	Spectral range, nm	Pressure, atm	Luminescent spectrum	Works cited
Mixtures of rare gases with N_2, CO_2				
He(Ne, Ar)-N_2 (^{210}Po α particles)	240–550	≤1.3	Bands 2^+N_2 and $1^-N_2^+$	[36]
Ne-N_2(CO_2) (^{239}Pu α particles)	300–820	0.1–6	Ne-N_2: bands 2^+N_2 and $1^-N_2^+$, Ne-CO_2: UV transitions of CO_2^+	[77]
He(Ne, Ar)-N_2 (^{210}Po α particles)	240–1,200	1	Bands 2^+N_2 and $1^-N_2^+$; bands of radicals CN, NO, OH, NH	[81]
He(Ar)-N_2 (^{252}Cf fission fragments)	200–500	0.2–6	Bands 2^+ N_2 and $1^-N_2^+$	[92, 93]
Mixtures of rare gases with halogen-containing substances				
^{3}He-Cl_2(F_2) (products of the reaction ^{3}He(n,p)^{3}H)	300–900	≤2	^{3}He-Cl_2: atomic lines ClI, ^{3}He-F_2: atomic lines of FI	[86]
He(Ne, Ar)-NF_3; He-SF_6(CCl_4, CCl_2F_2, $C_2F_3Cl_3$) (^{235}U fission fragments)	350–875	0.45–1.8	Relative intensities of 13 lines of FI and 10 lines of ClI	[94, 95]

In studies [92, 93], apart from the luminescence spectra of mixtures (Ar)-N_2, the time dependencies of luminescence and populating efficiencies of the states $N_2(C^3\Pi_u)$, $N_2^+(B^2\Sigma_u^+)$ were also determined.

Mixtures of Rare Gases and Metal Vapors

Mixtures of rare gases with metal vapors are considered to be promising media for NPLs (see Chap. 3, Sect. 3.1), thus the studies that investigated their spectral-luminescence characteristics are of great interest. Unfortunately, the number of such studies is not great (see Table 4.14), and the information obtained refers to active media operating at comparatively low temperatures (up to ~800 K).

Table 4.14 Experimental research on radioluminescent characteristics of mixtures of rare gases and metal vapors

Gas medium (excitation method)	Spectral range, nm	Pressure, atm (temperature, K)	Basic results	Works cited, Ref.
^{3}He-Cd(Hg, Cs), ^{3}He-Kr-Hg (reaction products $^3He(n,p)^3H$)	360–660	1 (≤720)	Relative intensities of 9 lines of CdI and CdII, 6 lines of HgI and HgII, 17 lines CsI and CsII	[96, 97]
^{3}He-Cd (products of reaction $^3He(n,p)^3H$)	270–860	0.86 (630)	Relative intensities of 27 lines of CdI and CdII	[98]
^{3}He-Cd(Zn, Hg, Na, K, Cs) (products of reaction $^3He(n,p)^3H$)	300–900	1 (≤770)	Survey spectra and discussion of level populating mechanisms	[99, 100]
^{3}He-Cs (products of reaction $^3He(n,p)^3H$)	300–900	0.86 (≤770)	Registered 25 lines of CsI and 29 lines of CsII	[101]
He-Cd (fission fragments of ^{252}Cf)	200–600	1.5–5 (≤670)	Analysis of kinetics of 12 lines of CdII, clarifica tion of the kinetics of processes in the mixture He-Cd	[102, 103]
^{3}He-Na (products of the reaction $He(n,p)^3H$)	300–850	0.3–1 (840–940)	Analysis of the possibility of creating an effective source of spontaneous radiation	[104]
(He)-Xe(Kr)-Hg with impurities N_2, H_2 and D_2 (α particles of ^{239}Pu and ^{210}Po; products of the reaction $^3He(n,p)^3H$)	200–830	0.8 (≤650)	Determination of the mechanism of formation of inverse population on the transitions 7^3S_1-$6^3P_{0,1,2}$ of a mercury atom	[105, 106]

References

1. L.I. Gudzenko, S.I. Yakovlenko, *Plasma Lasers (Plazmennyye lazery)* (Atomizdat, Moscow, 1978)
2. V.L. Bychkov, A.V. Yeletskiy, High-pressure beam plasma (Puchkovaya plazma vysokogo davleniya). In *Plasma Chemistry* (*Khimiya plazmy*), vol. 12 (Energoatomizdat, Moscow, 1985), pp. 119–158.
3. A. Dalgarno, Path lengths and energy losses (Probegi i poteri energii). In *Atomic and Molecular Processes* (*Atomnyye i molekulyarnyye protsessy*) (Mir, Moscow, 1964), pp. 540–556.
4. T.J. Moratz, M.J. Kushner, Fission fragment pumping of a neon plasma. J. Appl. Phys. **63**(5), 1796–1798 (1988)
5. T.J. Moratz, T.D. Saunders, M.J. Kushner, Heavy-ion versus electron-beam excitation of an excimer laser. J. Appl. Phys. **64**(8), 3799–3810 (1988)
6. A.V. Tyukavkin, Energy distribution of electrons in He when excited by ions (Energeticheskoye raspredeleniye elektronov v He pri vozbuzhdenii ionami). Fizika Plazmy **25**(1), 96–100 (1999)
7. A.P. Budnik, I.V. Dobrovolskaya, Specific features of kinetics of active media of gas lasers excited by fission fragments (Osobennosti kinetiki aktivnykh sred gazovykh lazerov, vozbuzhdayemykh oskolkami deleniya). Kvantovaya elektronika **24**(6), 506–510 (1997)
8. A.V. Karelin, O.V. Simakova, Kinetics of an active medium of a multiwave IR xenon laser with mixtures of He and Ar pumped by a hard ionizer. 2. Nuclear pumping (Kinetika aktivnoy sredy mnogovolnovogo IK lazera na ksenone v smesyakh s He i Ar, nakachivayemykh zhestkim ionizatorom. 2. Yadernaya nakachka). Kvantovaya elektronika **28**(2), 129–135 (1999)
9. W.H. Ellis, K. Imani, *Kinetic studies of fission fragment, $^3He(n,p)T$, and gamma ray produced plasmas*. Proc. Second Symp. on Uranium Plasmas: Research and Applications, Atlanta, 1971, pp. 45–52.
10. H.E. Wilhelm, Recombination in columns. J. Chem. Phys. **47**(11), 4356–4365 (1967)
11. V.P. Konovalov, E.Y. Son, Degradation spectra of electrons in gases (Degradatsionnyye spektry elektronov v gazakh). In *Plasma Chemistry* (*Khimiya plazmy*), vol. 14 (Energoatomizdat, Moscow, 1987), pp. 194–227.
12. M.J. Kushner, Response times and energy partitioning in electron-beam-excited plasmas. J. Appl. Phys. **66**(6), 2297–2306 (1989)
13. J. Bretagne, G. Delouya, J. Godart, V. Puech, High-energy electron distribution in an electron-beam-generated argon plasma. J. Phys. D **14**(7), 1225–1239 (1981)
14. F. Kannari, W.D. Kimura, High-energy electron distribution in electron beam excited Ar/Kr and Ne/Xe mixtures. J. Appl. Phys. **63**(9), 4377–4387 (1988)
15. E. Suzuki, Y. Hatano, Electron thermalization processes in rare gases with the ramsauer minimum. J. Chem. Phys. **84**(9), 4915–4918 (1986)
16. J.M. Warman, M.C. Sauer, An investigation of electron thermalization in irradiated gases using CCl_4 as an electron energy probe. J. Chem. Phys. **62**(5), 1971–1981 (1975)
17. F. Kannari, W.D. Kimura, Low-energy electron distribution in electron-beam-excited XeCl laser mixtures. J. Appl. Phys. **64**(2), 500–506 (1988)
18. J. Bretagne, J. Godart, V. Puech, Low-energy electron distribution in an electron-beamgenerated argon plasma. J. Phys. D **15**(11), 2205–2225 (1982)
19. C.J. Elliott, A.E. Greene, Electron energy distribution in e-beam generated Xe and Ar plasmas. J. Appl. Phys. **47**(7), 2946–2953 (1976)
20. V.A. Nikerov, G.V. Sholin, *Kinetics of Degradation Processes (Kinetika degradatsionnykh protsessov)* (Energoatomizdat, Moscow, 1985)
21. R.A. Walters, R.T. Schneider, Electron energy distribution function of fission fragment excited gases. Trans. Am. Nucl. Soc. **17**, 1–2 (1973)

22. R.H. Lo, G.H. Miley, Electron energy distribution in a helium plasma created by nuclear radiations. IEEE Trans. Plasma Sci. **PS-2**(4), 198–205 (1974)
23. Y.I. Dymshits, V.G. Neverov, V.G. Khoroshev, Calculation of outputs of primary products of irradiation of rare gases by fast electrons (Raschet vykhodov pervichnykh produktov oblucheniya inertnykh gazov bystrymi elektronami). Khimiya vysokikh energiy **16**(3), 201–208 (1982)
24. D.M. Bartell, G.S. Hurst, E.B. Wagner, Time-dependent studies of vacuum-ultraviolet emissions from helium. Phys. Rev. A **7**(3), 1068–1078 (1973)
25. G.R. Shipman, R.A. Walters, Excitation of helium caused by fission fragments. Trans. Am. Nucl. Soc. **17**, 1 (1973)
26. M.Y. Veselovskaya, V.R. Solovyev, Energy costs of ionization and excitation of atoms in beam plasma of Ar/Xe and Ne/Xe mixtures (Energeticheskiye tseny ionizatsii i vozbuzhdeniya atomov v puchkovoy plazme smesey Ar/Xe and Ne/Xe). Fizika Plazmy **21** (4), 344–349 (1995)
27. R.L. Platzman, Total ionization in gases by high-energy particles: an appraisal of our understanding. Int. J. Appl. Rad. Isotop. **10**(2/3), 116–127 (1961)
28. D.B. Rees, C.B. Leffert, D.J. Rose, Electron density in mixed gas plasmas generated by fission fragments. J. Appl. Phys. **40**(4), 1884–1896 (1969)
29. H.W. Schmitt, R.B. Leachman, Ionization vs energy relation for fission fragments. Phys. Rev. **102**(1), 183–185 (1956)
30. G.D. Alkhazov, An ionization cascade of non-relativistic electrons in helium (Ionizatsionnyy kaskad nerelyativistskikh elektronov v gelii). ZhTF **41**(12), 2513–2523 (1971)
31. J.A. Blauer, T.T. Yang, C.E. Turner, D.A. Copeland, Excimer kinetics and multiline model for the electron-beam pumped XeF (*B-X*) laser. Appl. Optics **23**(23), 4352–4367 (1984)
32. Y.I. Sytsko, S.I. Yakovlenko, Kinetics of ionization and excitation of gas by a hard source (Kinetika ionizatsii i vozbuzhdeniya gaza zhestkim istochnikom). Fiz. Plazmy **2**(1), 63–71 (1976)
33. W.P. Jesse, J. Sadauskis, Absolute energy to produce an ion pair by beta particles from ^{35}S. Phys. Rev. **107**(3), 766–771 (1957)
34. T.E. Bortner, G.S. Hurst, Ionization of pure gases and mixtures of gases by 5-MeV alpha particles. Phys. Rev. **93**(6), 1236–1241 (1954)
35. K. Saito, S. Sasaki, H. Tawara, T. Sanami, E. Shibamura, Simultaneous measurements of absolute numbers of electrons and scintillation photons produced by 5.49 MeV alpha particles in rare gases. IEEE Trans. Nucl. Sci. **50**(6), 2452–2459 (2003)
36. W.R. Bennett, Optical spectra excited in high pressure noble gases by alpha impact. Ann. Phys. **18**(3), 367–420 (1962)
37. G.S. Hurst, M.H. Nayfeh, J.P. Judish, C.H. Chen, E.B. Wagner, J.P. Young, Kinetics of He (2^1S) using resonance ionization spectroscopy. Phys. Rev. Lett. **35**(17), 1154–1156 (1975)
38. A.A. Radtsig, B.M. Smirnov, *Parameters of Atoms and Atomic Ions (Parametry atomov i atomnykh ionov)* (Energoatomizdat, Moscow, 1986)
39. Ye.D. Poletayev, Yu.B. Dorofeyev, P.P. Dyachenko, A.P. Kopay-Gora, A.A. Mavlyutov, A.I. Miskevich, B.S. Salamakha, Direct excitation of $2p^53p$ levels of NeI by heavy charged particles in high-pressure He-Ne mixtures (Pryamoye vozbuzhdeniye $2p^53p$ urovney NeI tyazhelymi zaryazhennymi chastitsami v He-Ne smesyakh vysokogo davleniya) (1990), Preprint of FEI, No. 2158, Obninsk.
40. Ye.D. Poletayev, Yu.B. Dorofeyev, P.P. Dyachenko, A.P. Kopay-Gora, A.A. Mavlyutov, A.I. Miskevich, B.S. Salamakha, Radiating characteristics of pure neon and He-Ne high-pressure mixtures during excitation by nuclear particles (Izluchatelnyye kharakteristiki chistogo neona i He-Ne smesey vysokogo davleniya pri vozbuzhdenii yadernymi chastitsami). ZhTF **62**(2), 1–8 (1992)
41. J.W. Wilson, R.J. DeYoung, W.L. Harries, Nuclear-pumped ^{3}He-Ar laser modeling. J. Appl. Phys. **50**(3), 1226–1235 (1979)

42. V.V. Ryzhov, A.G. Yastremskiy, Specific features of generation of low-temperature plasma on ionization of gas mixtures by an electron beam (Osobennosti generatsii nizkotemperaturnoy plazmy pri ionizatsii smesey gazov elektronnym puchkom). ZhTF **49**(10), 2141–2145 (1979)
43. A.D. Andreyev, S.V. Makarov, Y.V. Novoselov, V.V. Ryzhov, I.Y. Turchanovskiy, Modeling of processes in a He-Cd mixture with ion and electron pumping (Modelirovaniye protsessov v He-Cd smesi pri ionnoy i elektronnoy nakachakh). ZhTF **62**(3), 18–23 (1992)
44. D. Smith, A.G. Dean, J.C. Plumb, Three body conversion reactions in pure rare gases. J. Phys. B **5**(11), 2134–2142 (1972)
45. A.M. Voinov, S.P. Melnikov, A.A. Sinyanskiy, On the lasing mechanism of recombination lasers operating on transitions of rare gas atoms (O mekhanizme generatsii rekombinatsionnykh lazerov na perekhodakh atomov inertnykh gazov). Pisma v ZhTF **15**(19), 56–59 (1989)
46. P. Mansbach, J. Keck, Monte-Carlo trajectory calculations of atomic excitation and ionization by thermal electrons. Phys. Rev. **181**(1), 275–289 (1969)
47. L.P. Pitayevskiy, Recombination of electrons in a single-atom gas (Rekombinatsiya elektronov v odnoatomnom gaze). ZhETF **42**(5), 1326–1329 (1962)
48. V.A. Ivanov, Dissociative recombination of molecular ions in an rare gas plasma (Dissotsiativnaya rekombinatsiya molekulyarnykh ionov v plazme inertnykh gazov). Uspekhi fizicheskikh nauk **162**(1), 35–70 (1992)
49. A.B. Yeletskiy, V.D. Kulagin, The kinetics of processes in a plasma created by an electron beam in a dense rare gas (Kinetika protsessov v plazme, sozdavayemoy elektronnym puchkom v plotnom inertnom gaze). Fizika Plazmy **5**(1), 98–105 (1979)
50. F.C.M. Coolen, N. Van Schaik, Electron density and electron temperature in a neon plasma generated by 20-MeV protons. Physica **93C**(1), 121–130 (1978)
51. A.M. Boychenko, V.I. Derzhiyev, A.G. Zhidkov, A.V. Karelin, A.V. Koval, O.V. Sereda, S.I. Yakovlenko, Kinetic models of certain plasma lasers pumped by a hard ionizer (Kineticheskiye modeli nekotorykh plazmennykh lazerov, nakachivayemykh zhestkim ionizatorom). In *Visible and UV Plasma Lasers (IOFAN Proceedings)* (*Plazmennyye lazery vidimogo i UF diapazonov (Trudy IOFAN)*), vol. 21 (Nauka, Moscow, 1989), pp. 44–115.
52. A.M. Voinov, S.P. Melnikov, A.A. Sinyanskiy, A kinetic model of recombination lasers operating on transitions of the xenon atom. I. Laser mechanism and parameters of plasma (Kineticheskaya model rekombinatsionnykh lazerov na perekhodakh atoma ksenona. I. Mekhanizm generatsii i parametry plazmy). ZhTF **60**(10), 100–106 (1990)
53. A.M. Voinov, L.Y. Dovbysh, V.N. Krivonosov, S.P. Melnikov, I.V. Podmoshenskiy, A.A. Sinyanskiy, Low-threshold nuclear-pumped lasers operating on the transitions of atomic xenon (Nizkoporogovyye lazery s yadernoy nakachkoy na perekhodakh atomarnogo ksenona). Doklady AN SSSR **245**(1), 80–83 (1979)
54. A.M. Voinov, L.Y. Dovbysh, V.N. Krivonosov, S.P. Melnikov, I.V. Podmoshenskiy, A.A. Sinyanskiy, Infrared laser with nuclear pumping on mixtures of He + Xe and Ar + Xe (Infrakrasnyy lazer s yadernoy nakachkoy na smesyakh He + Xe i Ar + Xe). Pisma v ZhTF **7**(16), 1016–1020 (1981)
55. J.P. Riola, J.S. Howard, R.D. Rudel, R.F. Stebbings, Chemiionization reactions involving metastable helium atoms. J. Phys. B **7**(3), 376–385 (1974)
56. M. Bourene, J. Le Calve, Effets d'Additifs sur la Postluminescence de He_2 dans l'Helium Excite par une Impulsion Intense d'Electrons. J. Phys. (Paris) **32**(1), 29–33 (1971)
57. C.L. Chen, Atomic processes in helium-krypton and helium-xenon mixtures. Phys. Rev. **131** (6), 2550–2555 (1963)
58. C.B. Collins, F.W. Lee, Measurement of the rate coefficients for the bimolecular and termolecular ion-molecule reactions of $Ar^{+}{}_2$ with selected atomic and molecular species. J. Chem. Phys. **71**(1), 184–191 (1979)
59. F. Kannari, A. Suda, M. Obara, T. Fujioka, Theoretical simulation of electron-beamexcited xenon-chloride (XeCl) lasers. IEEE J. Quant. Electron. **QE-19**(10), 1587–1600 (1983)

60. H. Brunet, A. Birot, H. Dijols, J. Galy, P. Millet, Y. Salamero, Spectroscopic and kinetic analysis of the VUV emissions of argon and argon-xenon mixtures. II. Energy transfer in Ar-Xe mixtures. J. Phys. B **15**(17), 2945–2967 (1982)
61. S.A. Lawton, J.B. Richards, L.A. Newman, L. Specht, T.A. DeTemple, The high-pressure neutral infrared xenon laser. J. Appl. Phys. **50**(6), 3888–3898 (1979)
62. R. Deloche, P. Monchicourt, M. Cheret, F. Lambert, High-pressure helium afterglow at room temperature. Phys. Rev. A **13**(3), 1140–1176 (1976)
63. P.M. Dehmer, S.T. Pratt, Photoionization of ArKr, ArXe, and KrXe and bond dissociation energy of the rare gas dimer ions. J. Chem. Phys. **77**(10), 4804–4817 (1982)
64. M.S.B. Munson, J.L. Franklin, F.H. Field, A mass spectrometric study of homonuclear and heteronuclear rare gas molecule ions. J. Phys. Chem. **67**(7), 1542–1548 (1963)
65. S.P. Melnikov, A.A. Sinyanskiy, Kinetics of IR nuclear-pumped lasers operating on transitions of krypton and argon atoms (Kinetika IK lazerov s yadernoy nakachkoy na perekhodakh atomov kriptona i argona). ZhTF **62**(6), 159–169 (1992)
66. M.H. Voth, E.S. Kenney, Microwave diagnosis of a plasma generated in a reactor and its use as a reactor power monitor. Nucl. Inst. Meth. **63**(1), 45–52 (1968)
67. A.K. Bhattacharya, J.T. Verdeyen, F.T. Adler, L. Goldstein, Microwave investigation of plasmas produced in a reactor. J. Appl. Phys. **38**(2), 527–534 (1967)
68. G.A. Batyrbekov, Y.D. Kuznetsov, Z.S. Takibayev, Concentration of charge carriers of plasma obtained in a penning mixture upon irradiation by uranium fission fragments and reactor radiation (Kontsentratsiya nositeley zaryada plazmy, poluchennoy v smesi Penninga pri obluchenii oskolkami deleniya urana i reaktornym izlucheniyem). Zh. T. F. **44**(5), 1022–1024 (1974)
69. G.A. Batyrbekov, Y.D. Kuznetsov, B.Y. Pelmenshteyn, Z.S. Takibayev, Some characteristics of plasma of the isotopic gas He3 generated by products of the reaction He3(n, p)H3 in a nuclear reactor (Nekotoryye kharakteristiki plazmy izotopnogo gaza He3, generirovannoy produktami reaktsii He3(n, p)H3 v yadernom reaktore). ZhTF **46**(11), 2315–2317 (1976)
70. G.A. Batyrbekov, Y.D. Kuznetsov, B.Y. Pelmenshteyn, Z.S. Takibayev, An investigation of the parameters of inert gas plasma formed by uranium fission fragments in a nuclear reactor (Issledovaniye parametrov plazmy inertnykh gazov, obrazovannoy oskolkami deleniya urana v yadernom reaktore). ZhTF **46**(11), 2318–2320 (1976)
71. A.A. Ovsyannikov, Basic spectral methods of diagnostics of low-temperature plasma (Osnovnyye spektralnyye metody diagnostiki nizkotemperaturnoy plazmy), in *Studies on Low-Temperature Plasma Physics and Chemistry (Ocherki fiziki i khimii nizkotemperaturnoy plazmy)* (Nauka, Moscow, 1971), pp. 386–410
72. V.N. Ochkin, *Spectroscopy of Low-Temperature Plasma (Spektroskopiya nizkotemperaturnoy plazmy)* (Moscow, Fizmatlit, 2006)
73. J.C. Guyot, G.H. Miley, J.T. Verdeyen, Metastable densities in noble-gas plasmas created by nuclear radiations. J. Appl. Phys. **42**(13), 5379–5391 (1971)
74. G.N. Gerasimov, B.Y. Krylov, A.V. Loginov, A.S. Shchukin, Ultraviolet radiation of excited molecules of rare gases (Ultrafioletovoye izlucheniye vozbuzhdennykh molekul inertnykh gazov). Uspekhi fizicheskikh nauk **162**(5), 123–159 (1992)
75. A.M. Boychenko, V.F. Tarasenko, Y.A. Fomin, S.I. Yakovlenko, Wideband continua in rare gases and their mixtures with halogenides (Shirokopolosnyye kontinuumy v inertnykh gazakh i ikh smesyakh s galogenidami). Kvant. Elektron. **20**(1), 7–30 (1993)
76. J.B. Birks, *The theory and practice of scintillation counting* (Pergamon Press, Oxford, 1964)
77. A.B. Dmitriyev, V.S. Ilyashenko, A.I. Miskevich, B.S. Salamakha, The luminescence of neon and certain of its mixtures at high pressures (Lyuminestsentsiya neona i nekotorykh ego smesey pri vysokikh davleniyakh). Optika i spektroskopiya **43**(6), 1165–1168 (1977)
78. G.A. Batyrbekov, E.G. Batyrbekov, V.A. Danilychev, M.U. Khasenov, The efficiency of populating of $3p$-levels of neon when excited by a hard ionizer (Effektivnost zaseleniya $3p$-urovney neona pri ego vozbuzhdenii zhestkim ionizatorom). Optika i spektroskopiya **68** (6), 1241–1245 (1990)

79. P.P. Dyachenko, Yu.B. Dorofeyev, Ye.D. Poletayev, Ye.A. Seregina, Subthreshhold diagnostics of active media for lasers with direct nuclear pumping (Podporogovaya dignostika aktivnykh sred dlya lazerov s pryamoy yadernoy nakachkoy) (1990), Preprint of FEI, No. 2070, Obninsk.
80. P.P. Dyachenko Yu.B. Dorofeev, Ye.D. Poletaev, E.A. Seryogina, V.V. Korobkin, *Subthreshold diagnostics of active media for direct nuclear pumped lasers*. Proc. Int. Conf. LASERS'90, McLean, 1991, pp. 835–842.
81. S. Dondes, P. Hartek, C. Kunz, A spectroscopic study of alpha-ray-induced luminescence in gases. Radiat. Res. **27**(2), 174–210 (1966)
82. G.A. Batyrbekov, S.K. Kunakov, Y.A. Piotrovskiy, A.V. Redkin, Y.A. Tolmachev, Measurement of absolute intensities of spectral lines of optical radiation of ^{3}He(n, p)^{3}H plasma formed on irradiation of ^{3}He by a neutron beam of a VVR-K stationary nuclear reactor (Izmereniye absolyutnhkh intensivnostey spektralnykh liniy opticheskogo izlucheniya plazmy ^{3}He(n, p)^{3}H, obrazuyushcheysya pri obluchenii ^{3}He neytronnym potokom statsionarnogo yadernogo reaktora VVR-K). Teplofizika vysokikh temperatur **26**(4), 633–638 (1988)
83. G.A. Batyrbekov, E.G. Batyrbekov, V.A. Danilychev, M.U. Khasenov, The influence of helium on the efficiency of populating of *3p*-levels of neon atoms (Vliyaniye geliya na effektivnost zaseleniya 3*p*-urovney atomov neona). Kvantovaya elektronika **17**(9), 1175–1180 (1990)
84. P.E. Thiess, G.Y. Miley, *Spectroscopic and probe measurements of excited state and electron densities in radiation induced plasmas*. Proc. Second Symp. on Uranium Plasmas: Research and Applications, Atlanta, 1971, pp. 90–98.
85. S. Arai, R.F. Firestone, Evidence for the formation of neutral rare-gas molecules by electron-beam pulses. J. Chem. Phys. **50**(10), 4575–4589 (1969)
86. R.J. De Young, W.R. Weaver, Spectra from nuclear-excited plasmas. J. Opt. Soc. Am. **70**(5), 500–506 (1980)
87. F.P. Boody, Nuclear-driven fluorescence pumped solid-state lasers, Ph. D. Dissertation, University of Missouri-Columbia, 1991.
88. F.P. Boody, M.A. Prelas, *Absolutely calibrated spectra of nuclear-driven rare gases, 400–950 nm*. Proceedings of the Specialist Conference "The Physics of Nuclear-Excited Plasma and the Problems of Nuclear-Pumped Lasers," Obninsk, 1993, vol. 2, pp. 149–155.
89. V.V. Gorbunov, V.D. Grigoryev, L.Y. Dovbysh, S.P. Melnikov, A.V. Sinitsyn, A.A. Sinyanskiy, V.M. Tsvetkov, Luminescence spectra of dense gases in the region of 350– 875 nm during excitation by uranium fission fragments. I. Rare gases and their mixtures (Spektry lyuminestsentsii plotnykh gazov v oblasti 350–875 nm pri vozbuzhdenii oskolkami deleniya urana. I. Inertnyye gazy i ikh smesi), *Proceedings of RFYaTS-VNIIEF (Trudy RFYaTS-VNIIEF)*, (6), 148–173 (2004).
90. A.A. Abramov, V.V. Gorbunov, S.P. Melnikov, A.Kh. Mukhamatullin, A.V. Sinitsyn, A.A. Sinyanskiy, V.M. Tsvetkov, Luminescence spectra of rare gases and their mixtures in the near IR region at excitation by uranium fission fragments (Spektry lyuminestsentsii inertnykh gazov i ikh smesey v blizhney IK oblasti pri vozbuzhdenii oskolkami deleniya urana), *Proceedings of RFYaTS-VNIIEF* (*Trudy RFYaTS-VNIIEF*), (8), 180–201 (2005).
91. A.A. Abramov, V.V. Gorbunov, S.P. Melnikov, A.K. Mukhamatullin, A.A. Pikulev, A.V. Sinitsyn, A.A. Sinyanskiy, V.M. Tsvetkov, Luminescence of nuclear-induced rare-gas plasmas in near infrared spectral range. Proc. S. P. I. E. **6263**, 279–296 (2006)
92. Ye.D. Poletayev, Yu.A. Dyuzhov, A.V. Mironov, *Measurement of luminescence characteristics of an Ar-N2 mixture excited by fission fragments* (Izmereniye lyuminestsentnykh kharakteristik Ar-N2 smesi, vozbuzhdayemoy oskolkami deleniya), Proceedings of the 2nd International Conference "Physics of Nuclear-Excited Plasma and Problems of Nuclear-Pumped Lasers," Arzamas-16, 1995, vol. 1, pp. 134–141.
93. Yu.A. Dyuzhov, Ye.D. Poletayev, *Kinetics of excitation of the bands of the 2d positive system of the* N_2 *molecule and bands of the 1st negative system of the molecular nitrogen ion* $N^+{}_2$ *in nitrogen and Ar-*N_2*, Ne-*N_2 *high-pressure mixtures during excitation of the medium by fission*

fragments (Kinetika vozbuzhdeniya polos 2-y polozhitelnoy sistemy molekuly N_2 i polos 1-y otritsatelnoy sistemy molekulyarnogo iona azota $N^+{}_2$v azote i Ar-N_2, Ne-N_2 smesyakh vysokogo davleniya pri vozbuzhdenii sredy oskolkami deleniya), Proceedings of the 3rd International Conference "Problems of Nuclear-Pumped Lasers and Pulsed Reactors," Snezhinsk, 2003, pp. 64–68.

94. V.V. Gorbunov, V.D. Grigoryev, S.P. Melnikov, A.V. Sinitsyn, A.A. Sinyanskiy, V.M. Tsvetkov, *A search for lasing transitions during excitation of mixtures of helium and halogen-containing gases by uranium fission fragments* (Poisk lazernykh perekhodov pri vozbuzhdenii oskolakami deleniya urana smesey geliya s galogenosoderzhashchimi gazami). Proceedings of the 3rd International Conference "Problems of Nuclear-Pumped Lasers and Pulsed Reactors," Snezhinsk, 2003, pp. 191–198.
95. V.V. Gorbunov, V.D. Grigoryev, L.Y. Dovbysh, S.P. Melnikov, A.V. Sinitsyn, A.A. Sinyanskiy, V.M. Tsvetkov, Luminescence spectra of dense gases in the range of 350– 875 nm during excitation by uranium fission fragments. II. Mixtures of rare gases with halogen-containing materials (Spektry lyuminestsentsii plotnykh gazov v oblasti 350–875 nm pri vozbuzhdenii oskolkami deleniya urana. II. Smesi inertnykh gazov s galogenoso-derzhashchimi veshchestvami), *Proceedings of RFYaTS-VNIIEF* (*Trudy RFYaTS-VNIIEF*), (6), 174–185 (2004).
96. A.B. Dmitriyev, V.S. Ilyashenko, A.I. Miskevich, B.S. Salamakha, Excitation of lasing transitions in vapor-metal gas mixtures by products of neutron nuclear reactions (Vozbuzhdeniye produktami neytronnykh yadernykh reaktsiy lazernykh perekhodov v parometallicheskikh gazovykh smesyakh). ZhTF **52**(11), 2235–2237 (1982)
97. A.B. Dmitriyev, V.S. Ilyashenko, A.I. Miskevich, B.S. Salamakha, Spectroscopy of ^{3}He-Hg and ^{3}He-Kr-Hg high-pressure plasma excited by products of the reaction ^{3}He(n, p)T (Spektroskopiya ^{3}He-Hg and ^{3}He-Kr-Hg plazmy vysokogo davleniya, vozbuzhdayemoy produktami reaktsii 3He(n, p)T). Optika i spektroskopiya **47**(1), 63–66 (1979)
98. A.P. Kopay-Gora, A.A. Mavlyutov, A.I. Miskevich, B.S. Salamakha, The influence of wall atomization of cadmium on the processes of populating of levels in a high-pressure ^{3}He-Cd-plasma (Vliyaniye pristenochnogo raspyleniya kadmiya na protsessy zaseleniya urovney v ^{3}He-Cd-plazme vysokogo davveniya). Optika i spektroskopiya **67**(3), 526–532 (1989)
99. A.I. Mis'kevich, Visible and near-infrared direct nuclear pumped lasers. Laser Physics **1**(5), 445–481 (1991)
100. A.B. Dmitriyev, A.I. Miskevich, B.S. Salamakha, Excitation of potassium and sodium vapors by products of the neutron nuclear reaction ^{3}He(n, p)T (Vozbuzhdeniye parov kaliya i natriya produktami neytronnoy yadernoy reaktsii ^{3}He(n, p)T). Optika i spektroskopiya **61**(5), 939–945 (1986)
101. A.A. Mavlyutov, A.I. Miskevich, B.S. Salamakha, *On prospects for use of an He-Cs vapor-metal mixture in nuclear-pumped lasers* (O perspektivakh ispolzovaniya He-Cs parometal-licheskoy smesi v lazerakh s yadernoy nakachkoy). Proceedings of the Specialist Conference "The Physics of Nuclear-Excited Plasma and the Problems of Nuclear-Pumped Lasers," Obninsk, 1993, vol. 3, pp. 214–223.
102. Yu.B. Dorofeyev, Ye.D. Poletayev, P.P. Dyachenko, M.D. Tarasko, Time distributions of luminescence at cadmium ion transitions with excitation of a He-Cd medium by fission fragments (Vremennyye raspredleniya lyuminestsentsii na perekhodakh iona kadmiya pri vozbuzhdenii He-Cd sredy oskolkami deleniya) (1993), Preprint of FEI, No. 2311, Obninsk.
103. Ye.D. Poletayev, P.P. Dyachenko, M.D. Tarasko, Yu.A. Dyuzhov, *Time distributions of luminescence at cadmium ion transitions during excitation of a He-Cd medium by fission fragments* (Vremennyye raspredeleniya lyuminestsentsii na perekhodakh iona kadmiya pri vozbuzhdenii He-Cd sredy oskolkami deleniya), Proceedings of the 2nd International Conference "Physics of Nuclear-Excited Plasma and Problems of Nuclear-Pumped Lasers," Arzamas-16, 1995, vol. 1, pp. 121–133.

104. M.A. Prelas, L-T Lin, J.T. Bahns, *Progress in nuclear-driven flashlamps*. Int. Embedded Topical Meeting on Physics, Safety, and Applications of Pulse Reactors, Washington, 1994, pp. 155–162.
105. G.A. Batyrbekov, E.G. Batyrbekov, V.A. Dolgikh, I.G. Rudoy, A.M. Soroka, A.B. Tleuzhanov, M.U. Khasenov, Kinetics of excited states of Hg during pumping by ionizing radiation (Kinetika vozbuzhdennykh sostoyaniy Hg pri nakachke ioniziruyushchim izlucheniyem) (1987), Preprint of the Institute of Nuclear Physics, Academy of Sciences of the Kazakh SSR, No. 3-87, Alma-Ata.
106. G.A. Batyrbekov, E.G. Batyrbekov, V.A. Dolgikh, I.G. Rudoy, A.M. Soroka, A.B. Tleuzhanov, M.U. Khasenov, The luminescence of mixtures of mercury and rare gases with molecular additives on excitation by ionizing radiation, (Lyuminestsentsiya smesey rtuti i inertnykh gazov s molekulyarnymi dobavkami pri vozbuzhdenii ioniziruyushchim izlucheniyem). Zh. Prikl. Spektrosk. **49**(5), 770–774 (1988)

Chapter 5
Lasing Mechanisms and Kinetic Models of NPLs

5.1 Introduction

This chapter examines the kinetics of plasmochemical processes and the lasing mechanisms of various types of gas NPLs pumped using neutrons from pulsed nuclear reactors. The lasers considered here typically have Continuous Wave, i.e. CW operation (or, more accurately, quasi-CW) lasing mechanisms, because the characteristic times of basic plasma processes are much less than the durations of pulsed reactor pumping pulses ($\geq$50 μs). To determine the lasing mechanism, it is necessary to ascertain the processes of populating and quenching both of the upper and of the lower lasing levels. Information about the specific features and the constants of the most important plasmochemical processes is contained in a number of studies (see [1–4], for example).

5.2 Mechanisms of Excitation of Laser Media. Ultimate Efficiency

The active media of gas NPLs are mostly binary mixtures *A*–*B*, where *A* is a buffer gas with high ionization and excitation potential, and *B* is a lasing additive with lower ionization and excitation potential, the concentration of which is 1–10 %. In this case, the energy deposited in the buffer gas *A* is transported virtually without losses to the lasing additive *B* at the expense of the processes of charge-transfer, the Penning processes, and transfer of excitation, while the quenching of excited atoms B^* by their own atoms is insignificant [5].

Populating of the upper laser levels of the atoms and ions of *B* can occur at various stages of the sequence of relaxation processes (see Fig. 4.16). The first stage involves the processes of energy transfer from ions and atoms of the buffer gas *A* to atoms of the additive *B*:

S.P. Melnikov et al., *Lasers with Nuclear Pumping*,
DOI 10.1007/978-3-319-08882-2_5

1. The charge-transfer process $A^+(A_2^+) + B \rightarrow (B^+)* + A(2A)$. Depending on the pressure of the buffer gas, atomic ions A^+ or molecular ions A_2^+ will take part in this reaction.
2. The Penning reaction $A* + B \rightarrow (B^+)* + A + e$. The reaction takes place in the case when the energy of the excited atom $A*$ is more than the ionization potential of the atom B.
3. The process of excitation transfer $A* + B \rightarrow B* + A$.

The upper lasing levels can be populated by means of any of these three processes. For example, the first two are fundamental in populating the lasing levels of He-Cd and He-Zn NPLs that radiate on transitions of the excited ions Cd^+ and Zn^+ (see Chap. 3, Sect. 4.4), while in an Ar-N_2 UV laser excited by electron and ion beams [6–8], the levels $C^3\Pi_u$ of the molecule N_2 are populated as a result of the last process.

The processes of recombination play an important role among plasma processes; the main ones are three-body recombination of atomic ions $A^+(B^+) + 2e(e,A) \rightarrow A*$ $(B*) + e(A)$ and dissociative recombination of molecular ions $A_2^+(B_2^+, AB^+)$ $+ e \rightarrow A*(B*) + A(B, A)$.

Lasing levels can also be populated as a result of recombination processes. For example, as a result of three-body recombination, inverse population may occur between the excited states of the ion Sr^+ ($\lambda = 430.5$ nm) [9]. When atmospheric pressure lasers are pumped by nuclear radiation and the specific power deposition is comparatively low ($q < 5 \times 10^3$ W/cm^3, concentration of electrons $n_e \leq 10^{14}$ cm^{-3}), the processes of dissociative recombination of molecular ions B_2^+ or AB^+ with electrons whose rate constants for heavy rare gases are sufficiently high—$10^{-7} - 10^{-6}$ cm^3/s [10]—are of interest. As a result of dissociative recombination, one of the atoms is formed in an excited state; the number of such states is not great, which assures selectivity of populating of the upper lasing levels. First considered in study [11], such a populating mechanism occurs in NPLs operating on transitions of the atoms Ar, Kr, Xe, Ne, and Hg.

A high rate of depopulating (or quenching) of lower lasing levels is necessary for efficient operation of lasers in CW mode. In CW gas-discharge low-pressure lasers, as a rule the depopulating of these levels is owed to radiative decay. In NPLs operating with the gas medium at atmospheric pressures, such depopulating occurs basically through inelastic collisions: with atoms of the medium (NPLs operating on transitions of Ar, Kr, and Xe atoms), with electrons of the plasma (certain metal vapor NPLs), and as a result of Penning reactions when an atom in a lower lasing state collides with atoms or molecules of an additional additive (NPLs operating on transitions $3p - 3s$ of the Ne atom). In study [12], which contains the results of investigation of A-NF_3 lasers (A = He, Ne, Ar) during excitation by an electron beam and transverse discharge, harpoon reactions $A* + NF_3 \rightarrow AF* + NF_2$ having high rate constant values up to $\sim 10^{-9}$ cm^3/s were used for depopulating of lower levels. We note that the idea of using inelastic collisions for selective depopulating of excited states of atoms was put forward by V.A. Fabrikant as far back as 1940 [13, 14].

The mechanisms for forming inverse population of levels in the main types of NPLs reviewed here are briefly described in Table 5.1.

Table 5.1 Mechanisms of formation of an inverse population in NPLs

Active medium	Lasing wavelength, μm	Populating of upper lasing levels	Depopulating of lower lasing levels
IR lasers operating on transitions nd-(n + 1)p of atoms B = Xe, Kr and Ar (n = 5,4,3 for Xe, Kr, Ar, respectively)			
He-Xe,Kr, Ar; Ar-Xe; He-Ne(Ar)-Xe; Xe, Kr	26 lines in range of 1.15–3.51	$B_2^+ + e \rightarrow B^*(nd) + B$; $AB^+ + e \rightarrow B^*(nd) + A$	Quenching in collisions with atoms *A* and *B*
Visible-range laser operating on 3p-3s transitions of Ne atom			
He-Ne-*M*; Ne-*M* (M = Ar, Kr, Xe, H_2)	0.585, 0.703, 0.725	$Ne_2^+ + e \rightarrow Ne^*(3p) + Ne$	Penning reaction: $Ne^*(3s) + M \rightarrow M^+ + e + Ne$
Metal vapor rare gas lasers			
He-Cd	Cd^+ ion (0.442, 0.534, 0.538, 0.807, 0.853)	$He^+ + Cd \rightarrow (Cd^+)^* + He$ $He_2^+ + Cd \rightarrow (Cd^+)^* + 2He$ $He^* + Cd \rightarrow (Cd^+)^* + He + e$	Radiative decay and/or collisional quenching
He-Cd	Cd atom (1.43, 1.65)	$Cd_2^+ + e \rightarrow Cd^* + Cd$	Radiative decay and/or collisional quenching
He-Zn	Zn^+ ion (0.748)	$He^+ + Zn \rightarrow (Zn^+)^* + He$ $He_2^+ + Zn \rightarrow (Zn^+)^* + 2He$ $He^* + Zn \rightarrow (Zn^+)^* + He + e$	Radiative decay
He-Hg	Hg^+ ion (0.615)	$He^+ + Hg \rightarrow (Hg^+)^* + He$	Radiative decay and/or collisional quenching
He-Xe-Hg-H_2	Hg atom (0.546)	$Hg_2^+ + e \rightarrow Hg^* + Hg$	Quenching by H_2 molecules
Lasers operating on transitions of C, N, O, and Cl atoms			
He(Ne,Ar)–CO, CO_2, N_2, O_2, Cl_2, CCl_4	C (0.833, 0.940, 1.45); N (0.859, 0.863, 0.905, 0.939); O (2.65, 2.76); Cl (1.59, 2.45)	Lasing mechanism not established. The following processes for populating upper levels were proposed: (1) Dissociative excitation: $A^* + R_2 \rightarrow R^* + R$ (A^* is a metastable rare gas atom, R_2 is a molecular gas); (2) Ion–ion recombination: $A^+(A_2^+) + R \rightarrow R^* + A(2A)$; (3) Three-body recombination $R^+ + e + M \rightarrow R^* + M$ ($M = e, A$)	
Laser operating on transitions of molecular N_2^+ ion ($B \rightarrow X$)			
He-N_2-H_2	0.391, 0.428	$He_2^+ + N_2 \rightarrow N_2^+(B) + 2He$	$N_2^+(X) + H_2 \rightarrow N_2H^+ + H$

(continued)

Table 5.1 (continued)

Active medium	Lasing wavelength, μm	Populating of upper lasing levels	Depopulating of lower lasing levels
Laser operating on vibrational transitions of CO molecule			
CO	5.1–5.6	Lasing mechanism not established. Recombination of $(CO)_2^+$ ions or clustered $CO^+(CO)_n$ ions with electrons as well as direct excitation of CO_2 molecules by plasma electrons were considered as processes of populating of upper laser levels	

As noted in Chap. 4, when gases are irradiated with nuclear particles, a plasma is formed with the recombination type of non-equilibrium, which leads to preferred populating of excited states "top-down," owing to the processes of recombination and other relaxation processes. Therefore, for NPLs it is advisable to introduce the ultimate efficiency (η_{max}) as the ratio of the energy of laser quantum $h\nu$ to the energy expenditures for formation of one active particle (ion or excited atom of buffer gas) [5]:

$$\eta_{max} = (1 + f*/f^+)h\nu/w_i = (f^+ + f*)h\nu/q, \tag{5.1}$$

where w_i is the energy cost of formation of an ion–electron pair (see Chap. 4, Sect. 4.2); f^+, $f*$ are the formation rates of ions and excited atoms when the medium is irradiated by nuclear particles (for rare gases $f*/f^+ = 0.53$ [15]).

Formula (5.1) was written for the case when both ions and excited atoms of the buffer gas take part in the formation of upper laser levels. If only ions of the buffer gas take part in populating the upper laser levels, the formula for η_{max} has a simpler form:

$$\eta_{max} = h\nu/w_i = f^+h\nu/q. \tag{5.2}$$

In contrast to the quantum efficiency of gas discharge lasers, the ultimate efficiency for NPLs does not depend on the location of the upper laser level relative to the ground state.

The calculation results of ultimate efficiencies from formula (5.1) for a number of active media are given in Table 5.2. When some of these mixtures (He-Xe, Ar-Xe, He-Ne-*M*, He-Cd, He-Zn) were used, lasing was observed during nuclear pumping, while for other mixtures lasing was obtained in the afterglow of the gas discharge [16]. The value of η_{max} grows with an increase in the energy of the laser photon or with a reduction in w_i. Naturally, the real efficiency of the nuclear-pumped laser will depend not only on η_{max} but also on many other parameters:

Table 5.2 Ultimate efficiencies of laser mixtures in the recombination plasma mode

Atom or ion	λ, nm	Mixture	η_{max}, %
Xe	2,651	He-Xe	1.6
	1,731	He-Xe	2.4
	1,731	Ar-Xe	3.8
Ne	585	He-Ne-*M*	7.2
Cd^+	442	He-Cd	9.5
	325	He-Cd	13
Zn^+	748	He-Zn	5.6
Ca^+	371, 374	He-Ca	11
		Ne-Ca	13
		Ar-Ca	18
Sr^+	417, 431	He-Sr	10
		Ar-Sr	16
Hg	546	Xe-Hg-*M*	15

the correlation of rates of relaxation plasma processes, the selectivity of the populating process of the upper laser level, and the rate of depopulation of the lower laser level. Thus, the choice of active media with a high value of η_{max} is a necessary, but not sufficient, condition for creation of efficient NPLs.

In the final stage of the relaxation processes, metastable states B^m are formed in the *A–B* mixture, which are destroyed as a result of formation of the excimer molecules B_2* and subsequent radiative decay. However, under some conditions, the process of associative ionization $B^m + B^m \rightarrow B^+ + B + e$ can become the basic channel of destruction of the states B^m. In this case, atomic ions B^+ form again, and can again participate in populating the upper laser levels, and consequently the ultimate efficiency will be greater. According to estimates [17], for the He-Ar mixture, the ultimate efficiency as a result of this recirculation may be about twice as high.

Sequential population of excited states "top-down" in a nuclear-excited plasma makes it possible to organize a cascade of laser transitions. In this case, it becomes possible to obtain simultaneous lasing in different spectral ranges and to increase the efficiency. One can propose several lasing cascade schemes [5]: (a) a cascade of laser transitions between excited levels of an atom (ion) when the lower laser level of one transitions is simultaneously the upper level of the transition below; (b) sequential lasing: first operating on transitions of the ion, and then (after recombination of the ion), operating on transitions of the atom; (c) when using multicomponent mixtures, sequential lasing on transitions of atoms (ions) of individual components.

As an example, Fig. 5.1 shows one possible scheme of cascade lasing. When the ternary mixture He-Ne-Xe is used, simultaneous lasing can occur in the visible spectral range at the transition $3p$-$3s$ of the Ne atom ($\lambda = 585.3$ nm) and in the IR range at the transitions $5d$–$6p$ of the Xe atom ($\lambda = 1.73$, 2.65, or 2.03 μm). In this

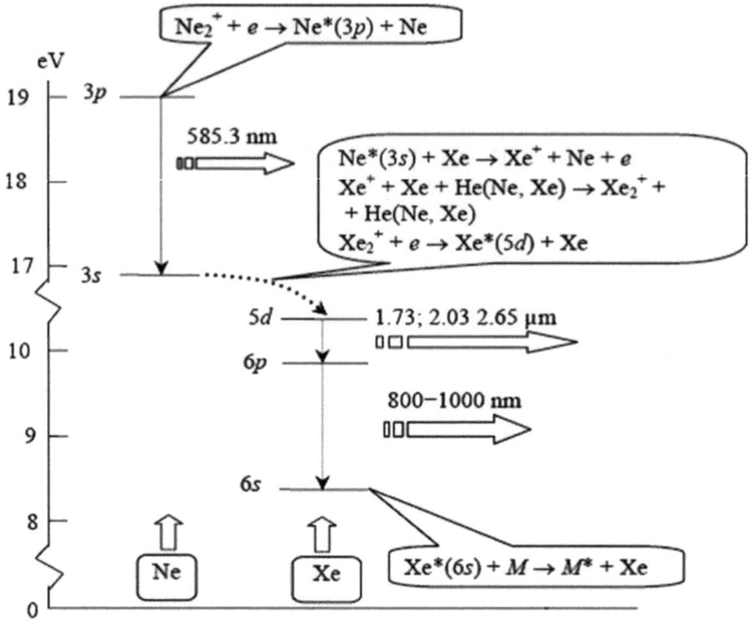

Fig. 5.1 Diagram of cascade lasing in He-Ne-Xe NPLs

case, xenon is used to depopulate the lower lasing 3*s*-level of the Ne atom and is simultaneously a laser component. But if we take the additive *M*, which efficiently quenches the long-lived 6*s* state of the Xe atom in collisions, then a cascade of laser transitions is possible between the levels 5*d*–6*p* and 6*p*–6*s* of the Xe atom. We note that in study [18], when the ternary mixture He-Ar-Xe is excited by an electron beam, one version of cascade lasing is obtained: simultaneous lasing at transitions of the Ar atom ($\lambda = 1.79$ μm) and Xe atom ($\lambda = 2.03$ μm).

5.3 Family of Lasers Operating on IR Transitions of Rare Gas Atoms

Lasers on Infra-Red (IR) transitions of Xe, Kr, and Ar atoms have the highest energy parameters among the NPLs (see Chap. 3, Sect. 3.1). Apart from high efficiency ($\eta_l \leq 2$–3 %), the chief merits of these lasers are the low laser thresholds, the full restoration of the medium after radiative influence, and the possibility of obtaining lasing over a wide range of the spectrum (1–3.5 μm). The record holder in terms of advantages is the laser operating on transitions *5d-6p* of the Xe atom ($\lambda = 1.73$, 2.03, and 2.65 μm).

From the survey of results of experimental investigations provided in Sect. 3.1 of Chap. 3, one can draw the following conclusions:

1. All the most intensive laser lines belong to the transitions nd-$(n+1)p$ of atoms Xe, Kr, and Ar ($n = 5$, 4, and 3 for Xe, Kr, and Ar, respectively). Analogous results in laser spectrums are also obtained with electron beam excitation.
2. Maximal energy parameters were registered for an Xe laser at the 1.73, 2.03, and 2.65 μm lines, which start from the common upper laser level $5d[3/2]_1{}^0$; these lines are the most powerful when using various buffer gases (He, Ar, or He-Ar and Ne-Ar mixtures).
3. For lasers operating on transitions of the Xe, Kr, and Ar atoms excited by nuclear radiation and electron beams, one observes a qualitative similarity of experimental dependencies of energy parameters on the mixtures' pressure and composition.
4. For the most intensive laser lines at excitation by nuclear radiation and electron beams, efficiencies were obtained that are close to ultimate values, which testify to the high populating selectivity of the upper *nd* levels.

Considering these circumstances, as well as the similar structure of energy levels of the Xe, Kr, and Ar atoms (Fig. 3.1), one can conclude that the basic processes of populating *nd* levels are similar or even identical, so that NPLs operating on IR transitions of Xe, Kr, and Ar atoms should be viewed as one family.

The characteristics of IR NPLs operating on transitions of rare gas atoms were investigated in detail for roughly 40 years in a wide range of experimental conditions (see Chap. 3, Sect. 3.1), and a large number of studies [19–39] were dedicated to their theoretical modeling. The processes leading to depopulating of the lower laser $(n+1)$ p levels can be considered well established: collisional quenching during collisions with atoms of the active medium and electrons (at high specific power depositions). However, to date the discussion on the mechanisms for populating the upper laser levels is not yet concluded. The variety of proposed mechanisms for populating *nd* levels (see Table 5.3) is explained by difficulties in registering and researching radiation in the IR range of the spectrum, uncertainties in the rate constants of many plasmochemical reactions, and by significant differences in the experimental conditions in which these lasers were studied.

Populating of Upper Lasing Levels

Let us examine in more detail the possible processes of populating of upper laser levels (see Table 5.3). The first models [19, 20] for NPLs operating on mixtures ^{3}He-Xe and ^{3}He-Ar were based on preferential populating of *nd* levels of Xe and Ar atoms as a result of the process of collisional-radiative recombination: $B^+ + e + e \rightarrow B^*(nd) + e$ ($B^+ = Xe^+, Ar^+$) (process (1)). Individual characteristics of lasers were calculated in [19, 20]; a detailed comparison with experimental results was not made. Process (1) was also included in kinetic models developed later for lasers

Table 5.3 Populating mechanisms of upper laser *nd* levels of atoms $B = \mathrm{Xe}$, Kr, Ar

Process No.	Process	Works cited								
		[19, 20]	[21, 22]	[23]	[24]	[25]	[26]	[27]	[28–32]	[33–39]
1	$B^+ + e + e(A) \rightarrow B^*(nd) + e(A)$ ($B^+ = \mathrm{Xe}^+, \mathrm{Ar}^+$; $A = \mathrm{He}$)	+	–	–	–	–	–	–	+	–
2	$B_2^+ + e \rightarrow B^*(nd) + B$ ($B = \mathrm{Xe}$, Kr, Ar)	–	–	–	–	–	–	+	+	+
3	$AB^+ + e \rightarrow B^*(nd) + A$ ($AB^+ = \mathrm{ArXe}^+, \mathrm{HeXe}^+, \mathrm{HeKr}^+, \mathrm{HeAr}^+$)	–	+	–	+	–	–	+	+	–
4	$(\mathrm{Xe}_2^+)^* + e \rightarrow \mathrm{Xe}^*(5d) + \mathrm{Xe}$	–	–	–	–	+	+	–	–	–
5	$\mathrm{Ar}^* + \mathrm{Xe} \rightarrow \mathrm{Xe}^*(5d) + \mathrm{Ar}$	–	–	+	–	+	–	+	+	–
6	$\mathrm{Xe}(6s, 6s') + e \rightarrow \mathrm{Xe}^*(5d) + e$	–	–	+	–	–	–	–	+	–

Note: "+" means taking this process into account in a model; $(\mathrm{Xe}_2^+)^*$ is the excited state of the Xe_2^+ ion (vibrational or electron). Processes (5) and (6) were not considered in study [31]

operating on the mixtures He-Xe [29, 30, 32] and He-Ar [31], the supposition was made that up to 60 % of the total reaction flux (1) may come to the *nd* levels. As is known (see [10], for example), process (1) is not selective, so as a result it is possible for atoms to appear in different excited states. In this regard, the assumption of a significant contribution of process (1) to populating of the *nd* levels of atoms Xe and Ar, advanced in studies [19, 20, 29–32], is doubtful. Moreover, the rate constant k_{cr} of process (1) greatly depends on the temperature of the electrons ($k_{cr} \sim T_e^{-4,5}$), so that He-Xe(Ar) lasers are hardly able to operate in a wide range of experimental conditions in the case of populating *nd* levels through this process.

In study [23], other populating mechanisms of upper laser levels were proposed for Ar-Xe NPLs: a) process (5), energy transfer in inelastic collisions $\mathrm{Ar}^* + \mathrm{Xe} \rightarrow \mathrm{Xe}^*(5d) + \mathrm{Ar}$; b) process (6), populating of *5d* levels of the Xe atom by electron impact from states *6 s*. Process (5) was also included in the kinetic models [25, 27–30, 32], and process (6) in models [28–30, 32]. Process (5) cannot provide efficient pumping of *5d* levels, since as a result of this process, highly excited Xe levels are formed initially (states *7d*, *8d*, etc.), while populating of *5d* levels is possible only as a result of subsequent radiative-collision processes with a probability of a few percent [40]. In this regard, the assumption of [28–30, 32] regarding populating of *5d* levels of the Xe atom through process (5) with a probability of 20–30 % is too optimistic. As for process (6), experimental studies [41, 42] show that there is no marked contribution of this process to populating of the *nd* levels.

The hypothesis regarding populating of *nd* levels of the Xe, Kr, and Ar atoms through dissociative recombination of heteronuclear molecular ions of ArXe^+ or

HeB^+ ($B = Xe$, Kr, Ar) with electrons is the most widespread. In a number of models (see [21, 22, 24], for example), this process is viewed as the only one, while in others [27–32] it is viewed as one of the main ones, with a populating probability of up to 40 % [28, 29]. In the majority of studies [22, 24, 27–32], it was proposed that populating of *nd* levels occurs directly as a result of process (3). Study [21] proposes for the Ar-Xe mixture a more complex two-stage mechanism of populating of the *5d* levels of the Xe atom—formation, in the process of recombination $ArXe^+ + e$, of high states *7p*, *7s* of the Xe atom, with subsequent collisional transitions $Xe^*(7p, 7s) + Ar(Xe) \rightarrow Xe^*(5d) + Ar$.

In binary *A–B* mixtures, heteronuclear molecular ions AB^+ are formed in three-body collisions:

$$B^+ + B + A \rightarrow B_2^+ + A \tag{5.3.a}$$

$$B^+ + B + A \rightarrow AB^+ + B; \tag{5.3.b}$$

$$B^+ + A + A \rightarrow AB^+ + A. \tag{5.4}$$

and are destroyed as a result of the processes:

$$AB^+ + A \rightarrow B^+ + 2A. \tag{5.5}$$

$$AB^+ + B \rightarrow B_2^+ + A. \tag{5.6}$$

$$AB^+ + \rightarrow B^+ + A + e. \tag{5.7}$$

Information about rate constants of processes (5.3)–(5.7) is extremely limited. The rate constants for these ions used in various models (see, for example, information for the $ArXe^+$ in Table 4.11) are estimates, sometimes quite rough, so that it is virtually impossible to predict the contribution of dissociative recombination of ions AB^+ to population of *nd* levels. Unknown constants most often were estimated in the process of fitting the results of calculation of NPL characteristics to the experimental data.

The destruction of heteronuclear ions largely depends on their bond energy, which decreases with an increase in the difference in masses of the atoms making up the ion [43, 44], and equals 0.14, 0.050, 0.030, and 0.027 eV for $ArXe^+$, $HeXe^+$, $HeKr^+$, and $HeAr^+$, respectively [44]. The estimates cited in study [10] provide the following values for the rate constants of process (5.5): for $ArXe^+$ ions, $k_{5.5} \approx 5 \times 10^{-11}$ cm³/s, while for the HeB^+ ions ($B = Xe$, Kr, Ar), $k_{5.5} \geq 10^{-10}$ cm³/s. Heteronuclear ions are also destroyed by plasma electrons as a result of process (5.7); moreover, the rate constants of such processes can reach 10^{-7}–10^{-6} cm³/s [10].

The characteristic time of process (5.5), for example for $HeXe^+$ ions when the atmospheric pressure of the He-Xe mixture is $\sim 3 \times 10^{-10}$ s, while the characteristic time of the process of dissociative recombination of this ion under the most favorable conditions ($T_e = 300$ °K, $n_e \leq 10^{15}$ cm^{-3}) will be $\geq 10^{-7}$ s. Consequently, heteronuclear ions HeB^+ are effectively destroyed as a result of collisional processes, so their recombination cannot make a marked contribution to formation of excited atoms. The latter circumstance is confirmed by the results of spectroscopic research on dissociative recombination of heteronuclear ions in the afterglow of gas-discharge plasma, in which "despite the large number of studies of these particles, it has not yet been possible to discover the formation of excited atoms associated with their recombination" [10].

An additional argument against a significant contribution from dissociative recombination of heteronuclear ions to the population of the upper laser levels comes from the results of experiments studying the influence of the medium temperature on laser characteristics. It follows from Table 3.7 (see Chap. 3, Sect. 3.1) that the temperature at which the output laser power decreases by half differs insignificantly for the mixtures Ar-Xe, He-Xe, and He-Ar, and is 350–450 °K. If populating of the upper laser levels occurred with the participation of the heteronuclear ions $ArXe^+$ (mixture Ar-Xe) or $HeXe^+$, $HeAr^+$ (mixtures He-Xe, He-Ar), for the mixtures He-Xe and He-Ar, the reduction in energy parameters would have to occur at substantially lower temperatures than for the mixture Ar-Xe, owing to the more efficient destruction of $HeXe^+$ and $HeAr^+$ ions as the temperature increases due to reaction (5.5). In addition, in experiments [45, 46] with mixtures He-Xe and He-Ar, the specific energy depositions were ~ 1 J/cm^3 per pulse, and the temperature of the medium by the end of the pumping pulse reached 800–1,000 °K, but lasing occurred throughout the entire pulping pulse (see Fig. 2.10, for example).

Thus construction of the models [22, 24] based on the reaction of dissociative recombination of ions $\mathrm{He}B^+$ with electrons, or its insertion in models [29–32], where the contribution of this reaction to populating of *nd* levels is estimated at 15–25 % [31], is invalid.

In the majority of kinetic models, the process of dissociative recombination of molecular ions $B_2{}^+$ with electrons was viewed as a loss channel, populating the lower laser $(n+1)\ p$ levels. This assumption was initially based on the results of spectroscopic investigations (see review [10] and the literature cited there), which registered intensive radiation from $(n+1)\ p$ levels. However, it should be noted that spectroscopic investigations were carried out in the spectral range of $\lambda \le 1{,}000$ nm using photoelectric multipliers for registration, so IR transitions *nd*-$(n+1)\ p$ could not be observed. Consequently, one can assume that the radiation from $(n+1)$ p levels registered in these studies is the consequence of preliminary IR transitions *nd*-$(n+1)\ p$.

In study [47], based on the measured intensities of spectral lines, a conclusion was drawn regarding preferential population as a result of the process $\mathrm{Xe}_2^+ + e$ of level $6p[5/2]_2$ in comparison with level $5d[3/2]_1{}^0$ which are the lower and upper levels of the laser transition respectively, with $\lambda = 1.73$ μm (see Fig. 3.1). The authors [47] drew this conclusion from calculation of the ratio of the recombination fluxes Γ, populating these levels:

$$\gamma = \frac{\Gamma_{5d}}{\Gamma_{6p}} = \frac{I_{5d}A_{6p}\lambda_{5d}\tau_{6p}}{I_{6p}A_{5p}\lambda_{6p}\tau_{5d}}, \quad (5.8)$$

where I_{5d} and I_{6p} are the intensities of the 1.73 μm and 0.992 μm spectral lines measured in [47], λ_{5d} and λ_{6p} are the wavelengths of these lines, A_{5d} and A_{6p} are the probabilities of radiative transitions, and τ_{5d} and τ_{6p} are the lifetimes of the levels. In [47] it is assumed that τ_{5d} is equal to the radiative lifetimes of the level 5*d* [3/2] 1^0, which is about 200 ns. With these assumptions, the authors of [47] obtained $\gamma \leq 0.2$ for $P_{\mathrm{Xe}} = 5$–10 Torr.

Study [48] reevaluated correlation (5.8), taking into account the processes of collisional quenching. From results of experiments [49], it follows that quenching of the level 5*d* [3/2]1^0 by Xe atoms in ground state already become noticeable at $P_{\mathrm{Xe}} \approx 0.1$ Torr. This means that for $P_{\mathrm{Xe}} \approx 0.1$ Torr, the radiative lifetimes of level $5d[3/2]_1{}^0$ are roughly equal to the lifetimes of this level through the process of collisional quenching. Hence, it follows that the rate constant of the quenching process is ~ 2×10^{-10} cm^3/s^{-1} (similar values of this constant were obtained in kinetic models [21, 25]). Therefore, for $P_{\mathrm{Xe}} = 10$ Torr, taking into account collisional quenching, τ_{5d} ~ 10 ns, and accordingly, the ratio γ ~ 5. Thus if one allows for the reduction in the lifetimes of the level $5d[3/2]_1{}^0$, then based on the spectral lines measured in [47], one can infer the opposite to the conclusion drawn by the authors [47]: the level $5d[3/2]_1{}^0$ is populated primarily due to the dissociative recombination $\mathrm{Xe}_2^+ + e$.

This inference is confirmed by the information of study [50], which concludes, based on experimental investigations of the process $\mathrm{Xe}_2^+ + e$, that there was preferential populating of the *5d* level of the Xe atom in comparison with the *6p* levels. A significant role of the process $\mathrm{Xe}_2^+ + e$ in populating levels *5d* is also noted in study [51]. And finally, analysis [52] of the kinetics of an Ar-Xe laser ($\lambda = 1.73$ μm) showed that the probability of populating of the level 5*d* $[3/2]_1{}^0$ through the process $\mathrm{Xe}_2^+ + e$ can reach 90 %.

In 1979, VNIIEF offered the hypothesis of selective populating of *nd* levels of Xe, Kr, and Ar atoms through dissociative recombination of molecular ions $B_2^+ + e \rightarrow B^*(nd) + B (B = \mathrm{Xe, Kr, Ar})$, published later in [33]. Based on this hypothesis, kinetic models were developed for calculation of NPL characteristics [34–39]. Populating of *nd* levels of Xe and Ar atoms through the processes of dissociative recombination of molecular ions B_2^+ with electrons was also taken into account (along with other processes) in kinetic models [27–32]. The efficiency of populating of *nd* levels was taken as equal to 15–20 % with regard to the total pumping flux.

A somewhat different lasing mechanism, proposed in the study [25] for Ar-Xe NPLs and later used in a kinetic model of a He-Xe NPL [26], was based also on preferential populating of the *5d* levels of the Xe atom as a result of dissociative recombination of molecular ions Xe_2^+. However, in this case dissociative recombination of the ion $(\mathrm{Xe}_2^+)^*$ in an excited state (electronic or vibrational) was proposed as the basic populating channel. According to this model, the excited $(\mathrm{Xe}_2^+)^*$ ions are formed in the reaction $ArXe^{+}(HeXe^{+}) + Xe \rightarrow (Xe_2^{+})^* + Ar(He)$, while the Xe_2^{+} in the ground state are a result of the reaction $Xe^{+} + Xe + Ar(He) \rightarrow Xe_2^{+} + Ar(He)$. Authors of [25, 26] proposed that recombination of the Xe_2^{+} ion in the ground state leads to populating of the levels $6p$ and $6p'$. To test the reliability of this model, which contains a large number of unknown constants, additional research is necessary, particularly calculations of the characteristics of a Xe laser in various experimental conditions.

Thus the most efficient populating of the *nd* levels of Xe, Kr, and Ar atoms, which are the upper laser levels in rare gas mixture NPLs, most likely occurs through the processes of dissociative recombination $B_2^{+} + e \rightarrow B^*(nd) + B$ ($B =$ Xe, Kr, Ar). Unfortunately, direct experimental proof of such inference is still absent, except perhaps for the information contained in study [50]. To resolve this problem, spectral-luminescence studies are needed, analogous to those performed using the single-photon spectrometry method in the visible and UV spectral ranges [53, 54]. However, IR photodetectors of the same high sensitivity as photoelectronic multipliers are needed in order to carry out such investigations.

Formation of the Laser Radiation Spectrum

In quasi-CW NPLs operating on transitions nd-$(n+1)\,p$ of Xe, Kr and Ar atoms, the depopulating of the lower $(n+1)\,p$ levels occurs as a result of collisional quenching in collisions with atoms. The efficiency of quenching, which depends on the pressure and composition of the gas mixtures, significantly influences not only the energy characteristics of the NPLs, but also the laser spectrum. The laser characteristics can be influenced not only by the quenching processes of the *nd* and $(n+1)\,p$ levels by atoms of the medium, but also by the quenching and mixing of these levels by plasma electrons, the influence of the latter being especially appreciable for $q \geq 10^3$ W/cm^3.

Let us examine the spectral characteristics of the most studied NPLs operating on transitions of the Xe atom (see Fig. 3.1a). Experiments showed that the laser spectra depend not only on the parameters of the cavity mirrors and the specific power depositions, but also on the type of buffer gas, the pressure, and the ratio of mix components. The list of laser lines and transitions for the different mixtures compiled on the basis of Sect. 3.1 in Chap. 3 is given in Table 5.4. The spectra of the Xe laser in experiments with electron beams were basically similar [57]; in addition, the 3.43 μm laser line, belonging to the transition $7p\,[5/2]_2$-$7s[3/2]_1{}^0$ [24], was observed.

The level $5d\,[3/2]_1{}^0$, from which the most intensive laser lines 1.73; 2.03 and 2.65 μm begin, is a resonance level. The radiation time of the isolated Xe atom in this state is equal to 0.58 ns [55], but with taking into account the imprisonment of the resonance radiation, about 200 ns [36].

Table 5.4 Laser transitions in xenon NPLs

Laser transition	λ, μm	$\tau_j{}^r$, ns	$\tau_i{}^r$, ns	$A_{ji} \times 10^6$, s^{-1}	Mixes in which lasing was observed
$5d[3/2]_1{}^0$–$6p[5/2]_2$	1.73	0.58	33.0	0.30 [55]; 0.48 [56]	**He-Xe, Ne-Xe, Ar-Xe, He-Ar-Xe, Ne-Ar-Xe**
$5d[3/2]_1{}^0$–$6p[3/2]_1$	2.03	0.58	29.9	2.46 [55]; 2.4 [56]	**He-Xe**, Ne-Xe, Ar-Xe, **He-Ar-Xe, Ne-Ar-Xe**
$5d[5/2]_3{}^0$–$6p[5/2]_3$	2.48	565	25.5	0.48 [55]	He-Xe, Ar-Xe
$5d[5/2]_2{}^0$–$6p[5/2]_2$	2.63	667	33.0	0.74 [55]; 0.62 [56]	Ar-Xe, **Kr-Xe**
$5d[3/2]_1{}^0$–$6p[1/2]_0$	2.65	0.58	22.4	1.27 [55]; 1.6 [56]	**He-Xe, Ne-Xe, Ar-Xe, He-Ar-Xe, Ne-Ar-Xe**
$5d[5/2]_2{}^0$–$6p[5/2]_3$	2.81	667	25.5	0.034 [55]	Kr-Xe
$5d[5/2]_2{}^0$–$6p[3/2]_1$	3.37	667	29.9	0.68 [55]; 0.59 [56]	Ar-Xe
$5d[7/2]_3{}^0$–$6p[5/2]_2$	3.51	1,190	33.0	0.74 [55]; 0.61 [56]	**Xe, He-Xe**
$7p[1/2]_1$-$7s[3/2]_2{}^0$	3.65	500	49.5	1.8 [55]	**He-Xe**

Note: Bold print shows the media in which these laser lines were predominant. A_{ji} is the probability of a radiative transition [55, 56]; τ_j and τ_i are the radiative lifetimes of the upper and lower laser levels [55]

From the data shown in Tables 3.1, 3.4, and 3.5 (see Chap. 3, Sect. 3.1), it follows that when a He-Xe mixture is used, the 1.73, 2.03, 2.65, and 3.65 μm lines are the most intensive. In the Ar-Xe mixture, lasing occurs primarily at the 1.73 and 2.65 μm lines, in the mixture Kr-Xe at the 2.63 μm line, and in pure Xe at the 3.51 μm line. All of the laser lines noted above (except for 3.65 μm) belong to the *5d*-*6p* transitions. In the Ne-Xe mixture, low-power lasing was observed at the 1.73, 2.03, and 2.65 μm lines [58].

According to the model proposed in study [33], regardless of the type of buffer gas, the level $5d[3/2]_1{}^0$ of the Xe atom is populated initially; it is the upper level for the three most intensive lasing lines: 1.73, 2.03, and 2.65 μm. Populating of this level occurs selectively with an efficiency close to 100 %, through the dissociative recombination process $Xe_2{}^+ + e$. The absence of 1.73, 2.03, and 2.65 μm lines in the Kr-Xe mixture and pure Xe is explained by the high rates of collisional quenching of the $5d[3/2]_1{}^0$ level by Kr and Xe atoms.

The weaker 2.48, 2.63, 3.37, and 3.51 μm laser lines are also present in He-Xe, Ar-Xe, and Kr-Xe mixtures and pure Xe; these begin with the levels $5d[5/2]_3{}^0$, $5d[5/2]_2{}^0$, and $5d[7/2]_3{}^0$.

Populating of these levels occurs as a result of collisional intra-multiplet transitions during collisions with He, Ar, Kr, and Xe atoms in ground states. In the He-Xe mixture, the 3.65 μm laser line, belonging to one of the *7p*-*7s* transitions, was also observed [59]. It is possible that the appearance of this line is related to populating

Table 5.5 Rate constants of processes of collisional quenching of 6*p* levels of the Xe atom (in units of 10^{-11} cm^3/s)

	Quenching atom				
Level	Xe	Kr	Ar	Ne	He
$6p[1/2]_0$	0.58 [60]; 0.59 [61]	11.0 [63]	14 [60]; 20 [61]	3.4 [64]	2.0 [64]
$6p[3/2]_2$	8.2 [60]; 10.1 [61]; 8.7 [62]	2.2 [63]	4.7 [60]; 4.0 [61]; 3.7 [62]	0.019 [62]; 0.05 [64]	0.17 [62]; 0.17 [64]
$6p[3/2]_1$	12.8 [60]; 11.5 [62]	1.2 [63]	0.2 [62]; 1.0 [63]	0.033[62]; 0.039 [64]	7.4 [62]; 7.5 [64]
$6p[5/2]_3$	5.3 [60]; 7.6 [62]	4.0 [63]	5.1 [62]; 2.5 [63]	0.22 [62]; 0.3 [64]	1.0 [62]; 0.7 [64]
$6p[5/2]_2$	9.6 [60]; 11.6 [61]; 10.1 [62]	4.5 [63]	8.2 [60]; 8.2 [61]; 8.0 [62]; 8.6 [63]	0.57[62]; 0.80 [64]	0.95 [62]; 0.90 [64]
$6p[1/2]_1$	13.3 [60]; 18.1 [62]	4.0 [63]	0.74 [62]; 0.6 [63]	0.035 [62]; 0.3 [64]	4.0 [62]; ≤0.2 [64]

of the 7*p* level as a result of the processes of collisional-radiative (three-body) recombination and subsequent cascade transitions. Indeed, experimental data [59] show that the optimal concentration of Xe for this line is much lower than for the 2.03 and 3.51 μm lines. The absence of the 3.65 μm line in mixtures with heavy buffer gases (Ar, Kr) may be explained by the fact that in this case, processes of the three-body recombination were suppressed because of the high electron temperature.

There are no reliable data in the literature on the rate constants for processes of collisional quenching of the 5*d* levels of the Xe atom, which makes analysis of experimental results and calculation of laser characteristics difficult. Such information exists for lower 6*p* levels (Table 5.5).

Laser spectra depend on the rates of processes of collisional quenching of lower 6*p* levels. From a comparison of the NPL energy characteristics shown in Sect. 3.1 of Chap. 3 and the data of Table 5.5, it is clear that the most powerful lasing is observed for lines with high rate constants of the processes of collisional quenching of lower 6*p* levels: the 2.03 and 2.65 μm lines for the He-Xe mixture and 1.73 and 2.65 μm lines for the Ar-Xe mixture. The addition of small amounts of He to the Ar-Xe mixture, which is of little consequence on the specific power deposition, leads to the appearance of powerful lasing at the 2.03 μm line and elimination of competition of the 1.73 and 2.03 μm lines (see Fig. 3.2), which may be explained by efficient quenching of the $6p[3/2]_1$ level by He atoms.

In the Ne-Xe mixture, almost all of the 6*p* levels were quenched with inadequate efficiency, so the energy characteristics for NPLs operating on this mixture are not great. As was shown in study [58], small additions of He or Ar to the Ne-Xe mixture led to a sharp increase in output power at the 2.03 and 2.65 μm lines.

Kinetic Models

In order to calculate the characteristics of any laser, it is necessary to know the populating rates of the upper and lower laser levels, R_j and R_i; the lifetimes of these levels τ_j and τ_i, with allowance for the different processes of their collisional quenching; the transition probability A_{ji}; and the line width $\Delta\nu_c$. The rates of populating R_j and R_i are determined from solving the system of kinetic equations which constitute a balance of the rates of formation and decay of individual components of the plasma (see Chap. 4, Sect. 4.3), with allowance for the processes of populating of the laser levels. If the above parameters of gas laser media are known, it is possible to determine the small-signal gain α_0, saturation intensity I_s, and output laser intensity I_{out} (for example, see [35, 36]):

$$\alpha_0 = \frac{\lambda_{ji}^2 A_{ji}}{4\pi^2 \Delta\nu_c}\left(R_j\tau_j - R_i\tau_i\frac{g_j}{g_i}\right), \quad I_s = \frac{4\pi^2 hc\Delta\nu_c}{\lambda_{ji}^3 A_{ji}\left(\tau_j + \tau_i g_j/g_i\right)},$$

$$I_{out} = I_s\left(\alpha_0 l - \beta l - \ln\sqrt{1/r_1 r_2}\right)\frac{\ln\sqrt{1/r_1 r_2}}{\beta l + \ln\sqrt{1/r_1 r_2}}$$

where g_j and g_i are degeneracies of the upper and lower laser levels, h is Plank constant, c is light velocity, β is distributed losses, $\Delta\nu_c$ is line width, / is length of active medium, and r_1 and r_2 are reflectivities of cavity mirrors.

For NPLs operating on the IR transitions of Xe, Kr, and Ar atoms, the poor accuracy in determining the rate constants of a number of basic plasma processes is a serious problem when computing R_j and R_i. In some cases, such constants from different literature sources differ a few times, and sometimes they differ by an order of magnitude. Approximately the same difficulties take place for A_{ji}, τ_j, τ_i, and $\Delta\nu_c$. Let us examine the problem in more detail using the example of a xenon NPL.

Transition Probability A_{ji} Virtually all kinetic models of xenon NPLs use calculated data on the probabilities of radiative transitions 5*d*-6*p* of the Xe atom (see Table 5.4). In a recently published study [65], an attempt was made to determine A_{ji} from measured intensities of spontaneous radiation for the 1.73, 2.03, and 2.65 μm lines during excitation of pure xenon ($P_{Xe} = 5$ Torr) by a nanosecond electron beam. The precision in determining A_{ji} was not great—values of A_{ji} for the 1.73, 2.03, and 2.65 μm lines were equal to 4×10^5, 3.4×10^6, and 7.9×10^5 s^{-1}, respectively.

Lifetimes τ_j *and* τ_i *of Laser Levels* To determine τ_j and τ_i, it is necessary to know the radiative lifetimes of the levels (see Table 5.4) and the rate constants of the processes of collisional quenching. The rate constants of the quenching processes of the lower 6*p* levels by different atoms are rather well known (see Table 5.5). For the upper 5*d* levels, such data are limited and frequently are estimates obtained as a result of numerical modeling. Table 5.6 provides information for the $5d[3/2]_1^0$ level, obtained on the basis of the best match of calculation results with experimental data [21, 23, 25, 32, 36] or from analysis of plasma processes in the

Table 5.6 Rate constants for the processes of collisional quenching of the $5d[3/2]_1{}^0$ level of the Xe atom by rare gas atoms

Quenching atom	He	Ne	Ar	Xe
Rate constant, 10^{-11} cm^3/s (calculated estimate)	0.01 [32]; 0.2 [36];1.5 [66]	–	0.1 [21]; 0.25 [23]; < 0.1 [25]; 0.1 [32]; 3 [67]	30 [21]; 10 [25]; 4 [32]; 20 [36]; 1 [66]
Rate constant, 10^{-11} cm^3/s (measurement)	–	<0.06 [51]	~ 0.1 [51]	5.5 ± 0.5 [51]

Note: Revised data of study [36] are shown, which were obtained using more reliable data on $\Delta\nu_c$

afterglow of a gas-discharge plasma [66, 67]. Study [51] was probably the only experimental study that determined a number of rate constants for quenching processes of the $5d[3/2]_1{}^0$ level by Xe, Ne, and Ar atoms and by certain molecules (N_2, CO, H_2, etc.), based on measurements of the intensity of resonance vacuum UV line 119.2 nm (transition $5d[3/2]_1{}^0 \rightarrow {}^1S_0$). The results of this research for Xe, Ne, and Ar atoms are also shown in Table 5.6.

At high specific power depositions ($q \geq 1$ kW/cm^3), when the concentration of electrons $n_e \geq 10^{13}$ cm^{-3}, there may be a reduction in the lifetime and change in the populations of the excited states owing to the processes of their quenching and mixing in collisions with plasma electrons. The rate constants of such processes with the participation of excited states of the Xe atom (6*s*, 6*p*, 5*d*, etc.), obtained by calculation, are contained in studies [32, 68], for example.

Line Width $\Delta\nu_c$ For high-pressure gas lasers (including gas NPLs), the broadening of the luminescent line is homogeneous and due primarily to collisions with gas atoms. Calculations of the parameters of xenon NPLs were carried out using values of $\Delta\nu_c$ obtained as a result of selection in the course of numerical modeling of laser characteristics [29], from calculations using the Lennard-Jones potential [69], or from the results of measurement [70, 71] (Table 5.7).

Then let us briefly examine some of the most well-developed kinetic models of NPLs operating on IR transitions of Xe, Kr, and Ar atoms, which were tested from experimental results obtained in the pumping of active media both by nuclear radiation and fast electron beams.

VNIIEF Kinetic Models VNIIEF developed "small" stationary models [33–39] to determine the lasing mechanisms of NPLs based on IR transitions of Xe, Kr, and Ar atoms and preliminary calculations of laser characteristics. As demonstrated by analysis of kinetic plasma processes, for an adequate description of the operation of lasers operating on binary mixtures He-*B* (*B* = Xe, Kr, or Ar) or Ar-Xe, it suffices to include 10–15 basic reactions in the model, which were selected beforehand as a result of analysis of the characteristic times of plasma processes (see Tables 4.10 and 4.11).

It was presumed that populating of upper laser *nd* levels of Xe, Kr, and Ar atoms occurs selectively, solely through reactions of dissociative recombination of molecular ions B_2^+ with electrons. The percentage of recombination flux populating these

Table 5.7 Collisional broadening of the 1.73, 2.03, and 2.65 μm lines of the Xe atom by He, He, Ar, and Xe atoms (half-width of the luminescent line $\Delta\nu_c$, GHz/atm)

	Quenching atom			
λ, μm	He	Ne	Ar	Xe
1.73	48 [29]; 15 [70]	1.0 [70]	48 [29]; 72 [69]; 15 [70]; 15 [71]	80 [29]
2.03	65 [29]	–	65 [29]; 63 [69]; 75 [71]	90 [29]
2.65	38 [29]	–	82 [29]; 43 [69]	43 [29]

levels was determined in the process of modeling, and was taken as 100, 50, and 40 % respectively for the Xe_2^+, Kr_2^+, and Ar_2^+ ions. Apart from the reactions cited in Tables 4.10 and 4.11, the calculation models included the processes of collisional quenching of *nd* levels. For level $5d[3/2]_1{}^0$ of the Xe atom, the constants of the processes of collisional "quenching" by Xe, Ar, and He atoms according to the revised data are 2×10^{-10}, 1×10^{-13}, and 2×10^{-12} cm^3/s, respectively.

The validity of the model was tested by calculations of the energy and threshold characteristics of NPLs using the mixtures He-Xe ($\lambda = 2.65$ μm), Ar-Xe ($\lambda = 1.73$ μm) [34–37], He-Kr ($\lambda = 2.52$ μm), and He-Ar ($\lambda = 1.79$ μm) [38, 39], excited by uranium fission fragments, products of the nuclear reaction $^3He(n,p)^3H$, and electron beams as a function of pressure, mixture composition, and power deposition. The calculated curves for energy and threshold characteristics agree satisfactorily with experimental data. Table 5.8 provides a comparison of the results of calculations and experimental data in optimal modes with respect to pressure and composition.

Kinetic Models of the University of Illinois Initially the model was developed for an Ar-Xe laser ($\lambda = 1.73$, 2.03, 2.63, 2.65 μm) and included around 100 plasmochemical reactions [21]. To test the model, data obtained from the pumping of Ar-Xe laser with electron beam [69, 78] and uranium fission fragments [79] were used. Later on this model was improved and was used to calculate the characteristics of NPLs using mixtures of He (Ne)-Ar-Xe ($\lambda = 1.73$, 2.03, 2.63, 3.37 μm) excited by uranium fission fragments [80, 81].

A rather complex two-stage process was proposed as the main mechanism for populating 5*d* levels of the Xe atom. First, as a result of dissociative recombination $ArXe^+ + e$, the states 7*p* and 7*s* of the Xe atom are populated, and then Xe atoms in the states $5d[3/2]_1{}^0$ and $5d[5/2]_2{}^0$ are formed in collisional processes Xe*(7*p*,7 *s*) + Ar(Xe). In these models it is assumed that level $5d\ [3/2]_1{}^0$ is populated during collisions with Ar atoms with an efficiency of 70 %, and level $5d\ [5/2]_2{}^0$—with Xe atoms with an efficiency of 50 % [21].

In these models, much attention is attended to processes of the so-called collisional mixing of levels 6*s*, 6*p*, and 5*d* by plasma electrons, which reduce the population inversion of laser transitions. Calculations have shown that such processes lead to a cessation of lasing if the degree of plasma ionization exceeds some critical value $(0.8–1.0) \times 10^{-5}$ [82]. At atmospheric pressure, this corresponds to an electron density of about 2×10^{14} cm^{-3}. In the opinion of the authors [21, 82], it is

Table 5.8 A comparison of calculated and experimental parameters of rare gas mixture lasers

Atom	Mixture	λ, μm	W^e_{out}, W	η_l^e, %	$\Phi_{th}^e \times 10^{-14}$, $cm^{-2} \times s^{-1}$	W^c, W	η_l^c,%	$\Phi_{th}^c \times 10^{-14}$, $cm^{-2} \times s^{-1}$	Pumping method
Xe	He-Xe	2.65	2,000	0.6	–	2,200	0.65	–	Fission fragments [45]
	He-Xe	2.65	300	1.5	1.0	260	1.2	1.2	Fission fragments [72]
	Ar-Xe	1.73	500	2.2	0.17	700	2.2	0.15	Fission fragments [73]
	^{3}He-Xe	2.65	40	0.05	–	42	–	–	$^3He(n,p)^3H$ [74]
	He-Xe	2.03	1.9×10^6	–	–	2.1×10^6	2.1	–	Electron beam [24, 75]
	Ar-Xe	1.73	8.3×10^6	1.5	–	1.3×10^7	2.5	–	Electron beam [24, 75]
Kr	He-Kr	2.52	110	0.6	2.7	105	0.52	2.9	Fission fragments [76]
Ar	He-Ar	1.79	390	0.6	3.2	450	0.44	3.5	Fission fragments [77]

Note: W^c, η_l^c, Φ_{th}^c are calculated parameters; W^e, η_l^e, Φ_{th}^e are experimental parameters

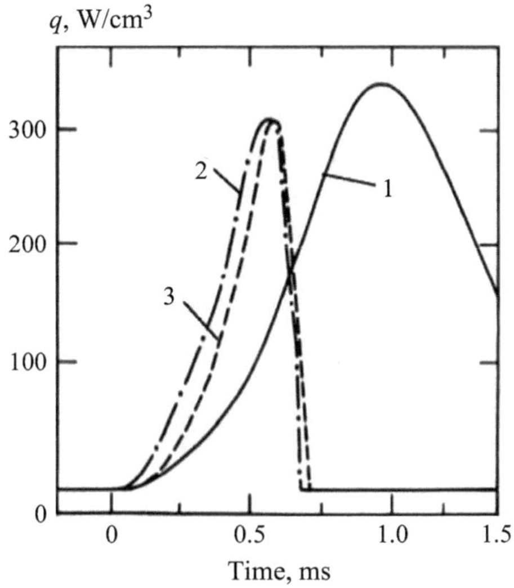

Fig. 5.2 (1) Specific power deposition (1) and output power at the 1.73 μm line (2, 3) at excitation of an Ar-Xe mixture (0.68 atm; Ar:Xe = 99.5:0.5) by uranium fission fragments; (2) experiment [79]; (3) calculation [21]

these processes that cause the early cessation of lasing observed in experiments [79] (Fig. 5.2).

Associates of the University of Illinois also examined the lasing mechanism of NPLs operating on the transition of an Ar atom with $\lambda = 1.79$ μm and carried out calculations of some of its characteristics [22]. It was proposed that the main process of populating of the upper laser level $3d[1/2]_1$ of the Ar atom in the mixture He-Ar is dissociative recombination of heteronuclear ions $HeAr^+$ with electrons, while the processes of collisional destruction of $HeAr^+$ ions in collisions with atoms were not included in the model. Earlier it was shown that such a populating mechanism is most likely mistaken, because of the low bond energy of the $HeAr^+$ ion (0.027 eV [44]), which roughly coincides with the energy of thermal motion of atoms at room temperature. Therefore, $HeAr^+$ ions will be effectively destroyed in the collisions (5.5) and (5.6).

IOFAN Kinetic Models The non-stationary models of xenon lasers [27–30, 32] developed at IOFAN are the most detailed. Their most recent modifications [29, 30, 32] include several hundred plasmochemical reactions with the participation of the ions He^+, He_2^+, He_3^+, Ar^+, Ar_2^+, $ArXe^+$, Xe^+, Xe_2^+, the excited atomic and molecular states He*, He_2*, Ar*, Ar**, ArXe*, Ar_2*, Xe_2*, and all of the excited states $5d$, $6p$, $6s$, and $6s'$ of the Xe* atom. The numerical modeling was carried out for the mixtures He-Xe, Ar-Xe, and He-Ar-Xe, and pure Xe. As the authors of [32] state, they were able to implement complete optimization of laser characteristics of when simultaneous lasing on six lines belonging to the transitions $5d$-$6p$ occurred.

Calculations under the conditions of pumping of active media with an electron beam showed that for all laser lines, the highest energy characteristics are observed for the Ar-Xe mixture, while the maximal efficiencies (4.5 %) and specific energy release (19 J/l) are achieved at the 1.73 μm line. The intervals of the parameters at which such high energy characteristics can be obtained are determined.

As was shown in Table 5.3, the authors of these models assume that virtually all plasma processes ever discussed participate in populating the upper laser levels. Thus, to populate the level $5d[3/2]_1{}^0$, 40 % and 15 % of the fluxes of dissociative recombination of molecular ions, respectively, $ArXe^+$ and Xe_2^+, 20 % of excited atoms formed as a result of the $Ar^* + Xe$ reaction, and about 40 % of the reaction of three-body recombination of the atomic Xe^+ ions are expended.

IOFAN associates also modeled an argon laser at the 1.27, 1.79, and 2.40 μm lines [31]. The model was tested with experimental data on pumping a He-Ar mixture laser ($\lambda = 1.79$ μm) with electron beams (FIAN) and uranium fission fragments (VNIIEF, VNIITF, Sandia). In the opinion of the authors of [31], populating of the upper laser level occurs by means of two processes: dissociative recombination of $HeAr^+$ ions (15–25 % of the total pumping flux) and Ar_2^+ ions (5–20 % of the total pumping flux). The model also considered processes associated with the presence of the impurities N_2, O_2, H_2, and H_2O in the He-Ar mixture, which makes it possible to more precisely consider the real experimental conditions. The inclusion of these impurities in the model was necessary to the authors of [31], in order to explain the behavior of experimental curves [77] in the region of partial pressures of argon ≤ 5 Torr.

5.4 Lasers Based on 3*p*-3*s* Transitions of the Ne Atom

Visible-range NPLs ($\lambda = 585.3$, 703.2, and 724.5 nm) on 3*p*-3*s* transitions of the Ne atom were studied at VNIIEF, VNIITF, and the Sandia Laboratories (see Chap. 3, Sect. 3.2). In contrast to NPLs operating on the nd-$(n+1)p$ transitions of Xe, Kr, and Ar atoms, in neon NPLs the main processes leading to the formation of an inverse population of levels were established rather reliably [4, 83–92]. Already in the first studies of neon lasers excited by nuclear radiation and electron beams there are nearly identical notions of the lasing mechanism, which may be explained by the extensive spectroscopic information about the processes of populating and relaxation of 3*p* and 3*s* states of the Ne atom (see [10], for example). In particular, populating of the level $3p'[1/2]_0$ owing to dissociative recombination of molecular ions Ne_2^+ with electrons was proven earlier in study [93], where additional Doppler broadening of the 583.3-nm line in the afterglow phase was observed, and the suggestion of using the Penning reaction to depopulate the lower laser states was demonstrated as early as 1970 (for example, see [1]). Because the lower 3*s* states of the Ne atom are depopulated in the Penning reactions

$$\text{Ne} * (3s) + M \rightarrow M^{+} + e + \text{Ne}, \tag{5.9}$$

sometimes such lasers are called Penning lasers. When nuclear pumping was used, the admixtures $M = \text{Ar}$, Kr, Xe, and H_2 were used to depopulate the $3s$ levels.

During the development of theoretical models, fundamental attention was paid to the transition $3p'[1/2]_0$-$3s'[1/2]_1{}^0$ of the Ne atom ($\lambda = 585.3$ nm), where lasing was most efficient when using the ternary mixture He-Ne-M (a diagram of Ne atom levels with lasing transitions is shown in Fig. 3.3). Formation of an inverse population in the He-Ne-M mixture occurs as a result of the following sequence of basic plasma processes:

$$\text{He} \xrightarrow[\text{particle}]{\text{nuclear}} \text{He}^{+} \xrightarrow{2\text{He}} \text{He}_2^{+} \xrightarrow{\text{Ne}} \text{Ne}^{+} \xrightarrow{\text{Ne, He}} \text{Ne}_2^{+} \xrightarrow{e}$$

$$\xrightarrow{e} \text{Ne} * \left(3p'[1/2]_0\right) \xrightarrow{h\nu(585{,}3)} \text{Ne} * \left(3s'[1/2]_1^0\right) \xrightarrow{M} M^{+} + \text{Ne} + e.$$

In this sequence there are three stages at which process competition takes place that influences the population of the upper lasing level:

(a) Competition of charge-transfer processes

$$\text{He}_2^{+} + \text{Ne} \rightarrow \text{Ne}^{+} + 2\text{He}, \tag{5.10}$$

$$\text{He}_2^{+} + \text{M} \rightarrow M^{+} + 2\text{He}; \tag{5.11}$$

(b) Competition of dissociative recombination

$$\text{Ne}_2^{+} + \text{e} \rightarrow \text{Ne} * (3p) + \text{Ne} \tag{5.12}$$

and the charge-transfer process

$$\text{Ne}_2^{+} + M \rightarrow M^{+} + 2\text{Ne}; \tag{5.13}$$

(c) Competition of radiative transition at the 585.3-nm line and the quenching processes of level $3p'[1/2]0$ owing to the Penning reaction, as well as in collisions with He and Ne atoms. Therefore the populating efficiency of level $3p'[1/2]_0$ will depend on the type of quenching admixtures M, the pressure and composition of the mixture, and the electron concentration (specific power deposition).

If a rare gas ($M = \text{Ar}$, Kr, or Xe) is used as the quenching admixture, then in the second phase the losses are not significant, because the processes (5.13) for these atoms have low rate constants ($\leq 10^{-13}$ cm^3/s [94]). Replacement of Ar(Kr,Xe) by H_2 leads to a marked competition of the processes in the second phase, because the rate constant of the process (5.13) for H_2 molecules is quite high—4.2×10^{-11} cm^3/s [95]. Therefore in NPLs based on $3p$-$3s$ transitions of the Ne atom, it is expedient to use hydrogen as the quenching admixture only at high specific power depositions

($q \geq 1$ kW/cm^3) [96]. We note that in the model [90, 91], the rate constant of process (5.13) with participation of the H_2 molecule, determined during the process of numerical modeling, proved to be three orders of magnitude less (1.3×10^{-13} cm^3/s). This low value contradicts the above data of [95] and the results of measurement [84] of the luminescence intensity at $\lambda = 585.3$ nm in the He-Ne-Ar mixture as a function of the concentration of H_2. As the data in [84] show, for $q \approx 1$ kW/cm^3 a twofold reduction in intensity (reduction in population of the upper lasing level by a factor of 2) occurs at H_2 pressure of about 1.5 Torr, while the influence of H_2 is determined precisely by process (5.13). The rate constant of this process, estimated from the conditions of the experiment [84], is no less than 5×10^{-11} cm^3/s, which agrees with the data of study [95].

In studies [89, 90], the models included the ternary charge-transfer reaction, which for $M = \text{Ar}$ has the view:

$$\text{Ne}_2^+ + \text{Ar} + R \rightarrow \text{Ar}^+ + 2\text{Ne} + R, \tag{5.14}$$

where R is the third particle (in this case, He, Ne, or Ar). The rate constant of this reaction in model [90] for $R = \text{Ne}$ was taken as equal to 5×10^{-32} cm^6/s. The model [89] used a significantly higher value of this constant (3.5×10^{-30} cm^6/s), which agrees with the data of [97]. The question of the need to include process (5.14) in the models remains open for now because the results of luminescence investigations [84] testify to an insignificant influence of this process on the concentration of Ne_2^+ ions.

To calculate lasing characteristics, it is necessary to determine the populations of the 3*p* levels of the Ne atom. To this end, usually a system of kinetic equations is solved which constitutes a balance of the rates of populating and relaxation processes for each of ten 3*p* levels. Of the basic processes, it is necessary to allow for populating of these levels through reaction (5.12) and the collisional intra-multiplet transitions between them, the radiative and collisional relaxation of the 3*p* levels during collisions with the He and Ne atoms, as well as the Penning reaction for each of these levels with the participation of *M* atoms.

Reaction (5.12) is the main populating process of 3*p* levels, and in most of the studies it was assumed that all of these levels are populated during recombination of molecular ions Ne_2^+ in the ground vibrational state. A somewhat different mechanism was considered in studies [87, 88], where it was assumed that level $3p'[1/2]_0$ (upper among the 3*p* levels) is populated with the participation of vibrationally excited ions Ne_2^+. The percentage of dissociative recombination flux of (5.12) reaching level $3p'[1/2]_0$, according to various information, is around 7 % at a low neon pressure [10], 3–17 % when neon pressure is reduced from 760 to 10 Torr [87, 88], and 7–10 % for 20–100 Torr [98]. In ternary mixtures He-Ne-*M* with a large content of helium, the efficiency of populating level $3p'[1/2]_0$ is increased [87, 88], which the authors explain by the formation of vibrationally excited ions Ne_2^+ in the processes:

$$Ne + 2He \rightarrow HeNe^+ + He, \quad (5.15)$$

$$HeNe^+ + Ne \rightarrow Ne_2^+(\upsilon > 0) + He. \quad (5.16)$$

The relationship between the processes of dissociative recombination of rare gas molecular ions with the kinetics of populating of their vibrational levels was studied in the afterglow of a pulsed gas discharge (see [10], for example).

Lasing was also observed in neon NPLs at the 703.2- and 724.5-nm lines, which start from level $3p[1/2]_1$. This level is effectively populated both through process (5.12), and as a result of intramultiplet collisional transitions from the higher 3*p*-states.

One more mechanism of populating of 3*p* levels, proposed in study [53] and based on the investigation of the kinetics of stopping fission fragments in neon and a He-Ne mixture, consists of direct excitation of these levels by nuclear particles. However, as was demonstrated by experimental data [89], for He-Ne-Ar and Ne-Ar mixtures excited by an electron beam with a short duration of 3 ns, pulses of luminescent and laser radiation appear with a delay of up to 200 ns with respect to the pumping pulse, while its value is inversely proportional to the helium pressure. This delay can arise only in the case when laser states are populated not as a result of direct excitation, but owing to subsequent plasma processes. The authors [89] explain the onset of delay to electron "cooling" processes to temperatures at which the rate of the recombination process (5.12) becomes significant and sufficient to achieve the laser threshold.

The population of 3*p* levels of the Ne atom is markedly influenced by collisional intra- and intermultiplet transitions during collisions with Ne [53, 99, 100] and He [101, 102] atoms. It is noted in studies [99, 100] that inter-multiplet transitions predominate for the $3p[1/2]_1$, which is the lowest among all the 3*p* states. At high specific power depositions, the processes of collisional relaxation of 3*p* levels during collisions with plasma electrons are added to these processes. For neon NPLs, the latter are less significant than, for example, for NPLs based on transitions 5*d*-6*p* of the Xe atom. In study [89] the conclusion was drawn that the effect of collisional mixing of the levels by electrons in a neon laser is not significant up to electron concentrations of 6×10^{15} cm^{-3}.

Populating of the lower lasing level $3s'[1/2]_1{}^0$ occurs through radiative and collisional intermultiplet transitions from 3*p* states. Depopulating of level $3s'[1/2]_1{}^0$ and three other 3*s* levels is carried out mainly through the Penning reaction (5.9), as well as through the reactions of associative ionization, intra-multiplet relaxation, and three-body processes with the formation of excimer molecules Ne_2*. The processes of depopulating of 3*s* levels were considered most fully in studies [103, 104]. Table 5.9 shows the rate constants of Penning processes for all four 3*s* levels. We note for comparison that the rate constants of Penning processes for the atoms Ne*($3p'[1/2]_0$) with the participation of Ar and H_2 are 5.3×10^{-11} and 4.6×10^{-11} cm^3/s [105], respectively.

Table 5.9 Rate constants of Penning processes (in units of 10^{-10} cm^3/s)

Level	Ar	Kr	Xe	H_2
$3s'[1/2]_1^0$ ($^1P_1$)	29 [100]	27 [100]	36 [100]	1.0 (estimate from [90])
$3s'[1/2]_0^0$ ($^3P_0$)	1.05 [103]; 1.1 [104]	1.07 [103]; 1.13 [104]	1.43 [103]; 1.1 [104]	0.65 [104]
$3s[3/2]_1^0$ ($^3P_1$)	0.96 [103]; 1.6 [104]	0.56 [103]; 1.1 [104]	1.34 [103]; 1.7 [104]	0.70 [104]
$3s[3/2]_2^0$ ($^3P_2$)	1.35 [103]; 0.97 [104]	1.35 [103]; 0.75 [104]	1.35 [103]; 0.9 [104]	0.65 [104]

The most detailed kinetic models for NPLs based on $3p$-$3s$ transitions of the Ne atom [88–90] include up to 450 plasmochemical reactions [90]. For example, model [90] considered the atoms, molecules, and ions He*, Ne*, Ar*, Ar**, Ne_2*, Ne_2**, He_2*, $HeNe^*$, Ar_2*, He^+, Ne^+, Ar^+, He_2^+, Ne_2^+, Ar_2^+, He_3^+, Ne_3^+, Ar_3^+, and $HeNe^+$, as well as the kinetics of population and relaxation of individual levels belonging to the groups of states Ne*($3s$,$3s'$), Ne*($3p$,$3p'$), Ne*($4s$), and Ne ($5s$). The calculations were primarily carried out for the transition $3p'[1/2]_0$- $3s'[1/2]_1^0$ of the Ne atom ($\lambda = 585.3$ nm). The kinetics of the neon NPL at the 703.2- and 724.5-nm lines (Ne-Kr mixture) were considered, evidently, only in studies [87, 88]. Experimental data obtained in pumping of a neon laser with nuclear radiation and electron beams were used for testing of the models.

The main differences of the models lie not in the quantity of plasma processes included in them, but in the use or absence of certain ones (for example, reaction (5.14)), differences in the rate constants for a number of important processes, and the probabilities of population of $3p$ levels through the reaction of dissociative recombination (5.12). One of the basic results of numerical modeling was the conclusion that the maximal efficiency of neon NPLs is no more than 0.5 % [90, 91].

5.5 Lasers Based on Rare Gas Mixtures with Metal Vapors

Mixtures of rare gases with metal vapors are of interest for NPLs in connection with the possibility of obtaining lasing in the visible and UV ranges, while the ultimate efficiency of such lasers can be 10–20 % because of the high energy of the laser quantum and/or the possibility of using Ar, Kr, and even Xe as the buffer gases, with low energy costs of formation of ion–electron pairs. In addition, populating of upper laser levels of ions of certain metals can occur even at the first phase of relaxation processes, directly as a result of charge-transfer and Penning reactions. This circumstance makes it possible to lower the energy losses in comparison with schemes in which the populating of laser levels is implemented through longer sequences of plasma processes.

Table 5.10 Optimal calculated parameters of NPLs operating on metal ion transitions [106]

Mixture	λ, nm	*P*, atm	q_{opt} kW/cm³	Optimal temperature of medium, °K	$\Phi_{th} \times 10^{-15}$, $cm^{-2} \times s^{-1}$	η_l, %
He-Cd	325.0	2–3	2	660–700	20	0.2
	441.6	3–5	2	660–690	0.5	0.7
	533.7; 537.8	1–1.5	2.5	750–800	1	0.13
He-Zn	610.3	2–4	10	930–1,020	3	0.4
	747.9	3–5	2	750–790	1	0.08
	758.8	~2	2	830–900	8	<0.1
He-Hg	615.0	1–2	2	500–550	7	0.02
He-Mg	448.1	2–3	1	1,000–1,020	1	0.2
	921.8	1–2	0.3	1,000	2	0.03
He-Sr	430.5	3–4	2.5	950	40	0.15
He-Ca	373.7	2–4	10	950	20	0.11
He-Ba	490.1	2–4	5	1,150–1,200	50	0.02

Note: q_{opt} is the optimal specific power deposition

This section briefly examines the laser mechanisms and kinetic models both of operating NPLs and of certain active media on which lasing has not yet been obtained. One can become more closely acquainted with these questions in the survey and original studies cited below. Nearly all of the theoretical studies were dedicated to the kinetics of the active media of NPLs based on the transitions of ions and atoms of group II metals on the Periodic Table. Among these studies, the first of which was monograph [1], one can distinguish a cycle of theoretical investigations by IOFAN associates (see reviews [4, 106]), dedicated to analysis and calculations of the kinetics of active media of NPLs based on the transitions of Cd, Zn, Hg, Mg, Sr, Ca, Ba and Be ions. The kinetic models took into account from 100 to 160 plasmochemical reactions. The basic results of the calculations performed for the experimental conditions of the pulsed EBR-L reactor (see Chap. 2, Sect. 2.5) are shown in Table 5.10.

Among the metal vapor NPLs, active media based on mixtures of helium with Cg, Zn, and Hg vapors have elicited the greatest interest.

Laser Media Based on Cd and Zn Vapors

The largest number of studies were dedicated to analysis of the lasing mechanisms and calculations of the characteristics of NPLs using a He-Cd mixture radiating at transitions of the Cd^+ ion ($\lambda = 441.6$, 533.7, and 537.8 nm) [4, 106–112]. A diagram of levels of the Cd ion and atom with laser transitions is shown in Fig. 3.5 (see Chap. 3, Sect. 3.3).

At present one can consider the basic populating processes of upper laser levels to be established:

- Levels $4f^2F^0{}_{5/2,7/2,}$ which are the upper levels for transitions with $\lambda = 533.7$ and 537.8 nm, are populated by means of charge-transfer processes

$$He^+ + Cd \rightarrow (Cd^+)* + He \quad (5.17)$$

with the formation of higher levels $6f$, $6\,g$, $8d$, $9\,s$, etc. [113] and subsequent cascade transitions in $4f$ states;

- Levels $5s^{22}D_{5/2,3/2}$, which are upper levels for transitions with $\lambda = 325.0$ and 441.6 nm, are populated by means of charge-transfer and Penning processes

$$He_2{}^+ + Cd \rightarrow (Cd^+)* + 2He \quad (5.18)$$

$$He * \left(2^{1.3}S\right) + Cd \rightarrow (Cd^+)* + He + e \quad (5.19)$$

As a result of processes (5.17)–(5.19), various states of the Cd^+ ion can arise. In a number of studies, for example [108, 109], it was proposed that by means of these processes, all energy-allowed states are populated with a probability proportional to the statistical weight of the given state. This approximation is inapplicable to charge-transfer processes, the probability of which is maximal if the potential energy of the starting ion, with an accuracy up to kT_g, is equal to the excitation energy of the final state of the Cd^+ ion. At present there is some information about partial rate constants of processes (5.17)–(5.19) for various states of the Cd^+ [110–115] ion. Study [112], based on examination of the populating kinetics of individual levels of the Cd^+ ion, concluded that the contribution of the charge-transfer reaction (5.18) to populating of levels $5s^{22}D_{5/2,3/2}$ and $6s^2S_{1/2}$ is roughly an order of magnitude greater than the Penning reaction (5.19).

Relaxation or quenching of laser levels can occur as a result of the ion conversion process

$$(Cd^+)* + Cd + He \rightarrow Cd_2{}^+ + He, \quad (5.20)$$

of radiative transitions, inelastic collisions with electrons, and (for levels above state $6s^2S_{1/2}$), the Penning process on its own atom

$$(Cd^+)* + Cd \rightarrow Cd^+ + Cd^+ + e. \quad (5.21)$$

Reaction (5.21) was not studied experimentally. In models [4, 106], its cross-section was assumed to be equal to the cross-section of polarization capture.

Among the transitions of the Cd^+ ion, much attention was given to the transition with $\lambda = 325.0$ nm, which is similar to the transition with $\lambda = 441.6$ nm [112, 116] in terms of its luminescence and other parameters. However, lasing under the conditions of nuclear pumping was obtained only in the second of them [116, 117], although for pumping with electron beams, lasing was observed at the 325.0-nm line both in the pulsed mode [118] and in the CW mode [119, 120], while the minimal specific power deposition at the laser threshold was just 10 W/cm^3 [120]. The reasons for the lack of lasing at the 325.0-nm line in nuclear pumping conditions have not as yet been finally ascertained. Possible reasons that have been considered include absorption of radiation at the 325.0-nm line by excited atoms Cd*($^3P_2$), quenching of level $5s^{22}D_{3/2}$ by helium atoms and plasma electrons and the influence of uncontrolled impurities [107, 110, 117, 120].

Neutralization of plasma in the He-Cd mixture occurs basically by means of dissociative recombination processes of Cd_2^+ molecular ions with electrons. As a result of recombination processes, the formation of an inverse population is possible between certain excited levels of the Cd atom. In experiments, lasing was registered at the lines 1.43 and 1.65 μm [116] and 361.0 nm [118] with pumping of the He-Cd mixture by uranium fission fragments [116] and an electron beam [118]. The kinetic model [107] included processes with the involvement of excited Cd* atoms, and an attempt was made to calculate some laser characteristics for the 1.65 μm line. In the opinion of the authors [107], the two-humped laser pulse observed at the 1.65 μm line may be explained by quenching of the upper laser level by electrons.

Differences in the results of calculations for various models for NPLs based on a He-Cd mixture are generally explained by the shortage of data on rate constants of many important processes. The maximal computed efficiency was obtained at the 441.6-nm line and is ≤ 1 % [106, 111].

Lasing during pumping of a He-Zn mixture by uranium fission fragments was observed at transition $4s^{22}D_{5/2}$-$4p^2P_{3/2}$ ($\lambda = 747.9$ nm) (see Table 3.8), and during pumping by an electron beam, it was observed at transition $5d^2D_{5/2}$-$5p^2P_{3/2}$ ($\lambda = 610.2$ nm) [121]. Lasing mechanisms of NPLs based on the transitions of Zn^+ and Cd^+ ions largely coincide. A kinetic model of NPLs based on a He-Zn mixture was examined in studies [4, 106, 121].

Just as for the He-Cd mixture laser, the main processes of populating the levels $4s^{22}D_{5/2,3/2}$ of the Zn^+ ion are the charge-transfer and Penning reactions of He_2^+ ions and metastable atoms He*($2^{1.3}S$) during interaction with Zn atoms, while for the higher states, it is the process of charge-transfer of He^+ ions on Zn atoms. Depopulating of lower laser levels with $\lambda = 747.9$ nm occurs due to radiative transitions, and that of the transition with $\lambda = 610.2$ nm by means of the Penning reaction on its own atom $(Zn^+)^* + Zn \rightarrow Zn^+ + Zn^+ + e$ and collisional quenching by electrons. The results of calculation of characteristics of He-Zn mixture NPLs in optimal modes are given in Table 5.10.

Hg Vapor Lasers

The mechanism of a laser operating on the transition $7p^2P_{3/2}$-$7s^2S_{1/2}$ of the Hg^+ ion ($\lambda = 615.0$ nm) was considered in studies [4, 106, 122]. Populating of the upper laser level $7p^2P_{3/2}$ occurs as a result of the charge-transfer process $He^+ + Hg \rightarrow (Hg^+)^* + He$. For low concentrations of Hg atoms, populating of the level $7p^2P_{3/2}$ can also occur by means of the Penning process with the participation of excited mercury atoms: $He^* + Hg^* \rightarrow (Hg^+)^* + He + e$ [122]. The basic quenching channels of the upper laser level are the Penning reaction on its own atom $(Hg^+)^* + Hg \rightarrow Hg^+ + Hg^+ + e$ and collisions with electrons [4, 106]. Calculations in [4, 106] show that at the 615.0-nm line, even in optimal modes, the energy laser characteristics are not great (see Table 5.10).

Significantly higher output energy parameters were obtained for NPLs operating on the transition 7^3S_1- $6^3P_2{}^0$ of the Hg atom Hg ($\lambda = 546.1$ nm) when the mixture He-Xe-Hg-H_2 was used (see Chap. 3, Sect. 3.3). The lasing mechanism and the possibility of pumping this laser with nuclear radiation were examined in the studies [122, 123] before lasing was obtained in experiments with the EBR-L pulsed reactor [116].

The most detailed kinetic models of the mercury laser at the 546.1-nm line are presented in studies [88, 124, 125]. The main populating channel of level 7^3S_1 is dissociative recombination $Hg_2^+ + e \rightarrow Hg^*(7^3S_1) + Hg$, and the populating efficiency of level 7^3S_1 by means of this process may reach 80 % [123, 124]. Formation of Hg_2^+ ions occurs in a sequence of processes: $Xe_2^+ + Hg \rightarrow Hg^+ + 2Xe$, $Hg^+ + Xe + M \rightarrow XeHg^+ + M$, $XeHg^+ + Hg \rightarrow Hg_2^+ + Xe$ (Xe is the buffer gas, M is a third particle). The presence of helium in the active medium leads to a reduction in the electron temperature and an increase in the rate of dissociative recombination of Hg_2^+ molecular ions with electrons. The use of an additional admixture (in this case H_2) for selective quenching of the lower levels $6^3P_2{}^0$ is a specific feature of this laser. We note that even before the appearance of the first lasers, the experiments of [14] were carried out on quenching of the levels $6^3P^0{}_{0,1,2}$ of the Hg atom during collisions with H_2 molecules, in which amplification was obtained at all lines of the mercury triplet: 546.1, 435.8, and 404.7 nm.

The lasing mechanisms of NPLs based on metal vapors, in which the necessary concentration of vapors (~ 1 Torr) was created by means of thermal evaporation of metals inside the laser cell, were examined above. A number of studies by MIFI associates (see [126, 127], for example), examine a different lasing mechanism, based on the formation of excited ions or atoms directly as a result of bombarding the metallic layer with charged particles. In the opinion of the authors [126, 127], this "atomization" mechanism of laser operation, which does not require a buffer gas, and forms excited particles directly in the upper laser states, makes it possible to increase laser efficiency by a factor of five to seven. To confirm this hypothesis, studies [126, 127] cite data that was obtained in experiments with the VIR-2 M pulsed reactor at irradiation by a neutron flux of two laser cells that have thin cadmium layers deposited to the walls and are filled with ^3He at a pressure of

1.1 atm. As the authors of [126, 127] state, the output laser power at the 441.6-nm line was ≤ 3 mW, with anomalously low temperatures of laser cells of 460 and 510 °K. At such temperatures, the concentration of cadmium vapors due to thermal evaporation was very low, $\leq 2 \times 10^{-4}$ Torr.

Based on a very limited number of experiments, one cannot draw an unequivocal conclusion regarding the presence of lasing effect in studies [126, 127], because signals from the photodetectors were too low and scarcely exceeded the level of electrical and radiation noise. If one assumes that lasing did indeed occur, then as the calculations in [128] show, the results of this experiment can be entirely explained based on the traditional kinetic model, if one allows for the effect of atomization of the cadmium layer by products of the nuclear reaction $^3He(n,p)^3H$ and subsequent injection of atomized cadmium atoms into the active medium.

5.6 Lasers Based on Transitions of C, N, O, and Cl Atoms

Lasers based on transitions of C, N, O, and Cl atoms (see Chap. 3, Sect. 3.4) have certain common features. First of all, in the active media of these lasers, consisting of the buffer gases He, Ne, and Ar, and the molecular additives CO, CO_2, N_2, O_2, Cl_2, CCl_4, etc., lasing occurs at IR transitions of the atoms making up the belonging to molecules. Second, the optimal concentrations of molecular additives are very low, ~ 10^{-2} Torr, which is comparable with the concentration of molecular impurities in highly pure rare gases. We note that lasing was earlier observed at many transitions of the C, N, O, Cl atoms in low-pressure gas-discharge lasers [129].

Lasing mechanisms of NPLs based on C, N, O, and Cl atoms have not as yet been finally established owing to the variety of plasma processes in active media which contain a large number of neutral particles (atoms, molecules, radicals), various types of positive and negative ions, as well as to the lack of information regarding the rate constants of many of them. The processes that are suggested as populating for laser levels include dissociative excitation during collision of a metastable atom of buffer gas with molecules [130, 131], collisional-radiative recombination of atomic ions [131, 132], as well as recombination of positive and negative ions [133].

Despite the sizeable lack of information noted above, IOFAN associates proposed quite detailed kinetic models for NPLs based on transitions of Cl [134], C, N [135], and O [136] atoms. It was proposed that the basic channels of populating of levels in lasers operating on C, N, and O atoms are the reactions of three-body recombination $C^+(N^+,O^+) + e + M \rightarrow C^*(N^*,O^*) + M$ (M is buffer gas atoms or electrons). In a laser operating on transitions of the Cl atom, along with the three-body recombination reactions, processes of dissociative recombination of molecular ions Cl_2^+ and CCl_2^+, and ion–ion recombination of an excited ion $(CCl_2^+)^*$ with a negative ion Cl_2^- were suggested. Unknown constants were determined during adjustment of calculation results to experimental data.

5.7 Molecular Lasers

CO Laser

The lasing mechanism of NPLs based on vibrational-rotational transitions of the CO molecule ($\lambda = 5.1$–5.6 μm) is yet to be determined. One possible channel for populating upper laser levels is the recombination process $(CO)_2{}^+ + e \rightarrow CO^*(\nu) + CO$ [137, 138]. The $(CO)_2{}^+$ ions are formed in three-body processes $CO^+ + 2CO \rightarrow (CO)_2{}^+ + CO$. However, in study [139] this mechanism of populating laser levels is called into doubt due to the lack of a time correlation between the recombination rate of $(CO)_2{}^+$ ions and the output laser power.

Excitation of the vibration levels of the CO molecule can occur by means of collisions between CO molecules with plasma electrons. In study [140], based on calculation of the electron spectrum formed in molecular gas under the effects of ionizing radiation, it was shown that the efficiency of a nuclear-pumped CO laser cannot exceed 0.5 %, and consequently carbon monoxide is not a promising medium for NPLs.

In a later study [141], dissociative recombination of the cluster ions $CO^+(CO)_n$ with formation of electronic-excited molecules CO* and subsequent collisions of these molecules with CO molecules in the ground state was proposed as the main mechanism for formation of $CO^*(\nu)$ molecules.

$N_2{}^+$ Laser

The most detailed kinetic model of NPLs based on the transitions $B \rightarrow X$ of the molecular ion N_2^+ ($\lambda = 391.4$; 427.8 nm) was presented in study [142]. The model includes 52 plasmochemical reactions among 19 plasma components. Populating of the upper laser level for a N_2^+ laser occurs by means of the charge-transfer reaction $He_2^+ + N_2 \rightarrow N_2^+\ (B) + 2He$, while depopulating of the lower X levels is as a result of the reaction $N_2^+\ (X) + H_2 \rightarrow N_2H^+ + H$.

As a result of calculation of the small-signal gain as a function of the pressure and composition of the He-N_2- H_2 mixture, it was concluded that it was possible to achieve lasing at $\lambda = 391.4$ and 427.8 nm with $q = 1$–3 kW × cm^{-3}, which is entirely attainable in experiments with EBR-L and SPR III fast pulsed reactors. Later on, in experiments with the EBR-L reactor, such lasing was indeed registered (see Chap. 3, Sect. 3.5).

Other Media using Molecule Transitions

Some studies examined lasing mechanisms and calculated the characteristics for active media that until recently did not lead to the making of NPLs, for example, active media using transitions of the 1st negative system of the CO^+ molecule ($\lambda =$ 219 nm) [143, 144]. From calculations of the characteristics of a CO^+ laser it follows that in the mixture He-CO-H_2, (hydrogen is used to depopulate the lower laser levels) for $q = 1$–2 kW/cm^3, it is possible to obtain quasi-CW lasing at $\lambda = 219$ nm with rather high efficiencies: $\eta_l = 3$–4 % [143]; 0.7 % [144]. It should be noted that in the experiments of [145] on pumping a He-CO mixture at atmospheric pressure with an electron beam, stimulated radiation was registered at transitions of the CO^+ molecule with $\lambda = 247$ nm (1st negative system) and $\lambda = 395$ and 421 nm (Baldet-Johnson system).

Authors of model [144] discussed the possibility of appearance lasing on the line of the O_2^+ ion with $\lambda = 160$ nm during excitation of the He-O_2-H_2 mixture by fission fragments and an electron beam. We could not establish the affiliation of this line with an electronic-vibrational transition of the O_2^+ ion (see [146], for example). As noted in study [147], as a result of the charge-transfer process $He_2^+ + O_2$, there is selective populating of the state $b^4\Sigma_g{}^-$ of the O_2^+ ion, from which the transitions of the first negative system of oxygen (500–850 nm) begin.

References

1. L.I. Gudzenko, S.I. Yakovlenko, *Plasma Lasers (Plazemennyye lazery)* (Atomizdat, Moscow, 1978)
2. V.L. Bychkov, A.V. Yeletskiy, High-pressure beam plasma (Puchkovaya plazma vysokogo davleniya). In *Chemistry of Plasma (Khimiya plazmy)*, Moscow: Energoatomizdat. (12), pp. 119–158, (1985)
3. B.M. Smirnov, *Ions and Excited Atoms in Plasma (Iony i vozbuzhdennyye atomy v plazme)* (Atomizdat, Moscow, 1974)
4. A.M. Boychenko, V.I. Derzhiyev, A.G. Zhidkov, A.V. Karelin, A.V. Koval, O.V. Sereda, S.I. Yakovlenko, *kinetic models of certain plasma lasers pumped by a hard ionizer* (Kineticheskiye modeli nekotorykh plazmennykh lazerov, nakachivayemykh zhestkim ionizatorom), Proceedings of IOFAN (Trudy IOFAN), 1989, vol. 21, pp. 44–115
5. S.P. Melnikov, A.A. Sinyanskiy, *On the ultimate efficiency of nuclear-pumped gas lasers* (O predelnom KPD gazovykh lazerov s yadernoy nakachkoy), Proceedings of the Industry Conference The Physics of Nuclear-Excited Plasma and the Problems of Nuclear-Pumped Lasers, Obninsk, 1993, vol. 2, pp. 133–148
6. S.K. Searles, G.A. Hart, Laser Emission at 3577 and 3805 Å in Electron-Beam Pumped Ar-N_2 Mixtures. Appl. Phys. Lett. **25**(1), 79–82 (1974)
7. V.M. Bystritskiy, A.N. Didenko, A.V. Kozhevnikov, Y.E. Krasik, A.M. Prokhorov, S.S. Sulakshin, An efficient Ar-N_2-laser pumped by a high-current ion beam (Effektivnyy Ar-N_2-lazer s nakachkoy silnotochnym ionnym puchkom). Kvantovaya elektronika **7**(9), 2006–2008 (1980)
8. B.M. Berkeliyev, V.A. Dolgikh, I.G. Rudoy, A.E. Sevastyanov, A.M. Soroka, Laser based on a 2^+ system of nitrogen at excitation powers up to 20 kW/cm^3 (Lazer na 2^+- sisteme azota pri

moshchnostyakh vozbuzhdeniya do 20 kW/cm^3). Kvantovaya elektronika **17**(9), 1135–1139 (1990)
9. S.M. Babenko, S.I. Yakovlenko, Possibility of nuclear pumping of a He-Sr laser. Phys. Lett., **76A**(3, 4), 237–239, (1980)
10. V.A. Ivanov, Dissociative recombination of molecular ions in an rare gas plasma (Dissotsiativnaya rekombinatsiya molekulyarnykh ionov v plazme inertnykh gazov), *Uspekhi fizicheskikh nauk*. **162**(1), 35–70, (1992)
11. B.M. Smirnov, Molecular ion gas laser (Gazovyy lazer na molekulyarnykh ionakh), *Doklady AN SSSR*. **183**(3), 554–557, (1968)
12. M.I. Lomayev, D.Y. Nagornyy, V.F. Tarasenko, A.V. Fedenev, G.V. Kirillin, Lasing on atomic transitions of rare gases in mixtures with NF_3 (Generatsiya na atomarnykh perekhodakh inertnykh gazov v smesyakh s NF_3). Kvantovaya elektronika **16**(10), 2053–2056 (1989)
13. V.A. Fabrikant, Mechanism of gas-discharge radiation (Mekhanizm izlucheniya gazovogo razryada). In *Electron and Ion Instruments (Proceedings of VEI)* (*Elektronnyye i ionnyye pribory (Trudy VEI)*), (41), 236–239, (1940)
14. F.A. Butayeva, V.A. Fabrikant, On a medium with a negative absorption coefficient (O srede s otritsatelnym koeffitsientom pogloshcheniya). In *Investigations into Experimental and Theoretical Physics* (Issledovaniya po eksperimentalnoy i teoreticheskoy fizike), Moscow: Izd. AN SSSR, pp. 62–70, (1959)
15. D.B. Rees, C.B. Leffert, D.J. Rose, Electron Density in Mixed Gas Plasmas Generated by Fission Fragments. J. Appl. Phys. **40**(4), 1884–1896 (1969)
16. I.G. Ivanov, Y.L. Latush, M.F. Sem, *Metal Vapor Ion Lasers (Ionnyye lazery na parakh metallov)* (Energoatomizdat, Moscow, 1990)
17. B.M. Berkeliyev, V.A. Dolgikh, I.G. Rudoy, A.M. Soroka, On the ultimate efficiency of lasers based on *d-p* transitions of heavy rare atoms (O predelnoy effektivnosti lazerov na *d-p* perekhodakh tyazhelykh inertnykh atomov). Kvantovaya elektronika **17**(12), 1537–1538 (1990)
18. B.M. Berkeliyev, V.A. Dolgikh, I.G. Rudoy, A.M. Soroka, Simultaneous efficient lasing in the near IR range of the spectrum on *d-p* transitions of Ar and Xe (Odnovremennnaya effektivnaya generatsiya v blizheney IK oblasti spectra na *d-p* perekhodakh Ar i Xe). Pisma v ZhTF **17**(21), 80–82 (1991)
19. J.E. Deese, H.A. Hassan, Direct nuclear pumping by a volume source of fission fragments. A. I. A. A. J. **16**(10), 1030–1034 (1978)
20. J.W. Wilson, R.J. DeYoung, W.L. Harries, Nuclear-Pumped ^{3}He-Ar Laser Modeling. J. Appl. Phys. **50**(3), 1226–1235 (1979)
21. M. Ohwa, T.J. Moratz, M.J. Kushner, Excitation mechanism of the electron-beam-pumped atomic xenon ($5d \rightarrow 6p$) laser in Ar/Xe mixtures. J. Appl. Phys. **66**(11), 5131–5145 (1989)
22. J.W. Shon, M.J. Kushner, excitation mechanism and gain modeling of the high-pressure atomic Ar laser in He/Ar mixture. J. Appl. Phys. **75**(4), 1883–1890 (1994)
23. K.S. Klopovskiy, A.V. Lukyanova, A.T. Rakhimov, N.V. Suetin, Numerical modeling of an atomic xenon laser (Chislennoye modelirovaniye lazera na atomarnom ksenone). Kvantovaya elektronika **16**(2), 205–211 (1989)
24. N.G. Basov, V.A. Danilychev, A.Y. Dudin, D.A. Zayarnyy, N.N. Ustinovskiy, I.V. Kholin, A.Y. Chugunov, Electroionization IR laser operating on Xe atoms (Elektroionizatsionnyy IK lazer na atomakh Xe). Kvantovaya elektronika **11**(9), 1722–1736 (1984)
25. I.V. Novobrantsev, V.R. *Solovyev, Kinetic processes in the active medium of a nuclear-pumped Ar-Xe laser* (Kineticheskiye protsessy v aktivnoy srede Ar-Xe lazera s yadernoy nakachkoy), Proceedings of the Specialist Conference The Physics of Nuclear-Excited Plasma and the Problems of Nuclear-Pumped Lasers, Obninsk, 1992, vol. 1, pp. 327–336
26. K.F. Grebenkin, A.L. Zherebtsov, *Excitation mechanisms of a recombination nuclear-pumped He-Xe laser* (Mekhanizmy vozbuzhdeniya rekombinatsionnogo He-Xe lazera s yadernoy nakachkoy), Proceedings of the 2nd International Conference Physics of Nuclear-

Excited Plasma and Problems of Nuclear-Pumped Lasers, Arzamas-16, 1995, vol. 1, pp. 184–192

27. V.I. Derzhiyev, A.G. Zhidkov, O.V. Sereda, S.I. Yakovlenko, Modeling of a multiwave Ar-Xe laser (Modelirovaniye mnogovolnovogo Ar-Xe-lazera). Proc IOFAN (Trudy IOFAN) **21**, 139–141 (1989)
28. A.V. Karelin, On the excitation mechanism of ir transitions of the xenon atom in a nuclear-pumped laser (O mekhanizme vozbuzhdeniya IK perekhodov atoma ksenona v lazere s yadernoy nakachkoy). Kvantovaya elektronika **25**(7), 619–621 (1998)
29. A.V. Karelin, O.V. Simakova, The kinetics of the active medium of a multiwave IR xenon laser in mixtures with He and Ar pumped by a hard ionizer. 1. Electron-Beam Pumping (Kinetika aktivnoy sredy mnogovolnovogo IK lazera na ksenone v smesyakh s He i Ar, nakachivayemykh zhestkim ionizatorom. 1. Elektronnopuchkovaya nakachka). Kvantovaya elektronika **28**(2), 121–128 (1999)
30. A.V. Karelin, O.V. Simakova, Kinetics of active media of a multiwave xenon laser in mixtures with He and Ar pumped by a hard ionizer. 2. Nuclear pumping (Kinetika aktivnykh sred mnogovolnovogo IK lazera na ksenone v smesyakh s He i Ar, nakachivayemikh zhestkim ionizatorom. 2. Yadernaya nakachka). Kvantovaya elektronika **28**(2), 129–135 (1999)
31. D.N. Babichev, A.V. Karelin, O.V. Simakova, K. Tomizava, Kinetic model of the active medium of an He-Ar laser with hard ionizer pumping (Kineticheskaya model aktivnoy sredy He-Ar-lazera s nakachkoy zhestkim ionizatorom). Kvantovaya elektronika **31**(3), 209–217 (2001)
32. A.V. Karelin, O.V. Simakova, Total optimization of an Xe laser with electron beam pumping at transitions with $\lambda = 1.73$, 2.03, 2.65, 2.63, 3.37 and 3.51 μm (Polnaya optimizatsiya Xe-lazera s elektronno-puchkovoy nakachkoy na perekhodakh s $\lambda = 1.73$, 2.03, 2.65, 2.63, 3.37 i 3.51 μm. Kvantovaya elektronika **34**(1), 29–36 (2004)
33. A.M. Voinov, S.P. Melnikov, A.A. Sinyanskiy, On the lasing mechanism of recombination lasers on the transitions of rare gas atoms (O mekhanizme generatsii rekombinatsionnykh lazerov na perekhodakh atomov inertnykh gazov). Pisma v ZhTF **15**(19), 56–59 (1989)
34. A.M. Voinov, S.P. Melnikov, A.A. Sinyanskiy, A kinetic model of recombination ir high-pressure lasers based on transitions of the xenon atom (Kineticheskaya model rekombinatsionnykh IK lazerov vysokogo davleniya na perekhodakh atoma ksenona), *Izvestiya AN SSSR. Ser. fizicheskaya*. **54**(10), 2040–2044, (1990)
35. A.M. Voinov, S.P. Melnikov, A.A. Sinyanskiy, A kinetic model of recombination lasers based on transitions of the xenon atom. I. Lasing mechanism and parameters of a plasma (Kineticheskaya model rekombinatsionnykh lazerov na perekhodakh atoma ksenona. I. Mekhanizm generatsii i parametry plazmy). ZhTF. **60**(10), 100–106, (1990)
36. A.M. Voinov, S.P. Melnikov, A.A. Sinyanskiy. A kinetic model of recombination lasers based on transitions of the xenon atom. II. Lasing characteristics (Kineticheskaya model rekombinatsionnykh lazerov na perekhodakh atoma ksenona. II. Lazernyye kharakteristiki). ZhTF. **60**(10), 107–113, (1990)
37. S.P. Melnikov, A.A. Sinyanskiy, Calculation of threshold characteristics of a nuclear-pumped laser operating on transitions of the xenon atom (Raschet porogovykh kharakteristik lazera s yadernoy nakachkoy na perekhodakh atoma ksenona). Pisma v ZhTF **17**(12), 78–82 (1991)
38. S.P. Melnikov, A.A. Sinyanskiy, *A lasing mechanism of nuclear-pumped IR lasers operating on transitions of krypton and argon atoms* (Mekhanizm generatsii IK lazerov s yadernoy nakachkoy na perekhodakh atomov kriptona i argona), Proceedings of the Specialist Conference The Physics of Nuclear-Excited Plasma and the Problems of Nuclear-Pumped Lasers, Obninsk, 1992, vol. 1, pp. 265–274
39. S.P. Melnikov, A.A. Sinyanskiy, Kinetics of nuclear-pumped IR lasers operating on transitions of krypton and argon atoms (Kinetika IK lazerov s yadernoy nakachkoy na perekhodakh atomov kriptona i argona). ZhTF **62**(6), 159–169 (1992)

40. S.A. Lawton, J.B. Richards, L.A. Newman, L. Specht, T.A. DeTemple, The high-pressure neutral infrared xenon laser. J. Appl. Phys. **50**(6), 3888–3898 (1979)
41. A.Yu. Dudin, D.A. Zayarnyy, L.V. Semenova, N.N. Ustinovskiy, I.V. Kholin, A.Yu. Chugunov, Electron-beam pumped lasers on Xe, Kr, and Ar mixtures with twocompnent buffer gases (Lazery s elektronnopuchkovoy nakachkoy na smesyakh Xe, Kr i Ar s dvukhkomponentnymi bufernymi gazami). Kvantovaya elektronika **18**(11), 1290–1294 (1991)
42. P.J.M. Peters, Q.-C. Mei, W.J. Witteman, Pressure-dependent optical delay time measurements in coaxial electron beam pumped Ar:Xe laser. Appl. Phys. Lett. **54**(3), 193–195 (1989)
43. D. Hausamann, H. Morgner, The heteronuclear rare gas ions. A simple model for the determination of the potential curves. Mol. Phys. **54**(5), 1085–1099 (1985)
44. P.M. Dehmer, S.T. Pratt, Photoionization of ArKr, ArXe, and KrXe and bond dissociation energy of the rare gas dimer ions. J. Chem. Phys. **77**(10), 4804–4817 (1982)
45. A.M. Voinov, L.Ye. Dovbysh, A.T. Kazakevich, V.N. Krivonosov, S.P. Melnikov, I.V. Podmoshenskiy, A.A. Sinyanskiy, Infrared nuclear-pumped lasers operating on ArI, KrI, and XeI transitions (Infrakrasnyye lazery s yadernoy nakachkoy na perekhodakh ArI, KrI i XeI), Pisma v ZhTF. **5**(7), 422–424, (1979)
46. A.V. Bochkov, E.P. Magda, V.V. Mironenko, V.M. Murzin, G.S. Sofiyenko, *Investigations into the parameters of lasers operating on transitions of rare gas atoms in conditions of high-energy loading* (Issledovaniya parametrov lazerov na perekhodakh atomov inertnykh gazov v usloviyakh vysokoy energeticheskoy nagruzki), Proceedings of the 3rd International Conference Problems of Nuclear-Pumped Lasers and Pulsed Reactors, Snezhinsk, 2003, pp. 134–139
47. B.N. Kolokolov, A.A. Kudryavtsev, V.A. Romanenko, A spectroscopic investigation of recombination populating of the $5p^5 6p$ and $5p^5 5d$ states of the Xe Atom (Spektroskopicheskoye issledovaniye rekombinatsionnogo zaseleniya $5p^5 6p$- and $5p^5 5d$sostoyaniy atoma Xe). Optika i spektroskopiya **67**(3), 500–504 (1989)
48. S.P. Melnikov, A.A. Sinyanskiy, On recombination populating of the states $5p^5 6p$ and $5p^5 5d$ of the xenon atom (O rekombinatsionnom zaselenii sostoyaniy $5p^5 6p$ i $5p^5 5d$ atoma ksenona). Optika i spektroskopiya **71**(3), 545–546 (1991)
49. C.C. Davis, T.A. King, Decay rates of infrared emitting laser levels in neutral xenon. J. Quant. Spectrosc. Radiat. Transfer **13**(9), 825–835 (1973)
50. X.K. Hu, J.B.A. Mitchell, R.H. Lipson, Resonance-enhanced multiphoton-ionization photoelectron study of the dissociative recombination and associative ionization of $Xe^+{}_2$. Phys. Rev. A **62**(5), 052712 (2000)
51. V.A. Alekseev, D.W. Setser, Generation and kinetics studies of $Xe(5d[3/2]_1)$ resonance state atoms. J. Phys. Chem. **103**, 8396–8403 (1999)
52. J.P. Apruzese, J.L. Giuliani, M.F. Wolford, J.D. Sethian, G.M. Petrov, D.D. Hinshelwood, M.C. Myers, D.M. Ponce, F. Hegeler, T. Petrova, Experimental evidence for the role of $Xe^+{}_2$ in pumping the Ar-Xe infrared laser. Appl. Phys. Lett. **88**(12), 121120 (2006)
53. Ye.D. Poletayev, Yu.B. Dorofeyev, P.P. Dyachenko, A.P. Kopay-Gora, A.A. Mavlyutov, A.I. Miskevich, B.S. Salamakha, Radiative characteristics of pure neon and high-pressure He-Ne mixtures when excited by nuclear particles (Izluchatelnyye kharakteristiki chistogo neona i He-Ne smesey vysokogo davleniya pri vozbuzhdenii yadernymi chastitsami), ZhTF. **62**(2), 1–8, (1992)
54. Yu.A. Dyuzhov, Ye.D. Poletayev, *The kinetics of exciting bands of the 2d positive system of the N2 molecule and bands of the 1st negative system of the $N^+{}_2$ nitrogen ion in high-pressure nitrogen and ar-n_2 and ne-n_2 mixtures during excitation of the medium by fission fragments* (Kinetika vozbuzhdeniya polos 2-y polozhitelnoy sistemy molekuly N_2 i polos 1-y otritsatelnoy sistemy molekulyarnogo iona azota $N^+{}_2$v azote i ArN_2, Ne-N_2 smesyakh vysokogo davleniya pri vozbuzhdenii sredy oskolkami deleniya), Proceedings of the 3rd International Conference "Problems of Nuclear-Pumped Lasers and Pulsed Reactors, Snezhinsk, 2003, pp. 64–68.

55. M. Aymar, M. Coulombe, Theoretical transition probabilities and lifetimes in KrI and XeI spectra. Atom. Data Nucl. Data Tables **21**(6), 537–566 (1978)
56. A. Dasgupta, J.P. Apruzese, O. Zatsarinny, K. Bartschat, C.F. Fischer, Laser transition probabilities in XeI. Phys. Rev. A **74**(1), 012509 (2006)
57. O.V. Sereda, V.F. Tarasenko, A.V. Fedenev, S.I. Yakovlenko, High-power IR lasers operating on transitions of the XeI atom (Moshchnyye IK lazery na perekhodakh atoma XeI). Kvantovaya elektronika **20**(6), 535–558 (1993)
58. A.I. Konak, S.P. Melnikov, A.A. Sinyanskiy, Quasi-CW nuclear-pumped laser based on a Ne-Xe-(He, Ar) mixture (Kvazinepreryvnyy lazer s yadernoy nakachkoy na smesi Ne-Xe-(He, Ar)). ZhTF **62**(3), 24–29 (1992)
59. C.R. Mansfield, P.F. Bird, J.F. Davis, T.F. Wimett, H.H. Helmick, Direct nuclear pumping of a ^{3}He-Xe laser. Appl. Phys. Lett. **30**(12), 640–641 (1977)
60. J.K. Ku, D.W. Setser, Collisional deactivation of Xe($5p^{5}6p$) states in Xe and Ar. J. Chem. Phys. **84**(8), 4304–4316 (1986)
61. M.R. Bruce, W.B. Layne, C.A. Whitehead, J.W. Keto, Radiative lifetimes and collisional deactivation of two-photon excited Xenon in argon and xenon. J. Chem. Phys. **92**(5), 2917–2926 (1990)
62. W.J. Alford, State-to-state rate constants for quenching of xenon $6p$ levels by rare gases. J. Chem. Phys. **96**(6), 4330–4340 (1992)
63. J. Xu, D.W. Setser, Deactivation rate constants and product branching in collisions of the Xe ($6p$) states with Kr and Ar. J. Chem. Phys. **92**(7), 4191–4202 (1990)
64. J. Xu, D.W. Setser, Collisional deactivation studies of the Xe(6p) states in He and Ne. J. Chem. Phys. **94**(6), 4243–4251 (1991)
65. I.A. Denezhkin, P.P. Dyachenko, V.P. Semenov, Time constants of radiative transitions from level $5d[3/2]_1$ of a xenon atom (Postoyannyye vremeni radiatsionnykh perekhodov s urovnya $5d[3/2]_1$ atoma ksenona). Optika i spektroskopiya **106**(4), 557–562 (2009)
66. N.B. Kolokolov, A.A. Kudryavtsev, A.G. Nikitin, Investigation into the processes of populating of $5d[3/2]_1$ and $6p[5/2]_2$ levels of xenon in the recombination plasma of a He + Xe mixture (Issledovaniye protsessov zaseleniya $5d[3/2]_1$ i $6p[5/2]_2$ urovney ksenona v rekombiniruyushchey plazme smesi He + Xe). Optika i spektroskopiya **69**(3), 527–533 (1990)
67. N.B. Kolokolov, A.A. Kudryavtsev, A.G. Nikitin, Investigation into the processes of populating of excited levels of xenon in recombination plasma of an Ar-Xe mixture Issledovaniye protsessov zaseleniya vozbuzhdennykh urovney ksenona v rekombiniruyushchey plazme smesi Ar-Xe). Optika i spektroskopiya **69**(5), 998–1002 (1990)
68. G.M. Petrov, J.L. Giuliani, J.P. Apruzese, A. Dasgupta, T. Petrova, K. Bartschat, D. Rose, Electron kinetics of the E-beam pumped Ar-Xe laser. J. Phys. D Appl. Phys. **40**(15), 4532–4543 (2007)
69. A. Suda, B.L. Wexler, B.J. Feldman, K.J. Riley, Measurements of gain, saturation, and line competition in an electron beam pumped high-pressure Ar/Xe laser. Appl. Phys. Lett. **54**(14), 1305–1307 (1989)
70. G.A. Hebner, G.N. Hays, measured pressure broadening and shift rates of the 1.73 μm ($5d$ $[3/2]_1$-$6p[5/2]_2$) transition of xenon. Appl. Phys. Lett. **59**(5), 537–539 (1991)
71. A.Yu. Dudin, D.A. Zayarnyy, L.V. Semenova, N.N. Ustinovskiy, I.V. Kholin, A.Yu. Chugunov, Dynamics of gain and lasing of an Ar-Xe electron-beam-pumped laser (Dinamika usileniya i generatsii lazera na smesi Ar-Xe s elektronno-puchkovoy nakachkoy). Kvantovaya elektronika **20**(7), 669–676 (1993)
72. A.M. Voinov, L.Ye. Dovbysh, V.N. Krivonosov, S.P. Melnikov, I.V. Podmoshenskiy, A.A. Sinyanskiy, Low-threshold nuclear-pumped lasers operating on transitions of atomic xenon (Nizkoporogovyye lazery s yadernoy nakachkoy na perekhodakh atomarnogo ksenona). Doklady AN SSSR. **245**(1), 80–83, (1979)
73. A.M. Voinov, L.Ye. Dovbysh, V.N. Krivonosov, S.P. Melnikov, I.V. Podmoshenskiy, A.A. Sinyanskiy, Nuclear-pumped IR laser operating on He + Xe and Ar + Xe mixtures,

(Infrakrasnyy lazer s yadernoy nakachkoy na smesyakh He + Xe i Ar + Xe), Pisma v ZhTF. **7** (16), 1016–1020, (1981)

74. N.W. Jalufka, Nuclear-pumped Lasing of ^{3}He-Xe at 2.63 μm. Appl. Phys. Lett. **39**(7), 535–536 (1981)
75. N.G. Basov, A.Y. Chugunov, V.A. Danilychev, I.V. Kholin, N.N. Ustinovsky, Powerful electroionization laser on Xe infrared atomic transitions. IEEE J. Quantum Electron. **QE-19** (2), 126–128 (1983)
76. A.M. Voinov, L.Ye. Dovbysh, V.N. Krivonosov, S.P. Melnikov, I.V. Podmoshchenskiy, A.A. Sinyanskiy, A high-pressure helium-krypton laser excited by uranium fission fragments (Geliy-kriptonovyy lazer vysokogo davleniya, vozbuzhdayemyy oskolkami deleniya urana). ZhTF. **52**(7), 1346–1350, (1982)
77. A.M. Voinov, V.N. Krivonosov, S.P. Melnikov, I.N. Mochkayev, A.A. Sinyanskiy, Quasi-CW nuclear-pumped laser operating on transitions of the argon atom (Kvazinereryvnyy lazer s yadernoy nakachkoy na perekhodakh atoma argona). Kvantovaya elektronika **18**(2), 177–179 (1991)
78. A. Suda, B.L. Wexler, K.J. Riley, B.J. Feldman, Characteristics of the high-pressure Ar-Xe laser pumped by an electron beam and electron-beam sustained discharge. IEEE J. Quantum Electron. **26**(5), 911–921, (1990)
79. W.J. Alford, G.N. Hays, Measured laser parameters for reactor-pumped He/Ar/Xe and Ar/Xe lasers. J. Appl. Phys. **65**(10), 3760–3766 (1989)
80. W.J. Alford, G.N. Hays, M. Ohwa, M.J. Kushner, The effect of He addition on the performance of the fission-fragment excited Ar/Xe atomic xenon laser. J. Appl. Phys. **69**(4), 1843–1848 (1991)
81. J.W. Shon, M.J. Kushner, G.A. Hebner, G.N. Hays, Predictions for gain in the fission-fragment-excited atomic xenon laser. J. Appl. Phys. **73**(6), 2686–2694 (1993)
82. M. Ohwa, M.J. Kushner, Energy loading effects in the scaling of atomic xenon Laser. IEEE J. Quantum Electron. **26**(9), 1639–1646 (1990)
83. A.Y. Aleksandrov, V.A. Dolgikh, O.M. Kerimov, Y.F. Myznikov, I.G. Rudoy, A.M. Soroka, Basic mechanisms of inversion formation at 3*p*-3*s* transitions of neon (Osnovnyye mekhanizmy obrazovaniya inversii na 3*p*-3*s* perekhodakh neona). Kvantovaya elektronika **14**(12), 2389–2395 (1987)
84. A.Y. Aleksandrov, V.A. Dolgikh, I.G. Rudoy, A.M. Soroka, Kinetics of a high-pressure electron beam-excited laser on the 'yellow' neon line ("Kinetika vozbuzhdayemogo elektronnym puchkom lazera vysokogo davleniya na 'zheltoy' linii neona"). Kvantovaya elektronika **18**(9), 1029–1033 (1991)
85. V.I. Derzhiyev, A.G. Zhidkov, A.V. Koval, S.I. Yakovlenko, Kinetic model of a penning Ne-laser on a beam He-Ne-Ar and Ne-H_2-plasma" (Kineticheskaya model penningovskogo Ne-lazera na puchkovoy He-Ne-Ar- i Ne-H_2-plazme). Kvantovaya elektronika **16**(8), 1579–1586 (1989)
86. A.M. Voinov, V.N. Krivonosov, S.P. Melnikov, A.I. Pavlovskiy, A.A. Sinyanskiy, Quasi-CW Lasing at 3*p*-3*s* transitions of the neon atom during excitation of rare gas mixtures by uranium fission fragments (Kvazinepreryvnaya generatsiya na perekhodakh 3*p*-3*s* atoma neona pri vozbuzhdenii smesey inertnykh gazov oskolkami deleniya urana). Doklady AN SSSR **312**(4), 864–867 (1990)
87. G.A. Batyrbekov, E.G. Batyrbekov, V.A. Danilychev, *kinetics of active laser media at 3p-3s transitions of the neon atom with pumping by an external ionization source* (Kinetika aktivnykh sred lazerov na 3p-3s perekhodakh atoma neona s nakachkoy istochnikom vneshney ionizatsii). Proceedings of the Specialist Conference The Physics of Nuclear-Excited Plasma and Problems of Nuclear-Pumped Lasers, Obninsk, 1992, vol. 1, pp. 348–358
88. E.G. Batyrbekov, Direct and Combined Nuclear Pumped Lasers. Preprint of IAENNC, Republic of Kazakhstan, Almaty (1994)
89. W.J. Shon, R.L. Rhoades, J.T. Verdeyen, M.J. Kushner, Short pulse electron beam excitation of the high-pressure atomic Ne laser. J. Appl. Phys. **73**(12), 8059–8065 (1993)

90. A.V. Karelin, S.I. Yakovlenko, A kinetic model of the He-Ne-Ar-H_2 laser with pumping by hard ionizing radiation (Kineticheskaya model He-Ne-Ar-H_2-lazera s nakachkoy zhestkim ioniziruyushchim izlucheniyem). Kvantovaya elektronika **22**(8), 769–774 (1995)
91. A.V. Karelin, V.F. Tarasenko, A.V. Fedenev, S.I. Yakovlenko, On the ultimate efficiency of a penning neon plasma laser (O predelnom KPD penningovskogo plazmennogo lazera na neone). Kvantovaya elektronika **23**(4), 299–302 (1996)
92. K. Sakasai, T. Kakuta, M. Nakazawa, Numerical simulation of a nuclear pumped ^{3}He-Ne-Ar gas laser for its optimization. Jap. J. Appl. Phys. **37**(9A), 4806–4811 (1998)
93. T.R. Connor, M.A. Biondi, Dissociative recombination in neon: spectral line-shape studies. Phys. Rev. **140**(3A), 778–791 (1965)
94. R. Johnsen, J. Macdonald, M.A. Biondi, Thermal energy charge transfer rates for Ne^+, $Ne_2{}^+$, Ar^+, $Ar_2{}^+$ ions with Kr and Xe atoms. J. Chem. Phys. **68**(6), 2991–2992 (1978)
95. A. Morozov, R. Krücken, A. Ulrich, J. Wieser, T. McCarthy. Energy transfer processes in neon-hydrogen mixtures excited by electron beams. J. Chem Phys. **123**, 234311, (2005)
96. A.I. Konak, S.P. Melnikov, V.V. Porkhayev, A.A. Sinyanskiy, Characteristics of a nuclear-pumped laser operating on 3*p*-3*s* transitions of the neon atom (Kharakteristiki lazera s yadernoy nakachkoy na perekhodakh 3*p*-3*s* atoma neona). Kvantovaya elektronika **22**(3), 225–230 (1995)
97. C.B. Collins, F.W. Lee, Measurement of the rate coefficients for the bimolecular and termolecular ion-molecule reactions of $Ne_2{}^+$ with selected atomic and molecular species. J. Chem. Phys. **72**(10), 5381–5389 (1980)
98. L.W.G. Steenhuijsen, N. Van Schaik, L.C.A.M. Van de Nieuwenhuyzen, F.H.P. Verspaget, Measurements of the production of neon 2*p* atoms by dissociative recombination. J. Phys. (Paris), **40**(7) 95–96, (1979)
99. R.S.F. Chang, D.W. Setser, Radiative lifetimes and collisional deactivation rate constants of excited Ne($2p^5 3p$) states. J. Chem. Phys. **72**(7), 4099–4110 (1980)
100. F. Fujimoto, C. Goto, K. Fukuda, Elimination of the excitation anisotropy effects in the laser-based determination of atomic collision rate coefficient—depopulating collisions of neon $2p^5 3p$ atoms. J. Phys. Soc. Jap. **53**(2), 574–581 (1984)
101. V.M. Baran, G.L. Kononchuk, A.V. Yakunov, Transitions between components of the fine structure of neon during inelastic collisions of neon and helium atoms (Perekhody mezhdu komponentami tonkoy struktury neona pri neuprugikh stolknoveniyakh atomov neona i geliya). Zhurnal prikladnoy spektroskopii **41**(1), 128–134 (1984)
102. M.P.I. Manders, J.P.J. Driessen, H.C.W. Beijerinck, B.J. Verhaar, Collision-induced intramultiplet mixing for Ne**$\{(2p^5)(3p)\}$ + He: experiments and quantum calculations. Phys. Rev. A **37**(9), 3237–3253 (1988)
103. D.A. Zayarnyy, I.V. Kholin, A.Y. Chugunov, Deactivation of 3*s* levels of a neon atom during collisions with neon, argon, krypton, and xenon (Dezaktivatsiya 3*s*-urovney atoma neona pri stolknoveniyakh s neonom, argonom, kriptonom i ksenonom). Kvantovaya elektronika **22**(3), 233–238 (1995)
104. A. Yokoyama, Y. Hatano, De-excitation rate constants of Ne($^3P_2$,$^3P_1$, and $^3P_0$) by atoms and molecules as studied by the pulse radiolysis method. Chem. Phys. **63**(1–2), 59–65 (1981)
105. M.L. Burshteyn, V.A. Komarovskiy, A.N. Fedorov, S.V. Yurgenson, An investigation of quenching of excited 2*p* levels of neon by molecular hydrogen and argon atoms (Issledovaniye tusheniya vozbuzhdennykh 2*p*-urovney neona molekulyarnym vodorodom i atomami argona). Optika i spektroskopiya **71**(2), 240–242 (1991)
106. A.V. Karelin, S.I. Yakovlenko, Kinetics of active media of high-pressure lasers based on metal vapors (Kinetika aktivnykh sred lazerov vysokogo davleniya na parakh metallov). Kvantovaya elektronika **20**(7), 631–650 (1993)
107. A.V. Karelin, R.V. Shirokov, Kinetics of active media of a nuclear-pumped laser operating on transitions of the cadmium atom (Kinetika aktivnoy sredy lazera s yadernoy nakachkoy na perekhodakh atoma kadmiya). Kvantovaya elektronika **25**(10), 917–921 (1998)

108. A.V. Parashchuk, G.L. Yudin, Kinetics of formation of a laser-active helium plasma of nuclear origin. plasmochemical processes and laser kinetics (Kinetika formirovaniya lazerno-aktivnoy geliyevoy plazmy yadernogo proiskhozhdeniya. Plazmokhimicheskiye protsessy i lazernaya kinetika), *Preprint FEI*, No. 1746, Obninsk (1985)
109. A.I. Miskevich, A kinetic model of a nuclear-pumped laser operating on cadmium vapors (Kineticheskaya model lazera s yadernoy nakachkoy na parakh kadmira). ZhTF **57**(9), 1767–1775 (1987)
110. S.V. Makarov, Y.N. Novoselov, V.V. Osipov, Modeling of a high-pressure helium-cadmium laser excited by a pulsed electron beam (Modelirovaniye geliy-kadmiyevogo lazera vysokogo davleniya, vozbuzhdayemogo impulsnym elektronnym puchkom). Kvantovaya elektronika **17**(8), 974–978 (1990)
111. N.M. Barysheva, N.V. Bochkova, K.F. Grebenkin, Analysis of kinetic models of a high-pressure helium-cadmium laser, (Analiz kineticheskikh modeley geliy-kadmiyevogo lazera vysokogo davleniya), Voprosy atomnoy nauki i tekhniki. Ser. Teoreticheskaya i prikladnaya fizika (2), 12–18, (1991)
112. Ye.D. Poletayev, P.P. Dyachenko, Yu.A. Dyuzhov, M.Z. Tarasko, Investigation of kinetics of populating of levels of the cadmium ion during excitation of a He-Cd medium by heavy charged particles (Issledovaniye kinetiki zaseleniya urovney iona kadmiya pri vozbuzhdenii He-Cd-Sredy tyazhelymi zaryazhennymi chastitsami). Optika i spektroskopiya. **84**(2), 186–192, (1998)
113. P. Baltayan, J.C. Pebay-Peyroula, N. Sadeghi, Determination of the rate constants for population of the individual Cd^{+*} levels in the thermal penning and charge-transfer reactions of $He^*(2^3S)$ and He^+ with cadmium. J. Phys. B **18**(17), 3615–3628 (1985)
114. S. Inaba, T. Goto, S. Hattori, Determination of the penning excitation cross-sections of individual Cd(II) states for He metastable atoms using penning electron spectroscopy. J. Phys. D **15**(1), 35–40 (1982)
115. F. Ranjbar, H.H. Harris, J.J. Leventhal, CdII excitation in Slow He^+-Cd and $He_2{}^+$-Cd collisions. Appl. Phys. Lett. **31**(6), 385–387 (1977)
116. E.P. Magda, *Analysis of experimental and computational-theoretical studies on nuclear-pumped lasers performed at VNIITF* (Analiz ekperimentalnykh i raschetno-teoreticheskikh rabot po lazeram s yadernoy nakachkoy, vypolnennykh vo VNIITF), Proceedings of the Specialist Conference The Physics of Nuclear-Excited Plasma and the Problems of Nuclear-Pumped Lasers, Obninsk, 1992, vol. 1, pp. 65–74
117. N.M. Barysheva, A.V. Bochkov, N.V. Bochkova, K.F. Grebenkin, V.A. Kryzhanovskiy, E.P. Magda, A.E. Neznakhina, Relaxation of the $5s^{22}D_{3/2}$ state of the CdII Ion during collision with He Atoms in a He-Cd medium excited by uranium fission fragments (Relaksatsiya sostoyaniya $5s^{22}D_{3/2}$ iona CdII pri stolknovenii s atomami He v He-Cd srede, vozbuzhdayemoy oskolkami deleniya urana). Optika i spektroskopiya **73**(1), 87–90 (1992)
118. S.P. Bugayev, F.G. Goryunov, S.Y. Nagornyy, V.S. Skakun, V.F. Tarasenko, A.V. Fedenev, UV lasing during electron-beam pumping of a Ne-Cd Mixture (UF generatsiya pri nakachke smesi Ne-Cd elektronnym puchkom). Optika i spektroskopiya **65**(3), 744–747 (1988)
119. V.A. Makeyev, Y.N. Novoselov, M.Y. Starovoytov, V.V. Uvarin, UV lasing on the cadmium ion excited by a microsecond electron beam (UF generatsiya na ione kadmiya, vozbuzhdayemaya mikrosekundnym elektronnym puchkom). Pisma v ZhTF **19**(9), 10–13 (1993)
120. Y.N. Novoselov, V.F. Tarasenko, V.V. Uvarin, A.V. Fedenev, The influence of impurities and power deposition on lasing characteristics of the high-pressure He-Cd laser (Vliyaniye primesey i moshchnosti nakachki na kharakteristiki generatsii He-Cd lazera vysokogo davleniya. Kvantovaya elektronika **23**(3), 211–216 (1996)
121. A.V. Karelin, S.Y. Nagornyy, V.F. Tarasenko, S.I. Yakovlenko, He-Zn beam laser at $\lambda = 610$ nm with penning and electron collisional purification (Puchkovyy He-Zn lazer na $\lambda = 610$ nm s penningovskoy i elektronnoy stolknovitelnoy ochistkoy). Pisma v ZhTF **16**(15), 52–55 (1990)

122. G.A. Batyrbekov, E.G. Batyrbekov, V.A. Dolgikh, I.G. Rudoy, A.M. Soroka, A.B. Tleuzhanov, M.U. Khasenov, Kinetics of excited states of Hg during ionizing radiation pumping (Kinetika vozbuzhdennykh sostoyaniy Hg pri nakachke ioniziruyushchim izlucheniyem), *Preprint of the Nuclear Physics Institute of the Kazakh SSR Academy of Sciences* (*Preprint Instituta yadernoy fiziki AN Kazakhskoy SSR*), No. 3–87, Alma-Ata (1987)
123. G.A. Batyrbekov, E.G. Batyrbekov, V.A. Dolgikh, I.G. Rudoy, A.M. Soroka, A.B. Tleuzhanov, M.U. Khasenov, On the possibility of creating a quasi-CW laser operating on *7s-6p* transitions of mercury with pumping by ionizing radiation (O vozmozhnosti sozdaniya kvazinepreryvnogo lazera na *7s-6p*-perekhodakh rtuti pri nakachke ioniziruyushchim izlucheniyem). Kvantovaya elektronika **14**(6), 1216–1218 (1987)
124. G.A. Batyrbekov, E.G. Batyrbekov, M.U. Khasenov, *A kinetic model of a laser operating on triplet (7^3S_1-6^3P) transitions of the mercury atom with direct nuclear pumping (Kineticheskaya model lazera na tripletnykh (7^3S_1-6^3P) perekhodakh atoma rtuti s pryamoy yadernoy nakachkoy)*, Proceedings of the Specialist Conference "The Physics of Nuclear-Excited Plasma and the Problems of Nuclear-Pumped Lasers", Obninsk, 1992, vol. 1, pp. 282–291
125. N.M. Barysheva, N.V. Bochkova, A.A. Kosorukova, E.P. Magda, *Kinetic model of the nuclear-pumped He-Xe-Hg-H_2 Laser* (Kineticheskaya model He-Xe-Hg-H_2 lazera s yadernoy nakachkoy), Proceedings of the 3rd International Conference "Problems of Nuclear-Pumped Lasers and Pulsed Reactors, Snezhinsk, 2003, pp. 218–224
126. A.A. Mavlyutov, A.I. Miskevich, B.S. Salamakha, *Plasma atomization of metallic films and use of this effect in nuclear-pumped lasers* (Plazmennoye raspyleniye metallicheskikh plenok i ispolzovaniye etogo effekta v lazerakh s yadernoy nakachkoy), Proceedings of the Specialist Conference The Physics of Nuclear-Excited Plasma and the Problems of Nuclear-Pumped Lasers, Obninsk, 1993, vol. 3, pp. 67–76
127. A.I. Miskevich, Generation of stimulated radiation at the $5s^{22}D_{5/2}$-$5p^2P_{3/2}$ transitions of CDII Ions with atomization of metalic cadmium by nuclear particles (Generatsiya stimulirovannogo izlucheniya na perekhode $5s^{22}D_{5/2}$-$5p^2P_{3/2}$ ionov CdII pri raspylenii metallicheskogo kadmiya yadernymi chastitsami). Kvantovaya elektronika **32**(9), 803–808 (2002)
128. A.V. Karelin, The mechanism of generation of stimulated radiation at the $5s^{22}D5/2$- $5p^2P3/2$ transition of CdII at anomalously low temperatures of the active medium (Mekhanizm generatsii stimulirovannogo izlucheniya na perekhode $5s^{22}D_{5/2}$-$5p^2P_{3/2}$ CdII pri anomalno nizkikh temperatuprakh aktivnoy sredy). Kvantovaya elektronika **33**(4), 315–316 (2003)
129. R. Beck, W. Englisch, K. Gürs, *Table of Laser Lines in Gases and Vapors* (Springer, New York, 1980)
130. J.B. Atkinson, J.H. Sanders, Laser action in carbon and nitrogen atoms following dissociative excitation transfer. J. Phys. B. **1**(6), 1171–1179 (1968)
131. M.A. Prelas, G.H. Miley, Dynamics of the nuclear and electrically pumped 1.45-μm atomic carbon laser in mixtures of helium + CO and helium + CO_2. Nucl. Technol./Fusion **1**(3), 402–413 (1981)
132. G.W. Cooper, J.T. Verdeyen, Recombination-pumped atomic nitrogen and carbon afterglow lasers. J. Appl. Phys. **48**(3), 1170–1175 (1977)
133. M.J. Kushner, Nuclear-reactor pumped lasers excited by ion-ion neutralization. J. Appl. Phys. **54**(1), 39–47 (1983)
134. A.V. Karelin, O.V. Simakova, Kinetics of the active medium of a nuclear-pumped laser operating on ir transitions of the chlorine atom (Kinetika aktivnoy sredy lazera s yadernoy nakachkoy na IK perekhodakh atoma khlora). Kvantovaya elektronika **24**(11), 993–997 (1997)
135. A.V. Karelin, O.V. Simakova, Kinetics of active media of nuclear-pumped lasers operating on transitions of carbon and nitrogen (Kinetika aktivnykh sred lazerov s yadernoy nakachkoy na perekhodakh atomov ugleroda i azota). Kvantovaya elektronika **25**(4), 319–323 (1998)

136. A.V. Karelin, O.V. Simakova, Kinetics of the active medium of a nuclear-pumped laser operating on transitions of the oxygen atom (Kinetika aktivnoy sredy lazera s yadernoy nakachkoy na perekhodakh atoma kisloroda). Kvantovaya elektronika **25**(9), 779–782 (1998)
137. D.A. McArthur, P.B. Tollefsrud, Measurements of optical gain in CO Gas excited only by fission fragments. IEEE J. Quantum Electron. **QE-12**(4), 244–253 (1976)
138. N.W. Jalufka, F. Hohl, A direct nuclear-pumped ^{3}He-CO laser. Appl. Phys. Lett. **39**(2), 139–142 (1981)
139. R.T. Schneider, F. Hohl, Nuclear-pumped lasers. Adv. Nucl. Sci. Technol. **16**, 123–287 (1984)
140. L.I. Gudzenko, V.S. Malyshevskiy, S.I. Yakovlenko, On a CO laser pumped by a hard source (O nakachivayemom zhestkim istochnikom CO lazere). ZhTF **48**(10), 2150–2156 (1978)
141. V.A. Zherebtsov, On nuclear pumping of a carbon monoxide laser (O yadernoy nakachke lazera na okisi ugleroda). ZhTF **68**(7), 80–85 (1998)
142. N.M. Barysheva, N.V. Bochkova, K.F. Grebenkin, E.P. Magda, O.V. Skvortsova, *Kinetics of a Nuclear-Pumped Nitrogen Laser* (Kinetika azotnogo lazera s yadernoy nakachkoy), Proceedings of the Specialist Conference The Physics of Nuclear-Excited Plasma and the Problems of Nuclear-Pumped Lasers, Obninsk, 1992, vol. 1, pp. 337–347
143. M.U. Khasenov, V.A. Dolgikh, A.M. Soroka, *Kinetics of excitation of the CO first negative system by ionizing radiation* (Kinetika vozbuzhdeniya pervoy otritsatelnoy sistemy CO ioniziruyushchim izlucheniyem), Theses of Reports at the Specialist Conference The Physics of Nuclear-Excited Plasma and the Problems of Nuclear-Pumped Lasers, Obninsk, 1992, p. 56
144. N.M. Barysheva, N.V. Bochkova, A.A. Kosorukova, E.P. Magda, *Kinetics of nuclear-pumped lasers operating on electron ultraviolet transitions of the ions of molecules* (Kinetika lazerov s yadernoy nakachkoy na elektronnykh ultrafioletovykh perekhodakh ionov molekul), Proceedings of the 3rd International Conference Problems of Nuclear-Pumped Lasers and Pulsed Reactors, Snezhinsk, 2003, pp. 51–55
145. R.A. Waller, C.B. Collins, A.J. Cunningham, Stimulated emission from CO^+ pumped by charge transfer from $He^+{}_2$ in the afterglow of an E-beam discharge. Appl. Phys. Lett. **27**(6), 323–325 (1975)
146. G.G. Telegin, A.S. Yatsenko, *Optical spectra of atmospheric gases (Opticheskiye spektry atmosfernykh gazov)* (Novosibirsk, Nauka, 2000)
147. C.B. Collins, W.W. Robertson, Spectra excited in a helium afterglow. J. Chem. Phys. **40**(3), 701–712 (1964)

Chapter 6
Nuclear-Pumped Laser Devices Based on Gas Media

In Chap. 2 we examined rather simple NPL designs for searching for and studying characteristics of laser active media, but there are also more complex nuclear laser devices. As a rule, they are structural elements of multichannel reactor lasers operating in stationary or pulsed modes, and are intended to study the specifics of their operation. This chapter examines some of these devices, whether in operation or still being designed.

6.1 LM-4/BIGR Experimental Complex

As the results of experiments and calculations showed (see Chap. 8), it is necessary to fulfill two basic conditions to obtain lasing in stationary mode: provide the necessary specific power deposition of the laser medium, and organize flowing of the gas mixture in the laser cell so as to eliminate its overheating and reduce the optical non-conformities. Taking into account these conditions, VNIIEF designed and built laser module LM-4 [1–6] in 1994 for joint operation with the powerful BIGR pulsed reactor [7], which can, in one of its operating modes, generate neutron pulses with a duration of ≥ 1 s. The experimental complex was supposed to demonstrate the possibility of creating reactor laser-type units with a stationary operating mode.

The LM-4 laser module [1–6] is a stationary-operating gas NPL functioning in gas-flowing mode (Figs. 6.1 and 6.2). The module and device for circulation of the gas were placed on a cart (see Fig. 6.1) and moved to the reactor core on it. The LM-4 module consists of four identical laser channels (Fig. 6.3). Over the length of the uranium layers, the average thermal neutron fluence reached values of (5–7) $\times$ 10^{14} cm^{-2} with a duration of 0.5–1.7 s.

Laser channels of the module are included in a single gas loop and are separated from one another by heat exchangers (radiators), which cool the gas when it is passed through. The radiators are packages 100-cm long, comprised of thin

S.P. Melnikov et al., *Lasers with Nuclear Pumping*,
DOI 10.1007/978-3-319-08882-2_6

Fig. 6.1 LM-4 laser module on transport cart [1, 3–5]. Before the pulse, the module is moved down the track to the core of the BIGR reactor

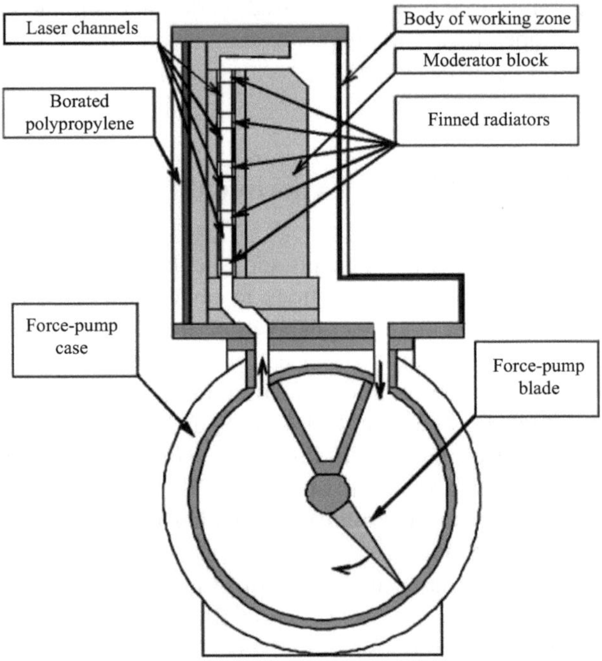

Fig. 6.2 Diagram of transverse section of LM-4 module [1, 4]

aluminum fins around 0.3-mm thick, and the same gaps between them. To equalize the gas dynamic perturbations of the gas flow, the edges of the fins were sharpened. The radiators do not have forced cooling, so their functionality is limited by their own heat capacity.

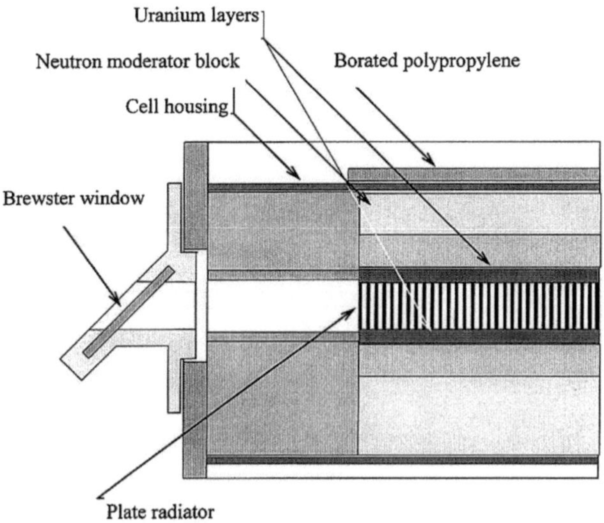

Fig. 6.3 Longitudinal section of laser channels of LM-4 Module [1, 4]

The active length of each channel, determined by the length of the uranium layers along the optical axis, is equal to 100 cm. The dimension of the channel in the direction perpendicular to the gas flow is 2 cm; in the direction parallel to the flow it is 6 cm. Layers of metallic ^{235}U about 5 mg/cm^2-thick are applied to 100×6 cm^2 plates. At their ends, the channels are closed by windows arranged at a Brewster angle. The experiments primarily used stable optical cavities consisting of a 100 % spherical mirror with a radius of curvature of 20 m and a flat semitransparent dielectric mirror. The distance between the mirrors is about 170 cm.

The gas-flowing system provides flow of the gas mixture sequentially through all the laser channels in the transverse direction relative to the channel axis at rates of 4–10 m/s. At a gas flow rate of 10 m/s, the duration of the stationary flow is 3–5 s. The gas-flowing device is a circular, single-blade force-pump capable of creating an overpressure of 80 Torr. The internal diameter of this unit is 80 cm, the blade length is 109 cm, and the angle of the full rotation of the blade is 270°.

To ensure the functionality of the LM-4, the following devices were developed: a transport system, which accomplishes and controls the module movement from the measurement area to the reactor hall, and fixes it close to the reactor core [1]; a system for degassing, filling, and flowing of the gas [1, 4]; the systems for remote adjustment of the cavity mirrors and withdrawal of laser radiation from the reactor hall to the measurement area [6, 8, 9]; an automated system for registration of parameters of the laser radiation, neutron flux, and gas mixture rate [1, 6, 10]; and a technological monitoring system for communications between the BIGR reactor and the module [1, 6].

The optical circuit scheme of the experiments is shown in Fig. 6.4. The route for withdrawal of laser radiation from the reactor hall includes aluminum 100 %

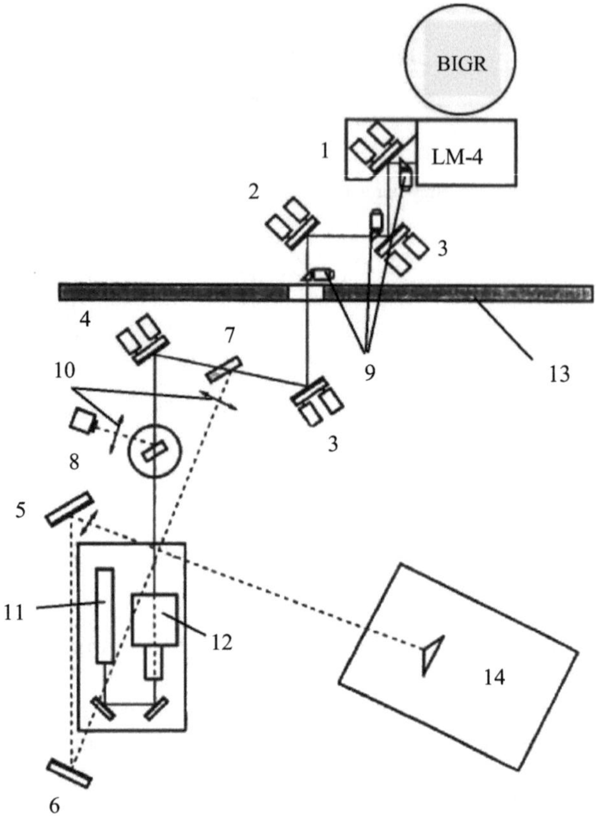

Fig. 6.4 Optical scheme for experiments on LM-4/BIGR complex [6]: (*1*) reference mirror; (*2–6*) flat aluminum mirrors; (*7*) beam splitter; (*8*) interferometer camera; (*9*) viewfinders; (*10*) lenses; (*11*) adjustment laser; (*12*) system for parallel shift of adjustment laser beam; (*13*) concrete shielding; (*14*) laser parameter registration unit

mirrors (parts 2–6) and a beam splitter (7). A lens system (10) was used to reduce the transverse cross section of the beam and construct a reduced image of the output mirrors of the laser channels at the input of the registration unit. Magnetostrictive motors with a feed pitch of around 1 μm were used to rotate some of the mirrors.

Depending on the experimental tasks, the registration unit included various instruments and methods: the photoresistors FSA-G1 and avalanche photodiodes LFD-2 for registration of the lasing pulse shape, laser pulse energy meters (IMO-2N and J50-710 of the Molectron Company), methods based on the Ragulsky wedge to measure radiation divergence, a lateral-shift interferometer with a Pulnix camera by the Spiricon Company to study the optical non-uniformities of the laser media, etc.

The first series of experiments from 1994 to 1995 demonstrated the possibility of NPL operation in stationary mode with a lasing duration of up to 1.5 s [2–5]. The choice of composition and pressure of the mixtures (He,Ne)-Ar-Xe ($\lambda = 1.73$, 2.03 μm) was made on the basis of experimental data obtained using the LUNA-2M setup (see Chap. 3, Sect. 3.1). Oscillograms of one of the experiments are shown in Fig. 6.5. Gas flowing was started roughly 0.5 s before the start of the neutron pulse. In the absence of gas flowing through the gas channels, there was no

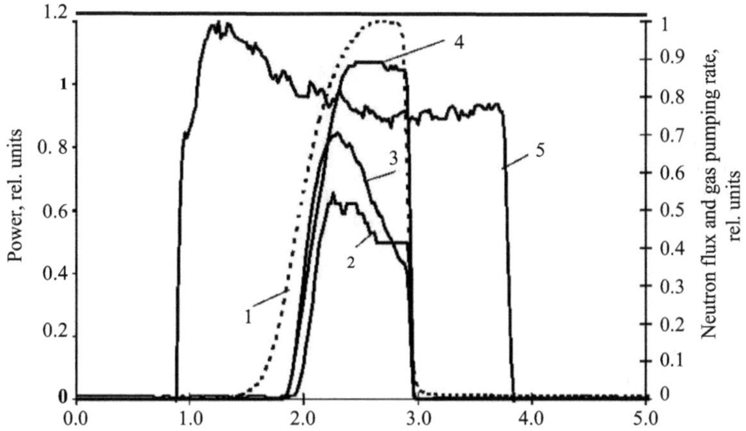

Fig. 6.5 Time dependencies of the neutron flux (*1*) and laser radiation from various channels of the LM-4 module (*2–4*) and rate of gas flow (*5*) for the Ar-Xe mixture (70:1, $\lambda = 1.73$ μm) at a pressure of 0.35 atm [2]

lasing. The lasing threshold for different mixtures was achieved with thermal neutron flux densities of $(1–5) \times 10^{14}$ cm^{-2} s^{-1}. When the lasing threshold was exceeded by a factor of 2–3, the lasing and pumping (neutron pulse) pulse shapes were roughly the same. The maximal lasing power was 15–20 W for each channel.

Before the second series of experiments (2001), the systems for module control, registration of radiation, and calibration of the optical scheme were improved. Experiments were aimed at studying the possibility of serial addition of laser channels (see Chap. 10, Sect. 10.5) and development of methods based on the lateral-shift interferometer to determine the optical non-uniformities of the laser medium [5, 6, 11, 12]. As a result of the experiments, the possibility of a serial scheme for adding laser channels was demonstrated. For example, for an Ar-Xe (100:1) mixture at a pressure of 0.5 atm, the output power at the 1.73 μm line when two channels were added proved to be 2.4 times more than for a single channel. The lasing duration at this time increased from 0.65 to 1.2 s.

Interferometric investigations of optical non-uniformities showed the correctness of conception about gas-dynamic and thermophysical processes in NPLs with transverse gas flowing (see Chap. 9, Sect. 9.3). Thus it follows from the interferograms obtained that an optical wedge forms in the longitudinal direction with respect to the gas flow, while in the transverse direction there is a positive distributed lens. The data of interferometric measurements can also be used to determine the energy deposition to the gas medium [13]. A comparison of the specific power depositions determined by the interferometric method and as a result of calculations using the measured number of fissions in uranium layers showed their approximate agreement, with a precision of 10–20 %.

Possible causes of insufficiently high energy parameters obtained using the LM-4 module include a number of design defects, in particular the presence of windows arranged at the Brewster angle in the laser channels, as well as the

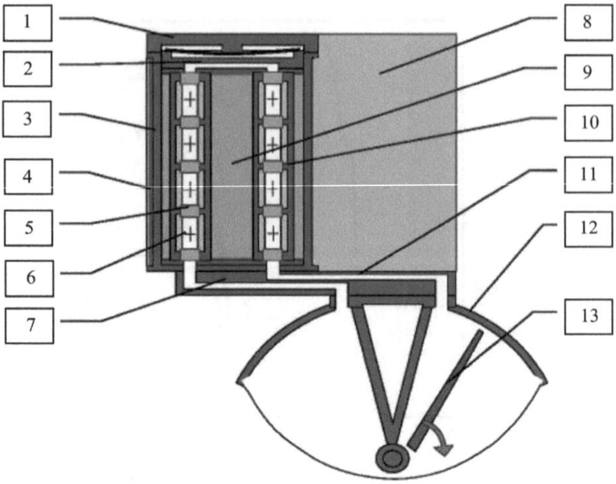

Fig. 6.6 Transverse section of LM-8 laser module: (*1*) cover of housing; (*2*) gas pipeline; (*3*) body of module's work zone; (*4*) shielding made of borated polypropylene; (*5*) radiator; (*6*) laser channel; (*7*) supplying gas pipeline; (*8*) graphite neutron reflector; (*9*) neutron moderator; (*10*) assemblage with four laser channels; (*11*) deferent gas pipeline; (*12*) force-pump housing; (*13*) force-pump blade

placement of the Plexiglas fast neutron moderator inside the sealed space to be evacuated. The latter circumstance can lead to a reduction in the purity of the gas mixture owing to its contamination by impurities adsorbed by the moderator. To eliminate these defects, the LM-8 eight-channel laser module [5, 14] was developed, in which the polyethylene moderator and graphite reflector are arranged outside the active laser volume. A diagram of the transverse section of the LM-8 module is shown in Fig. 6.6.

An optical scheme of the first series of experiments with the LM-8 laser module is shown in Fig. 6.7. The module allows some experiments with serial addition of two and four laser channels to be carried out. Channels 1, 8, and 3–6 are combined by optical cavities using reflecting mirrors. The numbering of the channels follows the direction of gas mixture flowing. The individual channels 2 and 7 are "control" channels.

In the first experiments [14] with the LM-8 module on the BIGR reactor, it was possible to serially add the radiation from four laser channels, and to obtain a relatively high efficiency of laser conversion ($\eta_l \approx 0.7$ %) at low neutron fluences of ~10^{13} cm^{-2}, which corresponds to a power deposition of just 0.4–0.6 W/cm^3.

All of the examined NPL designs and LM-4 and LM-8 modules are intended basically for operation close to the reactor core. The development of a laser module that is an elementary cell of the reactor core is the next phase in the path of development of the reactor laser. One such version, examined in study [15], was the 16-channel LM-16 module, surrounded by a graphite neutron moderator. Laser channels used in the LM-4 and LM-8 devices formed the basis of the LM-16 module. When such a module is used in the reactor core of a continuous stationary reactor, it is possible to resolve problems associated with the compatibility of the laser cell with the reactor core, and certain "laser" problems, for example addition of radiation of the laser channels and its withdrawal from the reactor core [15].

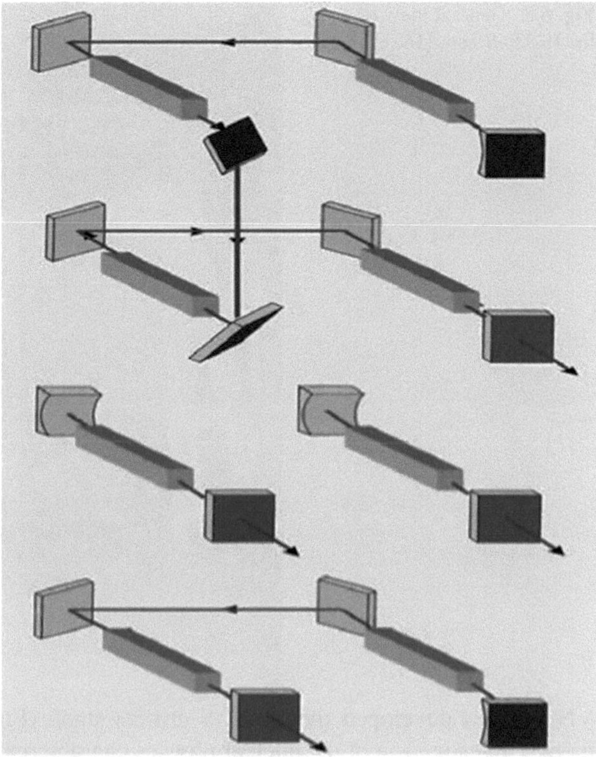

Fig. 6.7 Laser channel addition scheme in LM-8 module

6.2 Reactor Laser Model

At present VNIIEF is developing a nuclear physics facility—a physical model of a stationary reactor-laser (RL) with transverse flowing of the laser medium [16–19]. The facility includes an IKAR-500 pulsed reactor and LM-16 laser module.

The IKAR-500 core is a graphite matrix (a cube with 2,400-mm sides) with nine reach-through holes that are 500 × 500 mm in cross section, which accommodate the reactor modules. In the graphite matrix (between the modules) at the top and on the sides, zirconium channels are mounted to accommodate the reactor control systems. An LM-16 laser module can be placed in one of the reach-through holes instead of a reactor module. The planned energy release in the reactor core for a startup with a duration from fractions of a second to tens of seconds is 500 MJ; the expected energy of laser radiation is >20 kJ.

Calculations of the nuclear physics characteristics of the IKAR-500 type facilities are a very difficult problem, because the reactor core has significant dimensions, high porosity due to the presence of a large number of laser channels, and a clearly pronounced anisotropy of the leakage neutron field, which is not typical for traditional reactor designs. The calculations of reactor core characteristics of such facilities must therefore be confirmed by direct physical experiments. In this regard,

Fig. 6.8 General view of the IKAR-S core [16, 17]

VNIIEF has developed the IKAR-S critical stand (Figs. 6.8 and 6.9) for experimental investigation of the nuclear physics characteristics of the IKAR-500 reactor [18, 19]. The basic difference between the IKAR-S critical stand and the IKAR-500 reactor is the absence of full-fledged control elements.

Absorbing rods made of boron carbide, B_4C, are used as control elements for reactivity adjustment on the critical stand. Each of the control elements includes two absorbing rods. The reactivity control elements are functionally divided into four groups: two groups of reactivity control rods and two groups of emergency shutdown rods. The KNK-4 and KNK-15 chambers are used to register neutrons. Two plutonium-beryllium neutron sources are included as standard in the design of the critical stand.

Each reactor module consists of two fuel sections, which constitute a set of alternating layers of graphite and dispersed uranium-aluminum fuel elements (72 fuel elements in each section). A fuel element is a 5 × 60 × 900 mm uranium-aluminum plate containing 2.5 % uranium (90 % enriched in ^{235}U), confined in a vacuum-sealed case made of Zr-Nb alloy with walls 0.5-mm thick.

Recently performed experimental and computational investigations [19–21] of the nuclear-physical characteristics of the IKAR-S stand made it possible to determine the effective neutron multiplication factor at various stages of core assembling, and showed the possibility of forecasting the critical mass parameters of the RL.

It was initially assumed that the IKAR-S would consist of sections in accordance with version 1 (Fig. 6.10a), which most fully models the IKAR-500 reactor core,

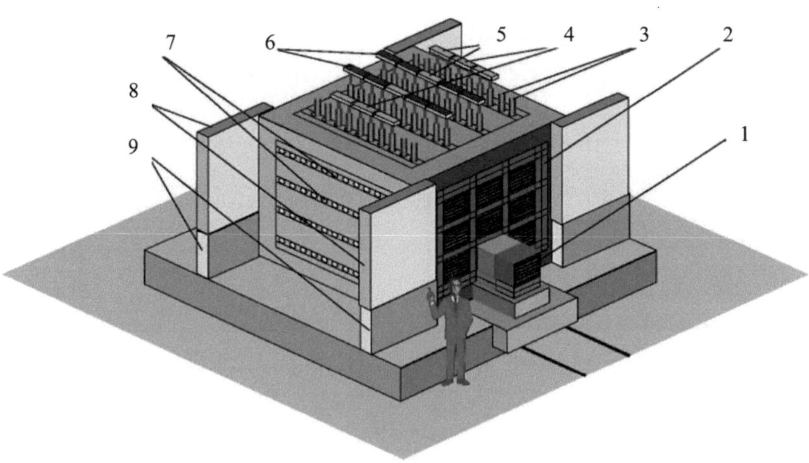

Fig. 6.9 Structural diagram of IKAR-S critical stand [19]: (*1*) reactor module of two fuel sections or LM-16 laser module; (*2*) graphite assembly; (*3*) neutron detector channels; (*4*) neutron source mechanisms; (*5*) control rods; (*6*) emergency shield rods; (*7*) horizontal channels; (*8*, *9*) shielding gates

reproducing the structure of the laser modules. The results of preliminary estimates using Nuclear Data Library (BAS) neutron constants for version 1 yielded an effective neutron multiplication factor $k_{ef} = 1.076$. The actually obtained value of $k_{ef} = 0.906$.

To bring IKAR-S to a critical state, the decision was made to increase the content of graphite in the sections while keeping the same number of fuel elements, which was supposed to lead to a shift of the neutron spectrum to the thermal range and a reduction in the core porosity, and accordingly, to a growth in k_{ef}. The corresponding calculations of k_{ef} were made using ENDF-B5 neutron constants, which were normalized to the experimental results obtained when IKAR-S was assembled per version 1. A diagram of the reactor sections, with a 46 % increase in the graphite content (version 2), is shown in Fig. 6.10b. In this case $k_{ef} = 0.988$ was obtained.

A further increase in reactivity was obtained when sections of the core's lower row were modified in accordance with Fig. 6.10c (version 3) by raising the fuel elements to the upper part of the section (closer to the center of the reactor core). According to calculations, the expected increase in k_{ef} was 0.012, while the value obtained experimentally was 0.004; that is, as a result of reconfiguration of sections in the module of the lower row, $k_{ef} = 0.992$ was obtained.

Some discrepancy of calculated and experimental results may be explained by inadequate allowance for the specific features of the reactor core in the computational model. In particular, when different versions of the reactor core were calculated, the significant influence of the quantity of aluminum (alloys) in the sections on the reactivity was noted; an increase in the quantity of aluminum (all other conditions remaining equal) led to a reduction in reactivity.

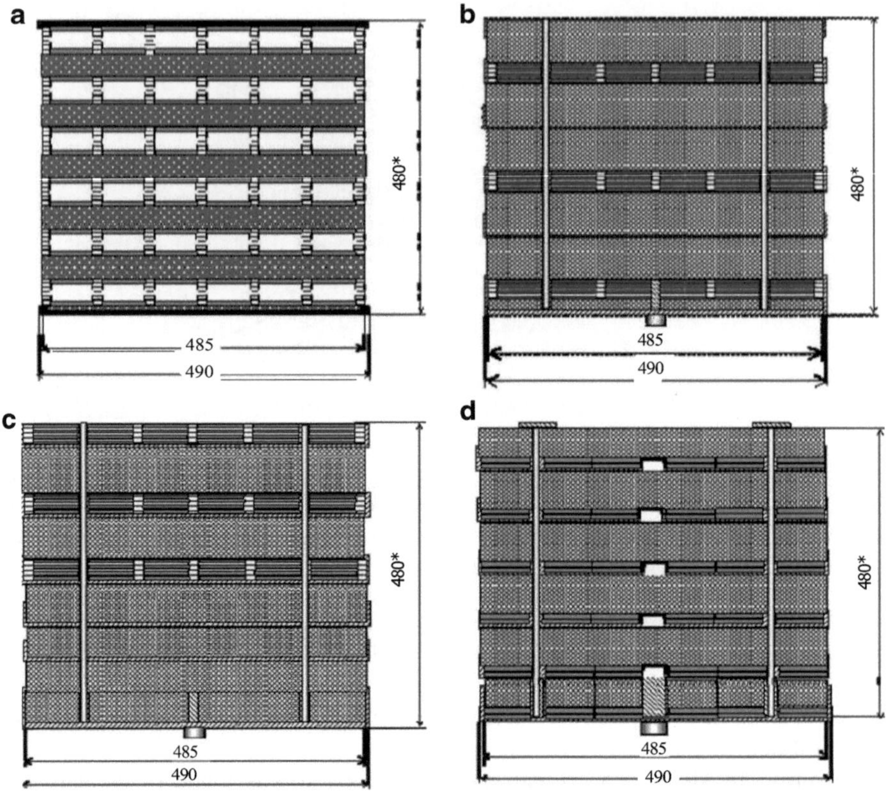

Fig. 6.10 Diagram of reactor section: (**a**) per version 1; (**b**) per version 2; (**c**) per version 3; (**d**) per version 4 of the studies and was examined in further experiments

In order to study this influence and achieve a critical state of the reactor core, the sections of version 2 of the central module were replaced with sections of version 4 (Fig. 6.10d), in which the aluminum details simulating radiators for cooling of the gas mixture were removed. The quantity of fuel elements and zirconium in the sections was not changed, while the quantity of graphite was increased by roughly 2 %. A value of $k_{ef} = 0.998$ was obtained.

After removal of the central source mechanism, a massive component of aluminum alloy, from the reactor core, the system achieved a critical state; moreover, the reactivity margin of the reactor core was $0.62\beta_{eff}$. To reduce the reactivity margin of the reactor core, a mockup of the temperature sensor was additionally (in mirror-image symmetry) placed in the horizontal channel of the core where the sensor is placed. As a result, the margin reactivity amounted to $0.35\beta_{eff}$. The reactivity was determined from the reactor acceleration period, which was 14 s. Figure 6.11 shows the configuration of the reactor core that was assembled as a result of the studies and was examined in further experiments.

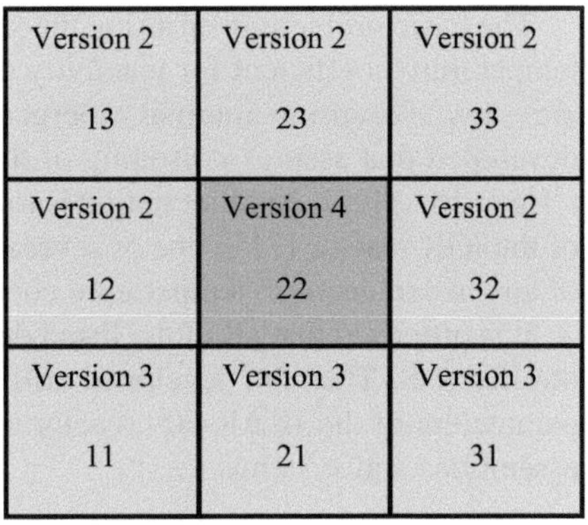

Fig. 6.11 Diagram of IKAR-S stand with nine modules in a critical state

In the course of the studies on physical startup of the critical stand, the characteristics (k_{ef}) were determined for roughly 40 different multiplying systems. The studies looked at how reactivity was affected by a reduction in reactor core porosity (by increasing the graphite content in the sections), an increase in the number of fuel elements in the central module, and a reduction in the quantity of aluminum alloy modeling the cooling radiators for the gas medium in the RL. Selection of the critical configuration of the critical stand core was accompanied by Monte Carlo calculations of k_{ef}. As new experimental data were obtained, the computational model was adjusted, which ultimately made it possible to quite accurately describe the multiplying properties of an IKAR type reactor core.

It was proposed that the IKAR-500 reactor be built on the basis of the IKAR-S critical stand. For this it was necessary to:

- select the reactor section configuration that would assure the required reactivity margin ($\sim 3\beta_{eff}$);
- ensure obtaining of a neutron pulse of the required (rectangular) shape;
- ensure safety of reactor operation.

The results of the physical startup showed that the proposed IKAR-500 reactor core configuration with sections per version 1 (Fig. 6.10a) did not assure the required reactivity margin when the existing fuel elements, with a content of 2.5 % uranium in a uranium-aluminum alloy, were used. One of the options to resolve the problem is to use "bare" fuel cores when the entire fuel section is sealed, which greatly reduces the mass of the zirconium alloy in the reactor core.

To assure the required pulse shape, it is proposed that a so-called reactivity modulator be used. This is a structure made of absorbent rods arranged in horizontal channels and controlled by linear stepper motors.

For fast transition of the reactor to a subcritical state, a system of fast-acting emergency shielding was developed based on the control elements existing on the IKAR-S stand. The absorbent rods are dropped to the lower position using compressed-air "guns" which reduces the reactor shutdown time to 0.2 s.

The basic problem in ensuring the safety of the IKAR-500 reactor is the lack of a temperature coefficient for reactivity damping when the characteristic pulse times are ~1 s. To ensure internal emergency shielding of the reactor, a system was developed that assures poisoning of the reactor core with a neutron-absorbing gas (^{3}He or BF_3) [22]. Another option is to use uranium-graphite fuel similar to the fuel of the IGR reactor [7] in one or several reactor modules. This ensures the presence of an "instantaneous" temperature coefficient for reactivity damping.

It is proposed that all of the listed developments be tested on the critical stand of the IKAR-S. Thus the developed critical stand makes it possible to determine the parameters of the IKAR-500 reactor and to develop the real design of its primary assemblies and systems.

6.3 Optical Nuclear Pumped Amplifier

In contrast to VNIIEF, where the concept of an autonomous stationary RL is being developed, since 1986, FEI (Obninsk) has been examining the concept of a high-power pulsed laser system based on an optical nuclear-pumped amplifier (OKUYaN) [23]. The "master oscillator–two-pass amplifier with phase conjugation" configuration underlay the work of the OKUYaN-based laser system [23, 24] (Fig. 6.12). The OKUYaN includes reactor and laser blocks. The reactor, which is

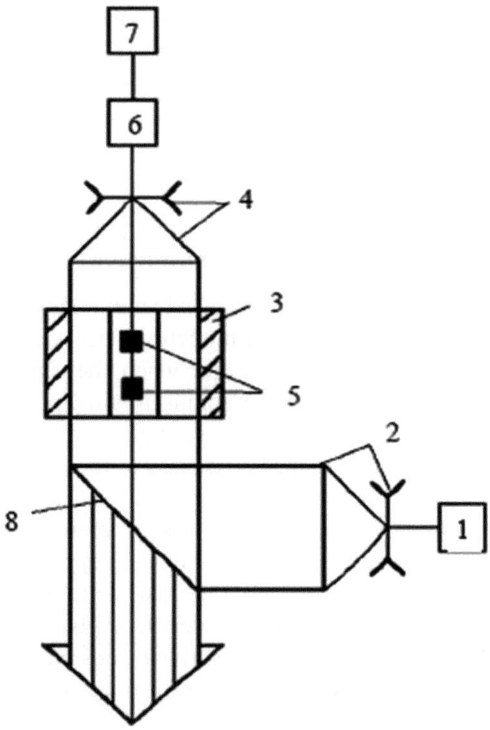

Fig. 6.12 Optical scheme of laser system based on OKUYaN [23, 24]: (*1*) master laser-oscillator; (*2*, *4*) telescope systems; (*3*) optical nuclear-pumped amplifier (OKUYaN); (*5*) pulsed reactor core; (*6*) Faraday cell; (*7*) phase conjugation cell; (*8*) polarizer

sometimes called an "initiating" reactor, presupposes use of a multicore fast-neutron pulsed reactor. The laser block constitutes a booster subcritical core in which neutrons are multiplied. We note that the first proposal to develop such devices appeared more than 20 years ago (see, for example, review [25]). Requirements for "initiating" reactors, their arrangement, and the results of calculation of the neutron physical characteristics, are considered in study [26].

At present there is an OKUYaN demonstration model operating at FEI, "Stand B" [23, 24, 27, 28]. It consists of two parts: the so-called "first workplace," consisting of a two-core BARS-6 reactor and an experimental section for studying the characteristics of individual lasing elements (cells), and the "second workplace." Organization of experiments for the "first workplace" was examined in Sect. 6.2.

The other part of Stand B (the "second workplace"), which includes the two-core BARS-6 reactor and laser block with a volume of around 2.5 m^3, was put in operation in 1999. The design and the neutron-physical characteristics of the BARS-6 reactor were considered in some detail in studies [23, 28]. The BARS-6 reactor, built on the basis of the BARS-5 reactor [7, 29] from the design documentation and with the technical assistance of VNIITF (Snezhinsk), has two reactor cores, which are arranged on a platform and can be moved along a rail track to one of the two "workplaces" or to the biological shield, into which the core of the reactor is moved for the period of preparation for experiments.

The relative arrangement of the BARS-6 reactor and laser block is shown in Fig. 6.13. The laser block (shown in Fig. 6.14) is a cylindrical structure with a diameter of 1.7 m and length of 2.5 m, with a longitudinal axial cavity for placement of two BARS-6 reactor cores. It consists of up to 800 laser elements, their simulators, and elements of the neutron reflector, made in the form of tubes 49-mm in diameter and 2.5-m long, filled with polyethylene. The space between the tubes of the laser block contains around 760 neutron moderator elements made from shaped polyethylene. The laser block, in neutron-physical relations, constitutes a deep subcritical core with a neutron multiplication coefficient $k_{ef} \leq 0.9$. Its neutron-physical characteristics were optimized in study [30].

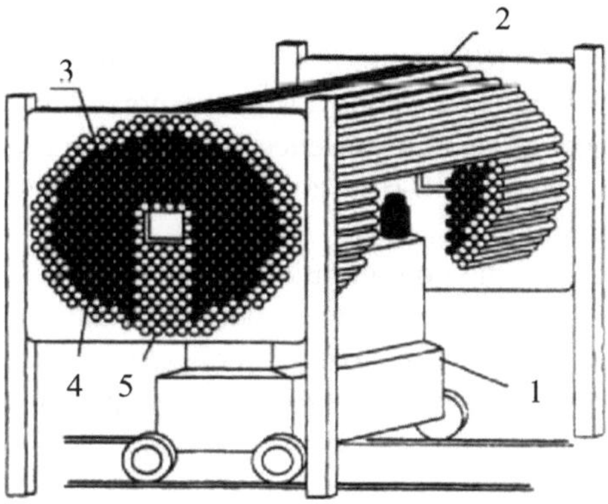

Fig. 6.13 Arrangement of the BARS-6 reactor and laser block at the "second workplace" [23, 24]: (*1*) two-core BARS-6 reactor; (*2*) laser block; (*3*) laser elements and its simulators; (*4*) outside neutron reflector; (*5*) inside neutron reflector

Fig. 6.14 Laser block [24]

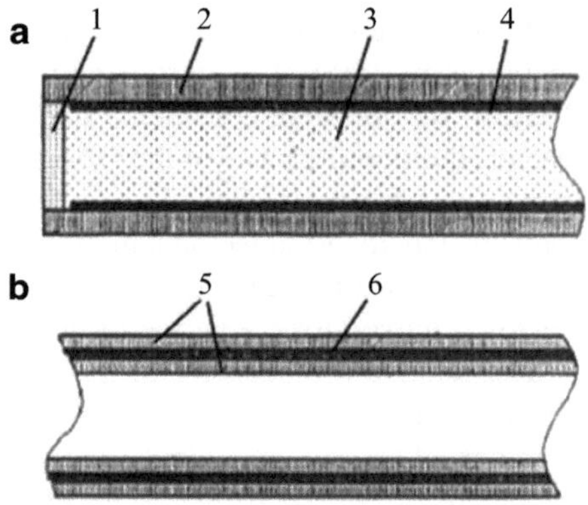

Fig. 6.15 Design of laser element (**a**) and its simulator (**b**) [23, 27, 28]: (*1*) optical window; (*2*) stainless steel tube; (*3*) laser medium; (*4*) metallic ^{235}U; (*5*) aluminum tubes; (*6*) $^{235}U_3O_8$

The laser element (Fig. 6.15a) is a stainless-steel tube with a diameter of 50 mm, wall thickness of 0.5 mm, and length of 2.5 m, which is coated on the inside with a 5 μm thick layer of metallic ^{235}U. The laser element simulator is in the form of two annular aluminum tubes, with the small space between them filled with $^{235}U_3O_8$ (Fig. 6.15b). It is a copy of a laser element in neutron-physical respects and contains the same amount of ^{235}U (32 g).

The specific features of neutron pulse generation in the reactor with the laser block are discussed in studies [23, 28]. The half-width neutron pulse duration for the BARS-6 reactor without the laser block is around 100 μs, while in the reactor with the laser block it is much more, 2 ms when there is an internal reflector present, and 20 ms in its absence. At the same time, the energy release in each reactor core of the BARS-6 is equal to 2×10^{17} fissions, and in the laser block, 2.2×10^{17} fissions.

For the configuration "reactor + laser block" with an internal reflector, experiments were carried out with the use of laser elements as part of a laser block [23, 31]. The main goal of the experiments is to study the operating modes of the "master oscillator–single-pass amplifier" configuration. In contrast to the amplifier element, the master oscillator has windows at the ends arranged at a Brewster angle, and a stable laser cavity consisting of spherical and flat dielectric mirrors. A telescope was used to allow passage of the beam of the master oscillator through the amplifier without losses. The laser medium was a He-Ar-Xe mixture ($\lambda = 2.03$ μm) at a pressure of 1.05 atm. The specific power deposition, averaged over the active volume, at the maximum of the neutron pulse for the amplifier element and the master oscillator were 40 W/cm^3 and 80 W/cm^3, respectively, at a half-width neuron pulse duration of 1.8 ms.

From the energy parameters of the laser radiation measured at the amplifier input and output, the small-signal gain (α_0) and saturation intensity (I_s) of the laser medium were determined: $\alpha_0 = 8.1 \times 10^{-3}$ cm^{-1}; $I_s = 92$ W/cm^2. Using these parameters, it is possible to determine the laser power efficiency $\eta_l = \alpha_0\ I_s/q \approx 2$ %, which agrees with the results obtained when NPLs were studied (see Table 3.5, and Chap. 3, Sect. 3.1). The results of these investigations allow one to hope for high energy parameters when there is full-scale operation of the entire OKUYaN laser block.

6.4 LIRA Laser-Reactor Setup

The design of the LIRA setup [32, 33] proposed by VNIITF associates is one version of a high-powered pulsed nuclear-laser facilities. It is presumed that the LIRA setup (like the OKUYaN) consists of two blocks: a rather large subcritical block made from laser elements in the form of long thin-walled pipes, and two relatively small "initiating" pulsed reactors disposed inside this block (Fig. 6.16).

The reactor block contains two DRAKON pulsed solution reactors operating synchronously (half-width of pulse duration is 2 ms, energy release in the core is 30 MJ). The ELIR reactor [29], which was operated for many years at VNIITF, was

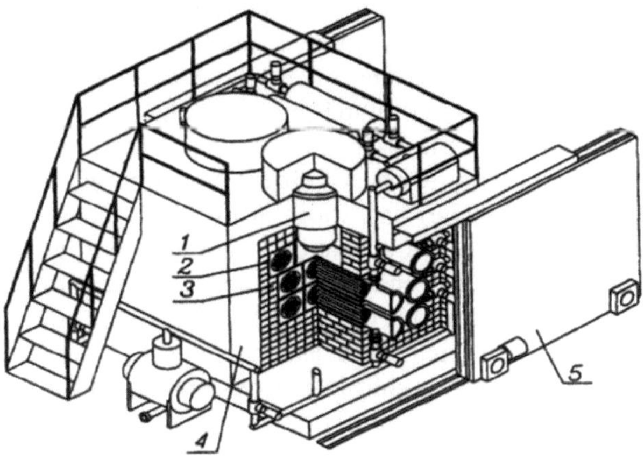

Fig. 6.16 LIRA setup scheme [32, 33]: (*1*) one of two "initiating" reactors; (*2*) laser block; (*3*) laser cell; (*4*) concrete shielding; (*5*) sliding door

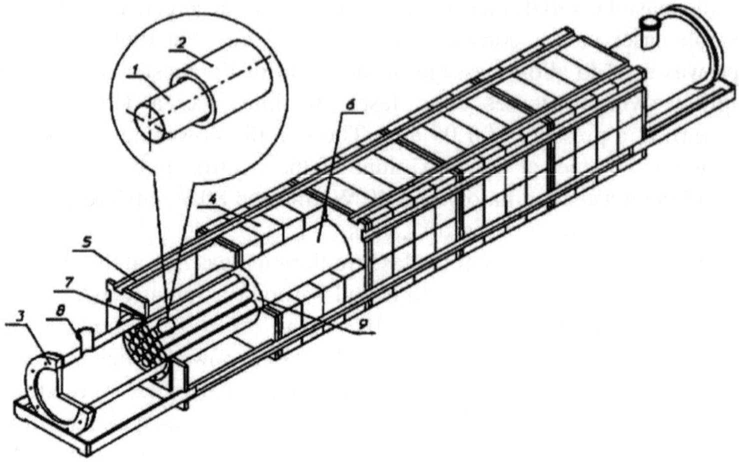

Fig. 6.17 Laser assembly of LIRA setup [33]: (*1*) laser element; (*2*) simulator; (*3*) optical window; (*4*) graphite neutron moderator; (*5*) fastening element; (*6*) aluminum tube; (*7*) flange; (*8*) exhaust pipe; (*9*) water neutron moderator

used as the basis of this reactor. The basic calculated characteristics of the DRAKON reactor are given in study [33].

The LIRA setup laser block consists of seven identical amplification laser assemblies, each of which consists of 19 laser elements. An eighth assembly includes eight laser elements, one of which is a master oscillator. The laser elements are enclosed in an aluminum tube 300-mm diameter, and are arranged it at intervals of 60 mm. The pipe is filled with water and surrounded by graphite, forming a laser assembly (Fig. 6.17). The outside dimensions of the assembly are 43 × 43 cm in cross section, with a length of 3.8 m. On the outside, the laser assemblies are surrounded by a neutron polyethylene moderator 10-cm thick.

Each laser element consists of three concentric tubes. A 6 μm thick layer of metallic ^{235}U is deposited to the inside surface of the first, innermost, tube, which is 48 mm in diameter. The two outer aluminum tubes have a 12 μm layer of ^{235}U between them and constitute a simulator, which is necessary to obtain a neutron multiplication coefficient in the laser block of $k_{ef} \leq 0.9$. The total mass of the uranium in the laser setup is around 20 kg.

To optimize the laser block design, calculations were carried out to study the effect on energy release of different neutron moderators in the assemblies (polyethylene, graphite) and water in the space between the laser elements and the absorbent screens. Calculations were also used to determine the temperature of the uranium layers and the gas pressure in the laser element after the pulse.

The optical circuit of the LIRA setup is shown in Fig. 6.18. One of the laser elements of the eighth laser assembly is a master oscillator. Laser radiation from the master oscillator is split into seven beams of identical power and directed to the other seven laser elements of this same assembly for preliminary amplification. Then the laser radiation is expanded using telescopes to a diameter of 300 mm and

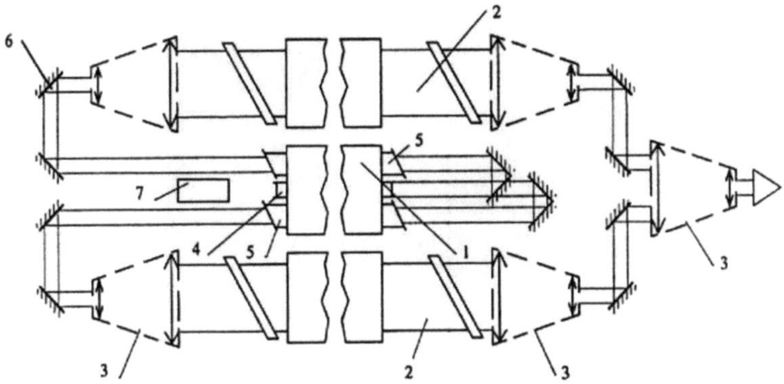

Fig. 6.18 LIRA setup optical channel scheme [32, 33]: (*1*) master oscillator assembly; (*2*) amplification laser assembly; (*3*) telescope; (*4*) master oscillator; (*5*) pre-amplification laser element; (*6*) rotary mirrors; (*7*) adjustment laser

Fig. 6.19 Multichannel laser assembly close to the BARS-5 reactor core [34]

goes to the other amplifier assemblies. After amplification, the beam aperture is reduced to 100 mm and output past withdrawn beyond the biological shielding.

In the opinion of the authors of [32, 33], when a He-Ar-Xe mixture ($\lambda = 2.03$ μm) is used as the laser medium, the specific energy deposition to the gas medium is about 1 J/cm^3, while the specific energy deposition is ~8 mJ/cm^3 for a lasing pulse duration of 5 ms. Under these conditions, the full energy of laser radiation of the LIRA setup can reach ~4.5 kJ.

In study [34], experimental investigations of a multichannel laser assembly on the BARS-5 pulsed reactor [29] were carried out (Figs. 6.19 and 6.20). The assembly consists of a package of 19 stainless steel tubes with an internal diameter of 15 mm and wall thickness of 0.2 mm. Each of the tubes held thin-walled aluminum tubes with a layer of $^{235}U_3O_8$ (layer thickness 2.5 mg/cm^2, layer length 1,050 mm). The specific power deposition of the gas mixtures He-Ar-Xe ($\lambda = 2.03$ μm) and He-Ar ($\lambda = 1.79$ μm) reached 5 kW/cm^3 at a helium pressure of 3 atm. The half-width pulse duration of the thermal neutrons was about 130 μs.

Fig. 6.20 Photo of end of laser assembly [34]

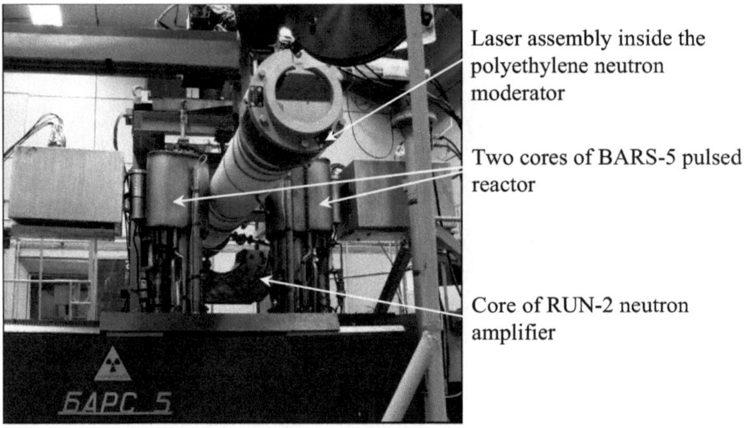

Fig. 6.21 Laser assembly inside the BARS-5 + RUN-2 facility [35]

Under these conditions, laser radiation energy from the entire assembly was 9.9 J (2.03 μm) and 10.3 J (1.78 μm). The output power was around 30 kW.

Higher energy parameters were recently obtained in experiments [35] when a multichannel laser assembly was irradiated by neutrons using the reactor facility BARS-5 + RUN-2 which consists of BARS-5 pulsed reactor and RUN-2 neutron amplifier (see reference [29]). One end of laser assembly was located between the two cores of the BARS-5 reactor and the other one was located inside the RUN-2 core (see Fig. 6.21).

The laser assembly in reference [35] consists of the 37 stainless steel tubes with a wall thickness of about 0.1 mm and a length of 1.5 m. The end of the laser assembly had approximately the same design as shown in Fig. 6.20. Inside the each tube, the ten aluminum cylinders were placed with the inner diameter of 19.5 mm which were covered by thin layers of $^{235}U_3O_8$ (with a layer thickness of 2.5 mg/cm^2). To

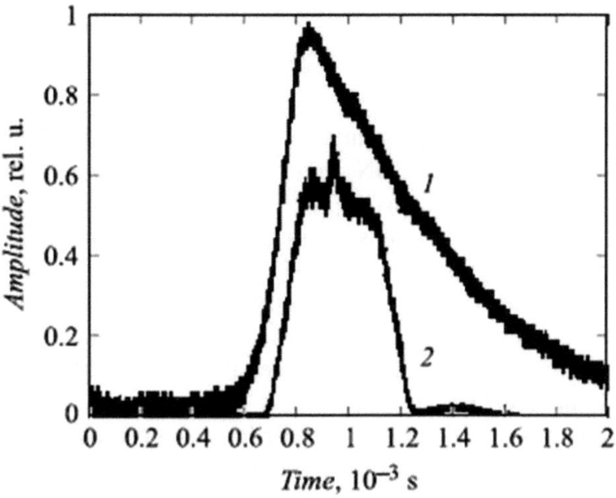

Fig. 6.22 Oscillorams of thermal neutron pulse (*1*) and laser pulse (*2*) with wavelength of 2.03 μm for the He-Ar-Xe (700:700:1) mixture at 4 atm pressure [35]

enhance the thermal-neutron flux, this laser assembly was located inside a cylindrical polyethylene neutron moderator with the wall thickness of 60 mm.

As described in reference [35], investigations were carried out using a He-Ar-Xe (700:700:1) mixture at 4 atm pressure. Lasing occurs on a transition in the Xe atom ($\lambda = 2.03$ μm). As an example of the result, Fig. 6.22 shows the oscillogram of one experiment. These experiments demonstrated record energy parameters for NPLs excited with help of pulsed reactors: laser output energy of 520 J with a laser pulse duration of about 400 μs (laser power output is 1.3 MW, $\eta_l \sim 3$ %). Such high laser energy parameters are attributed to the large active laser volume (16l) and high specific input power deposition (~1.5 kW/cm^3) which was distributed rather uniformly along the length of laser tubes.

Study [36] assessed the maximal possible energy parameters of pulsed nuclear-laser devices designed by an "initiating reactor–subcritical laser block" scheme. With allowance for the laser media efficiencies obtained up to now of $\eta_l \sim 1$ % (for example, for xenon NPLs) and the limitations associated with the operating modes of the laser element and the specific features of neutron pulse in such a system, it is possible to obtain a full laser radiation energy of ~2.5 MJ for a pulse duration of ~5 ms. Characteristic dimensions of a laser block that includes 10^4–10^5 laser elements with a total active volume of ~250 m^3 is 7–10 m. Three pulsed TRIGA- or ACPR-type reactors [7] with an energy release in the reactor core of 100–200 MJ can be used as "initiating" reactors.

6.5 Development of Laser Setups Based on Gas NPLs in the United States

In the United States, high-power NPL-based laser setups were developed in the scope of the FALCON program (Fission-Activated Laser CONcept) with Department of Energy support [37]. The basic executors of this program were the Sandia Laboratories with the support of the Idaho National Engineering Laboratory. Other participants included the Air Force Philips Laboratory, Los Alamos National Laboratory (LANL), Oak Ridge National Laboratory (ORNL), Mission Research Corporation, Babcock & Wilcox, W.J. Schafer Associates, Lockheed, Kaman, Science Research Labs, University of New Mexico, University of Illinois, and Texas A&M University. As is noted in study [37], the program examined two versions of high-power nuclear laser setups: a stationary RL with a laser radiation power of several megawatts, and a laser setup with a power of about 500 kW, which was pumped using a pool-type reactor. It is proposed that high-power NPL-based setups can be placed not only on land, but on ships as well.

Experimental investigations [38, 39] directed at modeling the pumping conditions of laser media in high-power nuclear-laser setups (such as large volumes of medium, long pumping pulses, effect of the medium-flowing and -heating on the quality of the laser beam) were carried out at Sandia Laboratories on an ACRR pool-type pulsed reactor. The description and characteristics of the ACRR reactor are provided, for example, in monograph [7]. The cylindrical core of the ACRR reactor, with a diameter of 80 cm and height of about 51 cm, is made of a mixture of UO_2-BeO oxides placed on the bottom of a tank of water at a depth of 7 m (Fig. 6.23). The duration of single pulses varied from 7 to 250 ms. At maximal energy release in the reactor core of 300 MJ, the neutron flux density in the central channel was 8.7×10^{17} cm^{-2} s^{-1} [38].

The first experiments with NPLs on the ACRR reactor were carried out in 1987, with irradiation of a $1.5 \times 7 \times 60$ cm laser cell in the central channel. An Ar-Xe mixture was used as the active medium. Placement of a laser cell with large dimensions in the central channel was difficult, because the channel has a diameter of just 22.8 cm. Therefore, in subsequent experiments, an ALEC (Advanced Laser External Cavity) was placed close to the surface of the core in a rectangular cavity with a cross-section of 20×132 cm^2 (Fig. 6.23).

Successful experiments with the ALEC device made it possible to obtain baseline data for study of stationary modes of NPL operation with flowing of the gas medium through the active volume. For this purpose, experiments were carried with FLE-1 and FLE-2 (Flowing Laser Experiments) devices, which were irradiated in the cavity of the FREC-II subcritical assembly [7], located close to the ACRR reactor core. Centrifugal air compressors were used to pump the gas. The active volume of the laser was ~5l, with specific power depositions of up to several tens of W/cm^3. The next phase of the investigations at Sandia involved experiments with the VSE (Volume Scaling Experiment) device, with an active volume of 50l, specific pumping power of 5 W/cm^3, and operating time of 0.2 s [39].

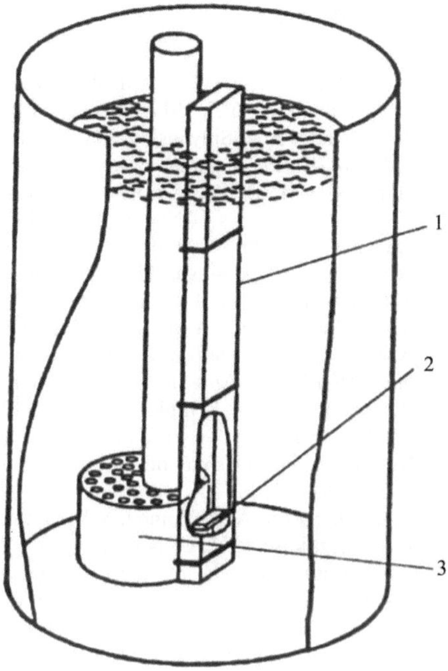

Fig. 6.23 ACRR reactor with ALEC laser device [38]: (*1*) ALEC device; (*2*) laser cell; (*3*) ACRR reactor core

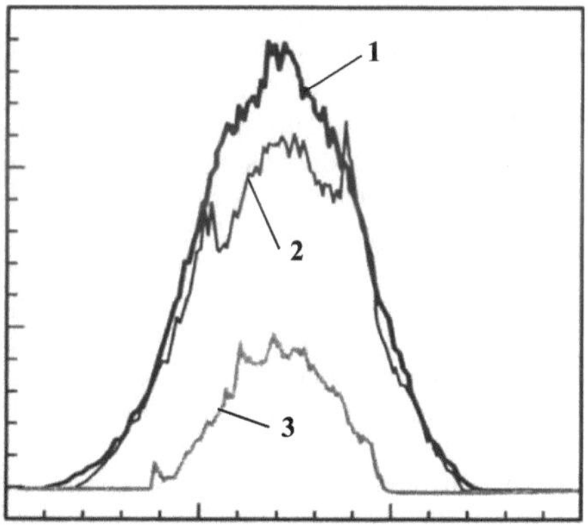

Fig. 6.24 Oscillograms of laser radiation for Ar-Xe mixture ($\lambda = 1.73$ μm) at argon pressures of 0.44 atm (*1*), 0.88 atm (*2*), and 1.25 atm (*3*) (specific power deposition 7 W/cm^3; scale division 2 ms) [40]

Unfortunately, studies [38, 39] provided only a schematic description of the conditions of experiments on the ACRR reactor, and there are virtually no data on the characteristics of the laser radiation. Only in study [40] is there some information on the results of the first experiments with the Ar-Xe mixture ($\lambda = 1.73$ μm) at various pressures and specific power depositions in the case of placement of the laser cell in the central channel of the reactor (Fig. 6.24).

References

1. V.N. Bogdanov, A.G. Vasilenko, V.A. Zherebtsov, V.V. Ivanov, M.I. Kuvshinov, A.Y. Lakhtikov, V.S. Mayornikov, A.B. Modenov, A.V. Nikulin, V.A. Nor-Arevyan, A.N. Pokalo, A.A. Sinyanskiy, I.G. Smirnov, S.L. Turutin, M.V. Khlestkov, V.V. Shirokov, *LM-4/BIGR experimental complex. Design and operating principle* (Eksperimentalnyy kompleks LM4/BIGR. Ustroystvo i printsip raboty). Proceedings of the 2nd International Conference "Physics of Nuclear-Excited Plasma and Problems of Nuclear-Pumped Lasers," Arzamas-16, 1995, vol. 2, pp. 172–178.
2. A.M. Voinov, A.A. Sinyanskiy, A.G. Vasilenko, O.A. Golubeva, L.Y. Dovbysh, V.V. Ivanov, M.I. Kuvshinov, B.V. Lazhintsev, A.Y. Lakhtikov, S.P. Melnikov, A.B. Modenov, V.A. NorArevyan, A.M. Pichugin, A.N. Pokalo, V.V. Porkhayev, Y.A. Pospelov, A.V. Sinitsyn, I.G. Smirnov, S.L. Turutin, M.V. Khlestkov, *LM-4/BIGR experimental complex. Obtaining lasing in an Ar-Xe mixture with a duration of more than 1 second* (Eksperimentalnyy kompleks LM4/BIGR. Polucheniye generatsii v smesi Ar-Xe s dlitelnostyyu boleye 1 s). Proceedings of the 2nd International Conference "Physics of Nuclear-Excited Plasma and Problems of Nuclear-Pumped Lasers," Arzamas-16, 1995, vol. 2, pp. 179–185.
3. A.A. Sinyanskii, A.M. Voinov, V.V. Gorbunov, L.E. Dovbysh, A.I. Konak, M.I. Kuvshinov, B.V. Lazhintsev, V.A. Nor-Arevyan, V.V. Porkhaev, A.V. Sinitsyn, I.G. Smirnov, S.L. Turutin, *Steady-state nuclear-pumped lasers with duration of 0.5–2 s*. Proc. Int. Conf. LASERS '96, McLean, 1997, pp. 100–106.
4. A.A. Sinyanskii, V.N. Krivonosov, Powerful nuclear-optical converters for applications in scientific investigations and national economy. Proc. SPIE **3574**, 440–448 (1998)
5. A.A. Sinyanskiy, *Multichannel nuclear-laser quasistationary setups in a BIGR reactor* (Mnogokanalnyye yaderno-lazernyye ustanovki kvazinepreryvnogo deystviya na reaktore BIGR). Proceedings of the 3rd International Conference "Problems of Nuclear-Pumped Lasers and Pulsed Reactors," Snezhinsk, 2003, pp. 377–387.
6. S.V. Patyanin, A.V. Lisenkov, A.A. Pikulev, A.A. Sinyanskiy, V.D. Grigoryev, Y.M. Limar, P.V. Sosnin, S.L. Turutin, V.M. Tsvetkov, Multichannel nuclear-pumped laser module based on noble gases (Mnogokanalnyy lazernyy modul na smesyakh blagorodnykh gazov s yadernoy nakachkoy). Opticheskiy zhurnal **69**(7), 33–39 (2002)
7. V.F. Kolesov, *Aperiodic Pulsed Reactors (Aperiodicheskiye impulsnyye reaktory)* (RFYaTs-VNIIEF, Sarov, 1999)
8. A.Y. Lakhtikov, V.S. Mayornikov, N.D. Rabion, A.A. Sinyanskiy, A.G. Stepanov. *Automated subsystem for optical tract calibration* (Avtomatizirovannaya podsistema yustirovki opticheskogo trakta), Proceedings of the 2nd International Conference "Physics of Nuclear-Excited Plasma and Problems of Nuclear-Pumped Lasers," Arzamas-16, 1995, vol. 2, pp. 77–80.
9. S.V. Patyanin, P.V. Sosnin, A.V. Lisenkov, V.M. Tsvetkov, V.A. Sharavin, *Calibration system of LM-4/BIGR multichannel laser module* (Sistema yustirovki mnogokanalnogo lazernogo modulya LM-4/BIGR). Proceedings of the 3rd International Conference "Problems of Nuclear-Pumped Lasers and Pulsed Reactors," Snezhinsk, 2003, pp. 353–357.
10. O.A. Golubeva, V.V. Gorbunov, L.Y. Dovbysh, A.M. Dyudyayev, Y.A. Pospelov, A.V. Sinitsyn, *A system for automated measurement of LM-4 laser module parameters* (Sistema avtomatizirovannykh izmereniy parametrov lazernogo modulya LM-4). Proceedings of the 2nd International Conference "Physics of Nuclear-Excited Plasma and Problems of Nuclear-Pumped Lasers," Arzamas-16, 1995, vol. 2, pp. 81–86.
11. S.V. Patyanin, A.A. Sinyanskiy, A.V. Lisenkov, L.V. Lvov, P.V. Sosnin, S.L. Turutin, V.M. Tsvetkov, V.A. Sharavin, *Investigation of lasing parameters in single and added channels of the LM-4 laser module* (Issledovaniye parametrov generatsii v odinarnom i slozhennom kanalakh lazernogo modulya LM-4). Proceedings of the 3rd International

Conference "Problems of Nuclear-Pumped Lasers and Pulsed Reactors," Snezhinsk, 2003, pp. 227–232.

12. A.A. Pikulev, L.V. Lvov, A.V. Lisenkov, V.D. Grigoryev, A.N. Korzenev, *Interferometric investigations on the JM-4/BIGR experimental complex* (Interferometricheskiye issledovaniya na eksperimentalnom komplekse LM-4/BIGR). Proceedings of the 3rd International Conference "Problems of Nuclear-Pumped Lasers and Pulsed Reactors," Snezhinsk, 2003, pp. 291–297.
13. A.A. Pikulev, Interference methods of determining the energy deposition in flowing-gas nuclear-pumped lasers (Interferentsionnaya metodika opredeleniya energovklada v prokachnykh lazerakh s yadernoy nakachkoy). Kvantovaya elektronika **31**(6), 500–504 (2001)
14. A.A. Sinyanskiy, Y.M. Limar, S.V. Patyanin, A.A. Pikulev, S.V. Sorokin, A.A. Sustatov, P.V. Sosnin, S.L. Turutin, V.M. Tsvetkov, N.N. Yurkov, *Nuclear-laser module LM-8. Investigation of laser characteristics* (Yaderno-lazernyy modul LM-8. Issledovaniye generatsionnykh kharateristik). Proceedings of the 4th International Conference "Physics of Nuclear-Pumped Lasers and Pulsed Reactors," Obninsk, 2009, vol. 1, pp. 137–145.
15. S.L. Turutin, A.A. Sinyanskiy, *A multichannel nuclear laser module* (Mnogokanalnyy yaderno-lazernyy modul). Proceedings of the 3rd International Conference "Problems of Nuclear-Pumped Lasers and Pulsed Reactors," Snezhinsk, 2003, pp. 421–425.
16. R.I. Ilkayev, *Physics of high-density energy* (Fizika vysokikh plotnostey energii). Fourth Practical Science Conference of the Russian Minatom "Using the Attainments of Fundamental Research in Nuclear Technologies," Moscow, 2003, pp. 123–137.
17. R.I. Ilkayev, *Research in the field of laser physics and high-temperature plasma* (Issledovaniya v oblasti fiziki lazerov i vysokotemperaturnoy plazmy). Fundamental Research of RFYaTs-VNIIEF in the Field of High-Density Energy (Fundamentalnyye issledovaniya RFYYaTs-VNIIEF v oblasti vysokikh plotnostey energii), Sarov, 2003, pp. 13–34.
18. A.M. Voinov, S.V. Vorontsov, V.N. Krivonosov, S.P. Melnikov, A.A. Pikulev, V.T. Punin, A.N. Sizov, A.A. Sinyanskiy, S.L. Turutin, *Studies fulfilled at VNIIEF to investigate the possibility of developing a reactor-laser* (Raboty, vypolnennyye vo VNIIEF, po issledovaniyu vozmozhnosti sozdaniya reaktora-lazera). Voprosy atomnoy nauki i tekhniki. Ser. Fizika yadernykh reaktorov 2007, (2), 19–28.
19. A.M. Voinov, G.V. Vlokh, S.V. Vorontsov, V.N. Krivonosov, V.A. Popov, A.A. Sinyanskiy, A.N. Sizov, S.V. Sorokin, V.I. Turutov, Y.M. Limar, A.G. Vasilenko, A.N. Kuznechenkov, I.A. Nikitin, I.A. Gorshkova, *IKAR-S critical stand. Results of physical startup* (Stend kriticheskikh sborok 'IKAR-S.' Rezultaty fizicheskogo puska). Voprosy atomnoy nauki i tekhniki. Ser. Fizika yadernykh reaktorov 2007, (2), 29–33.
20. A.M. Voinov, G.V. Vlokh, S.V. Vorontsov, A.K. Zhitnik, V.N. Krivonosov, V.A. Popov, A.A. Sinyanskiy, *Calculation and experimental determination of* k_{ef} *of multiplying systems modeling the core of a reactor-laser* (Raschet i eksperimentalnoye opredeleniye k_{ef} razmnozhayushchikh system, modeliruyushchikh aktivnuyu zonu reaktora-lazera). Proceedings of the 8th International Conference "VII Khariton Readings on the Problems of High-Density Energy Physics," Sarov, 2006, pp. 61–64.
21. A.M. Voinov, G.V. Vlokh, S.V. Vorontsov, A.K. Zhitnik, V.N. Krivonosov, V.A. Popov, A.A. Sinyanskiy, *Calculation and experimental determination of* k_{ef} *of multiplying systems modeling the core of the reactor-laser* (Raschet i eksperimentalnoye opredeleniye k_{ef} razmnozhayushchikh system, modeliruyushchikh aktivnuyu zonu reaktora-lazera). Proceedings of the 4th International Conference "Physics of Nuclear-Pumped Lasers and Pulsed Reactors," Obninsk, 2009, vol. 2, pp. 494–498.
22. A.M. Voinov, I.V. Kochetkova, V.N. Krivonosov, I.A. Nikitin, A.N. Sizov, Emergency protection element of a nuclear reactor (Organ avariynoy zashchity yadernogo reaktora), Patent RF SU 1810020 AZ kl. G21C 7/22. Inventions (Izobreteniya) **21**, 413 (1997)
23. P.P. Dyachenko, *On the status of research at the A.I. Leypunskiy state scientific center physics and power engineering institute on direct conversion of nuclear fission energy into the energy*

of laser radiation (O sostoyanii issledovaniy v GNTs RF-FEI im. A.I. Leypunskogo po pryamomu preobrazovaniyu energii deleniya yader v energiyu lazernogo izlucheniya). Proceedings of the 3rd International Conference "Problems of Nuclear-Pumped Lasers and Pulsed Reactors," Snezhinsk, 2003, pp. 5–30.
24. P.P. Dyachenko. *Nuclear-laser power engineering as a perspective direction in nuclear energy utilization*. Proceedings of the 8th International Conference on Emerging Nuclear Energy Systems, Obninsk, 1996, vol. 1, pp. 296–303.
25. R.T. Schneider, F. Hohl, Nuclear-pumped lasers. Adv. Nucl. Sci. Technol. **16**, 123–287 (1984)
26. A.V. Lukin, L.E. Magda, D.V. Khmelnitskiy, *Initiating reactors for a nuclear-pumped pulsed-periodic laser* (Zapalnyye reaktory dlya impulsno-periodicheskogo lazera s yadernoy nakachkoy). Voprosy atomnoy nauki i tekhniki. Ser. Fizika yadernykh reaktorov 1996, (2), 57–64.
27. A.V. Gulevich, P.P. Dyachenko, A.V. Zrodnikov, V.N. Kononov, Y.A. Prokhorov, V.Ya. Pupko, Energy model of a laser system pumped by a pulsed reactor (Energeticheskiy maket lazernoy sistemy s nakachkoy ot impulsnogo reaktora). Atomnaya energiya **80**(5), 361–365 (1996)
28. P.P. Dyachenko, O.A. Yelovskiy, Yu.A. Prokhorov, G.I. Fokin, P.A. Yakubov, A.A. Snopkov, 'Stand B' reactor-laser complex (Reaktorno-lazernyy kompleks 'Stend B'). Atomnaya energiya **88**(5), 337–342 (2000)
29. B.G. Levakov, A.V. Lukin, E.P. Magda, I.S. Pogrebov, A.A. Snopkov, V.A. Terekhin, *RFYaTs-VNIITF Pulsed Nuclear Reactors (Impulsnyye yadernyye reaktory RFYaTs-VNIITF)* (RFYaTs-VNIITF, Snezhinsk, 2002)
30. K.V. Berezhnoy, P.P. Dyachenko, O.A. Yelovskiy, O.F. Kukharchuk, Yu.A. Prokhorov, G.N. Fokin, P.A. Yakubov, *Optimization of neutron-physical characteristics of pulsed multicore reactor unit* (Optimizatsiya neytronno-fizicheskikh kharakteristik mnogozonnoy reaktornoy ustanovki impulsnogo deystviya). Proceedings of the 3rd International Conference "Problems of Nuclear-Pumped Lasers and Pulsed Reactors," Snezhinsk, 2003, pp. 443–448.
31. P.P. Dyachenko, Yu.A. Dyuzhov, A.A. Orlov, Ye.D. Poletayev, V.N. Smolskiy, O.Ye. Blagovestnyy, *Experimental investigation of laser characteristics of a nuclear-pumped optical amplifier* (Eksperimentalnoye issledovaniye lazernykh kharakteristik opticheskogo kvantovogo usilitelya s yadernoy nakachkoy). Proceedings of the 3rd International Conference "Problems of Nuclear-Pumped Lasers and Pulsed Reactors," Snezhinsk, 2003, pp. 140–145.
32. E.P. Magda, Powerful nuclear pumped lasers (review). Proc. SPIE **3574**, 93–103 (1998)
33. E.P. Magda, I.S. Pogrebov, I.S. Putnikov, A.V. Bochkov, A.I. Volkov, A.N. Maloyaroslavtsev, Yu.G. Gornitsyn, V.A. Vdovenko, Yu.M. Kononenko, Y.Z. Kandiyev, *LIRA laboratory laser-reactor unit* (Laboratornaya lazerno-reaktornaya ustanovka LIRA). Voprosy atomnoy nauki i tekhniki. Ser. Fizika yadernykh reaktorov 2001, (1/2), pp. 39–43.
34. S.A. Andreyev, A.A. Aradushkin, A.V. Bochkov, A.V. Zagidulin, D.V. Zaytsev, F.I. Krivenko, V.I. Litvin, E.P. Magda, V.V. Mironenko, A.S. Podymako, A.A. Snopkov, G.S. Sofiyenko, V.I. Cherashev, *A 10 J laser module* (Lazernyy modul na 10 Dzh). Proceedings of the 4th International Conference "Physics of Nuclear-Pumped Lasers and Pulsed Reactors," Obninsk, 2009, vol. 1, pp. 146–150.
35. A.V. Zagidulin, A.V. Bochkov, V.V. Mironenko, G.S. Sofienko, The 500-J nuclear-pumped laser (500-dzoulnyy lazer s yadernoy nakachkoy). Pisma v ZhTF **38**(23), 31–39 (2012)
36. E.P. Magda, A.V. Bochkov, A.V. Lukin, L.E. Magda, I.S. Pogrebov *On the ultimate characteristics of nuclear-pumped pulsed gas lasers* (O predelnykh kharakteristikakh impulsnykh gazovykh lazerov s yadernoy nakachkoy). Proceedings of the 3rd International Conference "Problems of Nuclear-Pumped Lasers and Pulsed Reactors," Snezhinsk, 2003, pp. 396–403.
37. J.R. Felty, R.J. Lipinski, D.A. McArthur, P.S. Pickard, DOE reactor-pumped laser program. Proc. SPIE **2121**, 2–9 (1994)
38. D.A. McArthur, G.N. Hays, P.S. Pickard, FALCON reactor-pumped laser technology for space power applications. Fus. Tech. **20**, 753–758 (1991)

39. D.E. Bodette, D.R. Neal, D.A. McArthur, P.S. Pickard, G.A. Hebner, R. Michie, *Reactor pumped laser research at the Sandia National Laboratories pulsed reactor facilities*. Int. Embedded Topical Meeting on Physics, Safety, and Applications of Pulse Reactors, Washington, DC, 1994, pp. 163–171.
40. D.A. McArthur, G.N. Hays, W.J. Alford, D.R. Neal, D.E. Bodette, J.K. Rice, Recent results on reactor-pumped laser studies at Sandia National Laboratories. Las. Interact. **8**, 75–86 (1988)

Chapter 7
Energy Deposition in Gas NPL Active Media

7.1 Comparison of Basic Methods of NPL Excitation

As was noted in Sect. 1.2 of Chap. 1, the nuclear reactions that can be used for pumping gas NPLs in experiments with pulsed reactors are:

$$^{3}\mathrm{He} + n \rightarrow {}^{3}\mathrm{H} + p + 0.76\,\mathrm{MeV} \tag{7.1}$$

$$^{10}\mathrm{B} + n \rightarrow {}^{7}\mathrm{Li} + \alpha + 2.38\,\mathrm{MeV}, \tag{7.2}$$

$$^{235}\mathrm{U} + n \rightarrow \text{fragments} + 168\,\mathrm{MeV}. \tag{7.3}$$

When reaction (7.1) is used, the laser medium is excited by protons and tritium ions. In this case, ^{3}He plays the role of a buffer component of the laser mixture and is simultaneously a volume pumping source. Calculations of the specific energy deposition to the gas medium when the reaction (7.1) is used were carried out in the studies [1, 2]. The gas pressure and the transverse dimension of the laser cell were varied (it was assumed that the length of the cell greatly exceeded its transverse dimension). It was shown that for each pressure value there is an optimal transverse dimension of the cell. When the dimensions are less than optimal, the losses of reaction (7.1) products on the walls of the cell become great. When the dimensions are greater than optimal, the decrease of the neutron flux in the direction from the cell boundary to the depth of the gas volume starts to have an effect. Some results of calculations are provided in Figs. 7.1 and 7.2. It should be noted that the model [2] makes it possible to perform calculations for different neutron spectra when more precise deceleration principles of charged particles are used. The results of the calculations in studies [1, 2] for the case of a mono-energetic flux of thermal neutrons differ by 10–15 % (Fig. 7.1).

The gaseous compound $^{235}UF_6$ can be used in the capacity of a volume pumping source along with ^{3}He. Calculations of the energy deposition for the mixture He-$^{235}UF_6$ for different pressures and concentrations of $^{235}UF_6$ are provided in

S.P. Melnikov et al., *Lasers with Nuclear Pumping*,
DOI 10.1007/978-3-319-08882-2_7

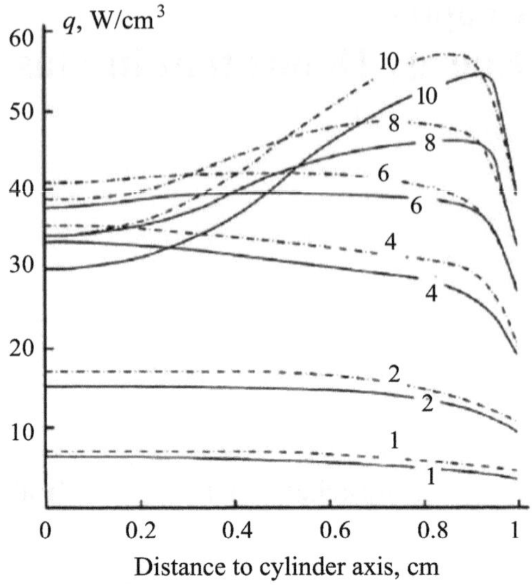

Fig. 7.1 Dependence of the specific power deposition on the distance from the axis of the cylindrical cell for a thermal neutron flux density of 10^{15} cm^{-2} × s^{-1} (cell radius 1 cm, neutron energy 0.0252 eV). Broken line curves are the results of study [1], solid curves are those of study [2]; *numbers* next to the curves indicate the ^{3}He pressure in atmospheres

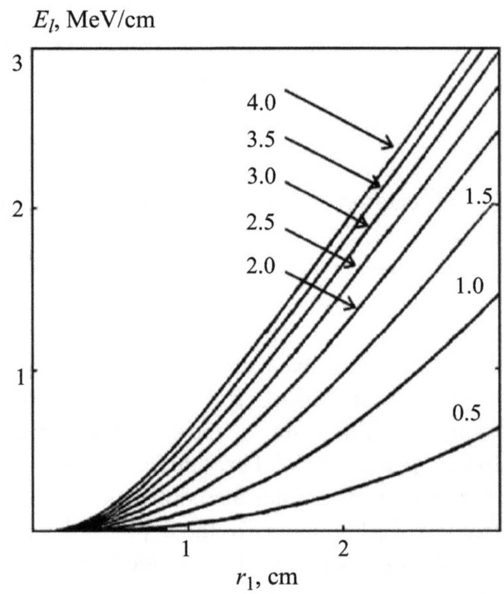

Fig. 7.2 Total absorbed energy E_l per unit of length of an infinite cylindrical tube depending on its radius r_1 with a thermal neutron fluence of 1 cm^{-2} [1]. *Numbers* next to the curves indicate the ^{3}He pressure in atmospheres

study [3]. However, the search for nuclear-pumped gas media based on UF_6 has not yielded a positive result in experiments to date. The issue involved is attributed to the high quenching rates of excited atoms by UF_6 molecules [4, 5].

Thin layers of materials containing the isotopes ^{10}B or ^{235}U (surface sources) are used to excite the active mixtures by means of the reactions (7.2) and (7.3).

However, in this case the efficiency of energy transfer to the gas mixture by the nuclear reaction products is limited to <50 % (for an ultra-thin layer) owing to the isotropy of their recession. For uranium layers, when conditions are optimal, this efficiency is 10–20 %. To our knowledge, the first calculations of the energy deposition from the surface source according to the reaction of (7.2) were carried out in study [6].

From the energy standpoint, the use of ^3He as the volume pumping source is as efficient as the use of thin uranium layers. Indeed, as an example let us examine two cylindrical cells of identical geometry. Let metallic ^{235}U be applied to the internal surface of one of them with an optimal thickness δ_U, equal to one half the range of the average fission fragment [7] in the uranium, that is, $\delta_U \approx 2.8 \times 10^{-4}$ cm. The cell is filled with a laser mixture that does not contain ^3He. Then the total energy transmitted to the gas by the fission fragment in a unit of time is

$$W = 2\pi r_1 \delta_U L_c \sigma_f N_U \Phi E_0 \varepsilon, \tag{7.4}$$

where E_0 is the total kinetic energy of the fission fragment of the uranium nuclei; L_c is the length of the cell; N_U is the density of the uranium nuclei in the active layer; r_1 is the internal radius of the cell; ε is the efficiency of energy transmission to the gas by fission fragment (ratio of the energy transmitted to the gas by fragments to the total kinetic energy of the fragments released in the uranium layer); σ_f is the cross section of uranium fission by the neutrons; and Φ is the neutron flux density.

Let the second cell not have a uranium coating, but be filled with a gas mixture whose basic component is ^3He. The total energy released in the gas volume of this cell per unit of time is

$$W_{He} = \pi r_1^2 L_c \sigma_{He} N_{He} \Phi E_{He} \varepsilon_{He}, \tag{7.5}$$

where N_{He} is the density of the ^3He; E_{He} is the kinetic energy of the reaction (7.1) products; ε_{He} is the efficiency of energy transmission by the products of reaction (7.1) to the gas; σ_{He} is the cross section of the reaction. From Eqs. (7.4) and (7.5) we have

$$\frac{W_{He}}{W} = \frac{r_1}{2\delta_U} \times \frac{\sigma_{He}}{\sigma_f} \times \frac{N_{He}}{N_U} \times \frac{E_{He}}{E_0} \times \frac{\varepsilon_{He}}{\varepsilon}.$$

To conduct a comparative assessment, we assume that the basic component of the gas mixture in the first cell is ^4He and the gas pressure in both cells is $\approx$2 atm; the transverse dimension of the cell $r_1 \sim 1$ cm. With such correlations between pressure (which at $T_0 = 300$ K corresponds to $N_{He} \approx 5 \times 10^{19}$ cm^{-3}) and transverse dimension of the cell, the decrease of the neutron flux owing to the interaction with ^3He is not great, and it can be ignored in estimates. On the other hand, the accepted correlation of the radius and pressure of a gas laser excited by fission fragment and not containing ^3He is close to optimal [7]; at the same time, $\varepsilon \approx 0.16$. The value of ε_{He} for these parameters is in the limits of $0.7 < \varepsilon_{He} < 1$ and is determined by the influence of the cell walls on the energy deposition of the reaction products. Assuming for thermal neutrons $\sigma_f = 582 \times 10^{-24}$ cm^2, $\sigma_{He} = 5{,}327 \times 10^{-24}$ cm^2, and considering that $N_U = 4.8 \times 10^{22}$ cm^{-3} (metallic uranium), $E_0 = 168$ MeV, $E_{He} = 0.76$ MeV, from the last equality we find $W_{He}/W \sim 0.5\ r_1 \varepsilon_{He} \sim 0.5\ r_1$ cm^{-1}. Consequently, in both cases, the energy depositions can be comparable in amount. However a serious restriction for pumping the gas medium with reaction (7.1) is necessary to use component ^{3}He in the capacity of the buffer component, greatly limiting the choice of gas mixtures for the laser media.

This limitation is eliminated when the reactions (7.2) and (7.3) are used. The total energy transferred to the gas per unit of time by products of the reaction (7.2) is

$$W_B = 2\pi r_1 \delta_B L_c \sigma_B N_B \Phi E_\alpha \varepsilon_B. \tag{7.6}$$

Here E_α is the kinetic energy of the α-particles; N_B is the density of the nuclei of ^{10}B in the active layer; δ_B is the thickness of the boron layer; ε_B is the efficiency of energy transition to the gas by α-particles; σ_B is the cross section of the reaction (7.2). From expressions (7.4) and (7.6) we find:

$$\frac{W}{W_B} = \frac{\delta_U \varepsilon \sigma_f N_U E_0}{\delta_B \varepsilon_B \sigma_B N_B E_B}.$$

An α-particle that appears as a result of a reaction (7.2) can travel from the boron layer into the gas mixture only if it was born at a distance from the coating-gas interface that is less than its stopping range in the coating (source) material. Therefore, the maximal effective thickness of the boron layer should not exceed the stopping range of an α-particle with an energy $E_\alpha = 2.38$ MeV which is about $\delta_B \approx 0.6 \times 10^{-4}$ cm with a density of boron nuclei of $N_B = 0.136 \times 10^{24}$ cm^{-3}. The maximal value of efficiency of the energy deposition, as already mentioned, cannot exceed 50 %, that is, $\varepsilon_B < 0.5$. Assuming laser cells with uranium or boron layers are fully identical to the above cell in regard to geometry and gas mixture composition, then with allowance for the fact that for thermal neutrons, $\sigma_B \approx 3{,}840 \times 10^{-24}$ cm^2, we find $W/W_B > 5.6$. The advantage of using uranium layers with their larger energy release is evident.

Because the absorption cross section for thermal neutrons by the isotopes ^{3}He and ^{10}B is quite large, the placement of laser cells with such isotopes in the core of the nuclear reactor can lead to a perceptible reduction in its core reactivity margin, and if the number of cells becomes too large, this may also bring the reactor to a subcritical state. But cells with ^{235}U layers can simultaneously play the role of fuel elements in the reactor core, avoiding a reduction in criticality. Furthermore, if the configuration is optimal, an autonomous reactor-laser may be formed with an adequate super-criticality margin, using a core entirely assembled from such laser cells (channels).

7.2 On Methods of Calculating the Energy Deposition of Fission Fragments

While conducting investigations associated with development of NPLs and the study of their characteristics, it becomes necessary to determine the total energy absorbed in the gas mixture, and the spatial distribution of the energy deposition through the gas volume. Such data make it possible to assess and to optimize the physical characteristics, including the laser efficiency, the threshold pumping power, the optimal pressure for the given geometry and composition of the gas mixture, the space–time structure of the optical inhomogeneities, and so forth. Both the total energy absorbed in the gas mixture and the specific energy deposition in each point depend on the geometry of the laser channel, the composition and density of the mixture, and the composition and thickness of the active layer containing the fissionable material. Usually lasers excited by fission fragments have the shape of an elongated circular cylinder or rectangular channel (see Chap. 2). In the first case, the active fissionable material is applied in a thin layer ($\delta_U = 1$–10 μm) to the internal surface of the cylindrical laser body; in the second, to two opposing parallel internal surfaces of the rectangular channel.

NPLs are characterized by transverse dimensions comparable to the range of the fragment in the gas laser mixture. Therefore, the profile of distribution of the specific energy deposition in the transverse cross section of the laser markedly depends on the transverse coordinates. The dimensions of the lasers in the direction of the optical axis of the system exceed many times over the range of the fission fragments in the utilized mixtures. This means that in the overwhelming portion of the volume of the laser channel, the longitudinal distribution of the specific energy deposition is determined only by the shape of the longitudinal distributions of the gas density and neutron flux irradiating the laser.

One of the chief deficiencies of early methods [8–14] of calculating the energy deposition of heavy ions, including fission fragments, was that the density of the excited gas was assumed to be uniform. In principle, the Monte Carlo method makes it possible to calculate the energy deposition with allowance for the non-uniformity of spatial distribution of the medium density.

Study [12] employs an alternate "two region" approach to address this problem. However, the spatial distribution of the energy deposition is given in histogram form [14] (calculation of the energy deposition is done for a finite selected volume in the neighborhood of the given point), which creates evident difficulties for restoration of a continuous dependence of the target value on coordinates. Breaking up each volume element into smaller ones, with the goal of clarifying the behavior of the spatial dependence, requires a significant increase in the calculating time and operational memory of the computer to retain the desired statistical accuracy of the calculation.

Matyev [15–17] developed an analytical method of calculating the energy deposition to media with an arbitrary distribution of density of the excited gas.

The space–time distribution of the specific power deposition to the gas $F\left(\vec{r},t\right)$ is determined as in [15, 16]

$$F\left(\vec{r},t\right) = E_0 q\left(\vec{r},t\right)\frac{R_1}{R_0} \times \frac{\rho\left(\vec{r},t\right)}{\rho_0} I\left(\vec{r},t\right). \tag{7.7}$$

Here q is the average fission power for the uranium-containing active layer per unit of volume of this layer; R_0, R_1 are the mean range of a fission fragment in the gas and in the material of the uranium-containing layer respectively; $\rho\left(\vec{r},t\right)$ is the density of the gas at the point in question at the time t; ρ_0 is the average (non-perturbed) gas density; and $I\left(\vec{r},t\right)$ is a dimensionless functional, defined by the distribution of gas density and the geometry of the cell. The correlations for $I\left(\vec{r},t\right)$ obtained in [15, 16] come to operations involving double integration. Although they have a very cumbersome appearance, it then becomes possible with comparatively little effort to conduct a precise calculation of the energy deposition in non-uniform media with a finite transverse dimension.

As in other studies [1–3, 8–14] dedicated to calculations of the energy deposition when there is slowing down of heavy ions, including fission fragments, in the gas, the dependence of the energy E_i of a decelerating particle of given mass on the range x in the model [15–17] is presented in the form

$$E_i = E_{i0}(1 - x)^n. \tag{7.8}$$

The exponent n determines the "slowing down law": for $n = 1$, one speaks of a "linear law" for slowing down; for $n = 2$, a "square law," for $n = 3/2$, a "3/2 law." The linear law is usually used to describe the deceleration of light ions—protons and α-particles [1]. It is rarely applied to fission fragments [10]. Usually $n = 2$ and $n = 3/2$ are used to describe the deceleration of fission fragments. Both values of the exponent n were discussed in study [18]. The square law of deceleration [18], which is also the one used in most studies, is considered the most adequate for fission fragments.

A comparison of results of calculations by the methods of [15–17] with the published results of calculations by the Monte Carlo method [14] shows their close identity. Figures 7.3 and 7.4 show the results of calculation of the efficiency for absorption of the kinetic energy of fission fragments in a gas medium. Results are for both cylindrical and flat uranium layers. The figures include the designations: ε, the ratio of fission fragment kinetic energy absorbed in the gas medium to their total kinetic energy, released in the uranium layer; $D_0 = d/R_0$, the normalized transverse dimension of the gas volume (d is the diameter in the case of a cylindrical uranium layer or thickness of gas in the case of a flat layer); and $D_1 = \delta_U/R_1$, the normalized thickness of the uranium-containing layer.

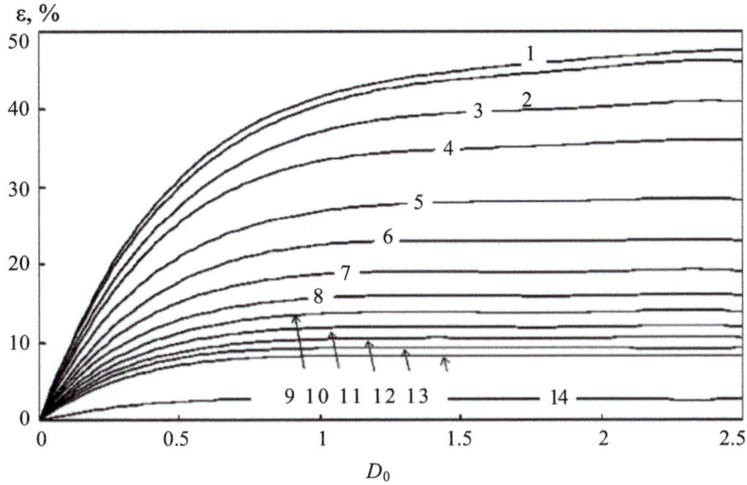

Fig. 7.3 The efficiency of an infinite cylindrical uranium layer depending on the normalized cylinder diameter for $D_1 = 0.001$ (*1*); 0.01 (2); 0.05 (*3*); 0.1 (*4*); 0.2 (*5*); 0.3 (*6*); 0.4 (*7*); 0.5 (*8*); 0.6 (*9*); 0.7 (*10*); 0.8 (*11*); 0.9 (*12*); 1.0 (*13*); 3.0 (*14*)

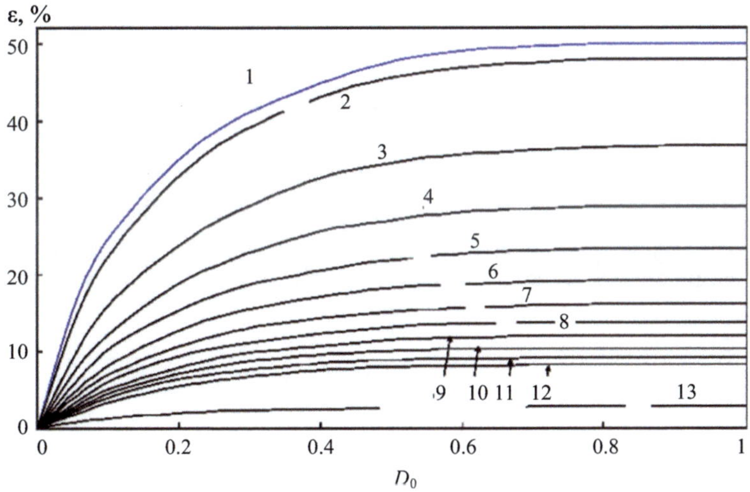

Fig. 7.4 The efficiency of an infinite flat uranium layer depending on the normalized thickness of the gas layer for $D_1 = 0.001$ (*1*); 0.01 (2); 0.1 (*3*); 0.2 (*4*); 0.3 (*5*); 0.4 (*6*); 0.5 (*7*); 0.6 (*8*); 0.7 (*9*); 0.8 (*10*); 0.9 (*11*); 1.0 (*12*); 3.0 (*13*)

The ranges of the uranium fission fragments in certain materials are provided in Table 7.1. To describe the mass spectrum of fission fragments, frequently an approximation is used, which assumes that when a ^{235}U nucleus is split by thermal neutrons, two fragments are formed. These are a light and heavy one, which have average masses of 96 amu and 140 amu, and kinetic energies of 100 MeV and

Table 7.1 Ranges of ^{235}U fission fragments in certain media

Medium	Density under normal conditions, g/cm^3	Fragment mass	Fragment range, mg/cm^2		
			Experiment [14]	Calculation per Eq. (7.9)	Calculation per Eq. (7.10)
He	1.78×10^{-4}	Medium light	1.3	2.10	1.85
		Medium heavy	1.1	1.63	
Ne	9.00×10^{-4}	Medium light	–	3.65	3.21
		Medium heavy	–	2.82	
Ar	1.78×10^{-3}	Medium light	4.4	5.02	4.41
		Medium heavy	3.9	3.88	
Kr	3.71×10^{-3}	Medium light	7.4	7.04	6.19
		Medium heavy	5.7	5.44	
Xe	5.85×10^{-3}	Medium light	8.6	8.73	7.67
		Medium heavy	6.7	6.74	
H_2	8.99×10^{-5}	Medium light	0.9	1.30	1.14
		Medium heavy	0.8	1.00	
Air	1.29×10^{-3}	Medium light	3.5	3.19	2.81
		Medium heavy	2.5	2.47	
H_2O	1.00	Medium light	3.0 (most probable range)	2.77	2.44
		Medium heavy		2.14	
Al	2.7	Medium light	4.0	4.16	3.66
		Medium heavy	3.5	3.22	
U	18.7	Medium light	12.6 (maximum range)	11.7	10.3
		Medium heavy		9.02	
UO_2	10.0–10.9	Medium light	8.0–8.7	9.04	7.94
		Medium heavy		6.98	
U_3O_8	7.3	Medium light	10.0 (maximum range)	8.45	7.42
		Medium heavy		6.53	

68 MeV respectively. An empirical relationship [14] can be used to determine the mass ranges R (mg/cm^2) of fission fragments in various media:

$$R = \left(0.0391 + 0.0202 A_s / Z_s^{1/2}\right) E_f^{2/3}, \tag{7.9}$$

where A_s and Z_s are the mass number and the charge of atoms of the medium; E_f, MeV is the initial kinetic energy of the fragment. Sometimes a simpler relationship is used, which makes it possible to determine the so-called mean mass range $\overline{R}$ without dividing the fragments into light and heavy [14]:

$$\overline{R} = 0.755 + 0.388 A_s \times Z_s^{-1/2}. \tag{7.10}$$

Prior to development of the computational methods [15–17] during the early computational investigations of the dynamics of development of optical inhomogeneities in sealed laser cells, the influence of changes in gas density on the profile of distribution of the energy deposition into the gas was taken into account by tying the specific energy deposition sources to Lagrangian coordinates [19–21]. But the distribution itself of these sources was determined by the method described in [14]. In this case, the power deposition of the decelerating fragments in a gas portion of fixed mass, calculated relative to the intensity of uranium nuclear fissions, does not depend on the time, and remains equal to its initial value in a non-perturbed gas. That approach provides a precise description of the spatial distribution of the energy deposition only for an infinitely extended flat geometry [16, 22]. For other types of geometry of the cell, in particular for cylindrical or rectangular laser channels, this technique does not adequately allow for the influence of redistribution of density on the energy deposition. Thus, in the case of cylindrical cells with an axially symmetrical distribution of medium density, in which the diameter is equal to the mean range of a fragment in a non-perturbed gas, or exceeds it, with an increase in the energy deposition, the gas particles are concentrated toward the axis of the channel, where when there are very large energy deposition (significantly exceeding those realistically attainable even in the most powerful pulsed reactors), a region inaccessible to the fragments may form [23]. In [23] this phenomenon is called "axial screening." Nonetheless, special investigations showed that for cylindrical cells, with pumping durations of $\tau_{1/2} \leq 3$ ms, fixation of the energy deposition in Lagrangian coordinates makes it possible to obtain gas temperature and density deviations for any point of the volume which are no more than 10 % of the corresponding values obtained in calculations with precise allowance for the influence of redistribution of density on the specific energy deposition [16, 21].

7.3 Influence of Inhomogeneities of Uranium-Containing Layers on Energy Deposition

One of the chief parameters of NPLs is the lasing energy (power), which is unambiguously associated with the parameter ε, the share of kinetic energy of fission fragments absorbed in the gas mixture with respect to their total kinetic energy released in the uranium layers. The thickness of the active layer usually is comparable with the fragment range in the material making up this layer. The methods developed by now [14–17] make it possible, for given parameters of laser cells (geometric dimensions, material of active layer, composition of gas mixture), to calculate ε as a function of gas density and thickness of the active layer, which is assumed constant. Within this approach the structure of the layer is considered to be uniform. But in fact, using the presently existing methods of thin layer's manufacturing, the above conditions (thickness invariability and uniformity) are difficult to achieve. A real layer is non-inform, and its surface is rough [24, 25]. The amplitude of the deviations in layer inhomogeneities from the average value of its thickness, and the frequency of their repetition, can depend directly on the technology of deposition of the fissile material on the substrates, and on the resistance of the layers to external environment factors, such as ejection of fuel micro-particles, erosion, corrosion and the like.

Study [26] showed that the presence of inhomogeneities in the uranium layer must lead to a reduction in the resultant efficiency in comparison with a uniform layer. This reduction must occur independently of whether the roughness of the active layer surfaces exerts a screening effect on the escape of fragments into the gas or not.

Periodic Symmetrical Inhomogeneities

Studies [14–16] propose determining the energy deposition of fission fragments (in particular the efficiency ε) through the normalized transverse size of the gas volume $D_0 = d/R_0$ and the normalized thickness of the uranium-containing layer $D_1 = \delta_U/R_1$, where d is the internal transverse size of the laser cell (diameter for a cylindrical cell or distance between uranium layers for a cell with plane-parallel arrangement of the uranium layers). Such an approach is convenient, because the properties of the media are included in the values R_0 and R_1, so the slowing of the fragments in different media, but with an identical value of normalized sizes, must be identical. We note that since the fragment energy losses per unit of range are proportional to the density of the decelerating medium, the normalized transverse size of the cell is proportional to the initial pressure of the gas medium.

Let $\varepsilon(D_1)$ be the share of fission fragment energy transmitted to the gas when the relative thickness of the active layer D_1 is unchanged. The total energy released in the system is proportional to D_1, while the energy absorbed in the gas, $E \sim D_1\varepsilon(D_1)$.

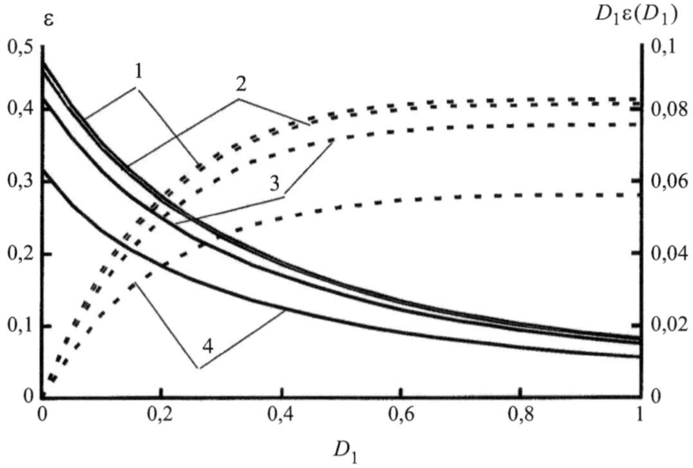

Fig. 7.5 Dependencies of efficiency ε (*solid curves*) and the parameter $D_1\varepsilon(D_1)$ (*broken curves*) on the normalized thickness of the uranium-containing layer: (*1*) $D_0 = 1$; (*2*) $D_0 = 0.6$; (*3*) $D_0 = 0.4$; (*4*) $D_0 = 0.2$

Figure 7.5 shows the dependencies, calculated using formulas [14, 16] of the energy deposition efficiency ε and the parameter $D_1\varepsilon(D_1)$ on the relative thickness D_1 for four values of the normalized width of the gas interval D_0 in a system comprised of two plane-parallel plates, where active layers of identical thickness are deposited on two surfaces facing each other.

Let us examine a simple one-dimensional model (depending only on the x coordinate along the surface of the layer),

$$D_1 = \overline{D}_1\{1 + \delta f(x)\}, \tag{7.11}$$

whose inhomogeneities are arranged with a specific period h, that is, $f(x+h) = f(x)$. Here $\overline{D}_1$ is the average value of relative thickness of the active layer, and δ is the dimensionless amplitude of inhomogeneities.

For a uniform layer, let the number of fragments entering the gas volume from a unit of surface be equal to N. If the maximal deviation $\delta R_1 \overline{D}_1$ of the thickness of the active layer from its average value does not exceed the range R_1 by an order of magnitude, that is,

$$\delta \overline{D}_1 \leq 1, \tag{7.12}$$

and the distance between the nearest maxima or minima of inhomogeneities markedly exceeds the fragment range in the material of the layer

$$h \gg R_1, \tag{7.13}$$

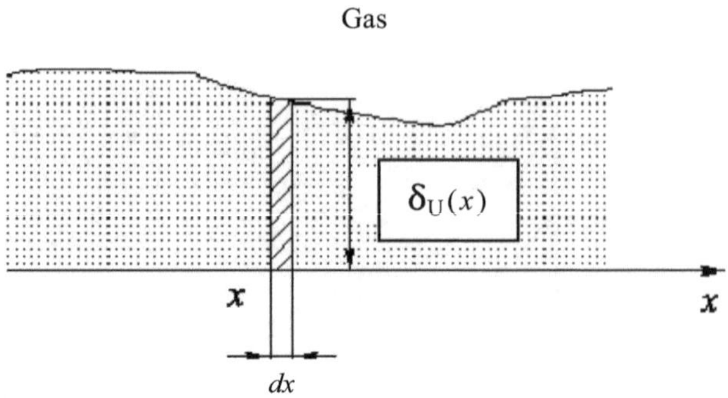

Fig. 7.6 One-dimensional model of uranium-containing layer of variable thickness

then the relative change in the number of fragments $\Delta N/N$, which escape into the gas owing to the inhomogeneities (that is, screening), can be estimated by the ratio $\Delta N/N \sim \alpha/\pi < <1$, where $\alpha \approx \delta R_1 \overline{D}_1/h \tilde{} R_1/h$. Hence it follows that the influence of the screening factor can be ignored.

The efficiency of energy deposition to the gas $\varepsilon(x)$ from a strip of thickness dx and height $\delta_U(x)$ (see Fig. 7.6) evidently must coincide with the energy deposition of an infinitely elongated uranium layer with a constant thickness ($\delta_U = \text{constant}$) equal to the height $\delta_U(x)$ of the selected strip. Accordingly, the total energy transmitted to the gas from the uranium strip in question is $dE(x) \sim D_1(x)\varepsilon(x)dx$.

If the flux distribution of neutrons initiating fission of the active nuclide can be considered unchanging within the limits of the spatial period of inhomogeneities, the effective value of the share of energy deposited into the gas by fragments emitted from the non-uniform layer must be determined by the expression:

$$\overline{\varepsilon} = \frac{\int_{-h/2}^{h/2} D_1(x)\varepsilon(x)dx}{\int_{-h/2}^{h/2} D_1(x)dx}, \tag{7.14}$$

where $\varepsilon(x)$ is the efficiency of the energy deposition from the uniform, infinitely elongated layer of constant thickness, equal to the thickness $\delta_U(x)$ of the real layer at point x.

Figure 7.7 shows the results of numerical calculations of the dependence $\overline{\varepsilon}$ on the amplitude of the inhomogeneities in a system, comprised of two flat parallel plates, for two simple models with $\overline{D}_1 = 0.5$:

(a) inhomogeneities in the form of alternating teeth of rectangular shape, the distance between which is equal to the width:

$$f(x) = \begin{cases} 1, & mh < x < (m+1/2)h, \\ -1, & (m+1/2)h < x < (m+1)h' \end{cases} \qquad |m| = 0, 1, 2, 3, \ldots;$$

(b) inhomogeneities of sinusoidal shape

$$f(x) = \sin \frac{2\pi x}{h}.$$

For cylindrical systems with an active layer deposited on the internal surface of the cell wall, the dependencies ε and εD_1 on D_1 and D_0, as well as $\overline{\varepsilon}$ on δ, D_1, and D_0 have the same appearance as for plane-parallel systems.

Direct calculations of $\overline{\varepsilon}$ for active layers with different types of periodic symmetrical one-dimensional inhomogeneities showed, as was to be expected, that the maximal reduction of the energy deposition to the gas occurs with inhomogeneities of the rectangular shape. Here, the efficiency of the energy deposition can decrease by 50 % in comparison with the efficiency of uniform layers. Inhomogeneities with a smoothed profile are less dangerous. Thus for inhomogeneities with a profile of the sinusoidal type, a reduction of the total energy deposition of $\leq$21 % is possible.

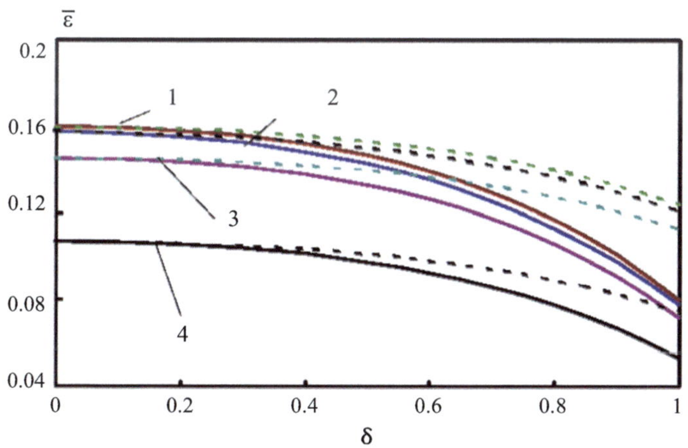

Fig. 7.7 Dependence of the share of energy of fission fragments effectively transmitted to the gas on the amplitude of the inhomogeneities: ($\overline{D}_1 = 0.5$): (*1*) $D_0 = 1$; (*2*) $D_0 = 0.6$; (*3*) $D_0 = 0.4$; (*4*) $D_0 = 0.2$; the *solid curves* are the inhomogeneities of the "a" type; the *broken curves* are inhomogeneities of the "b" type

Inhomogeneities in the Form of Craters

Clearly, in the case of non-symmetrical periodic inhomogeneities, including two-dimensional (that is, depending on two spatial coordinates), the maximal reduction in the energy deposition must be manifested when there is a rectangular profile of these non-uniformities. Let us examine the influence of the presence of craters with vertical walls and an arbitrary base (Fig. 7.8) on the efficiency of the energy deposition. Let us introduce the dimensionless parameters of the inhomogeneities:

$$\delta_+ = \frac{D_1^+ - \overline{D}_1}{\overline{D}_1}; \qquad \delta_- = \frac{\overline{D}_1 - D_1^-}{\overline{D}_1}; \qquad \delta = \frac{1}{2}(\delta_+ + \delta_-). \tag{7.15}$$

Let $\overline{D}_1(\delta_+ + \delta_-)$ be the depth of the craters; σ the area of their base; and S the equivalent area of the surface of the elementary cell of the active layer containing one crater, defined as the ratio of the total surface of the active layer to the total number of craters. As in the one-dimensional model, we shall assume that the maximal deviation of the thickness of the active layer from its average value does not exceed the range R_1 by an order of magnitude (that is, $\delta_\pm \overline{D}_1 \leq 1$), and the distance between the nearest craters markedly exceeds the fragment range in the layer material:

$$S^{1/2} \gg R_1. \tag{7.16}$$

From the conservation of the average thickness of the active layer $(S - \sigma)\delta_+\overline{D}_1 = \sigma\delta_-\overline{D}_1$ it follows that:

$$\delta_+ = 2\frac{\sigma}{S}\delta; \qquad \delta_- = 2\frac{S-\sigma}{S}\delta. \tag{7.17}$$

If $D_1^- = 0$, that is, the depth of the craters reaches the substrate, then $\delta_- = 1$. Hence

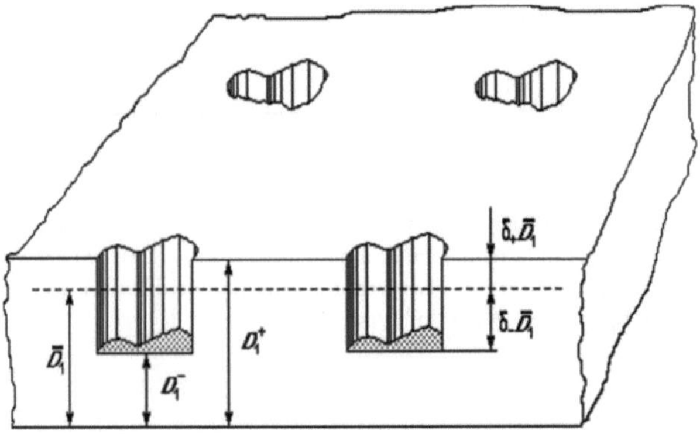

Fig. 7.8 Model of two-dimensional periodic non-uniformities of the crater type with vertical walls

in accordance with the second equation (7.17) we find that for a given ratio of areas σ/S, the value of the parameter δ cannot exceed

$$\delta_{\max} = \frac{S}{2(S-\sigma)}, \tag{7.18}$$

and the minimal value which the ratio σ/S can assume for a given δ is

$$\left(\frac{\sigma}{S}\right)_{\min} = 1 - \frac{1}{2\delta}.$$

The effective energy deposition in the two-dimensional case is determined by an equation analogous to Eq. (7.14), in which integration is performed over the area of the equivalent cell active layer surface S. For the crater model in question, this equation boils down to the simple expression

$$\overline{\varepsilon} = \frac{1}{\overline{D}_1}\left\{D_1^-\varepsilon\left(D_1^-\right)\frac{\sigma}{S} + D_1^+\varepsilon\left(D_1^+\right)\left[1-\frac{\sigma}{S}\right]\right\}. \tag{7.19}$$

Figure 7.9 shows the results of calculations for the model of craters with vertical walls when $D_0 = 0.5$ in a system comprised of two plane parallel plates.

Let us transform Eq. (7.19) using Eqs. (7.15) and (7.17):

$$\overline{\varepsilon} = \varepsilon\left(D_1^-\right)[1-2(1-\sigma/S)\delta]\frac{\sigma}{S} + \varepsilon\left(D_1^+\right)(1+2\delta\sigma/S)(1-\sigma/S).$$

Substituting the value of $\delta_{\max}$ from Eq. (7.18), which clearly corresponds to the minimal possible value of $\overline{\varepsilon}$, we get $\overline{\varepsilon}_{\min} = \varepsilon\left(D_1^+\right)$. Thus the limit minimally possible value of $\overline{\varepsilon}$ for a non-uniform layer is equal to the value of ε which has a uniform layer with a normalized thickness equal to the maximal thickness of the non-uniform layer. Such correlations are also not hard to write for the model of

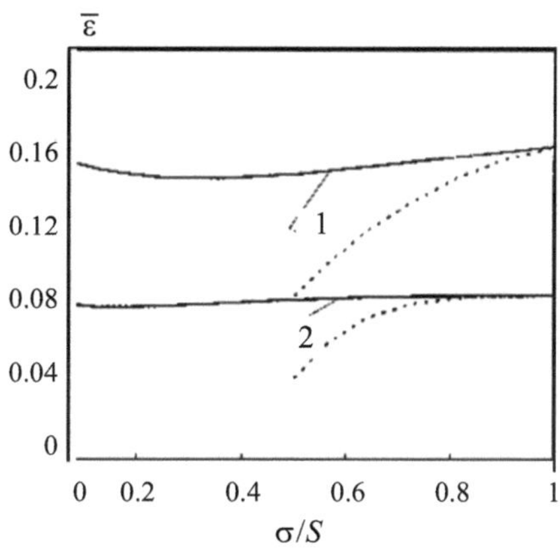

Fig. 7.9 Dependence of the share of energy of fission fragments efficiently transmitted to the gas on the relative area of the craters: (*1*) $\overline{D}_1 = 0.5$; (*2*) $\overline{D}_1 = 1$; the *solid line* is $\delta = 0.5$; the *dotted line* is $\delta = 1$

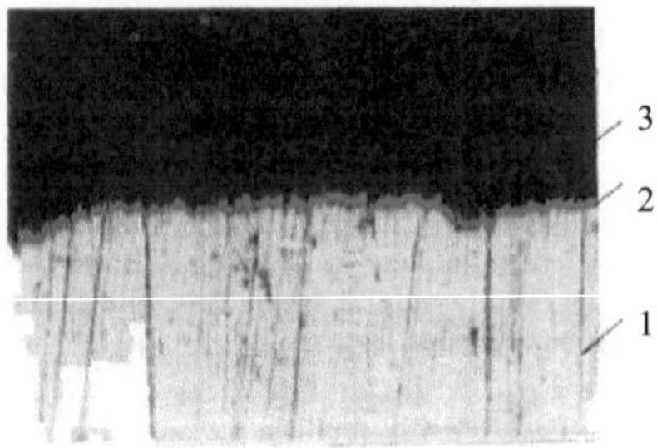

Fig. 7.10 Cross section of the layer of uranium oxide-protoxide: (*1*) substrate of aluminum; (*2*) layer of uranium oxide-protoxide; (*3*) epoxide resin

active layers with two-dimensional projections having vertical walls, which give results analogous to the crater model.

Of course, the models we have discussed are only an illustration of the significant influence of inhomogeneities; they cannot serve as a method providing precise results of calculation of $\overline{\varepsilon}$, because the spectrum of possible configurations of real inhomogeneities, even for one specific layer, is hard to convey mathematically. Figure 7.10 shows photograph [24] enlarged roughly 1,000 times of the cross section of one of the layers of the uranium oxide-protoxide used in NPL experiments at VNIIEF. The best method of allowing for the influence of inhomogeneities in determining $\overline{\varepsilon}$ is evidently to introduce the corrective multiplier $\overline{\varepsilon} = \beta\varepsilon\left(\overline{D}_1\right)$, which is found from a comparison of the experimentally determined efficiency of energy deposition to the gas with the computed efficiency $\varepsilon\left(\overline{D}_1\right)$, obtained on the assumption of uniformity of the active layer, $D_1 = \overline{D}_1$.

The circumstance that a maximal reduction in the energy deposition occurs for inhomogeneities having a rectangular profile in the cross section of the active layer makes it possible to determine the lower limit of the possible drop in the energy deposition to the gas, and to select acceptable values of the amplitude of the inhomogeneities. Analysis conducted on the results of calculations showed that if the amplitude of fluctuations of the inhomogeneities does not exceed one half the thickness of the active layer, then the maximal possible reduction in the energy deposition will be no more than 10 %. However, it should be noted that this conclusion is valid provided that the conditions (7.12) and (7.13), or (7.16), are met. Otherwise, a reduction in the energy deposition can prove more significant. The reason for a marked reduction in the energy deposition could be, for example, screening of the escape of fragments to the gas by the inhomogeneities themselves, when $h \leq R_1$.

In study [21], in several specific examples, the influence of periodic one-dimensional inhomogeneities of the active layer, determined by inhomogeneities of the surface of the substrate on which this layer was deposited, was examined computationally. The inhomogeneities had a rectangular profile. The width of the projections of the non-uniformities is equal to their height, while the period of the non-uniformities is equal to twice their height.

The surface of the uranium turned toward the gas was considered ideally smooth. Calculations of the energy deposition of the fragments were carried out by the Monte Carlo method, which automatically led to allowance for the screening factor of substrate projections in the active layer. As the material of the active layer, uranium dioxide UO_2 was considered ($R_1 \sim 8$ μm), and the substrate material was Al and W. The height of the inhomogeneities was varied within the limits of 2–10 μm. The results of these calculations agreed with the results discussed above: with the growth in amplitude of the inhomogeneities, the efficiency of the energy deposition drops, and when the amplitude corresponds to $\delta = 1$ in Eq. (7.11), it decreases by ~50 %. This last evidently testifies to a relatively weak influence of screening (at least for the examples considered).

In contrast to direct numerical calculations for a specific geometric profile of the active layer and its composition, the methods of (7.14) and (7.19) in the limits of the constraints (7.12) and (7.13) have greater universality. These methods used the formulas (7.11) and (7.14)–(7.19) and the methods of calculations from studies [14–16], in which the efficiency of the energy deposition of both $\overline{\varepsilon}$ and $\varepsilon(D_1)$ are determined by means of the normalized thickness of the active layer $D_1 = \delta_U/R_1$. Because the range R_1 is inversely proportional to the density of the decelerating medium, in the utilized approach, the inhomogeneities of the normalized thickness of active layers include both the inhomogeneities of the real thickness, provoked by roughness of the geometric surfaces bounding these layers, and the inhomogeneities of density of the layer material.

7.4 Results of Experiments to Determine the Energy Deposition

With expansion and deepening of investigations into the NPL lasing mechanism, as well as the specific features of onset and development of optical inhomogeneities in them, the question of how to determine the share of fission energy which is transmitted directly to the gas became increasingly more critical. Despite some differences in approaches, all of the calculation and theoretical work yields similar quantitative results. But in experimental investigations, lower values of efficiencies ε were obtained, in comparison with calculations. And the results of different authors differed markedly from one another [27–33]. The greatest discrepancy between experiment and calculation (by roughly double) was obtained in studies [30, 31]. If such a great difference did indeed occur, it could lead to the need for adjustment of the efficiency values, and possibly the kinetic models of the NPL (see Chap. 5). Therefore, analysis of the causes of the spread of the experimental results on the efficiency of the energy deposition by fission fragments and their discrepancy from the calculated values is very important. Apart from that, this parameter plays a significant role in calculations of the energy and mass-dimension properties of reactor-laser facilities.

Experiments When the Excitation Duration is ≤0.4 ms

In studies [30, 31], the identical method was used to determine the energy deposition to the gas. After irradiation of the laser cell with a short neutron pulse, damping pressure oscillations arose in the gas. From a comparison of the amplitude of the first peak of dependence of the pressure on the time with analogous amplitudes of dependencies obtained by means of computer simulation of gas dynamics (here the values of the energy transported to the gas are varied), the share of the energy of fission fragments transmitted to the gas was determined. In both studies, this percentage was roughly half the value obtained from direct calculation of the fragment energy deposition by the described methods. It is noted in [30], however, that direct measurements of the number of fission fragments emitted to the gas agree with the calculated values.

The conditions of conduct of experiments in these studies had some differences. For instance, in [30] a rectangular cell with plane-parallel arrangement of the layers containing uranium was used; the half-width of the neutron pulse was $\tau_{1/2} \sim 0.15$ ms; only argon was studied, with three values of initial pressure (0.7; 2.8; 4.14 atm). In [31], a cylindrical cell was used, with a uranium layer deposited on its internal surface (on a section with length 0.65 that of the total length). The half-width of the neutron pulse was $\tau_{1/2} \sim 0.4$ ms. Experiments were conducted with helium and neon with several values of pressure, in a range from 0.5 to 5 atm.

In study [31], the energy deposition was also evaluated from the pressure established after damping of the fluctuations. From the oscillograph curve shown in [31], it follows that the time of pressure establishment is ~10 ms. In the opinion of the authors [31], corrections allowing for the cooling of the gas through heat conduction were not great.

It should be noted that the calculations of [19, 20] (see also Chap. 8) of thermal and gas-dynamic processes in laser cells with the same internal diameter as in [31] (2.8 cm) showed that a relative reduction of pressure of ~20–35 % occurred in the gas due to heat removal from the walls of the cell after ~10 ms (from the start of the irradiating pulse with half-height duration of $\tau_{1/2} \sim 3$ ms). We also make mention of analogous calculations [34] for a rectangular cell with dimensions of $1 \times 7 \times 60$ cm, from which it follows that in ~4 ms, around 10 % of the deposited energy departs owing to thermal conduction through the cell walls.

As one possible reason for the discrepancy of data between theory and experiment, it is proposed that a significant portion of the deposited energy does not have time to be thermalized, and is carried to the cell walls in the form of luminescent radiation [30], in particular through transformation into UV radiation of excimer molecules [31]. Another reason for the reduction in the energy deposition might be the low quality of fabrication of the active layers. Unfortunately, study [31] provides no data on the specific energy deposition to the gas, which hampers analysis of the results of this study.

Experiments with Excitation Durations ≥ 3 ms

Next experimental measurements of the energy deposition to the gas obtained in investigation of optical inhomogeneities in NPLs [27] will be analyzed. In this study, in each variant of filling of the laser cells with a gas mixture, the average specific energy deposition was determined, but issues related to the efficiency of the energy deposition and its dependence on various conditions were not examined. Later on that analysis of the data [27] was done in studies [25, 35]. The results yielded much better agreement with theory than prior studies [30, 31], and matched the results mentioned in [32].

Cylindrical Layers. In the first series of experiments [27], the cell was formed by a cylindrical pipe with an internal diameter of 35 mm and length of 77 cm. It held an aluminum tube 57-cm long, with an internal diameter of 28 mm and wall thickness of 3 mm. A layer of uranium oxide-protoxide (90 % enriched ^{235}U) 5.9-mg/cm^2 thick was applied to the internal surface of the aluminum tube. The cell was filled with rare gases (He, Ne, Ar) and irradiated with a pulsed thermal neutron flux from the VIR-2M reactor with a half-height pulse duration of $\tau_{1/2} \approx 3$ ms. The distribution of the number of fissions per pulse along the length of the uranium layer, measured by the activation method using copper and uranium indicators, was found to be symmetrically relative to the cell center. In a reactor pulse, the average number of fissions along the length of the uranium layer, calculated per unit of uranium mass, was 1.5×10^{13} fissions/g. A change in the pressure in the course of the irradiation was measured using a DMI-6-2 inductive differential pressure sensor. The shape of the irradiating pulse and dependence of the pressure change on the time given are shown in Fig. 7.11.

The characteristic time of equalization of the pressure in the gas filling the cell is $\tau_p \sim L_c/2u_s$, where u_s is the speed of sound in the gas; $u_{He} \approx 10^3$ m/s; $u_{Ne} \approx 4.5 \times 10^2$ m/s; $u_{Ar} \approx 3 \times 10^2$ m/s. Thus even for argon, the characteristic time of pressure relaxation $\tau_p \sim 1$ ms is less than the half-width of the excitation pulse $\tau_{1/2}$. Therefore, the pressure in the cell in the course of irradiation by the

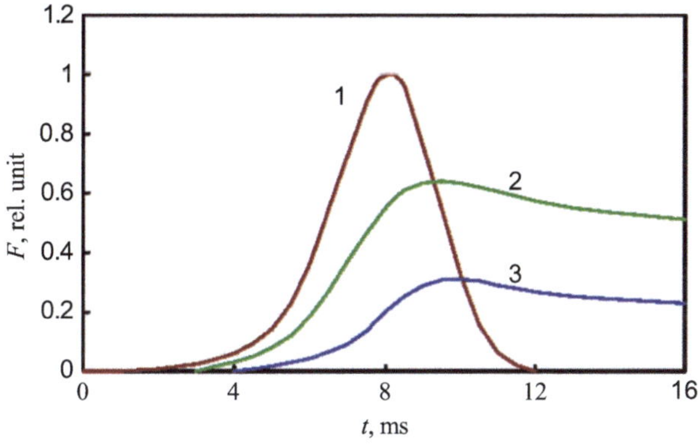

Fig. 7.11 Shape of neutron pulse F (*1*) and dependencies of increments of pressure in the cell Δp on time for neon (*2*) and helium (*3*) with $P_0 = 1$ atm

pulsed neutron flux with a duration of $\tau_{1/2}$ can be assumed in a first approximation to be uniform and it may be considered that redistribution of the gas density is in equilibrium.

This was confirmed by experiments [27], in which neither the interference method nor the pressure sensor registered acoustic oscillations in gas density and pressure (Fig. 7.11). Assuming that the gas is ideal, the pressure at each moment of time is uniform over the cell volume, and its walls are absolutely rigid, and using the energy conservation equation (with allowance for heat transfer, but ignoring the effect of viscosity), the constitutive equation, and the continuity equation, one can show (see Chap. 8, Sect. 8.2) that pressure at any moment of time may be described by the correlation:

$$P(t) = P_0 + \frac{\varepsilon}{V}(\gamma - 1)\int_0^t Q(\xi)d\xi + \frac{\gamma - 1}{V}\int_0^t \Pi(\xi)d\xi; \tag{7.20}$$

$$\Pi(t) = \oint k_g \nabla T\left(\vec{f}, t\right)d\vec{f},$$

where $d\vec{f}$ is an element of area of the internal surface of the cell; k_g is the coefficient of heat conductivity of the gas; P_0 is the initial pressure; Q is the intensity of energy release in the entire volume of uranium layers; V is the total gas volume in the cell; γ is the adiabatic index; and Π is the total heat flux through the surface bounding the cell volume.

Equation (7.20) essentially is a reflection of the energy conservation law and coincides with the correlation for an average pressure in the cell obtained in [34] for the more general case. If heat transfer through the cell walls can be ignored, then as in [22] we have

$$P(t) = P_0 + (\gamma - 1)q_c(t), \tag{7.21}$$

where q_c is the specific energy deposition to the gas averaged over the cell volume.

Based on the maximal pressure jump, Eq. (7.21) was used in experiments [27] to determine the average specific energy deposition for the cell volume, q_c, from which the specific energy deposition in the active portion of the gas volume V_U (that is, the part of the cell volume directly bounded by the uranium layer) was then calculated using the formula $q_A = q_c V/V_U$. From the formula

$$\varepsilon = \frac{q_c V}{q_U} = \frac{q_A V_U}{q_U}$$

(q_U is the total energy release in the uranium layers), it is not difficult to calculate the share of fission fragment energy transmitted to the gas. The results of

Table 7.2 Experimental data and results of calculation of ε for a cell with a cylindrical uranium layer

Gas	He				Ne	Ar		
P_0, atm	1	2	3	5	1	0.25	0.5	1
q_A, J/cm^3	0.07	0.14	0.16	0.18	0.15	0.07	0.13	0.16
D_0	0.243	0.486	0.729	1.217	0.700	0.254	0.508	1.018
ε	0.026	0.054	0.055	0.067	0.057	0.026	0.047	0.058

calculations are given in Table 7.2. This table also provides the corresponding values of the specific energy deposition q_A, the initial pressures in the cell [27], and the equivalent diameters $D_0 = d/R_0$ (d is the internal diameter of the active layer).

Figure 7.12 shows the experimental dependence $\varepsilon(D_0)$ and the calculated dependence obtained by the methods of [14–16]. As is evident from the figure, the calculation and experiment differ roughly by a factor of 1.5 in the high value range of D_0, and only for $D_0 \approx 0.25$ does the difference reach ~2.

Flat Layers. In the second series of experiments with helium and argon [27], an interference method was used to measure the optical inhomogeneities in a cell with plane-parallel arrangement of uranium layers as a part of the LUNA-2M studies (see Chap. 2, Sect. 2.4). In this experiment, the pressure was not measured directly. The presence of a large buffer volume in the cell led to marked gas displacement from the active region in the course of excitation. According to the reduction in gas density measured by the interference method in the pumping region by the methods of [27], the average change in gas temperature over the entire cell was determined. Then using the formula $q_c = c_V \times \Delta T$ (where c_V is the specific heat capacity of the gas with a constant volume) and with allowance for the correlation of the active and total volumes, the value of the average energy deposition for the active volume was determined.

The cell constituted an aluminum pipe 220-cm long and 8 cm in diameter, in which two flat aluminum plates (2,000 × 60 × 2 mm) were placed in parallel with one another with a separation of 2 cm, with layers of uranium oxide-protoxide (90 % enriched ^{235}U) 3.2 mg/cm^2 thick deposited on the inside surfaces. The VIR-2M reactor was also used as the neutron source. The distribution of fission reactions along the length of the uranium layer was measured by the same method as previously. Table 7.3 provides experimental values of the specific energy deposition in the volume between the uranium layers. The relative thickness of the gas region is defined as $D_0 = d/R_0$, where d is the distance between the uranium-containing layers.

Figure 7.13 provides experimental and calculated dependencies of $\varepsilon(D_0)$ for flat uranium-containing layers. The calculated dependence, as in [30], was obtained in an approximation of infinitely extended layers. In the case in point, the calculation and experiment differ roughly by a factor of 1.3–1.4, except for the region of small gas pressures ($D_0 \approx 0.08$), where the relative difference reaches ~2.

Somewhat later, VNIIEF researchers conducted experiments [29] in which the efficiency of the energy deposition in a cell with flat uranium layers was measured

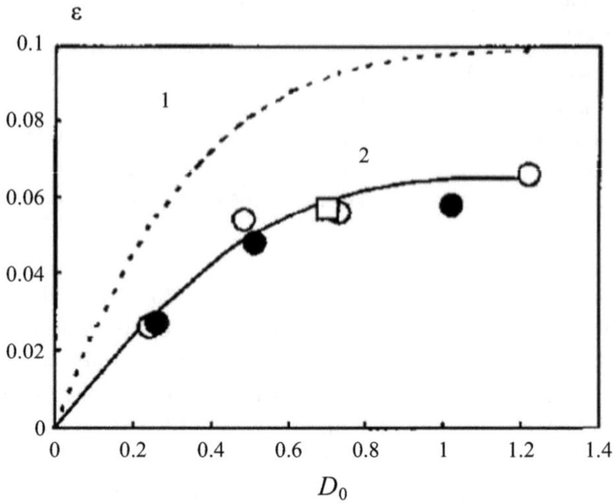

Fig. 7.12 Dependence of the efficiency of a cylindrical uranium layer on the equivalent cell diameter: (*1*) calculation by the methods of [14–16]; (*2*) experiment (*open circle* He, *open square* Ne, *filled circle* Ar)

Table 7.3 Experimental data and results of calculation of ε for a cell with flat uranium layers

Gas	He			Ar	
P_0, atm	0.5	2	5	0.25	0.5
q_A, J/cm^3	0.026	0.09	0.1	0.07	0.08
D_0	0.087	0.347	0.87	0.181	0.363
ε	0.031	0.108	0.120	0.084	0.096

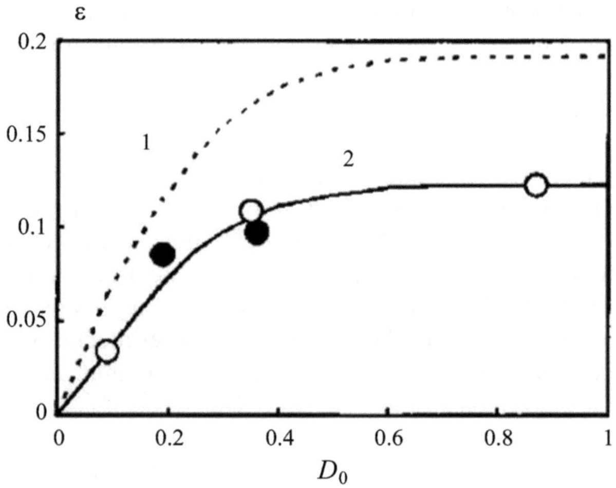

Fig. 7.13 Dependence of efficiency of flat uranium layers on the relative distance between them: (*1*) calculation for infinite plates by the methods of [14–16]; (*2*) experiment (*open circle* He, *filled circle* Ar)

from the pressure jump. In these experiments, flat uranium layers with an area of 20×6 cm^2 were arranged in a rectangular cell with a volume of $20 \times 6 \times 2$ cm^3, irradiated by a neutron pulse of the VIR-2M reactor ($\tau_{1/2} \sim 3$ ms). The cell was filled with helium. When the relative specific energy depositions were measured, a reduction was found in the experimental values in comparison with the theoretical ones, roughly by a factor of 1.5 for $D_0 \geq 0.6$ and by roughly double for $D_0 \leq 0.4$. The average specific energy deposition to the gas was ~0.25 J/cm^3 for $D_0 = 0.5$. Direct determination of the average specific energy deposition was done from the maximal pressure jump measured using a DMI-6 sensor. Apart from this, in the same series of experiments, the total energy output from the layer by fission fragments was determined under vacuum conditions using thermoresistors in the form of nickel wires arranged at a distance of 5 mm from the uranium layer. The results demonstrated fair agreement with theory: the measured output energy was just 11 % lower than calculated. This difference is comparable in size with the measurement error of the method utilized.

Analysis of Experimental Results. The experimental results [27, 29] obtained by recalculation using the formula (7.21) do not allow for the influences of heat outflow to the aluminum substrate and the cell walls. The deposition of this factor, estimated in [27] from the pressure drop in the cell after the end of the neutron pulse, can amount to 15–25 %. It was noted above that calculations [19, 20] of thermal and gas dynamic processes carried out for experiments [27] with the assumption of uniformity of fission density along the length of the cell showed that by the end of the irradiating pulse, the relative reduction in pressure in the cell owing to heat removal to the aluminum substrate (the third member in the right part of Eq. 7.20) reaches 20–35 % and depends on the type of gas used and its initial density. The maximal deviation of calculated pressures from experimental did not exceed 15 %.

Another factor which can lead to a reduction in the experimental energy deposition is the non-uniformity of the uranium-containing layer. Layers viewed in their cross sections are reminiscent in shape to something halfway between rectangular projections and a sinusoid (see Fig. 7.10). A study of photographs of the cross sections of uranium layers showed that the deviations in thickness of the layers from their average value could be 50–70 %. Figure 7.14 shows the results of calculation of dependence $\overline{\varepsilon}$ for the cylindrical layers used in experiments [27] on the dimensionless amplitude of inhomogeneities δ for two of the one-dimensional models shown in Sect. 7.3 of this chapter ("a" and "b"). Calculations show that for the above shape and amplitude in the thicknesses deviation of the uranium layers from an average value, these inhomogeneities can lead to a reduction of ε of 6–19 %.

Thus allowance for the heat outflow and the influence of the inhomogeneities in the uranium layers can markedly reduce or possibly even nullify the discrepancy between experimental and calculated values of specific energy deposition of fission fragments to the gas. We stress that in [27, 29], the energy deposition was found directly from the pressure jump, in contrast to [30, 31], where the energy deposition was determined by comparing the experimental curves of pressure oscillations with analogous oscillations curves obtained from calculating the dynamics of gas motion in the cell, while simultaneously selecting the specific energy deposition. The method used here could be implemented in view of the comparatively long duration of the irradiating pulse of $\tau_{1/2} \sim 3$ ms. With this long duration, the pressure in the

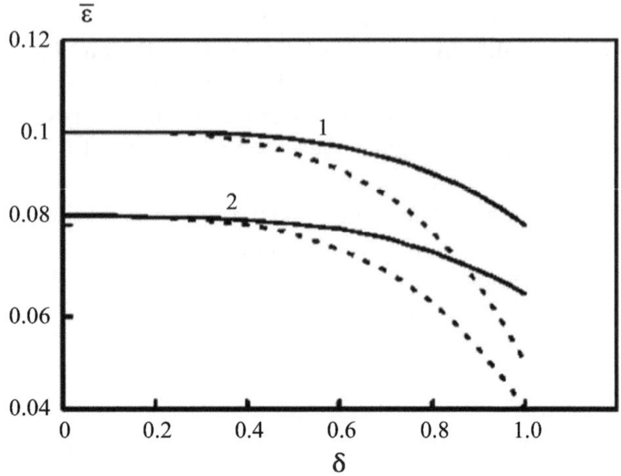

Fig. 7.14 Dependence of the efficiency of the energy deposition for a cylindrical uranium-containing layer on the amplitude of its inhomogeneities: (*1*) $D_0 = 1$; (2) $D_0 = 0.5$; *solid curves* are sinusoidal inhomogeneities; *broken curves* are inhomogeneities in the form of rectangular teeth

cell in the course of irradiation changes quasistatically, and at each moment is virtually uniform for the entire gas volume.

The increase in the discrepancy (with a reduction in the initial gas density) between the calculated efficiencies of the energy deposition and their values determined by recalculation using formula (7.21) with pressure measurement data can be explained using the following simple correlations. Let $F\left(\vec{r},t\right)$ be the specific power of the sources of energy release in the gas due to the deceleration of fission fragments at the point $\vec{r}$ at the moment t. Then the temperature increment at this point in the time δt can be defined as $\delta T\left(\vec{r},t\right) \sim F\left(\vec{r},t\right)\delta t/\rho c_n$, where c_n is the specific heat capacity of the polytropic process which occurs at the point $\vec{r}$ when the gas is heated by fission fragments. For very low energy deposition, the change in pressure in the cell can virtually be ignored; then $c_n \approx c_p$.

At high values of energy deposition, when starting at a certain moment the pressure markedly exceeds the initial pressure, and heat removal from the gas is negligibly small, allowing a stationary density profile to be established in the cell. This profile remains unchanged with a further delivery of energy to the gas [22, 36]. In this case, $c_n \approx c_v$.

Because specific energy losses of fission fragments per unit of range are proportional to the density of the nuclei of the decelerating medium [14], one can consider that the specific power deposition for variable gas density in the cell is approximately described by the equation $F\left(\vec{r},t\right) = F_0\left(\vec{r},t\right)\rho\left(\vec{r},t\right)/\rho_0$, where $F_0\left(\vec{r},t\right)$ is the specific power deposition at the point with coordinate $\vec{r}$ at the time t for a uniform non-perturbed distribution of gas density ρ_0.

The source function can be represented as $F_0 = W_r\left(\vec{r}\right)\Phi(t)$, where $W_r\left(\vec{r}\right)$ is the spatial distribution of specific sources in a uniform gas, calculated relative to a unit of neutron flux. Then the gas temperature at the point $\vec{r}$ at the time t, with weak influence of heat conductivity is

$$T\left(\vec{r},t\right) \sim T_0 + \frac{W_r\left(\vec{r}\right)}{\rho_0 c_n} \int_0^t \Phi(\xi) d\xi. \qquad (7.22)$$

The heat flow from the gas to the cell wall through a unit surface is equal to $\vec{g}_T = -k_g \times \operatorname{grad} T$. Calculations [19, 20] and experiments [27] showed that a narrow gas region forms close to the cell walls with the deposited uranium layer, from which heat is intensively removed into the wall. The temperature of the latter varies weakly and remains virtually equal to the initial gas temperature T_0. The characteristic dimension l of this region is determined by the correlation $l \sim \sqrt{at}$, where $a = k_g/\rho_0 c_p$ is the coefficient of thermal diffusivity of the gas. In the remaining part of the cell volume, the gas temperature increases in the course of irradiation, and its profile is very close to that of Eq. (7.22). Based on this, we can estimate the temperature gradient near the cell wall as

$$\operatorname{grad} T \sim \frac{T\left(\vec{r}_m, t\right) - T_0}{\sqrt{at}}.$$

where $\vec{r}_m$ is the coordinate corresponding to the maximal temperature close to the wall.

Because the wall region of intensive heat removal is rather narrow, we can consider the energy deposition $W_r\left(\vec{r}_m\right)$ in Eq. (7.22) to be approximately equal to its value directly on the wall, W_w. Thus,

$$g_T \sim \frac{\sqrt{k_g c_p} W_w}{\sqrt{\rho_0} c_n \sqrt{t}} \int_0^t \Phi(\xi) d\xi.$$

The total amount of heat passing through a unit area of the surface is

$$G(t) \sim \frac{\sqrt{k_g c_p} W_w}{\sqrt{\rho_0} c_n} \int_0^t \frac{d\xi}{\sqrt{\xi}} \int_0^\xi \Phi(\eta) d\eta.$$

To determine the qualitative connection of the considered parameters, it suffices to represent the time dependency of the pulsed neutron flux in the form

$$\Phi(t) = \begin{cases} \Phi_m t/\tau_n, & 0 \le t \le \tau_n; \\ \Phi_m(2 - t/\tau_n), & \tau_n \le t \le 2\tau_n; \\ 0, & t \ge 2\tau_n. \end{cases}$$

As a result, for the period of time when the gas is still cooling, we find

$$G(t) \sim \frac{\sqrt{k_g c_p} W_w \Phi_m \tau_n^{3/2}}{\sqrt{\rho_0} c_n} f(t),$$

where

$$f(t) = \begin{cases} 0.2(t/\tau_n)^{3/2}, & 0 \le t \le \tau_n; \\ 1.07 - 0.2(t/\tau_n)^{5/2} + 1.33(t/\tau_n)^{3/2} - 2(t/\tau_n)^{1/2}, & \tau_n \le t \le 2\tau_n; \\ 2(t/\tau_n)^{1/2} - 1.95, & t > 2\tau_n. \end{cases}$$

Considering that the neutron fluence during the pulse is equal to $\varphi = \Phi_m \tau_n$, by the moment of the pulse end and after it ceases, the total quantity of heat passing through a unit surface area may be approximately represented by

$$G(t) \sim \frac{\sqrt{k_g c_p} W_w \varphi}{c_n \sqrt{\rho_0}} \left(2\sqrt{t} - 1.95\sqrt{\tau_n}\right). \tag{7.23}$$

This dependence well reflects the experimental curves of the pressure drop in the cell after the pulse. It also shows that with a reduction in the initial density (or D_0, because $D_0 = d/R_0 \sim \rho_0$), the influence of the heat removal both on the pressure measured in experiments [27, 29] and the average gas temperature, and on the formation of acoustic oscillations in studies [30, 31] must grow. It should be noted that the value of W_w also depends on ρ_0 and consequently on D_0. However, this is a comparatively weak dependence and manifests itself only in the narrow region of $0 \le D_0 \le 0.4$, where the function W_w increases roughly by a factor of 2 with a decrease in D_0. For a relatively thick gas interval of $D_0 > 0.4$, the value of W_w remains practically unchanged. Thus, using the results of [15, 16], it is not hard to show that when the uranium layers are in a plane-parallel arrangement, W_w for $0 \le D_0 \le 1$ can be approximately described by the correlation $W_w \sim W_0(1 - 0.5D_0^2 + D_0 \ln D_0)$, where W_0 is equal to the value of W_w for $D_0 = 0$.

Both the data of experiments [27, 29] on the deviation of ε from the theoretical value, and the results of direct calculation of $G(t)$ in direct calculations of gas behavior [20] confirm that the dependence of the heat removed through the cell walls on the initial gas density is indeed close to Eq. (7.23).

Further, when the energy deposition from flat uranium layers is calculated, one should allow for the influence of edge effects. Direct measurements [29] confirmed the presence of these effects. A marked reduction in the energy flux carried away by fragments is perceptible at a distance ~$R_0/3$ from the edge of the layer and at the edge itself is equal to half the value of the energy removed in the case of an infinitely extended layer.

One of the reasons for the discrepancy between the calculated (obtained for a non-perturbed medium) and experimental values of the efficiency of the energy deposition may be spatial redistribution of the gas density arising in the course of irradiation owing to the spatial non-uniformity of the energy deposition by fission fragments. In addition, the influence of dynamic and thermal relaxation of the gas may play a role. Thus the calculations carried out [30] showed that the values of gas densities in different points in the cell by the end of irradiation for a specific energy deposition of 0.58 J/cm^3 differ two times over. To check these assumptions, a series of experiments were carried out using a rectangular cell with a volume of $20 \times 6 \times 2$ cm^3 with the same configuration as in [29] (the same neutron source and same gas), but with a somewhat greater distance from the source. The latter circumstance made it possible to reduce the average specific energy deposition from 0.25 to 0.03 J/cm^3, and accordingly to reduce the density differentials through the cell volume. Measurements showed that the absolute discrepancy between the calculation for a non-perturbed medium in this experiment decreases by a factor of 1.5. This fact clearly testifies to a definite influence of the specific energy depositions on the experimental results.

Possibly a marked role is played by the effect, discussed in [28, 33], of removal of some of the energy injected into the gas, in the form of UV radiation of excimer molecules. According to [28, 33], this effect is manifested in pure gases which do not contain impurities. When insignificant concentrations (~1 %) of impurities (nitrogen, hydrogen, etc.) are added in the gas, the experimentally measured efficiency may increase by 40 %. When the concentrations of impurities are >1 %, an increase in its quantity has little influence on the energy deposition. In this regard, it should be noted that in experiments [27], the criterion for gas purity was only the requirement that the presence of impurities not affect the value of the energy deposition, and the presence of impurities in a quantity <1 % was not tested. Because additions of an active laser component (xenon, for example) in NPL mixtures based on rare gases as a rule exceed 1 %, there should not be a marked drop in the efficiency of fragment energy deposition through the formation of excimer molecules.

Study [37] identified the important role of rapidly cooled small regions of the cell, residual volumes of the gas filling lines, the gaps between the optical and structural elements, and so forth. These regions are at first glance of insignificant volume, and for that reason their effects usually were not considered when results were processed. The authors of [37] showed, in the example of a rectangular cell of the type of [30], irradiated with a ~10 ms pulse, that correct allowance for the influence of rapidly cooled regions (in this cell their total volume is 15 %) leads to a reduction in the calculated pressure jumps in the cell by roughly a factor of 2 in comparison with analogous jumps in calculations that do not allow for the influence of these regions. In addition, in contrast to the results without allowance for the effects of small regions, calculations of the dependence of pressure on time in the updated scheme agree well with experiment [37].

Thus the method of indirect determination of the energy deposition by means of recalculating the data on measurement of pressure jumps [27, 29–31] does not allow for the marked influence of rapidly cooled small regions, and partial removal of energy to the walls due to heat conductivity, which greatly depends on the initial gas density. Moreover, in computer simulation (in a one-dimensional approximation) of the oscillations of the gas pressure in the cell after excitation by short neutron pulses of duration $\tau_{1/2} \sim 0.15$ ms [30] and $\tau_{1/2} \sim 0.4$ ms [31], the comparison was done only with regard to the amplitude of the first peak of the pressure dependence on time. But in the behavior of the peaks following the first, as the data cited in [30] shows, there are significant differences between calculation and experiment. Allowance for heat removal (15–35 %), the influence of small, rapidly cooled regions (up to ~50 %), as well as the removal, by the excimer molecules of rare gases, of some of the energy injected by the fragments, in experiments with highly purified single-component media (up to ~40 %), allows us to say that there is no contradiction between the experimental and calculated values of energy deposition efficiency.

Thus according to presently existing data, evidently only the inhomogeneities of the uranium layers lead to a real reduction in the energy deposition of fission fragments to mixtures based on rare gases. Direct measurements [29] of the energy carried away by fragments from uranium layers under vacuum conditions conform within the limits of error to the corresponding calculated values of this parameter.

7.5 Optimization of the Energy Deposition

One way to raise the total efficiency of nuclear-laser facilities is to increase the efficiency of use of the nuclear fission energy generated in the uranium-containing layers of the laser channels. This efficiency depends on the thickness, composition, and degree of uniformity of the uranium-containing layers, the geometry of the lasers themselves, and the composition and density of the laser-active gas.

The basic problem of optimization is to force the fission fragments to inject as much energy as possible into the gas mixture. The energy absorbed in the gas is proportional to $D_1\varepsilon(D_0, D_1)$ (see Sect. 7.3 of this chapter). With an increase in D_1, the value of $D_1\varepsilon(D_0, D_1)$ at first increases nearly linearly (see Fig. 7.5), but as D_1 continues to increase, the rate of increase of $D_1\varepsilon(D_0, D_1)$ decreases, because segments of the uranium layer far from the gas inject more and more energy into the layer itself, and less and less into the gas. When $D_1 = 1$, $D_1\varepsilon(D_0, D_1)$ reaches a maximal value fission fragments from the most remote layers of uranium do not reach the gas at all. Saturation begins with $D_1 \approx 0.5$. The energy intensity in the substrates with the uranium layer with an increase in thickness of this layer grows linearly, while the energy injection in the gas, starting with $D_1 \geq 0.5$, grows insignificantly (≤ 6 %). Consequently, the thickness of the uranium layer $D_1 \sim 0.5$ can be considered optimal.

The value of ε falls monotonously with an increase in D_1. The drop continues even with $D_1 > 1$. Thus with regard to D_1, the functions $\varepsilon(D_0, D_1)$ and $D_1\varepsilon(D_0, D_1)$ behave in the opposite manner: the thicker the active layer, the less efficiently it is used, but the greater the total energy deposition to the gas.

With regard to the normalized transverse dimension of the laser channel D_0, the functions $\varepsilon(D_0, D_1)$ and $D_1\varepsilon(D_0, D_1)$ behave identically (see Fig. 7.5, for example), because both are proportional to the total energy absorbed in the gas. For low values of D_0, which correspond to small gas densities, the functions $\varepsilon(D_0, D_1)$ and $D_1\varepsilon(D_0, D_1)$ are small, because fragments intersect the laser channel with small energy losses. As D_0 increases, the functions in question also increase and for $D_0 \sim 1$ they reach saturation—then fragments injected in the gas transfer nearly all of the energy to it.

The characteristics of NPLs excited by fission fragments are determined by many factors, which also include the formation and development of optical inhomogeneities of the medium, that is, inhomogeneities of the gas density. Next we will examine the link between the efficiency of the energy deposition of fragments and the growing optical inhomogeneities. We shall limit ourselves to the mode of short irradiation of the gas in the laser channel, which is of practical importance, that is, to pulsed excitation of the cells or excitation of a portion of the gas when there is continuous flowing of the gas medium through the laser channel under the conditions of stationary irradiation with neutrons. The characteristic time of equalization of the gas temperature in the laser cell is $\tau_T \sim d^2/a \sim 1$ s. If the excitation time of the gas in the channel is sufficiently small ($\tau_{1/2} \leq 10^{-2}$ s), then heat exchange has a marked effect only on the thin wall layer of gas with a thickness of ~1 mm [19, 20, 25, 27, 38, 39]. A "passive zone," with a large positive density gradient, is formed here which is not part of the volume involved in lasing. In the remaining part of the channel, the processes of heat conductivity do not have time to markedly influence the distribution of gas density arising in the course of irradiation (see Chap. 8). The distribution of the gas temperature and density is determined by the distribution function of the specific fragment energy deposition $F\left(\vec{r}\right)$. Because this function drops with distance from the channel walls [14], the gas density increases with approach to the channel axis; a focusing gas lens is formed. During lasing, light beams both from the axial and from the peripheral regions of the active part of the laser volume oscillate around the optical axis of the system during their multiple passage of the laser cavity. Consequently, the axial region can have the greatest influence on the parameters of the laser radiation.

The degree of non-uniformity of the energy deposition of fission fragment can be determined by the ratio

$$\mu_1 = F(0)/F_1,$$

where $F(0)$ is the specific deposition at the center of the channel (minimal value), and F_1 is the maximal value of the specific energy deposition to the gas (directly at the surface of the uranium layer).

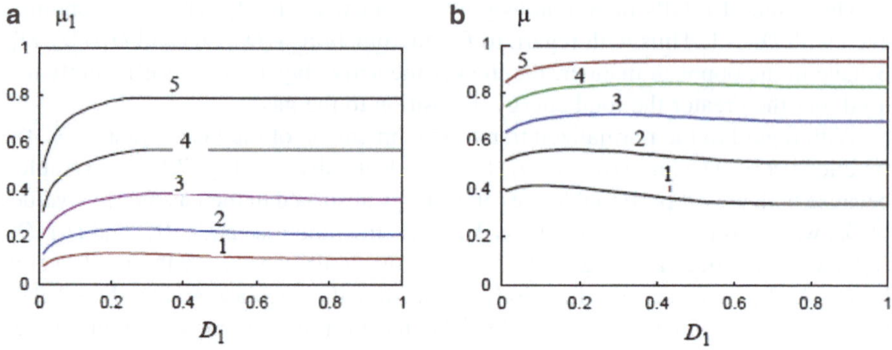

Fig. 7.15 Dependence of the coefficient of non-uniformity of the energy deposition μ_1 (a) and μ (**b**) on the normalized parameters D_1 and D_0: Curves 1–5 correspond to values of D_0 equal to: 1.0; 0.8; 0.6; 0.4; 0.2

The relative share of the energy deposited by fragments in the central part of the channel is characterized by the expression

$$\mu = F(0)/\langle F \rangle, \tag{7.24}$$

where $\langle F \rangle$ is the average value of the specific energy deposition along the channel section. Figure 7.15 shows the dependencies μ and μ_1 calculated from the correlations of studies [14, 16] for a system formed by two parallel infinitely extended flat uranium layers.

For all the clarity of the coefficient μ_1 for optimization of the energy deposition, it is more convenient to use the parameter μ. The fact is that the function $F\left(\vec{r}\right)$ has the greatest gradient and reaches the greatest value on the channel wall; however, it is in the wall region that the "passive zone" is formed such that light beams are bent on the channel wall.

The value of μ with growth in the relative width of the channel D_0 decreases monotonously, while the efficiency ε of the energy deposition grows monotonously, and starts to reach saturation in the region of $D_0 \geq 0.4$ [14]. Because the total energy deposited in the gas by the fragments is proportional to $D_1\varepsilon(D_0, D_1)$, the energy released by them in the axis of the laser channel can be characterized by the dimensionless parameter

$$\omega = D_1 \varepsilon \mu. \tag{7.25}$$

Figure 7.16 shows the dependence of the parameter ω on the normalized values D_0 and D_1 for a system of two plane-parallel infinitely extended layers. In the region of $D_0 \sim 0.4$, the parameter ω has a clearly pronounced maximum. With an increase in the thickness D_1 of the active layer, the parameter ω grows monotonously, reaching saturation at $D_1 = 1$. This increase is similar to the behavior of the dependence of the product $D_1\varepsilon(D_0, D_1)$ on D_1. Saturation starts in the region of $D_1 \sim 0.5$. Thus the

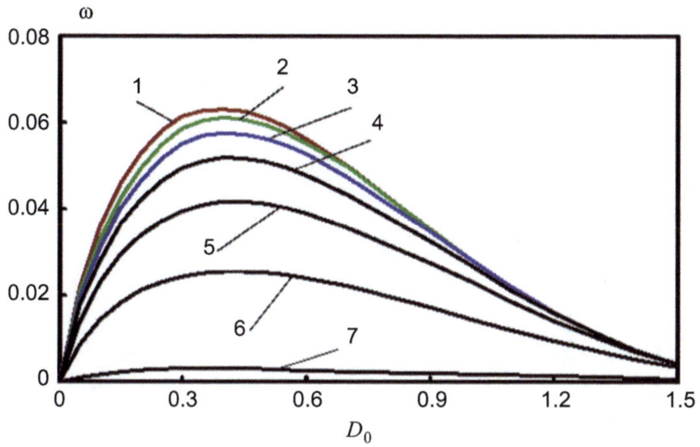

Fig. 7.16 Dependence of optimization parameter ω on the normalized width of the gas interval: Curves 1–7 correspond to values of D_1 equal to 1.0; 0.5; 0.4; 0.3; 0.2; 0.1; 0.01

behavior of the parameter ω also leads to the conclusion that an active layer thickness of $D_1 \sim 0.5$ can be considered optimal. Data of the dependence for ω, like the curves in Fig. 7.15, were calculated for uniform gas distribution in the cell and a square law of fragment deceleration, that is for $n = 2$ in Eq. (7.8).

In Fig. 7.17, the broken curve shows the dependence $\omega(D_0)$, obtained by calculation by the method [14], for a system formed by two parallel, infinitely extended flat uranium layers for $D_1 = 0.4$ (which corresponds to a uranium oxide-protoxide layer with a thickness $\delta_U = 3.2$ mg/cm^2, used on the LUNA-2M setup [40–43]). Correlations from [15, 16] make it possible to allow for the influence of edge effects arising due to the finite length b of the uranium layers in the direction perpendicular to the optical axis of the laser cell. In the experiments shown here, $b = 6$ cm, and the distance between the uranium layers is $d = 2$ cm. The results of calculation of ω with allowance for edge effects are shown in Fig. 7.17 by a solid curve; in addition, experimental values of laser output powers are marked on the graph for various mixtures [40–45]. For convenience in comparison, the values of the parameter ω, and laser output power J are calculated relative to their maximal values ω_{max} and J_{max}.

Deviations of experimental results from the curve ω/ω_{max} can be explained by differences in the kinetics of plasma-chemical processes and in the formation of an inverse population when there are variations in the partial composition of gas mixtures. The ω parameter itself is entirely independent of these processes. The correlation of its behavior to the experimental dependence $J(D_0)$ testifies to the significant influence not only of the indicated processes, but also of the regularities of the energy transfer from charged particles to the gas, thermo- and gas-dynamic processes, and geometry of the system. In addition, this parameter does not allow for more subtle effects. Thus, for example, the lower the initial pressure of the gas, which is equivalent to a lower value of D_0, the greater the volume of the near-wall

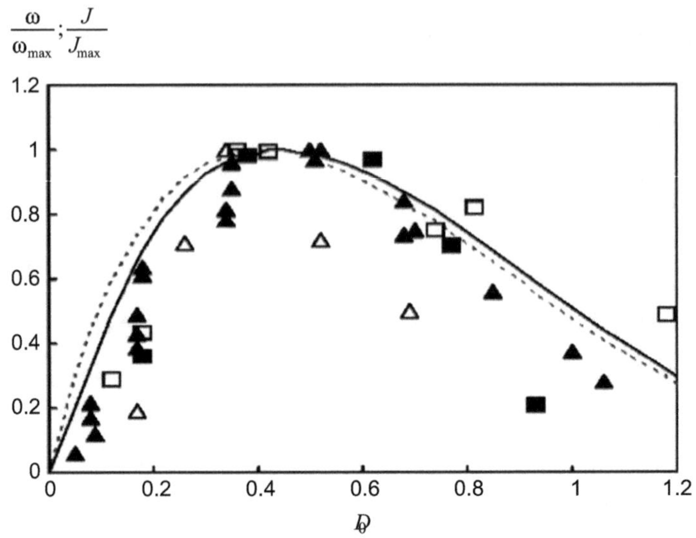

Fig. 7.17 Dependencies of the parameter ω of energy deposition optimization and the laser output power J on the normalized transverse dimension of the laser cell: the *broken curve* is the calculation for infinitely extended uranium layers; the *solid curve* is calculation for layers with a finite transverse dimension $b = 6$ cm; *filled triangle* (He-Xe) [40, 44]; *open triangle* (He-Kr) [41, 45]; *open square* (Ar-Xe) [42]; *filled square* (pure Xe) [43] experiment

passive zone (see Chap. 8), which at the moment of end of the neutron pulse can amount to ~30 % for $D_0 \sim 0.1$. Consequently, the volume of the active lasing region and the output laser energy will be less. Allowance for this factor would lead to improvement of the correlation between the curve ω/ω_{max} and the experimental points J/J_{max} for $D_0 < 0.4$.

Apart from calculations of the parameters μ, μ_1, and ω cited above for the "square law" of fragment deceleration, corresponding to $n = 2$ (see Sect. 7.2 of this chapter), for a deeper investigation analogous calculations were carried out for the "linear deceleration law" ($n = 1$) and the $n = 3/2$ law. These dependences behave similarly to the dependencies obtained for $n = 2$. Figure 7.18 shows the curves for each of the deceleration laws, reflecting the change in the optimal normalized transverse dimension D_{0max}, that is, the dimension corresponding to the maximum of the function ω, depending on the normalized thickness D_1 of the active layer for a system of two infinitely extended flat layers. A comparison of these curves with the experimental results in Fig. 7.17, according to which $D_{0max} \approx 0.4$–0.45, testifies more in favor of the "square law" of deceleration.

Investigations of an analogous type were also conducted for a cylindrical geometry of laser cells. As was to be expected, qualitatively the results in no way differ from the corresponding results for a plane geometry. Figure 7.19 provides a comparison of the dependencies of optimal transverse dimension D_{0max} for cylindrical and infinitely extended flat geometries calculated using the "square law" of deceleration.

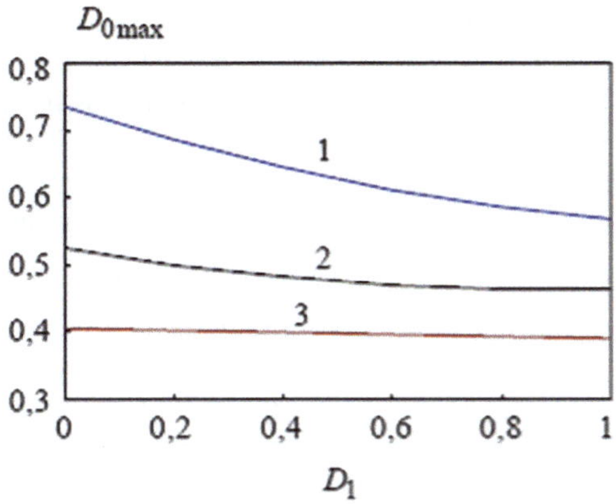

Fig. 7.18 Dependence of optimal normalized width for a system formed by two parallel infinitely extended flat uranium layers, on the thickness of the active layer for three deceleration laws: (*1*) "linear"; (*2*) "3/2 law"; (*3*) "square law"

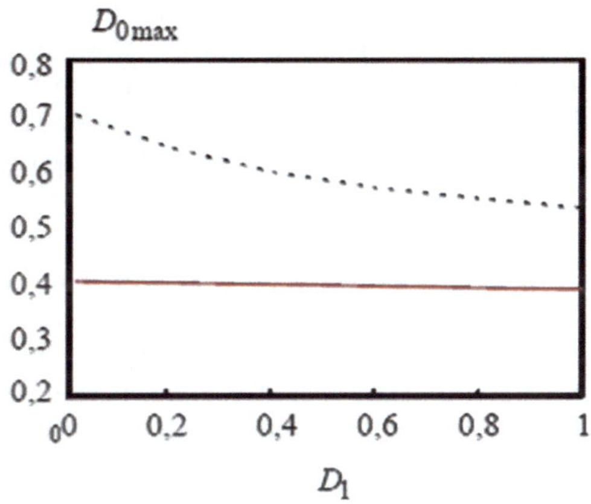

Fig. 7.19 Dependence of the optimal normalized width of the laser channel on the thickness of the active layer for a system formed by two parallel, infinitely extended flat uranium layers (*solid curve*) and for a cylindrical system (*broken curve*)

The calculations shown above for the optimization parameter ω were carried out with the assumption of a uniform gas distribution. For a closed system, formed by two infinite flat layers of uranium, the possible redistribution of the gas density in the course of irradiation, owing to the non-uniformity of the specific energy deposition function and under the influence of heat transfer processes, will not lead to a change either in the dependence itself, ω (D_0, D_1), or in the optimal value D_{0max}. Indeed, in such a system, the gas density is a function only of the transverse coordinate. The distribution of the specific sources $F\left(\vec{r}, t\right)$, fixed in Lagrangian coordinates, does not depend on redistribution of the density, and is equal to its value in a non-perturbed medium [16, 22, 36]. For a closed cylindrical cell, this distribution of sources for $D_0 \leq 0.5$, fixed in Lagrangian coordinates, also differs little from the true one [36]. The same also occurs for $D_0 \geq 0.5$, for irradiating times $\tau_{1/2} \leq 3$ ms [16].

Matters are a little more complicated when the cell has large buffer volumes, into which a considerable quantity of gas may flow in the course of irradiation when there are high power depositions. The gas flow leads to a reduction in its average density in the active part of the cell, which is equivalent to a reduction in D_0; thus the parameter ω varies in the course of irradiation. In this case, the question is at what moment of the excitation pulse is it necessary to achieve an optimal energy deposition? The very process of optimization requires performance of gas-dynamic calculations to determine the transformation in time of the specific energy deposition function $F\left(\vec{r}, t\right)$ over the entire length of the excitation pulse. However, if distortions of density through the cell volume are not too large and the quantity of gas flowing to the buffer volumes is not great, as occurred in experiments [40–43], optimization for non-perturbed gas as a reasonable zero-order approximation is reasonably accurate. This also applies to laser channels with continuous gas pumping, in which a constant mass expense of gas mixture is realized.

In conclusion of this section, it should be pointed out that other authors have also sought the criteria for energy deposition optimization. An essentially very similar method was used in study [46]: "Since in order to raise the efficiency of fission fragments, it is necessary to ensure the greatest escape of fragments from the fuel to the gas with the minimal non-uniformity of the energy release profile with respect to the thickness of the gas channel, it is possible to adopt as the channel optimization criterion" $\omega' = \varepsilon F(0)/\langle F \rangle$, that is, $\omega' = \varepsilon\mu$. At the same time, the "minimal energy intensity which facilitates cooling of the channel is reached in the case when" the value ω' of [46] "is maximal." This approach, in agreement with the results discussed above, shows that the optimal width of the gas channel formed by two infinitely extended plane-parallel uranium layers is equal to $D_{0\max} \approx 0.5$ for $n = 3/2$ and $D_{0\max} \approx 0.4$ for $n = 2$. However, "with an increase in the thickness of the fuel, the maximal value of ω' decreases quickly." Graphs of the dependencies $\omega'(D_0, D_1)$, similar in form to the graphs of ω in Fig. 7.16, are arranged in order opposite that of ω in the indicated figure [46]. The parameter ω' climbs steeply with a reduction in D_1, reaching a maximal value for $D_1 = 0$, which physically corresponds to a zero energy deposition because of the absence of uranium. The authors [46] did not allow for one significant detail. The coefficient ε is a relative value, reflecting the share of energy deposited into the gas relatively to the total energy of the fission fragments released in the active layer. In reality, the energy characteristics of the laser are determined by the energy deposition, which is characterized by the product $D_1\varepsilon(D_0, D_1)$, rather than $\varepsilon(D_0, D_1)$. Therefore, for optimization it is necessary to use the parameter ω, defined by Eq. (7.25), rather than the product $\varepsilon\mu$. The approach used in study [46] allowed its authors to optimize with respect to D_0 (the gas pressure), but they could not optimize with respect to D_1, because the coefficient of efficiency $\varepsilon\mu$ introduced in this way reaches a maximal value for $D_1 = 0$. From the academic standpoint, the material of the active layer is indeed utilized most fully in infinitely thin layers. However, the chief practical goal is still not to maximize some parameters that could be interpreted as the "fission fragment efficiency," but to inject as much energy as possible in the laser-active gas, and to use it to pump the laser.

We note that study [47] proposed using, in the capacity of optimization criterion, the product $\omega'' = \varepsilon F(0)/F_1 = \varepsilon\mu_1$. The results of calculation for the "square law" of deceleration show a maximum of ω'' in the region of $D_0 \approx 0.3$–0.4. This is somewhat below the value of $D_{0\max}$, determined by the aggregate of experimental results on the laser output power (see Fig. 7.17).

References

1. J.W. Wilson, R.J. DeYoung, Power density in direct nuclear-pumped ^{3}He lasers. J. Appl. Phys. **49**(3), 980–988 (1978)
2. A.A. Pikulev, The energy deposition in nuclear-pumped gas lasers based on helium-3 (Energovklad v gazovykh lazerakh s yadernoy nakachkoy na osnove geliya-3). ZhTF **76** (10), 91–96 (2006)
3. J.W. Wilson, R.J. DeYoung, Power deposition in volumetric $^{235}UF_6$-He fission-pumped nuclear lasers. J. Appl. Phys. **49**(3), 989–993 (1978)
4. R.J. DeYoung, Y.J. Shiu, M.D. Williams, Fission-fragment nuclear lasing of Ar(He)-Xe. Appl. Phys. Lett. **37**(8), 679–681 (1980)
5. R.J. DeYoung, W.R. Weaver, Spectra from nuclear-excited plasmas. J. Opt. Soc. Am. **70**(5), 500–506 (1980)
6. J.C. Guyot, G.H. Miley, J.T. Verdeyen, Application of a two-region heavy charged particle model to noble-gas plasmas induced by nuclear radiations. Nucl. Sci. Eng. **48**(3), 373–386 (1972)
7. A.N. Sizov, On the link of spatial non-uniformities of pumping with output powers of lasers excited by uranium fission fragments (O svyazi prostranstvennykh neodnorodnostey nakachki s moshchnostyu generatsii lazerov, vozbuzhdayemykh oskolkami deleniya urana). Lett. ZhTF **20**(9), 64–67 (1994)
8. P.J. Richards, B.A. Rubin, Irradiation of small volumes by contained radioisotopes. Nucleonics **6**(6), 42–49 (1950)
9. M. Steinberg, Deposition efficiency of fission-fragment energy. Nucleonics **21**(8), 151–153 (1963)
10. C.B. Leffert, D.B. Rees, F.E. Jamerson, Noble gas plasma produced by fission fragments. J. Appl. Phys. **37**(1), 133–142 (1966)
11. D.H. Nguen, L.M. Grossman, Ionization by fission fragments escaping from a source medium. Nucl. Sci. Eng. **30**(2), 233–241 (1967)
12. G.H. Miley, P.E. Thiess, A unified approach to two-region ionization-excitation density calculations. Nucl. Appl. **6**(5), 434–451 (1969)
13. A.K. Chung, M.A. Prelas, The transport of heavy charged particles in a cylindrical nuclear-pumped plasma. Nucl. Sci. Eng. **86**(3), 267–274 (1984)
14. V.T. Kazazyan, B.A. Litvinenko, L.P. Roginets, 4.A. Savushkin, *Physical Foundations of the Use of the Kinetic Energy of Fission Fragments in Radiation Chemistry (Fizicheskiye osnovy ispolzovaniya kineticheskoy energii oskolkov deleniya v radiatsionnoy khimii)* (Nauka i tekhnika, Minsk, 1972)
15. V.Y. Matyev, *Methods of calculation of the energy deposition of ions in non-uniform media* (Metody rascheta energovklada ionov v neodnorodnykh sredakh). Proceedings of the Specialist Conference "Physics of Nuclear-Excited Plasma and Problems of Nuclear-Pumped Lasers," Obninsk, 1993, vol. 2, pp. 79–88.
16. V.Y. Matyev, A.N. Sizov, *The influence of non-uniformities of the laser medium on gas-dynamic calculations of nuclear-pumped lasers*, Ibid, pp. 209–218.

17. V.Y. Matyev, The energy deposition of fission fragments in nuclear-pumped lasers. I. General method of calculation (Energovlad oskolkov deleniya v lazerakh s yadernoy nakachkoy. I. Obshchiy metod rascheta). ZhTF **71**(1), 72–78 (2001)
18. S. Kahn, R. Harman, V. Forgue, Energy distributions of fission fragments from uranium dioxide films. Nucl. Sci. Eng. **23**(1), 8–20 (1965)
19. A.N. Sizov, Y.N. Deryugin, *Calculation of spatial non-uniformities in cylindrical nuclear-pumped gas lasers* (Raschet prostranstvennykh neodnorodnostey v tsilindricheskikh gazovykh lazerakh s yadernoy nakachkoy). Tez. dokl. VI Vsesoyuz. konf. "Optika lazerov," Leningrad, 1990, p. 132.
20. A.N. Sizov, Y.N. Deryugin, Calculations of spatial non-uniformities in cylindrical nuclear-pumped gas lasers (Raschety prostranstvennykh neodnorodnostey v tsilindricheskikh gazovykh lazerakh s nakachkoy oskolkami deleniya). ZhTF **62**(9), 107–111 (1992)
21. M.G. Anuchin, K.F. Grebenkin, Y.Z. Kandiyev, Y.I. Cherepanova, Computational investigation of pumping of a gas medium by charged particles/nuclear reaction products (Raschetnoye issledovaniye nakachki gazovoy sredy zaryazhennymi chastitsami-produktami yadernykh reaktsiy). ZhTF **61**(1), 3–8 (1991)
22. J.R. Torczynski, On the motion of a gas experiencing range-dependent volumetric heating. J. Fluid Mech. **201**, 167–188 (1989)
23. J.R. Torczynski, R.J. Gross, Axis shielding in cylindrical nuclear-reactor-pumped lasers. J. Appl. Phys. **64**(9), 4323–4328 (1988)
24. G.B. Vlokh, L.L. Kazakov, N.S. Kosulin, A.A. Sinyanskiy, G.E. Filippov, V.N. Cherevatyuk, *Film energy releasing elements for nuclear-pumped lasers* (Plenochnyye energhovydelyayushchiye elementy dlya lazerov s yadernoy nakachkoy). Proceedings of the 2nd International Conference "Physics of Nuclear-Excited Plasma and Problems of Nuclear-Pumped Lasers," Arzamas-16, 1995, vol. 2, pp. 47–54.
25. V.V. Borovkov, G.B. Vlokh, B.V. Lazhintsev, V.A. Nor-Arevyan, A.N. Sizov, A.A. Sinyanskiy, G.E. Filippov, Energy deposition efficiency in lasers pumped by uranium fission fragments (Effektivnost energovklada v lazerakh s nakachkoy oskolkami deleniya urana). Kvantovaya elektronica **22**(3), 219–224 (1995)
26. A.N. Sizov, *Influence of non-uniformities of uranium-containing layers on the energy deposition to the active medium of nuclear-pumped lasers* (Vliyaniye neodnordnostey uransoderzhashchikh sloyev na energovklad v aktivnuyu sredu lazerov s yadernoy nakachkoy). Proceedings of the Specialist Conference "Physics of Nuclear-Excited Plasma and Problems of Nuclear-Pumped Lasers," Obninsk, 1993, vol. 3, pp. 83–88.
27. V.V. Borovkov, B.V. Lazhintsev, S.P. Melnikov, I.N. Mochkayev, V.A. Nor-Arevyan, A.A. Sinyanskiy, G.I. Fedorov, An investigation of optical non-uniformities in nuclear-pumped lasers (Issledovaniye opticheskikh neodnorodnostey v lazerakh s yadernoy nakachkoy). Izvestiya AN SSSR. Ser. fizicheskaya **54**(10), 2009–2015 (1990)
28. E.P. Magda, S.L. Mukhin, V.A. Kryzhanovskiy, A.V. Bochkov, Determination of the energy deposition to a gas medium (Opredeleniye energovklada v gazovuyu sredu). Pisma v ZhTF **19** (23), 54–57 (1993)
29. G.B. Vlokh, A.I. Konak, V.Y. Matyev, A.N. Sizov, A.A. Sinyanskiy, G.E. Filippov, *Measurement of energy release in a nuclear-pumped gas laser* (Izmereniye energovydeleniya v gazovom lazere s yadernoy nakachkoy). Proceedings of the Specialist Conference "Physics of Nuclear-Excited Plasma and Problems of Nuclear-Pumped Lasers," Obninsk, 1993, vol. 2, pp. 55–62.
30. J.R. Torczynski, R.J. Gross, G.N. Hays, G.A. Harms, D.R. Neal, D.A. McArthur, W.J. Alford, Fission-fragment energy deposition in argon. Nucl. Sci. Eng. **101**(3), 280–284 (1989)
31. M.G. Anuchin, K.F. Grebenkin, V.A. Kryzhanovskiy, E.P. Magda, Efficiency of pumping of gas media with fission fragments (Effektivnost nakachki gazovykh sred oskolkami deleniya). Pisma v ZhTF **18**(5), 92–95 (1992)
32. A.A. Androsenko, P.A. Androsenko, N.V. Gusev, P.P. Dyachenko, *Modeling of neutron-physical and gas-dynamic processes in an NPL* (Modelirovaniye neytronno-fizicheskikh i

gazodinamicheskikh protsessov v LYaN). Proceedings of the Specialist Conference "Physics of Nuclear-Excited Plasma and Problems of Nuclear-Pumped Lasers," Obninsk, 1993, vol. 2, pp. 23–40.

33. E.P. Magda, A.V. Bochkov, V.A. Kryzhanovskiy, S.L. Mukhin, *An investigation of the influence of nitrogen on the value of the measured pressure pulse in determining the energy deposition of fission fragments to helium and argon* (Issledovaniye vliyaniya azota na velichinu izmerennogo impulsa davleniya pri opredelenii energovklada oskolkov deleniya v geliy i argon). Proceedings of the 2nd International Conference "Physics of Nuclear-Excited Plasma and Problems of Nuclear-Pumped Lasers," Arzamas-16, 1995, vol. 1, pp. 452–456.
34. J.R. Torczynski, Heat transfer regimes in nuclear-reactor-pumped gas lasers. J. Thermophys. Heat Trans. **5**(3), 318–326 (1991)
35. V.V. Borovkov, G.B. Vlokh, B.V. Lazhintsev, V.A. Nor-Arevyan, A.N. Sizov, A.A. Sinyanskiy, G.E. Filippov, *Efficiency of the energy deposition of fission fragments to NPLs* (Effektivnost energovklada oskolkov deleniya v LYaN). Proceedings of the 2nd International Conference "Physics of Nuclear-Excited Plasma and Problems of Nuclear-Pumped Lasers," Arzamas-16, 1995, vol. 1, pp. 388–396.
36. V.Y. Matyev, *The gas dynamics of near-central regions in sealed NPL channels* (Gazodinamika okolotsentralnykh oblastey v germetichnykh kanalakh LYaN), Ibid, pp. 410–420.
37. J.R. Torczynski, D.R. Neal, Transient gas motion in nuclear-reactor-pumped lasers: computational and experimental results. Nucl. Sci. Eng. **113**(3), 189–206 (1993)
38. I.I. Borovkov, B.V. Lazhintsev, V.A. Nor-Arevyan, A.N. Sizov, A.A. Sinyanskiy, G.F. Fedorov, *Specific features of gas dynamics of flowing NPLs and problems of forming light-beam quality* (Osobennosti gazodinamiki prokachnykh LYaN i problemy formirovaniya kachestva svetovogo puchka). Proceedings of the 2nd International Conference "Physics of Nuclear-Excited Plasma and Problems of Nuclear-Pumped Lasers," Arzamas-16, 1995, vol. 1, pp. 399–406.
39. V.V. Borovkov, B.V. Lazhintsev, V.A. Nor-Arevyan, A.N. Sizov, A.A. Sinyanskiy, G.F. Fedorov, Specific features of gas dynamics of flowing lasers excited by uranium fission fragments (Osobennosti gazodinamiki prokachnykh lazerov, vozbuzhdayemykh oskolkami deleniya yader urana). Kvantovaya elektronika **22**(12), 1187–1191 (1995)
40. A.M. Voinov, L.Y. Dovbysh, V.N. Krivonosov, S.P. Melnikov, I.V. Podmoshenskiy, A.A. Sinyanskiy, Low-threshold nuclear-pumped lasers on atomic transitions of xenon (Nizkoporogovyye lazery s yadernoy nakachkoy na perekhodakh atomarnogo ksenona). Doklady AN SSSR **245**(1), 80–83 (1979)
41. A.M. Voinov, L.Y. Dovbysh, V.N. Krivonosov, S.P. Melnikov, I.V. Podmoshenskiy, A.A. Sinyanskiy, A high-pressure helium-krypton laser excited by uranium fission fragments (Geliy-kriptonovyy lazer vysokogo davleniya, vozbuzhdayemyy oskolkami deleniya urana). ZhTF **52**(7), 1346–1350 (1982)
42. A.M. Voinov, L.Y. Dovbysh, V.N. Krivonosov, S.P. Melnikov, I.V. Podmoshenskiy, A.A. Sinyanskiy, An infrared nuclear-pumped laser based on mixtures of He + Xe and Ar + Xe (Infrakrasnyy lazer s yadernoy nakachkoy na smesyakh He + Xe i Ar + Xe). Pisma v ZhTF **7**(16), 1016–1020 (1981)
43. A.I. Konak, S.P. Melnikov, V.V. Porkhayev, A.A. Sinyanskiy, Lasing on Xe atom transitions with excitation of a Kr-Xe mixture and pure xenon by uranium fission fragments (Generatsiya na perekhodakh atoma Xe pri vozbuzhdenii smesi Kr-Xe i chistogo ksenona oskolkami deleniya urana). Kvantovaya elektronika **22**(12), 1184–1186 (1995)
44. A.M. Voinov, S.P. Melnikov, A.A. Sinyanskiy, A kinetic model of recombination lasers based on transitions of the xenon atom (Kineticheskaya model rekombinatsionnykh lazerov na perekhodakh atoma ksenona). ZhTF **60**(10), 100–113 (1990)
45. S.P. Melnikov, A.A. Sinyanskiy, Kinetics of IR nuclear-pumped lasers on transitions of the krypton and argon atom (Kinetika IK-lazerov s yadernoy nakachkoy na perekhodakh atomov kriptona i argona). ZhTF **62**(6), 159–169 (1992)

46. A.V. Karelin, O.V. Sereda, V.V. Kharitonov, K.R. Chikin, F. Yu. Naumkin, Heating of a gas by fission fragments (Nagrev gaza oskolkami deleniya). Atom. Energ. **61**(1), 44–46 (1986)
47. A.V. Karelin, V.V. Kharitonov, K.R. Chikin, On direct conversion of nuclear energy into light radiation (O pryamom preobrazovanii yadernoy energii v svetovoye izlucheniye), in *Issues of Thermal Physics in Nuclear Power Units (Voprosy teplofiziki v yadernykh energeticheskikh ustanovkakh)* (Energatomizdat, Moscow, 1986), pp. 3–8

Chapter 8
Optical Inhomogeneities in Sealed NPLs

8.1 Inhomogeneities in Cylindrical Cells

The short ranges of fission fragments causes a specific energy deposition inhomogeneities which, along with excited gas heat exchange with laser cell walls, leads to spatial density redistribution, i.e., to the violation of active medium optical homogeneity. As noted in paper [1], with a reference to the 1980 report of A. N. Sizov and Yu. N. Deryugin, numerical and theoretical investigations of the dynamics of optical inhomogeneity origination and development in nuclear-pumped lasers first began at the VNIIEF. During first stage, these investigations only pertained to the phenomena occurring in sealed lasers excited by pulsed neutron fluxes. This important aspect of NPLs is discussed in this chapter.

Calculation Procedure

The neutron flux from the nuclear reactors used in experiments involving NPLs varies noticeably at distances that considerably exceed a laser cell's transverse dimensions [2–14]. Thus, it can be assumed that fission density distribution in a uranium layer, as well as gas temperature and density distribution at any given point within a cylindrical cell, are essentially only dependent upon the radial coordinate, r, and the axial coordinate, z, which coincides with the axis of symmetry.

More precisely, the nonstationary thermal gasdynamic processes taking place in a laser of the type under conditions only permit modeling to be performed using a computer. But even for a geometry that is dependent upon two spatial variables, such calculations prove to be quite difficult and require large outlays of computer time. Therefore, the first investigations of this type were conducted for alternatives involving dependence upon just one spatial variable; the primary objective of these investigations consisted of determining the scale of the inhomogeneities occurring

S.P. Melnikov et al., *Lasers with Nuclear Pumping*,
DOI 10.1007/978-3-319-08882-2_8

and establishing the regularities of their dynamics. For simplicity's sake, it was assumed that all physical parameters, including the distribution of specific energy release sources in the gas and in the uranium layer, are only dependent upon a single radial coordinate, r. Consequently, it was decided that the distribution of the thermal neutron fluxes initiating uranium nucleus fission is not dependent upon coordinates and equals the average value over the length of a cell.

It should be noted that the effect of viscosity forces is negligible. Actually, the order of magnitude of the gas velocity scale comes to $u \sim l/\tau_{1/2}$, where l is a typical cell dimension (a transverse dimension of $d \sim 1$ cm or a longitudinal dimension of $L \sim 1$ m), and; $\tau_{1/2}$ is the excitation pulse duration. The Reynolds number for the gas flow over the course of excitation is then $\mathrm{Re} = lu/\nu \sim 10^2$–$10^7$ (here v represents kinematic viscosity, which for rare gas mixtures over the temperature and pressure variation ranges actually used comes to ~10^{-1} cm^2/s), suggesting the negligible effect of viscosity on gas flow.

All the initial calculations were performed using the standard VNIIEF program described in study [15]. In ignoring viscosity, it becomes possible to undertake the numerical solution of the thermal gasdynamic problems described by:

the motion equation

$$\frac{\partial u}{\partial t} = r^{N-1}\frac{\partial}{\partial s}P = 0;$$

$$ds = \frac{1}{n}\rho d\left(r^N\right);$$

$$\frac{\partial r}{\partial r} = u,$$

the continuity equation

$$\frac{\partial}{\partial t}(1/\rho) = \frac{\partial}{\partial s}\left(r^{N-1}u\right),$$

and the energy equation

$$\frac{\partial E}{\partial t} + P\frac{\partial}{\partial t}(1/\rho) - \frac{\partial}{\partial s}\left[r^{2(N-1)}\rho k_i\frac{\partial T}{\partial s}\right] = F_\rho(\rho, T, s, t).$$

Here, E is the internal energy per unit mass of gas; F_ρ is the internal source power per unit mass; k_i is the thermal conductivity coefficient; u is matter velocity; P is pressure; r is the matter particle Eulerian coordinate; s is the matter particle Lagrangian coordinate; T is temperature; and ρ is density. The N parameter characterizes system geometry: $N = 1$ for a planar geometry, $N = 2$ for a cylindrical geometry, and $N = 3$ for a spherical geometry.

Over wide temperature and pressure variation limits, the rare gases that serve as NPL active media can be quite accurately described as ideal. Therefore, a state equation was given in the form

$$P = (\gamma - 1)\, c_V \rho T, \tag{8.1}$$

$$E = c_V T, \tag{8.2}$$

where $\gamma = c_p/c_V$; c_p, c_V represents specific heat capacities at a constant pressure and a constant volume, respectively.

At the contact boundaries of the gas and the uranium-containing active layer, as well as the active layer and the metal tube (the substrate) to which this layer is applied, the equality of temperatures and heat fluxes was postulated. The substrate's outer surface was considered to be heat-insulated. The heat capacity and thermal conductivity coefficients of the active layer and the substrate were approximated by a quadratic dependence upon temperature. Based on data [16], the dependences of the gas thermal conductivity coefficients upon temperature were described by a fourth-degree polynomial.

Over the course of irradiation, the gas distribution density in the cell is transformed with the passage of time. Accordingly, because the fragment energy losses per unit range are unequivocally linked to the density of the stopping medium, the specific energy deposition profile along the laser cell cross-section is also deformed. During early calculations, specific source distribution, F_ρ, in the gas was tied to Lagrangian coordinates in order to take this factor into account, and was given in the form

$$F_\rho(s, t) = F_{\rho 0}(s) q(t).$$

Here $F_{\rho 0}(s)$ is the specific power deposition at a point with a Lagrangian coordinate of s in the unperturbed gas (at the reference time) attributed to fission output per uranium layer unit volume, and $q(t)$ is the fission output per unit volume of this layer. The spatial distribution, $F_{\rho 0}(s)$, was determined using the method described in study [17]. A comparison [18, 19] of this approach to the technique in which the precise dependence of the F_ρ function on the gas density distribution at any given moment in time is used revealed that discrepancies in the density and temperature calculations using both procedures over typical durations of $\tau_{1/2} \leq 3$ ms do not exceed 10 % (also see Chap. 7, Sect. 7.2).

After specific energy deposition calculation techniques were worked out for the media with variable densities [19, 20] as mentioned above, the program was modified to make it possible to perform calculations with strict allowance for the effect of the density redistribution on the energy deposition.

Dynamics of Inhomogeneity Development in the Pulsed and Quasi-Stationary Excitation Modes

Numerical investigations of the dynamics of the development of inhomogeneities during NPL excitation by pulsed neutron fluxes from a TIBR reactor [21] with a duration (at half the exciting pulse height) of $\tau_{1/2} = 0.8$ ms were among the first performed. During experiments [4] on this reactor, the aluminum cylindrical cell employed had an inside diameter of 2.7 cm and an effective length of 57 cm. A ^{235}U oxide-protoxide layer with a thickness of 9 mg/cm^2 was applied to the cell's inner surface. The cell wall thickness was 2 mm, while the distance between the cavity mirrors equaled 1 m, and the average thermal neutron fluence per pulse over the cell's effective length was 4.2×10^{13} cm^{-2}. Curve 1 in Fig. 8.1 represents the time dependence of the thermal neutron flux density. Figure 8.2a reflects the results of computerized numerical calculations of the time evolution of the gas density distribution in a cell filled with an He-Xe mixture (a component ratio of 200:1) in the presence of an initial pressure of $P_0 = 1$ atm and an initial temperature of $T_0 = 300$°K [4]. These calculations demonstrate that the spatial distribution of temperature is inversely proportional to density distribution, while the pressure at any given moment in time is virtually uniform throughout the cell's cross-section. This behavior is expected, because the typical distribution time scale for the perturbation in the cell, $\tau_p \sim r_1 / u_s$ (u_s is sound velocity in the gas, and r_1 is the cell's inside radius), is significantly shorter than exciting pulse duration.

The results of numerical investigations [22, 23] revealed that the gas volume is divided into two parts. The equation $\partial \rho(r,t)/\partial r = 0$ defines the $r_A(t)$ interface. In the active region, $0 < r < r_A(t)$, the gas density gradient, and consequently the refractive index gradient, are negative. At $r_A(t) < r < r_1$ (the passive zone), the gradient is positive and far exceeds the absolute value of the gradient in the active region. A passive zone originates where heat removal to the cell wall plays a leading role,

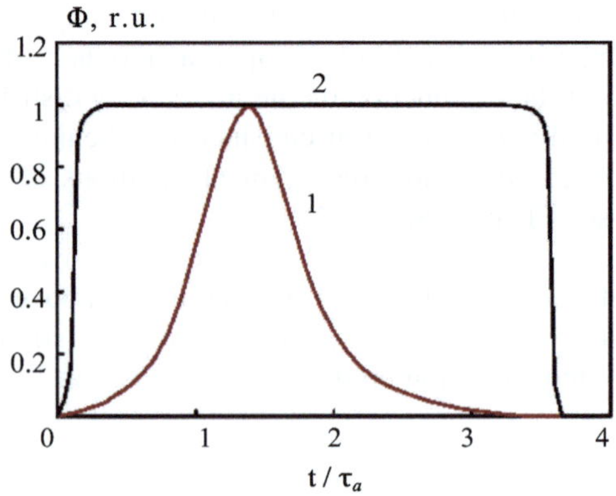

Fig. 8.1 Thermal neutron flux density time dependence: (1) pulse with a half-height duration of $\tau_n = 0.8$ ms ($\tau_a = 1$ ms); (2) pulse with a duration of $\tau_n \approx 1$ s ($\tau_a = 0.3$ s)

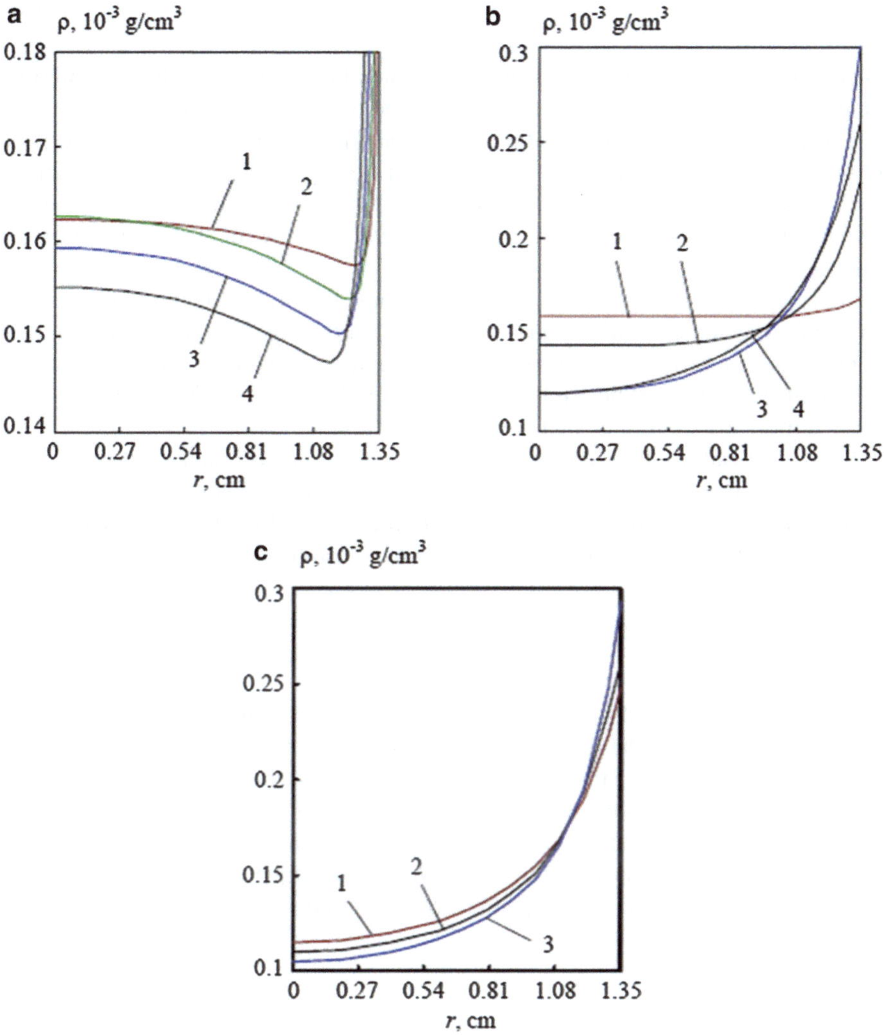

Fig. 8.2 Gas density distribution throughout the cross-section of a laser cell: (**a**) a pulse with a duration of $\tau_{1/2} = 0.8$ ms: (1) $t = 1$ ms; (2) $t = 1.4$ ms; (3) $t = 1.8$ ms; and, (4) $t = 2.2$ ms; (**b**) a pulse with a duration of $\tau_{1/2} \approx 1$ s: (1) $t = 0.03$ s; (2) $t = 0.06$ s; (3) $t = 0.1$ s; and, (4) $t = 1$ s; (**c**) the stationary irradiation mode: (1) coolant velocity, $u_l = 0.8$ m/s; (2) $u_l = 2$ m/s; and, (3) $u_l = 8$ m/s

because the heat capacity per unit volume appreciably exceeds the gas unit volume heat capacity. At $l = r_1 - r_A << r_1$, the time dependence of this zone's transverse dimension is well described by the correlation [22, 23]

$$l \sim \sqrt{at}, \tag{8.3}$$

where a represents gas thermal diffusivity. When long-duration pulses are present, the passive zone involves the entire gas volume over a time frame of $t \sim 0.1$s.

The results of similar calculations for a gas mixture of the same composition during excitation by a pulse with a duration of $\tau_{1/2} \approx 1$ s, a fluence of 4×10^{14} cm^{-2}, and a close-to-rectangular shape (see Curve 2, Fig. 8.1) are presented in Fig. 8.2b. The passive zone encompasses the entire gas volume in a few tenths of a millisecond.

Stationary Mode. Under stationary conditions, heat removal through the outer surface of a sealed cell must be arranged. Heat removal of this type can be realized by means of a gaseous or liquid heat-transfer agent washing over the outer surfaces of systems with internal heat release. As is generally known, a liquid heat-transfer agent is more efficient than a gas, so it is preferred.

Unlike the pulsed alternatives described above, a precise analytical solution to the problem of gas temperature and density distribution exists for the stationary mode. Under the conditions being examined, the temperature distribution along the cylindrical cell's tube wall is defined by the expression

$$T_w(r) = T_l - \frac{q_w}{k_w}\left\{r_1 \ln\frac{r}{r_2} - \frac{k_w r_1}{\alpha_l r_2}\right\}, \tag{8.4}$$

where k_w is the thermal conductivity coefficient of the tube material; q_w is the thermal flux density at the interface of the tube and the gas filling it; r_2 is the radius of the tube's inner surface that the heat-transfer agent washes over; $T_w\,(r)$ is the tube wall temperature at a point with a coordinate of r; T_l is the temperature of the external heat-transfer agent; and α_l is the coefficient of heat transfer from the tube's outer surface to the heat-transfer agent. We derive equation (8.4) by means of solving a thermal conductivity equation for a geometry with cylindrical symmetry

$$a_w \frac{1}{r}\frac{\partial}{\partial r}\left(r\frac{\partial T_w}{\partial r}\right) = \frac{\partial T_w}{\partial t}, \tag{8.5}$$

and with boundary and initial conditions of the type

$$k_w \frac{\partial T_w(r_2,t)}{\partial r} = -\alpha_l[T_w(r_2,t) - T_l(t)]; \frac{\partial T_w(r_1,t)}{\partial r} = -\frac{q_w(t)}{k_w}; T_w(r,0) = T_{w0}. \tag{8.6}$$

where a_w is the thermal diffusivity of the tube material.

System (8.5) and (8.6) describes the temperature space–time distribution in a cylindrical layer, $r_1 \le r \le r_2$, on the outer surface of which, with a coordinate of $r = r_2$, convective heat exchange occurs. On the layer's inner surface, at $r = r_1$, a thermal flux of $q_w(t)$ is specified.

When the a_w and k_w coefficients are constant, the solution to (8.5) and (8.6) takes the form:

$$T_w(r,t)=T_l(t)-\frac{q_w(t)}{k_w}\left\{r_1\ln\frac{r}{r_2}-\frac{k_1r_1}{\alpha_l r_2}\right\}-2\sum_{n=1}^{\infty}\frac{\Phi_n(r)}{\lambda_n\left\{r_2^2\left[1+\frac{\alpha_l}{\lambda_n k_w}\right]\Phi_n^2(r_2)-r_1\Phi_n^2(r_1)\right\}}\times$$
$$\times\left\{\frac{r_1}{k_w}\Phi_n(r_1)\left[q_w(t)-a_w\lambda_n\int_0^t q_w(\theta)\exp\{a_w\lambda_n(t-\theta)\}d\theta\right]+\frac{\alpha_l r_2}{k_w}\Phi_n(r_n)\times\right.$$
$$\left.\left[T_l(t)-a_w\lambda_n\int_0^t T_l(\theta)\exp\{a_w\lambda_n(t-\theta)\}d\theta-T_{w0}\exp\{-a_w\lambda_n t\}\right]\right\}. \tag{8.7}$$

The eigenvalues of λ_n are roots of the equation

$$k_w\sqrt{\lambda_n}\{N_1(\sqrt{\lambda_n}r_1)J_1(\sqrt{\lambda_n}r_2)-J_1(\sqrt{\lambda_n}r_1)N_1(\sqrt{\lambda_n}r_2)\}$$
$$=\alpha_l\{N_1(\sqrt{\lambda_n}r_1)J_0(\sqrt{\lambda_n}r_2)-J_1(\sqrt{\lambda_n}r_1)N_0(\sqrt{\lambda_n}r_2)\},$$

while the eigenfunctions that correspond to them are given by

$$\Phi_n(r)=N_1\left(\sqrt{\lambda_n}r_1\right)J_0\left(\sqrt{\lambda_n}r\right)-J_1\left(\sqrt{\lambda_n}r_2\right)N_0\left(\sqrt{\lambda_n}r\right).$$

Here J_ν, N_ν are the (first-order) Bessel and ν-order Neumann functions. It is not difficult to see that in the stationary case, i.e., at $\partial q_w/\partial t=\partial T_l/\partial t=0$ and $t\to\infty$, and the solution to (8.7) takes on the simple form of (8.4).

In the stationary mode, the equation for heat balance in the gas on a surface with a coordinate of r in the presence of problem cylindrical symmetry is written as

$$-2\pi rk_g\frac{dT}{dr}=2\pi\int_0^r r'F(r')dr', \tag{8.8}$$

where, as before, F is specific power deposition of gas; k_g is the thermal conductivity coefficient of the gas, which is very slightly dependent upon density for rare gases [24]. However, its dependence upon temperature is communicated with the same degree of accuracy as during the polynomial representation by the equation [24, 25]

$$k_g(T)=k_{g0}(T/T_{k0})^{\beta}. \tag{8.9}$$

The fragments transfer a portion of the uranium nucleus fission energy released in the uranium layer to the gas, while a portion $(1-\varepsilon)$ remains in the layer itself.

Due to this layer's low thickness ($\delta_U \sim 1$–10 μm), the layer can simply be regarded as a surface heat source with a surface power density of S_U when solving a thermal conductivity problem. In this case, the following boundary conditions must be satisfied at the gas and cell wall interface when $r = r_1$:

$$k_g(r_1)\frac{dT(r_1)}{dr} - k_w\frac{dT_w(r_1)}{dr} = S_U; \tag{8.10}$$

$$T(r_1) = T_w(r_1). \tag{8.11}$$

Source surface power density equals (see the designations in Chap. 7, Sect. 7.1)

$$S_U = (1-\varepsilon)N_U\sigma_f\Phi\delta_U E_0.$$

The balance of the energy released in a cylindrical gas volume of unit length and a radius of r_1 is written as

$$2\pi r_1\frac{\varepsilon}{1-\varepsilon}S_U = 2\pi\int_0^{r_1} xF(x)dx. \tag{8.12}$$

Integrating equation (8.8) with allowance for equation (8.9) and using the conditions of (8.10)–(8.12), we obtain

$$T(r) = \left\{\left[T_1 + \frac{S_U}{1-\varepsilon}\left(\frac{r_1}{\alpha_l r_2} - \frac{r_1}{k_w}\ln\frac{r_1}{r_2}\right)\right]^{(1+\beta)} - \right.$$
$$\left. - \left[\frac{1+\beta}{k_{g0}}T_{k0}^{\beta}\int_{r_1}^{r}\frac{dy}{y}\int_0^{y} xF(x)dx\right]\right\}^{\frac{1}{1+\beta}}, \quad 0 \le r \le r_1; \tag{8.13}$$

$$T_w(r) = T_l + \frac{S_U}{1-\varepsilon}\left\{\frac{r_1}{\alpha_l r_2} - \frac{r_1}{k_w}\ln\frac{r}{r_2}\right\}, \quad r_1 \le r \le r_2. \tag{8.14}$$

The heat transfer coefficient, α_l, can be determined using the formula [26, 27]

$$\mathrm{N_u} = 0,023\mathrm{Re}^{0,8} \times \mathrm{Pr}^{0.4} \tag{8.15}$$

through a Nusselt number of $\mathrm{N_u} = \alpha_l d_e / k_l$, a Prandtl number of $\mathrm{Pr} = \nu_l / a_l$, and a Reynolds number of $\mathrm{Re} = u d_e / \nu_l$, where d_e is the equivalent diameter of the heat-transfer agent channel, k_l is the thermal conductivity coefficient of the heat-transfer agent, u is its velocity, and ν_l is the heat-transfer agent's kinematic viscosity.

Using the expressions obtained, the temperature spatial distribution was calculated in a gas mixture of the same composition as in the previous section at three liquid heat-transfer agent circulation velocities. Water at a temperature of $T_l = 323$ K was selected as the heat-transfer agent. The equivalent diameter of the heat-transfer agent channel was $d_e = 2$ cm. Gas density spatial distributions were determined based on the temperature fields found using a state equation. All the calculations were performed for a thermal neutron flux density of $\Phi = 4 \times 10^{14}$ cm^{-2} s^{-1}. The calculation results (Fig. 8.2c) demonstrate that the density and refractive index gradients are positive in the entire gas volume in the stationary mode.

*Evaluation of Optical Characteristics***.** Reducing of radiation angular divergence and bringing it down to the diffraction limit insofar as possible is one of the most urgent problems in laser physics. Its solution is linked to the problem of the electromagnetic field distribution within a cavity. This field depends on the type of cavity used, the design and type of the laser itself, and the optical quality of the active medium filling the cavity (determined by refractive index spatial distribution, as well as this medium's gain and absorption factors). In the general case, wave optics techniques should be used to calculate the characteristics of a cavity filled with an active medium. However, the solution of a wave equation is difficult for a wide range of cavity configurations involving an inhomogeneous medium [28, 29]. Still, a number of important characteristics, especially when high Fresnel number values, $N_f = a_r^2/\lambda L_f$ (a_r is the transverse dimension of the cavity L_f is the distance between its mirrors, and λ is the radiation wavelength) are present, can also be determined using ray optics techniques [28–30]. For example, a formula in a paraxial approximation is derived in [31] for estimating the laser beam divergence with an inhomogeneous transverse refractive index distribution of laser medium

$$\vartheta \approx 2\sqrt{2\Delta n}, \tag{8.16}$$

Here Δn is the refractive index variation in the transverse direction within the confines of the lasing zone. This procedure provides quit good agreement with experiments [31, 32].

Moreover, for optical systems that can be represented in the form of a series of mirrors and elements equivalent to thin lenses, or for systems with a quadratic transverse refractive index dependence, and for combinations of these systems (see study [30] for an example), it has been shown that a ray approximation is in full agreement with the results provided by a Gaussian beam wave optics approximation. Gaussian beams play a significant role in laser optics: at $N_f >> 1$, they quite accurately describe the radiation field of a broad range of lasers with a stable cavity.

An important advantage of ray optics is visualization and the possibility of the comparatively simple qualitative and quantitative interpretations of the results obtained. In a geometric optics approximation, the equation that defines the shape of light rays takes the form [30, 33]:

$$\frac{d}{d\varsigma}\left(n\frac{d\vec{r}}{d\varsigma}\right) = \nabla n\left(\vec{r}\right),$$

or in a different version

$$\frac{d}{d\varsigma}\left(n\vec{\varsigma}\right) = \nabla n\left(\vec{r}\right), \tag{8.17}$$

where n is the refractive index; ς is the length of the ray arc from a certain fixed point to a point that is characterized by the radius vector, $\vec{r}(\varsigma)$; and, $\vec{\varsigma} = d\vec{r}/d\varsigma$ is a unit vector that is tangential to the ray trajectory; it determines the direction of light beam propagation, i.e., it coincides with the normal to the electromagnetic wave front.

Here, it is appropriate to note a peculiarity. Let us assume that $\vec{e}_z$ is the unit vector in the direction of the optical axis, z. For the variation of the scalar product of $\vec{e}_z \times \vec{n}\varsigma$ along the light ray, we have

$$\frac{d}{d\varsigma}\left(\vec{e}_z \cdot n\vec{\varsigma}\right) = \vec{e}_z \frac{d}{d\varsigma}\left(n\vec{\varsigma}\right).$$

The right-hand part of an equation based on (8.17) can be rewritten in the form of $\vec{e}_z \cdot \nabla n$. If ∇n is perpendicular to the z axis everywhere, then from the latter equation we get $\vec{e}_z \times n\vec{\varsigma} = \text{const}$; i.e., the scalar product of $\vec{e}_z \times n\vec{\varsigma}$ remains unchanged along the ray. Hence it follows that

$$\cos\vartheta = \frac{n_0 \cos\vartheta_0}{n\left(\vec{r}\right)}, \tag{8.18}$$

where n_0 is the refractive index at the point where the beam enters the gas volume under consideration; ϑ_0 is the angle between the z axis and the direction of the incoming ray and ϑ is the angle between the tangent to the beam at this point and the z axis.

Let us suppose that a medium's refractive index spatial distribution has a cylindrical or planar symmetry, and is only dependent on a coordinate perpendicular to the axis (or plane) of symmetry. Assume that in the region where the refractive index gradient is negative, a light ray passes parallel to the axis (or plane) of symmetry at a certain point at a distance of r_0 from it, i.e., at this point, $\cos\vartheta_0 = 1$. In this instance, whatever trajectory a ray traces as a result of the effect of the refractive index gradient, $n\left(\vec{r}\right)$, may not be less than $n_0 = n\,(r_0)$, since $\cos\vartheta \leq 1$ always, ray distance from the axis (or plane) of symmetry must not exceed r_0.

If the laser beam divergence with a gas density distribution that is only dependent upon a single transverse coordinate, as is the case in [31], is estimated as double the value of the maximum angle by which a ray intersecting an active lasing region can be deviated during propagation, formula (8.16) for estimating this divergence then follows from equation (8.18) in the $\vartheta << 1$ and $n - 1 << 1$ approximations.

In the paraxial approximation, $\vartheta << 1$, equation (8.17) for a cylindrically symmetric system with a transverse refractive index dependence takes the form [30]:

$$\frac{d^2r}{dz^2} = \frac{1}{n}\frac{dn}{dr}. \tag{8.19}$$

If a parabolic dependence approximates refractive index distribution

$$n(r) = n_\alpha\left(1 - \frac{\alpha_{n1}}{2}r^2\right),$$

which quite accurately communicates the real refractive index distribution in the bulk of an NPL's gas volume [2, 34–36], then from (8.19) at $n - 1 << 1$ and in the presence of the boundary conditions

$$r(z=0) = r_0; \quad \frac{dr(z=0)}{dz} = \vartheta_0,$$

we have [30]

$$r = r_0 \cos\left(\sqrt{\alpha_{n1}}z\right) - \frac{\vartheta_0}{\sqrt{\alpha_{n1}}}\sin\left(\sqrt{\alpha_{n1}}z\right).$$

Hence, in particular, it follows that the rays in a cavity formed by planar mirrors and filled with the inhomogeneous medium under consideration oscillate around the optical axis with a period of $z_0 = 2\pi/\sqrt{\alpha_{n1}}$.

Similarly, if the refractive index can be approximated by the dependence

$$n(r) = n_\alpha\left(1 + \frac{\alpha_{n2}}{2}r^2\right),$$

in the same approximations from equation (8.19), we have

$$r = r_0\mathrm{ch}\left(\sqrt{\alpha_{n2}}z\right) + \frac{\vartheta_0}{\sqrt{\alpha_{n2}}}\mathrm{sh}\left(\sqrt{\alpha_{n2}}z\right).$$

In the case of irradiation by a pulse with a duration of $\tau_{1/2} = 0.8$ ms at a moment in time of $t = 2.2$ ms (Curve 4, Fig. 8.2a) in an active region of $r < r_A$, then according to the well-known dependence $n - 1 = \rho C_n$ (C_n is the constant for a gas of this type), the gas refractive index distribution can be approximated by a parabola with a coefficient of $\alpha_{n1} \approx 2.4 \times 10^{-6}$ cm^{-2}, which yields $z_0 \approx 40$ m for the ray oscillation period. The displacement of a beam with initial parameters of $\vartheta_0 = 0$, $z = 0$, and $r_0 = 1$ cm after transmission the active section of a cell with a length of $L_c = 57$ cm equals $\Delta r \sim 10^{-6}$ cm. In the passive region, $r > r_A$, so a rough approximation yields $\alpha_{n2} \approx 0.77 \times 10^{-4}$ cm^{-2}. Given this α_{n2} value, following a single pass with a distance of $L_c = 57$ cm along the cell axis, rays with initial parameters of $\vartheta_0 = 0$, $z = 0$, and $r_0 > 1.2$ cm reach its wall. Furthermore, following several cavity passes, almost all rays with parameters of ϑ_0 called $= 0$, $z = 0$, and $r_0 > r_A$ reach the cell wall; therefore, the passive zone may also be a "dead zone." The divergence of output laser beam from an active region, $r < r_A$, estimated using formula (8.16), comes to $\vartheta \sim 3 \times 10^{-3}$ rad.

For the quasistatic irradiation mode, starting at $t \approx 0.1$ s, and for the stationary mode (Fig. 8.2b, c), the refractive index spatial distribution at $r > 0.6$ cm can be roughly approximated by a parabola with a coefficient of $\alpha_{n2} \geq 0.5 \times 10^{-4}$ cm^{-2}. At this α_{n2} value, a light ray that has initial parameters of $\vartheta_0 = 0$, $z = 0$, and $r_0 > 0.6$ cm is displaced toward the cell wall ($r_1 = 1.35$ cm), traversing a distance of $z < 2$ m, which almost corresponds to a single round trip of a cavity with planar mirrors.

Thus, in sealed gas lasers with fission fragment pumping in the stationary excitation mode, conditions created in the bulk of the gas volume that are not favorable for lasing. The same thing also occurs in a quasi-stationary excitation mode, starting at a time of $t \sim 10^{-1}$ s. Under these conditions, lasing can only originate in the region adjacent to the cell axis.

In pulsed excitation modes with a duration of $\tau_{1/2} \leq 10^{-2}$ s, the refractive index gradients in the bulk of the gas volume are negative and small, which provides the conditions needed to achieve lasing.

Comparison with Experiments. Already the results of the initial numerical and theoretical investigations of the inhomogeneities originating in NPLs have attracted much attention. Experiments using the interference procedure were carried out in 1982 to study inhomogeneities. The results of these investigations were later published in the unclassified press [2]. In the wake of these experiments, calculations [23] were performed that modeled them with the maximum possible approximation of actual conditions using the procedure described at the start of this chapter. The set of gases, their initial pressure and temperature values, and the irradiation intensity time dependence embedded in the calculations fully duplicated the corresponding experimental characteristics.

The conditions under which the experiments [2] were performed, and the parameters of the cell with a cylindrical uranium layer that was used, are presented in Sect. 7.4 of Chap. 7. The longitudinal cross-section of a cell with the plane in which the axis of symmetry is situated along with the relative distribution of the thermal neutron flux is presented in Fig. 8.3. The full internal volume of the cell (excluding the space occupied by the aluminum cylindrical substrate) was 561 cm^3. The volume of the cell's active section (i.e., the internal volume of the substrate tube, bound by its end faces) equaled 351 cm^3. We will call the difference in these

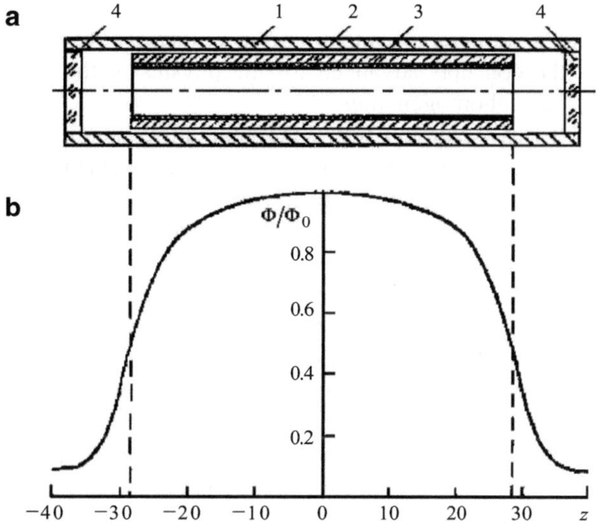

Fig. 8.3 Experimental cell: (a) longitudinal cross-section (1) outer cylindrical body; (2) aluminum cylindrical substrate; (3) uranium-containing layer; (4) quartz windows; (**b**) thermal neutral flux distribution over the length of the cell (Φ_0 is the flux density at a point of $z = 0$)

two volumes (i.e., the space between the windows in the cell's outer body and the end faces of the aluminum substrate, as well as the gap between this substrate and the cell's inner cylindrical surface), the "buffer" volume.

During the calculations, all physical parameters, including the distribution of specific energy sources in the gas and the uranium layer, were assumed to be dependent on a single coordinate, r, alone. The approach used corresponds to a one-dimensional calculation of the processes in a certain equivalent cell with an inside radius, r_1, and active volume equal to the inside radius and full volume of the real cell. An uranium layer of the same thickness and composition as those of the real cell stretches over the entire length of the equivalent cell, which equals the ratio of the real cell's full volume to the cross-sectional area of the active volume's gas column. Here, it is assumed that the thermal neutron flux distribution is not dependent on the coordinates but equals the average flux value over the length of the equivalent cell. This value is normalized in such a way that the total fissions occurring in the real and equivalent cells were the same.

When using an approach of this type no allowance is made for the possibility of the longitudinal motion of the gas and its partial seepage into the buffer volume. Also, certain density and temperature distributions averaged over the length of the cell are obtained as a result of the calculations. However, it should be remembered that over the course of experiments [2] aimed at investigating changes in density distributions along a cell's cross-section, an interference procedure was used. During this measurement the variation in the optical path of an He-Ne laser probing beam over the entire length of a cell, also including the end buffer volumes, is recorded. Experimental results obtained in this manner only make it possible to approximately evaluate the gas density changes averaged over the entire length of a cell [2]. Of course, it is impossible to regard a density distribution "averaged" along a beam's optical path in this manner as fully equivalent to the distribution obtained during calculations. Nevertheless, a comparison of the results of such calculations with a similar experiment permits a judgment to be made concerning the accuracy of the concepts advanced for the physics of the origination and development of optical inhomogeneities.

It was shown in Chap. 7 that uranium-containing layer inhomogeneities can lead to a decrease in the efficiency of fission fragment energy deposition into a gas. A series of photographs of the transverse sections of the layers used in the experiments (similar to the photos in Fig. 7.10) reveal that the thickness deviations in these layers from the average value are 50–70 %, while the shape of the inhomogeneities is reminiscent of something halfway between rectangular and sinusoidal protuberances. As analysis demonstrates (see Chap. 7, Sect. 7.7), such fluctuations necessarily lead to a 5–18 % decrease in energy deposition efficiency. During the calculations, an average value of 11 % was selected for this decrease, i.e., $\overline{\varepsilon} = 0.89\varepsilon$, where ε is the energy deposition efficiency for an ideal uranium layer. This correction factor was taken into account in the subsequent calculation of specific energy depositions.

Over the course of the experiments [2], changes were recorded in the pressure, $\Delta P(t) = P(t) - P_0$ (P_0 is the initial gas pressure in the cell), and the optical path of the He-Ne probing laser beam. The density changes of the media under investigation were recalculated using optical measurement data, which are unequivocally related to the changes in the optical path.

The dependences of pressure variation within the cell upon time and density spatial distribution at different moments in time obtained from the calculations are in good agreement with the experimental results. Some calculated (the solid lines) and experimental (the broken lines) dependences of pressure, $\Delta P(t)$, and relative density distribution, $\chi(r,t) = \rho(r,t)/\rho_0$, over the radius of a cell, as well as changes in relative density at the cell axis over time, are shown in Figs. 8.4, 8.5, 8.6, and 8.7. In addition, the shape of the neutron pulse is presented in Fig. 8.4.

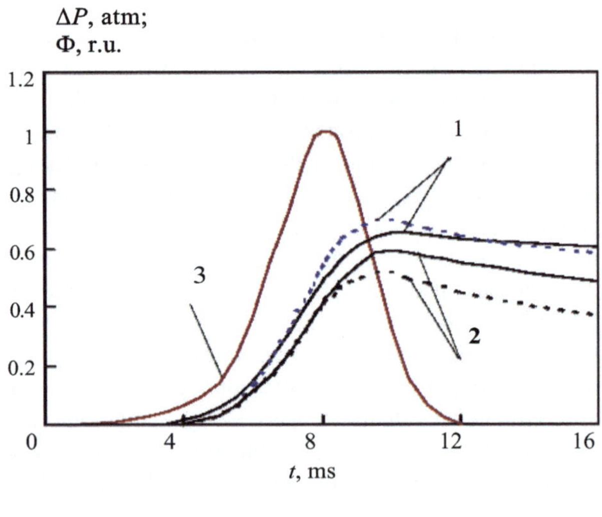

Fig. 8.4 Dependences of the pressure increase within a cell (1 and 2) and neutron flux variation (3) upon time: (1) He ($P_0 = 5$ atm); (2) He ($P_0 = 2$ atm); *solid lines* are calculated; *broken lines* are experimental

The characteristics obtained experimentally and as a result of calculations are presented in Table 8.1 ($F(1)/F(0)$ is the ratio of the specific energy deposition into the gas at a distance of $r = 1$ cm from the cell axis to its value at $r = 0$; ΔP_{max} is the maximum value of the pressure increase within the cell; and $\chi(0, t)$ is the relative density value at the cell axis at a moment in time of $t = 10$ ms.) With the exception of two calculations involving argon at pressures of $P_0 = 0.25$ and 0.5 atm, wherein the deviation in calculated relative density from the experimental value reached 20 %, this deviation was less than 2.5 % in all of the remaining calculations. The fact that the calculated relative density, $\chi(0, t)$, consistently exceeds the experimental value stands out. This is probably explained by the fact that no allowance was made in the calculations for the gas seepage into the buffer volume during the course of irradiation. However, the overrun is not large and does not lead to noticeable differences in the density profile's spatial distribution.

An estimate of the effect of gas fraction seepage into the buffer volume can be obtained from quite simple considerations. It was noted in Sect. 7.4 of Chap. 7

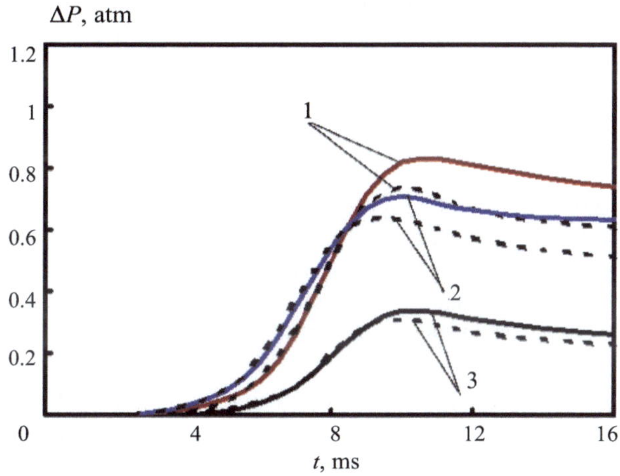

Fig. 8.5 Dependences of the pressure increase within a cell upon time: (1) Ar ($P_0 = 1$ atm); (2) Ne ($P_0 = 1$ atm); (3) He ($P_0 = 1$ atm); *solid lines* are calculated; *broken lines* are experimental

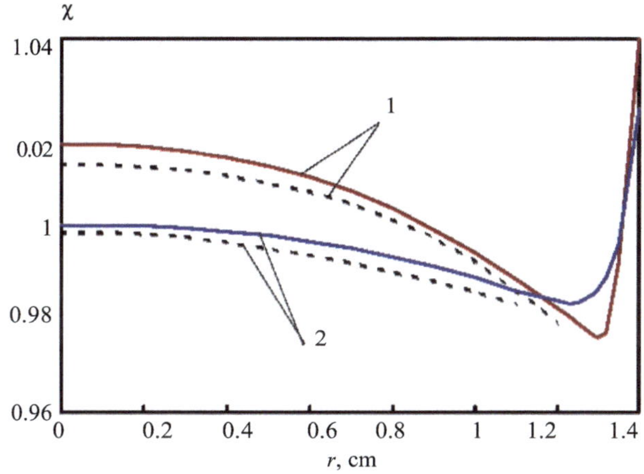

Fig. 8.6 Distribution of normalized gas density over the cell radius: (1) He ($P_0 = 5$ atm) for $t = 6.6$ ms; and (2) He ($P_0 = 2$ atm) for $t = 6.7$ ms; *solid lines* are calculated; *broken lines* are experimental

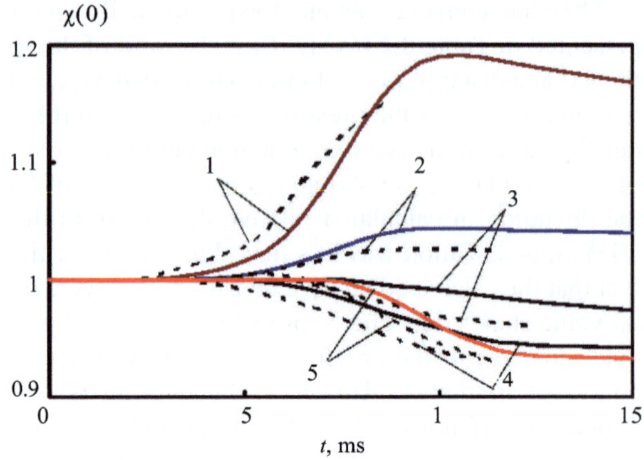

Fig. 8.7 Dependence of normalized density in the center of a cell upon time: (1) Ar ($P_0 = 1$ atm); (2) He ($P_0 = 5$ atm); (3) He ($P_0 = 2$ atm); (4) He ($P0 = 1$ atm); (5) Ne ($P_0 = 1$ atm); *solid lines* are calculated; *broken lines* are experimental

Table 8.1 Comparison of experimental and calculated characteristics

Gas		He				Ar			Ne
P_0, atm		1	2	3	5	0.25	0.5	1	1
$F(1)/F(0)$	Exper.	1.11	1.33	1.43	2.5	1.11	1.43	2.5	1.67
	Calcul.	1.10	1.21	1.45	3.1	1.10	1.22	2.15	1.41
ΔP, atm	Exper.	0.30	0.51	–	0.69	–	–	0.74	0.61
	Calcul.	0.34	0.58	0.68	0.65	0.41	0.68	0.83	0.72
$\chi(0, t)$	Exper.	0.939	0.969	0.980	1.026	0.790	0.870	1.16[a]	0.945
$t = 10$ ms	Calcul.	0.959	0.991	1.013	1.044	0.970	1.043	1.19	0.959

Note: [a]Only the $\chi(0, t)$ dependence up until a moment in time of $t \approx 9$ ms is presented in study [2] for Ar ($P_0 = 1$ atm). A value of $\chi(0, t) = 1.16$ was obtained by extrapolating the experimental results up until a moment in time of $t = 10$ ms

that the pressure within a cell over the course of irradiation by a pulsed neutron flux with a duration of $\tau_{1/2} \geq 3$ ms can, as a first approximation, be assumed to be homogeneous at any irradiation pulse moment, and that gas density redistribution can be regarded as having an equilibrium nature. In this instance, the gas density distribution averaged over the cross-section along the z coordinate, which coincides with the cell's optical axis, can be presented in the form

$$\overline{\rho}(z,t) = \frac{P(t)}{c_v(\gamma - 1)\overline{T}(z,t)}, \tag{8.20}$$

where $\overline{T}(z,t)$ is the effective average (for the cell cross-section) temperature value, determined by the equation

$$\overline{T}(z,t) = \left\{ \frac{2}{r_1^2} \int_0^{r_1} \frac{rdr}{T(r,z,t)} \right\}^{-1}, \tag{8.21}$$

at which time formula (8.20) holds true.

From the condition of gas mass constancy within a cell, we obtain

$$\frac{P}{c_v(\gamma - 1)} = \frac{\rho_0 L}{2} \left\{ \int_0^{L/2} \frac{dz}{\overline{T}(z,t)} \right\}^{-1}.$$

In this case it is noted that energy release distribution along the z axis is symmetrical relative to the geometric center of the cell. Thus,

$$\overline{\rho}(z,t) = \frac{\rho_0 L}{2\overline{T}(z,t)} \left\{ \int_0^{L/2} \frac{dz}{\overline{T}(z,t)} \right\}^{-1}. \tag{8.22}$$

For simplicity's sake, we assume that the $\overline{T}(z,t)$ dependence takes the form

$$\overline{T}(z,t) = \begin{cases} T_A, & 0 \le z \le L_A/2; \\ T_B, & L_A/2 \le z \le L/2, \end{cases} \tag{8.23}$$

where L_A is the length of the real cell's active volume. Such a dependence must approximately correspond to the real temperature distribution, as can be concluded from the type of thermal neutron flux distribution (see Fig. 8.3). Placing the distribution from (8.23) into (8.22), we obtain

$$\overline{\rho}(z,t) = \frac{\rho_0 L}{\overline{T}(z,t)} \left\{ \frac{L - L_A}{T_B} + \frac{L_A}{T_A} \right\}^{-1}. \tag{8.24}$$

From (8.23) and (8.24) for the relative variation of average density, we have

$$\frac{\Delta\rho_A}{\rho_0} = \frac{\overline{\rho}_A}{\rho_0} - 1 = -(L - L_A) \frac{\left(1 - \frac{T_B}{T_A}\right)}{L - L_A\left(1 - \frac{T_B}{T_A}\right)}. \tag{8.25}$$

The temperature determined using formula (8.21) can be used as the temperature, $T_A(t)$, for estimates. In this instance, the $T(r, t)$ dependence is taken from a calculation using the procedure described above.

For the approximate determination of the gas temperature in the buffer volume, it can be assumed that its compression over the course of gas expansion from the active volume occurs adiabatically, i.e.,

Table 8.2 Relative variation of the average density of the gas in a cell

Gas		He				Ar			Ne
P_0, atm		1	2	3	5	0.25	0.5	1	1
$\Delta\rho_A/\rho_0$, %	t_m, ms	4.7	4.6	3.7	1.5	14.6	13.1	8.8	6.4
	t_c, ms	8.5	7.3	5.6	3.2	25.4	21.2	13.6	12.8

$$T_B(t) = T_0\left(\frac{P}{P_0}\right)^{\frac{\gamma-1}{\gamma}}, \tag{8.26}$$

where P_0, T_0 are the initial pressure and temperature values in the cell. Such an approach is equivalent to ignoring heat removal on the cell wall within the buffer volume. In approximate calculations, this is warranted by the fact that, even in the active volume, ~30 % of the energy absorbed in the gas is carried away by means of similar heat removal. However, the temperature gradients in the buffer volume must be smaller than in the active volume, and accordingly, smaller than the share of energy carried off by means of heat transfer.

The results of a $\Delta\rho_A/\rho_0$ calculation for all the gas pressures that were studied in the experiments [2] are presented in Table 8.2. The moment in time t_m in the table corresponds to the irradiation pulse peak, while the moment in time t_c corresponds to the pulse termination.

The calculated and experimental results revealed that for the rare-gas NPLs the lower density difference values, and consequently the lower refractive index values take plact for gases with a lower atomic numbers. The difference values decrease as the initial pressure drops. The experiments confirmed that a narrow region with a large positive density gradient originates near a substrate with a uranium-containing layer. This region grows over time. A comparison of the calculated and experimental results makes it possible to reach a conclusion concerning the applicability of the proposed calculation model to the investigation of the optical properties of NPLs.

The experimental data [2] were later examined in study [37]. The calculations described above [23] were performed in accordance with a full gasdynamic program that quite correctly reflected heat transfer processes but made no allowance for gas seepage into the buffer volumes. A simplified analytical model was developed in [37] within which an attempt was made to take gas seepage into account while simultaneously introducing a correction factor for heat exchange between the gas and the cell's metal wall. The authors [37] proposed that viscosity and thermal conductivity are absent in the main part of the cell (not adjacent to the wall), that the gas is ideal, and that the gas pressure is homogeneous at any given moment of time. The thickness of the heat exchange zone adjacent to the wall, which is called the "passive" or "dead" zone, was taken to equal $l = \sqrt{at}$. Using the latter equation, the heat flux in the active and buffer parts of the cell wall was found, and then the correction factor for the calculated pressure was determined, which consists of the last addend in equation (7.20). The transverse specific energy deposition profile was assumed to be constant in Lagrangian coordinates. In the longitudinal direction within the active part of the cell, the specific energy deposition was regarded at homogeneous, while in the buffer part, it was considered to equal zero.

A comparison of calculation results [37] and experimental data [2] revealed that a somewhat better agreement was achieved between the calculated and experimental values in five of the six cases examined. Here, the maximum discrepancy between the calculation results from studies [23] and [37] is exhibited during exciting pulse termination. As far as density, the greatest discrepancy in these results (~5 %) is observed for neon. In the remaining cases, these discrepancies are 0.3–2.7 %. All these figures are perceptibly lower than the estimated value of the effect of neglecting gas seepage into the buffer volumes cited in Table 8.2.

In conclusion, the numerical and theoretical investigations of other authors concerning the development of transverse optical inhomogeneities should be mentioned [18, 38–40]. The authors [18, 38, 39] completely ignore heat exchange. Furthermore, in [39], the specific energy deposition profile, tied to Eulerian coordinates, is assumed to be constant over the entire duration of the excitation pulse. In studies [18, 38], above profile is tied to Lagrangian coordinates, which more properly corresponds with the physics of the processes occurring. In study [40], the dependence of the specific energy deposition profile upon variable density is precisely defined; however, simplified gasdynamic equations are used.

8.2 Density Inhomogeneities for Planar Uranium Layers

Calculations Based on the Approximation of Infinitely Extended Uranium Layers

If a cell's working volume has the shape of a parallelepiped bounded by planar walls, and uranium layers are deposited to the inner surfaces of opposing walls, the transverse dimension of which noticeably exceeds the distance between them, and the cell itself has no buffer volumes (a cell of this type was used, for example, in studies [35, 36, 41–43]), the transverse density distributions of the gas averaged along the length of the cell can then be calculated in an approximation of infinitely extended uranium layers. Using this approach, all the physical parameters are dependent only upon a single transverse coordinate. The equation system presented in the first section of this chapter for $N = 1$ describes the gas behavior in this approximation. Here, the specific source distribution, as indicated above, can be fixed in Lagrangian coordinates (see study [19]), which considerably simplifies the calculation procedure. The density inhomogeneity formation processes in cells of this type do not qualitatively differ from the processes occurring in cylindrical cells.

Calculation of Density Inhomogeneities For the Plane-Parallel Distribution of Uranium Layers with Finite Dimensions

The situation is significantly more complex if a cell has large buffer volumes and the dimension of the planar uranium layers in one direction is comparable to the distance between these layers. Two such cells are a part of the LUNA-2M setup [5–7, 14] (see Chap. 2, Sect. 2.4, and Chap. 7, Sect. 7.4). The cell's transverse cross-section is presented in Fig. 8.8. The z axis coincides with the system's optical axis and is oriented perpendicular to the plane of the figure. It is impossible to reduce calculations of the thermal gasdynamic processes of such a cell to a problem with one spatial variable. Any existing multidimensional gasdynamic program could be modified, or a special program could be created for purposes of the numerical calculations of such problems using a computer. However, large computational burden would be needed in order to perform numerical solutions; this is scarcely warranted because this type of laser cell design is far from optimum. For this reason, when conducting the numerical investigations described next, a series of model approximations based on the specific features of the experiments involving these cells was used.

Pressure. During the calculations, uranium nucleus fission density distributions and gas densities within the expanse of the uranium layers were assumed not to be dependent upon the longitudinal coordinate, z. In the presence of a cell body inside radius of $R \sim 4$ cm, the typical perturbation propagation time is $\tau_p \sim R/u_s$, which is considerably shorter than the exciting pulse duration in the experiments mentioned above, $\tau_{1/2} \sim 3$ ms. Consequently, the pressure at any given moment in time can be regarded as homogeneous, i.e., temperature distribution determines gas density distribution as a whole.

By virtue of the distinctive features noted, the geometry of a real cell with a fission density distribution that is not dependent upon the longitudinal coordinate, in purely gasdynamic terms, is virtually equivalent to a cell with a rectangular cross-section (Fig. 8.9). The heating of the metal body of a real cell and the plates to

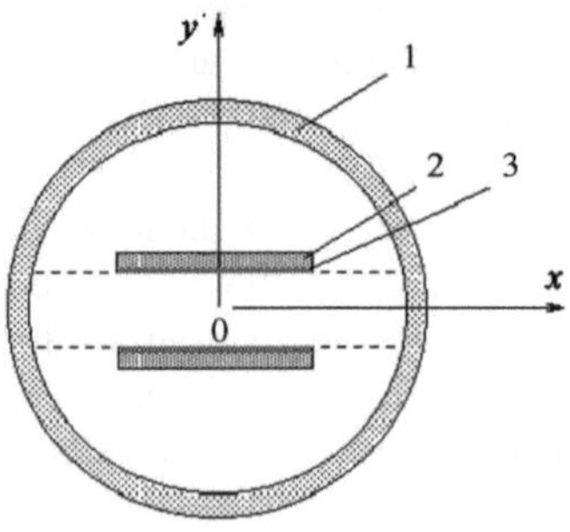

Fig. 8.8 Cross-section of a laser cell with planar uranium layer: (1) outer tube; (2) aluminum plates; (3) uranium oxide-protoxide layer

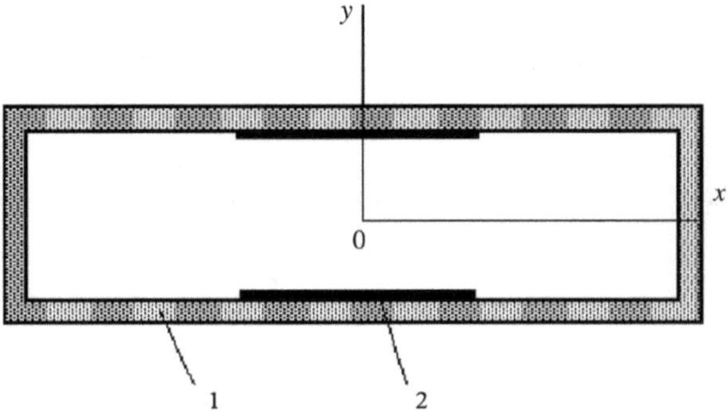

Fig. 8.9 Cross-section of an equivalent rectangular cell: (1) cell wall; (2) uranium oxide-protoxide

which a uranium layer has been deposited, as well as the plates in an equivalent cell with a rectangular geometry, in the presence of the energy depositions studied in this problem is negligible as compared to the heating of the gas; therefore, this factor's effect can be ignored during the transition from one geometry to another. The volume, mass, and initial pressure of the gas in the equivalent cell, the dimensions and composition of the uranium layers, and the distance between them are the same as in the initial geometry.

In ignoring the effect of viscosity forces, the equation for the energy of the gas in an Eulerian coordinate system takes the form:

$$\rho T\left(\frac{\partial S}{\partial t}+\vec{u}\nabla S\right)=F+\operatorname{div}\left(k_g\nabla T\right), \tag{8.27}$$

where $\vec{u}$ is the gas velocity, and S is the entropy of a unit mass of gas. Based on the well-known thermodynamic correlation,

$$dS=\frac{1}{T}\left(dE-Pd\rho/\rho^2\right)$$

together with an expression for the internal energy of an ideal gas,

$$E=c_vT,$$

equation (8.27) can be written in the form

$$c_v\rho\frac{\partial T}{\partial T}+c_v\rho\vec{u}\nabla T-\frac{P}{\rho}\frac{\partial\rho}{\partial t}-\frac{P}{\rho}\vec{u}\nabla\rho=F+\operatorname{div}\left(k_g\nabla T\right).$$

We rearrange this equation as:

$$c_V\rho\frac{\partial T}{\partial t} + c_v\,\mathrm{div}\left(\rho\vec{u}T\right) - c_vT\,\mathrm{div}\left(\rho\vec{u}\right) - \frac{P}{\rho}\frac{\partial\rho}{\partial t} - \frac{P}{\rho}\mathrm{div}\left(\rho\vec{u}\right) + P\,\mathrm{div}\vec{u}$$
$$= F + \mathrm{div}\left(k_g\nabla T\right),$$

after which, using the continuity equation

$$\frac{\partial\rho}{\partial t} + \mathrm{div}\,\rho\vec{u} = 0,$$

we obtain

$$c_V\frac{\partial}{\partial t}(\rho T) + c_v\mathrm{div}\left(\rho\vec{u}T\right) + P\,\mathrm{div}\vec{u} - \mathrm{div}\left(k_g\nabla T\right) = F. \tag{8.28}$$

We integrate (8.28) over the entire gas volume. Taking into account the fact that the pressure, based on the simplifications used, is homogeneous over the volume, we rearrange the integral from the second, third, and fourth summands on the left side according to the Gauss–Ostrogradski theorem into an integral over a closed surface,

$$\int_V\left\{c_v\mathrm{div}\left(\rho\vec{u}\right) + P\,\mathrm{div}\,\vec{u} - \mathrm{div}\left(k_g\nabla T\right)\right\}dV = \int\left\{c_V\rho\,\vec{u}T + P\vec{u} - k_g\nabla T\right\}d\vec{f}_s.$$

Assuming that the cell is closed, its walls are fixed, and the velocity near the inner surface of the walls equals zero, we deduce that the latter integral equals

$$\Pi(t) = \oint k_g\nabla T\left(\vec{f},t\right)d\vec{f}.$$

The integral for cell volume from the right side of equation (8.28) equals the total energy released into the gas per unit of time,

$$\int_V F(x,y,t)dV = \varepsilon Q(t)$$

(see the designations in Chap. 7, Sect. 7.4). According to the state equation for an ideal gas, we rearrange the first term of equation (8.28) in the form

$$c_V\frac{\partial}{\partial t}[\rho(x,y,t)T(x,y,t)] = \frac{1}{\gamma-1}\frac{\partial}{\partial t}P(t).$$

Thus, following integration for the entire gas volume, equation (8.28) is written as

$$\frac{V}{\gamma - 1}\frac{dP}{dt} = \varepsilon Q(t) + \Pi(t),$$

whence equation (7.20) ensues. A similar correlation was obtained in [45] for the pressure averaged over a cell.

Gas Density. The typical temperature equalization time in the gas between plates containing uranium is on the order of $\tau_T \sim d^2/a$. The thermal diffusivity for the gas mixtures used is $a \sim 1\ \text{cm}^2/\text{s}$. Hence, it follows that $\tau_T \sim 1$ s, i.e., $\tau_T >> \tau_{1/2}$. Therefore, heat exchange in the bulk of the gas volume during a neutron pulse plays a negligibly minor role, with the exception of the thin layers directly adjacent to the plates containing the uranium, where heat removal to the plates manifests itself in a noticeable manner. However, the thickness of these layers, even for the termination of a pulse with a duration of $\tau_{1/2} \sim 3$ ms reaches a total of $l \sim 10^{-1}$ cm (see the first section of this chapter). It is heat removal from these layers precisely that determines the last summand in equation (7.20). Due to the smallness of l, even for the pulse termination, this term accounts for a total of 15–25 % of the overall volume in a cell; therefore, in approximate calculations, we will ignore heat exchange in the entire gas volume and will calculate the pressure using formula (7.21). Other authors also proceed in a similar manner (see [18, 35–38, 46] for example).

In the approximations under consideration, the spatial distribution of the temperature increase associated with fairly small density variations is more or less similar to the spatial distribution of specific energy sources, which is confirmed both by calculations [23, 45] and by experiments [2]:

$$T(x, y, t) = T_0 + A(t)W_e(x, y,\ t), \tag{8.29}$$

where $A(t)$ is a proportionality factor that is dependent upon the integral of energy release within the cell, and $W_e\ (x,\ y,\ t)$ is the spatial distribution of the specific sources caused by fission fragment deceleration in the gas and standardized per unit of energy released.

The density distribution of the gas between the plates varies over time. Because the fission fragment energy losses are unequivocally related to the density of the stopping medium, the W_e function profile is also transformed over time. In the first approximation, this factor can be taken into account by tying specific energy release to Lagrangian coordinates. In studies [18, 19], it is shown that such a technique for pulses with a duration of $\tau_{1/2} \sim 3$ ms at any point in the volume of a cell will not lead to noticeable gas temperature and density deviations from the corresponding values obtained during calculations if precise allowance is made for the effect of density redistribution on specific energy deposition. Bearing these results in mind, together with the fact that the fragment energy losses in the medium are proportional to its density [17], we will describe the spatial distribution of the sources using the function

$$W_e(x,y,t) = \frac{\rho(x,y,t)}{\rho_0} W_{e0}(x,y), \tag{8.30}$$

where W_{e0} is the distribution of specific energy deposition in an unperturbed homogeneous medium.

Using the state equation for an ideal gas

$$P(t) = (\gamma - 1)c_v\rho(x,y,t)T(x,y,t),$$

then from (8.29) and (8.30) we obtain

$$\rho = \frac{T_0\rho_0}{2AW_{e0}}\left\{\left(1 + \frac{4APW_{e0}}{(\gamma-1)c_v\rho_0T_0^2}\right)^{1/2} - 1\right\}. \tag{8.31}$$

Taking the condition of the conservation of the total gas mass within a sealed cell into account, we have

$$\frac{T_0\rho_0P(t)}{2}\int_V \frac{dV}{U(t)W_{e0}(x,y)}\left\{\left[1 + \frac{4U(t)W_{e0}(x,y)}{(\gamma-1)c_v\rho_0T_0^2}\right]^{1/2} - 1\right\} = \rho_0V, \tag{8.32}$$

where $U(t) = A(t)P(t)$. Solving jointly equations (7.21) and (8.32), we can find the dependence of $A(t)$, and using it, we can determine the $\rho(x, y, t)$ profile from equation (8.31).

The source distribution function, $We_0\ (x, y)$, was taken into account using the technique described in [20, 46]. Under this approach, the share of energy, ε, deposited into the gas was regarded as constant and equal to its value in an unperturbed gas. It was assumed that cell length equals the dimension of the uranium layers along the z axis. Therefore, two fairly small cylindrical gas volumes occupying the space between the cell end windows and the parallel planes within which the edges of the plates containing the uranium lie were included in lateral (buffer) volumes of the equivalent rectangular cell not bounded by uranium layers (see Fig. 8.9).

Calculation Results. Calculations using the model described, in which an exciting neutron flux was specified based on the shape (see Figs. 7.11 and 8.4) and fluence of the pulses of a VIR-2 M reactor [21], were performed for three cell gas filling alternatives: He ($P_0 = 2$ atm), He ($P_0 = 5$ atm), and Ar ($P_0 = 0.5$ atm) at $T_0 = 300$K [2, 46]. Naturally, considering the approximate model, it would be incorrect to talk about the good accuracy of the calculations. One of the principal causes that might lead to the deviation of the calculation results from real gas behavior consists of the fact that the cell under consideration has a large free gas volume that is inaccessible for the fission fragments escaping from the uranium layers. The gas from the active region excited by the fission fragments is displaced into this volume during the pulse. Consequently, gas density in the active region is decreased, which will necessarily lead to the reduction of ε energy deposition

Table 8.3 Relative errors (in %) in the calculation of density changes, $\Delta\rho^*/\Delta\rho$,in a cell with planar uranium layers

t (ms)	He ($P_0 = 2$ atm)	He ($P_0 = 5$ atm)	Ar ($P_0 = 0.5$ atm)
8	5	1	20
11	20	10	70

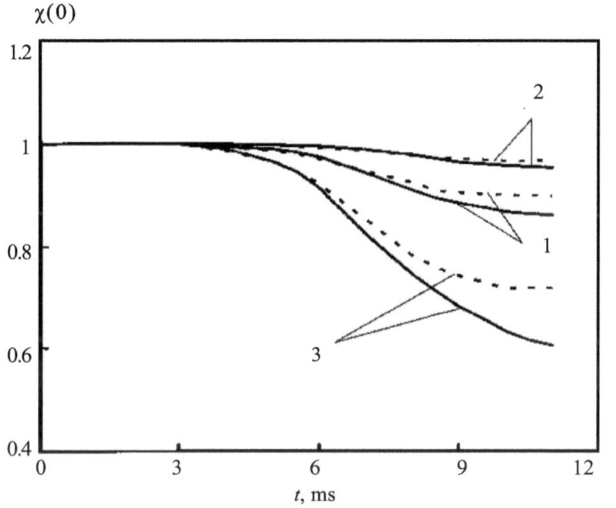

Fig. 8.10 Dependence of normalized density in the center of a cell upon time: (1) He ($P_0 = 2$ atm); (2) He ($P_0 = 5$ atm); (3) Ar ($P_0 = 0.5$ atm); *solid lines* are calculated; *broken lines* are experimental [2, 46]

efficiency in this region, but it remains constant in the model used over the entire pulse duration.

Gas density values averaged over the active volume were determined based on calculation results for several moments in time. Using these data, the corresponding energy deposition efficiencies were recalculated and the latter figures were used in conjunction with equations (7.21), (8.31), and (8.32) to estimate possible errors in the calculation of the $A(t)$ factors and density changes, $\Delta\rho^*$. The relative error values, $\Delta\rho^*/\Delta\rho$ where $\Delta\rho$ is the density variation in the center of the cell obtained during calculations using the procedure described above, are presented in Table 8.3 for moments in time that correspond to the excitation pulse peak ($t = 8$ ms) and pulse termination ($t \approx 11$ ms).

Analysis revealed that the seepage effect increases noticeably over the course of irradiation. This effect is most intensely manifested if a heavier gas (such as argon) is used. Therefore, it only makes sense to talk about the quantitative agreement of the calculation results with the real processes occurring in similar laser cells for time intervals up to the exciting pulse peak. Past the peak, the calculation results may have a perceptible divergence from real gas behavior.

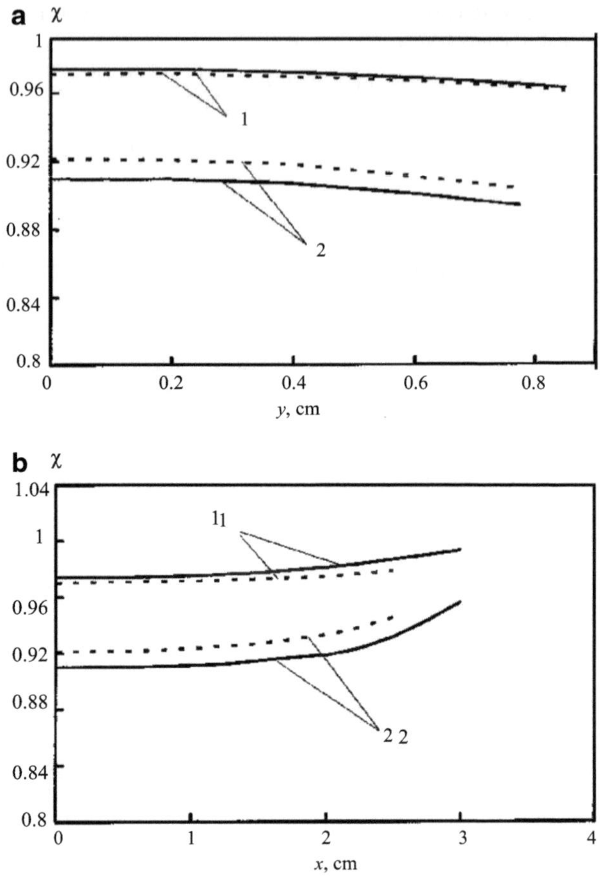

Fig. 8.11 Normalized gas density distribution for He ($P_0 = 2$ atm) in the $x = 0$ (**a**) and $y = 0$ (**b**) planes: (1) $t = 6$ ms; (2) $t = 8$ ms; *solid lines* are calculated; *broken lines* are experimental [2, 46]

The calculated and experimental dependences of the variation in relative gas density in the center of a cell upon time are presented in Fig. 8.10. Gas density distribution for the helium ($P_0 = 2$ atm) in the $x = 0$ (Fig. 8.11a) and $y = 0$ (Fig. 8.11b) planes is shown in Fig. 8.11. The calculation results are found to be in satisfactory agreement with the experimental data all the way up to the exciting pulse peak, while past the peak, the development of gas density variation over time is only qualitatively described. In addition, the calculations good enough reproduce the density gradient's spatial dependence. This fact made it possible to successfully use this calculation technique to analyze temporal changes in the laser cavity stability with the type of cell under consideration [34] (see Sect. 5 below in this chapter).

Later, in [46], the computational modeling of the experiments [2] was carried out in two steps. In the first approximation, in fact, as during the technique described above for the gas particles which present in the active volume prior to the start of irradiation, distribution of specific energy deposition was fixed in an unperturbed homogeneous medium, W_{e0}. Energy deposition into the gas outside the confines of this volume was ignored. Pressure was assumed to be homogeneous throughout this volume, while the gas was assumed to be nonviscous and thermally nonconductive. The distribution of fission fragment specific energy deposition was calculated using the same procedure [19, 20].

Unlike the technique described above, during which gas temperature distribution was simply assumed to be proportional to specific source distribution (here, the time-dependent proportionality factor was found by means of jointly solving equations (7.21) and (8.32)), as a starting point in [46], it was assumed that the energy absorbed by the gas mass component goes toward an increase in the internal energy of the gas and toward work based on the expansion of this component. Formally, this approach is more inherent, since it does not include a constant $A(t)$ which does not carry a specific physical "load,".

The average gas density in the active section of a cell's volume was determined in the first approximation based on density space–time distribution calculation results. Specific energy deposition distribution was then recalculated for an unperturbed medium characterized by a density that equaled the average density found in the first approximation. Using this figure, density and temperature distribution were recalculated, but this time with a correction factor for gas heat exchange with the uranium metal substrates and the cell walls in the buffer volume. As in study [37], the thickness of the passive zone adjacent to the walls and the substrate was *a priori* taken to be $l = \sqrt{at}$ (see also the first section of this chapter).

According to relative gas density calculation data [46], the difference in the second and first approximations comes to not more than ~5 %, though it is not always closer to the experiment. Calculation results [46] in the first approximation all the way up to the excitation pulse peak differ negligibly from the results presented in Figs. 8.10 and 8.11. The divergence becomes more noticeable past the peak. So, for helium at $P_0 = 2$ atm at a moment in time of $t = 11$ ms, the divergence of results for relative density at the optical axis ($x = 0$ and $y = 0$) reaches 5 %, while for argon at $P_0 = 0.5$ atm, this figure is 17 %.

8.3 Possibilities for NPL Parameter Optimization

Dynamics of Near-Wall Passive Zone Development

It was previously shown that a so-called passive zone is formed in direct proximity to the side wall of a laser cell with deposited uranium layer, from which heat is intensely removed to the wall. Large positive density and refractive index gradients up to $\nabla n \sim 10^{-4}$ cm^{-1} characterize a passive zone. Light rays that have a direction

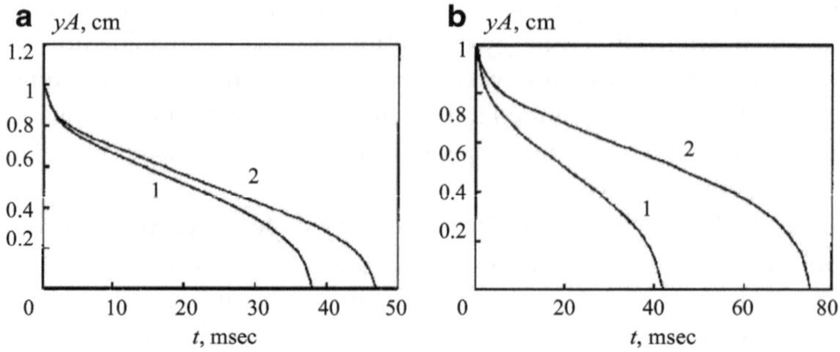

Fig. 8.12 Dependence of active lasing region and passive zone interface variation upon time in a cylindrical cell (1) and an infinitely extended plane-parallel cell (2): (**a**) a cell filled with a mixture based on He ($P_0 = 1.5$ atm); (**b**) a cell filled with a mixture based on Ar ($P_0 = 0.35$ atm)

parallel to the cell axis at any point in a passive zone are deflected toward its walls (at least, after several cavity passes). Consequently, this zone should not make a contribution to a laser beam's output energy. Passive zone size, l, increases over time. During the initial stage, when $l << d$ (d is the transverse dimension of the cell), passive zone size variation in time is adequately communicated by dependence (8.3). Outside the confines of this zone (in the midsection of the gas volume) the medium density and refractive index gradients are negative, during which $|\nabla n| \sim 10^{-6}$ cm^{-1}. Lasing develops exactly in this very region.

At $l \sim d$, the dependence of near-wall passive zone size noticeably deviates from the regularity expressed by (8.3). This zone's development is most demonstrably observed based on the example of a cell irradiated by a time-constant neutron flux. Calculation results for the dependence of active region and passive zone interface variation upon time are presented in Fig. 8.12. Two types of cells were examined: a cylindrical cell with an inside diameter of $d = 2$ cm and an infinitely extended planar cell, in which the distance between the uranium layers was also $d = 2$ cm. In Fig. 8.12, the distance in the direction of the y axis is reckoned from the axis of symmetry for the cylindrical cell and from the plane of symmetry for the planar cell. As in the experiments described in [47], the transverse dimension of the aluminum outer cell walls to which metallic uranium layers with a thickness of $\delta_U = 5$ mg/cm^2 were deposited equaled 0.5 cm. In both cases, the layers were coated with a protective aluminum film that had a thickness of $\delta_A = 0.5$ mg/cm^2. These are the same layers with a protective coating that were used during the experiments described in [47]. Calculations were performed for an Ar-Xe mixture (70:1) at a pressure of $P_0 = 0.35$ atm and an initial temperature of $T_0 = 293$ K (Fig. 8.12a), as well as for a He-Xe mixture (2000:1) at $P_0 = 1.5$ atm and the same temperature (Fig. 8.12b). In these cases, the cell's normalized transverse dimension equaled $D_0 = 0.26$. The results presented in Fig. 8.12 were obtained under the assumption that, from a moment in time of $t = 0$, neutron flux density remains constant

$\Phi = 3.5 \times 10^{14}$ cm^{-2}·s^{-1}. The results show that without gas flowing at irradiation times exceeding 41 ms for Ar and 38 ms for He in the cylindrical cell, or 74 ms (Ar) and 47 ms (He) in the infinitely extended planar cell, a heat removal zone characterized by a positive refractive index gradient spreads over the entire gas mixture volume. From a moment in time of $t \sim 15$ ms, a steeper approach of the active region boundary, $y_A(t)$, to the center is observed as compared to correlation (8.3). In addition to the specific role of gas-dynamic effects, this is explained by the nonlinear dependence of the gas heat conductivity coefficient upon temperature [16]. The effect of the geometric shape of the cells is also obvious.

Exponential Increase in Power Deposition

Laser output intensity is linked to the size of the working gas's active region that takes part in lasing: reducing the size of the passive zone in a gas mixture can increase laser emission intensity.

The typical temperature equalization time in a gas mixture with a transverse dimension of r_1 is

$$\tau_T \sim r_1^2/a. \tag{8.33}$$

If a laser cell is irradiated by a time-varying neutron flux, the typical rise time of which is $\tau < \tau_T$, it can then be anticipated that a similar procedure would make it possible to slow passive zone development. The time dependence of this type that is the simplest one actually feasible, and that is the most convenient for numerical and theoretical investigations, consists of an exponentially rising neutron flux density

$$\Phi(t) = \Phi_0 e^{t/\tau}. \tag{8.34}$$

The capabilities of this method were investigated based on the example of a helium-filled cylindrical cell with an internal radius of $r_1 = 1$ cm and a wall thickness of $\delta_w = 1$ mm (the wall material during the calculations was zirconium). During all the calculations, it was assumed that $\Phi_0 = 10^{13}$ cm^{-2} s^{-1}. The reduced transverse dimension values of the cell's metallic uranium layer and the tube were selected as $D_1 = 0.5$ и $D_0 = 0.5$ (which corresponds to $\delta_U = 2.775 \times 10^{-4}$ cm and $P_0 = 2.88$ atm) at an initial temperature of $T_0 = 300$ K. The calculated temperature profiles in the laser cell for several successive moments in time at a power deposition rise time of $\tau = 10$ ms are represented by the solid lines in Fig. 8.13. The passive zone involves the region between the cylindrical surface, $r = r_A$, on which the maximum temperature is reached and the cell's inner surface, $r = r_1$.

The dependences of the radial coordinate of the active region's outer boundary, r_A, upon the thermal neutron flux rise time, τ, and fluence, φ, which equals the

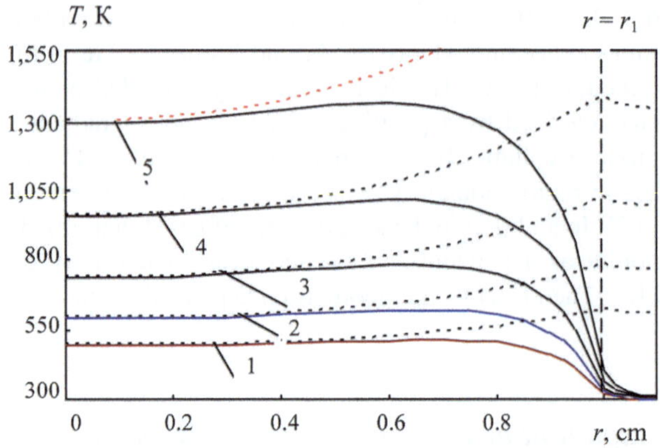

Fig. 8.13 Temperature distribution in a cell at an excitation power rise time of $\tau = 10$ ms ($D_0 = 0.5$): (1) $t = 44.6$ ms; (2) $t = 48.6$ ms; (3) $t = 52.6$ ms; (4) $t = 56.6$ ms; (5) $t = 60.6$ ms; the *solid lines* are without cell wall heating; the *broken lines* are with wall heating by the ^{235}U nuclei at $N_w = N_{ow}$)

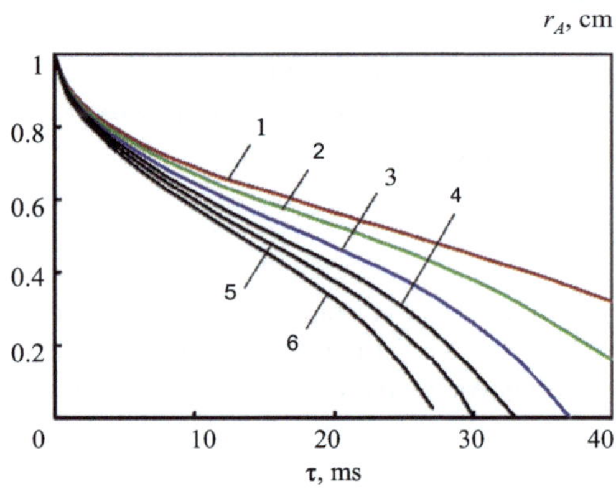

Fig. 8.14 Dependence of the radial coordinate of an active region's outer boundary upon the thermal neutron flux rise time, τ, and its fluence, φ, at $D_0 = 0.5$: Curves 1–6 correspond to the values $\varphi = (0.5, 1, 2, 3, 4, \text{and } 5) \times 10^{13}$ cm^{-2}

integral of expression (8.34), that were obtained as a result of calculations are shown in Fig. 8.14. Here, by way of comparison, it seems appropriate to recall that the fluence in the center of the laser cell during experiments performed on a VIR-2 M reactor reached 1.4×10^{13} cm^{-2} [2, 5–7, 14]. For the sake of convenience in comparing the results obtained, the relationship between the pumping power rise time, τ, thermal neutron fluence, φ, and the moment in time, t, by which this fluence is reached is reflected in the time scales in Table 8.4.

The results obtained make it possible to reach a conclusion: in order for at least half of the gas mixture volume in a cell with the parameters selected (which correspond to $r_A \approx 0.7$ cm) to be an active region, the neutron flux rise time must be no higher than $\tau = 7.5$ ms. Here, the time that an active region with this volume exists does not exceed 0.1 s.

Table 8.4 Relationship between the pumping power rise time, thermal neutron fluence, and the moment in time that it is reached

φ (10^{13} cm^{-2})	τ (ms)	t (ms)	φ (10^{13} cm^{-2})	τ (ms)	t (ms)
	5	23.08		5	31.99
	10	39.32		10	57.07
0.5	20	65.16	3.0	20	100.35
	30	86.15		30	138.45
	40	104.11		40	173.23
	5	26.52		5	33.43
	10	46.15		10	59.94
1.0	20	78.64	4.0	20	106.07
	30	106.08		30	147.01
	40	130.32		40	184.61
	5	29.97		5	34.54
	10	53.03		10	62.17
2.0	20	92.30	5.0	20	110.51
	30	126.44		30	153.66
	40	157.27		40	193.45

Variations in Initial Gas Pressure

Investigations revealed that if the initial thermal neutron flux value is on the order of, or higher than, $\Phi_0 = 10^{13}$ cm^{-2}·s^{-1}, passive zone development is then followed by a regularity of the type described in (8.3)

$$l \approx A_l \sqrt{at} \tag{8.35}$$

not only at $l << r_1$, but also at $l \leq 0.3\, r_1$, with an accuracy no worse than 20 %. If gas thermal diffusivity, a, in this instance is assumed to equal its value at the initial moment in time, the proportionality factor, A_l, will then fluctuate within limits of a few units. So, for a cell with planar layers (see Fig. 8.12), $A_l = 3.24$ for Ar and $A_l = 2.75$ for He. Here, expression (8.35) describes the $l(t)$ dependence derived from an accurate calculation all the way up to $l \approx 0.25d$, with an error no higher than 6 %. For a cylindrical cell, $A_l = 4.61$ for Ar and $A_l = 3.04$ for He, with this same accuracy.

Calculations of different excitation alternatives demonstrated that at $l < 0.3r_1$, the A_l factor is most highly dependent upon cell geometry (planar or cylindrical), and to a lesser extent, upon the shape and intensity of the exciting neutron pulse, uranium layer thickness, and gas mixture thermophysical properties, despite the fact that mixture characteristics are partially taken into account by the thermal diffusivity in correlation (8.35). This A_l factor behavior is explained by the fact that passive region development is determined not only by heat removal to the cell wall, but also by the gas motion, during which the occurrence of thermal and

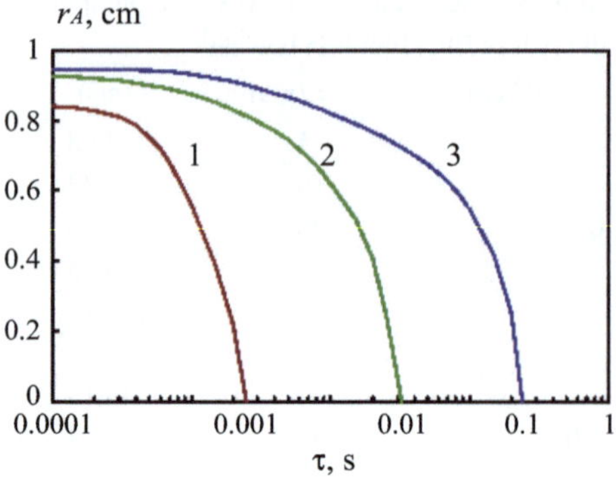

Fig. 8.15 Dependence of the active region's outer boundary upon pumping power rise time: (1) $D_0 = 0.1$; (2) $D_0 = 0.5$; 3.) $D_0 = 1$

gas-dynamic processes is dependent upon the power and spatial distribution of the energy release sources in the gas at any given moment in time.

The regularity noted makes it possible for the initial stage of passive zone development in cells with identical geometries to establish the approximate relationship between their dimensions and the thermal diffusivities of the gases filling the cells

$$l_i \sim l_j \left(a_i a_j\right)^{1/2}.$$

If two identical cells irradiated by identical neutron pulses are filled with mixtures that have identical compositions, but different initial densities, it then follows from the latter correlations that

$$l_i \sim l_j \left(\rho_{0j} / \rho_{0i}\right)^{1/2}. \tag{8.36}$$

Here, the determination of thermal diffusivity is used, $a = k_g / c_v \rho$.

The calculation results for the moment in time at which the fluence reaches $3 \cdot 10^{13}$ cm^{-2} are presented in Fig. 8.15 for a cylindrical cell with $r_1 = 1$ cm, filled with helium at different pressures, and irradiated by an exponentially rising neutron flux (8.34). Because the moment in time, t, and the fluence, φ, are linked by the formula $t = \tau \ln(1 + \varphi / \Phi_0 \tau)$, the τ value at given Φ_0 and φ corresponds to specified t value. The data in Fig. 8.15 confirm the existence of approximate dependence (8.36) for the values of $l < 0.3 r_1$.

Synchronous Cell Wall Heating

In study [48], it is proposed that cell wall heating be carried out at the same time as laser pumping in order to diminish, or totally eliminate, a passive zone. The simplest means for putting this process into effect is to inject ^{235}U nuclei into the cell walls: during the irradiation of a laser of this type by a neutron flux, cell body heating due to the nuclear fission of the uranium injected into its walls occurs synchronously with the escape of fragments from uranium-containing layer into the gas.

It is convenient to examine the synchronous heating process based on the example of a cylindrical laser cell with an active uranium layer that is directly deposited to its inner wall surface. Let us assume that the concentration of uranium nuclei in the cell wall material is such that, over the course of a neutron flux rise time that exceeds the typical temperature equalization time of the gas in the cell, τ_T, determined by means of correlation (8.33), there is no noticeable flow of thermal energy from the gas to the wall and vice versa. Thus, the correlation that defines the optimum concentration of uranium nuclei in the wall directly follows from gas, uranium layer, and wall temperature equality requirements at any given time during irradiation, as well as from the absence of a heat flow from the gas to the wall:

$$N_{w0} = N_U \frac{\left(r_1^2 - r_U^2\right)}{\left(r_2^2 - r_1^2\right)} \left(\frac{\varepsilon}{\varepsilon_w} - 1\right) \quad ; \qquad (8.37)$$

$$\varepsilon_w = \frac{c_v \rho_0 r_U^2}{c_v \rho_0 r_U^2 + c_U \rho_U \left(r_1^2 - r_U^2\right) + c_w \rho_w \left(r_2^2 - r_1^2\right)},$$

where c_U, c_w are the specific heat capacities of the uranium-containing layer and the cell wall material; r_U, r_1 are the internal and external uranium layer radius values; r_2 is the external radius of the tube wall; ε_w is the fraction of energy that the fission fragments transfer to the gas from all the energy released in the cell, including the energy released within the wall; and ρ_U, ρ_w are the uranium-containing layer and cell wall density density values.

In fact, when the above conditional requirements are present

$$\begin{aligned}
&(c_U V_U \rho_U + c_w V_w \rho_w)(T - T_0) = (1 - \varepsilon_w) E_w, \\
&c_v \rho_0 V (T - T_0) = \varepsilon_w E_w, \\
&\varepsilon_w E_w = \varepsilon E_U.
\end{aligned}$$

Here, E_w is the total energy released in the uranium layer and the cell wall; E_U is the energy released in the uranium layer; and V, V_U, V_w are the volumes of the gas, the uranium layer, and the tube wall, respectively. Taking into account the fact that energy release in the wall and the uranium layer is proportional to the concentration of uranium nuclei therein, it is not difficult to derive correlation (8.37) from the latter three equations.

In order to verify the effect of the uranium, series of calculations were performed that involved different uranium concentration values, N_w, in the wall for different neutron flux rise times, starting with $\tau = 4$ ms or higher. These calculations revealed that at $N_w = N_{w0}$, the passive zone completely disappears at any of the neutron flux rise times studied. The temperature profiles for several successive moments in time

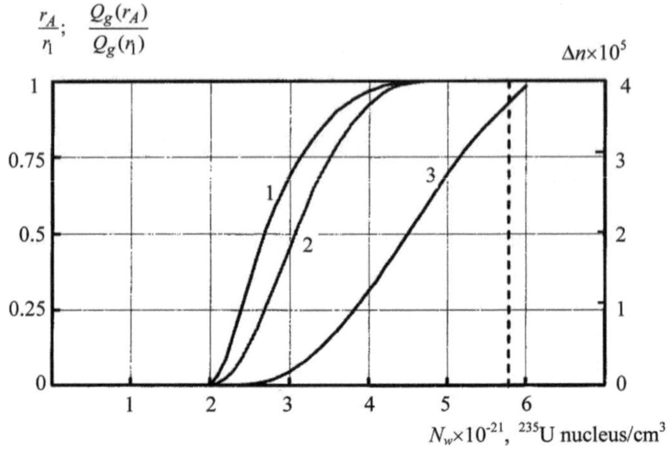

Fig. 8.16 Dependences of the radial coordinate of the active region's outer boundary, the fraction of the energy absorbed in this region, and the refractive index gradient in this same region upon the concentration of uranium nuclei in the cell wall: (1) r_A/r_1; (2) $Q_g(t_A)/Q_g(r_1)$; (3) Δn; the broken line is $N_w = N_{w0}$

at $N_w = N_{w0} = 4.95 \times 10^{21}$ cm^{-3} are designated by the broken line in Fig. 8.13. The calculated dependences of the radial coordinate of the active region's outer boundary and the fraction of fission fragment energy absorbed in this region, $Q_g(t_A)/Q_g(r_1)$, upon the total energy absorbed in the gas, as well as of the refractive index gradient in this region upon the N_w concentration value at $D_0 = 0.5$ a neutron flux rise time of $\tau = 0.04$ s, are presented in Fig. 8.16.

The calculations demonstrated that, despite a positive effect, the synchronous wall heating process has a specific drawback: over a time frame of $t < 0.1$ s (see Fig. 8.13), the cell wall and the gas are heated by more than $\Delta T = 1{,}000$ °K, which limits opportunities for using the process.

We note that injecting uranium nuclei into the cell wall material is virtually equivalent to simply increasing the uranium mass in the uranium layer by a value that equals its mass in the wall. In Fig. 8.13, $N_{ow} \approx 5 \times 10^{21}$ cm^{-3}, which at $r_1 = 1$ cm and in the presence of a wall thickness of $\delta_w = 1$ mm, is equivalent to increasing the thickness of a metallic uranium layer by ~40 times. In the presence of such an increase, the energy that the fragments carry into the gas (D_1 is increased from 0.5 to 20), which is proportional to $D_1\varepsilon(D_1)$ (see Chap. 7, Sect. 7.3), increases by 7 %. The energy $(1 - \varepsilon)$ that the fragments do not carry into the gas is released directly onto the inner surface of the cell wall; thus, just as when uranium is placed in the wall material, energy release therein is more homogeneous. In the example under consideration, the wall is made from zirconium, the thermal diffusivity of which is $a_w \approx 0.12$ cm^2/s; the typical temperature equalization time for a wall thickness of $\delta_w = 1$ mm comes to $\tau_w \sim \delta^2_w / a_w \sim 0.1$ s, which is an order larger than the neutron pulse rise time. Therefore, when all the necessary uranium is concentrated in the active uranium-containing layer, the cell wall's inner surface will be heated quite a bit more intensely than during uranium blending, which will lead to a decrease in the permissible fluence values and accordingly in irradiation time. However, it follows from the foregoing that in order to achieve the desired effect over fairly short rise time of $\tau < \tau_w$ when concentrating all the uranium in the active layer, its total quantity may be smaller than during its blending. It cannot be ruled out that under specific conditions, the version when all the uranium is concentrated in the uranium-containing layer may prove to be preferable. The question of the optimum placement location and the total quantity of the additional uranium must be separately considered in each given case based on actual structural solutions and the requirements imposed on laser characteristics.

8.4 Dynamics of NPL Cavity Stability Variations

In NPLs based on mixtures of rare gases excited by the fission fragments of uranium nuclei, the shape of the lasing pulse is usually close to that of the neutron pulse that initiates the fission reaction [4–8, 49]. However, during a number of experiments involving mixtures, the bulk of which were comprised of gases with high atomic numbers (for example, Ar and Kr), the lasing pulse consisted of an irregular series of small peaks on the crest of a common pulse; the total suppression of lasing followed by its restoration was sometimes observed [4, 34–36].

Various explanations exist for the irregular behavior of lasing with an increase in power deposition. For example, in study [50], lasing in an Ar-Xe mixture ($\lambda = 1.73$ μm) only occurred on the leading edge of the pumping pulse, which in the opinion of the authors of [51], was the result of Xe atom 6*s*, 6*p*, and 5*d* level collisional mixing by the plasma electrons.

Other attempts at explaining this effect are based on the hypothesis of the appearance of refractive index discontinuities that originate within low-intensity longitudinal sound and shock waves in the presence of comparatively short neutron pulses ($\tau \sim 10^{-3}$ s) due to a pumping inhomogeneity in the direction of the optical axis. In particular, the cause of these inhomogeneities may be the presence of gas regions directly adjacent to a laser cell's end windows that are not excited by the fragments. However, the absence of a correlation between the duration of the aforementioned micropeaks, the time intervals that separate them, the geometric dimensions of the cell, and sound velocity in the gas, as well as, more importantly, the disappearance of these peaks with a decrease in gas mixture density when other conditions remain unchanged, have given rise to doubts concerning the correctness of the proposed interpretation.

On the other hand, the peculiarities of the appearance and development of the transverse optical inhomogeneities that lead to the alteration of cavity stability can explain the absence of a time correlation between lasing and pumping pulses. A comparison of the calculated and experimental data cited in [35, 36] confirms this proposition. During the experiments, a rectangular laser cell was used that had two parallel planar uranium layers located close to one another. The calculation procedure was based on the use of the ray matrix technique to analyze the stability of a laser cavity filled with a medium that had a time-varying parabolic refractive index profile.

In studies [35, 36], during gasdynamic calculations of gas density redistribution in a laser cell that were performed while ignoring heat transfer processes, it was assumed that a gas motion only occurs in the direction normal to the uranium layer planes. Thus, the dynamics of altering cavity stability were examined in an approximation of the exclusive existence of one-dimensional inhomogeneities that were symmetrical relative to the cell's central plane. The inhomogeneity of density distribution (and, accordingly, the refractive index) obtained by means of calculations at each studied moment in time were approximated by a polynomial with a finite number of members. This polynomial's first two coefficients were used for the parabolic representation of the refractive index profile in the entire volume of the cell's active section. The calculations made it possible to identify the moments in time at which the loss of cavity stability and its subsequent restoration occur. They are in quite accurate agreement with the experimentally observed lasing cessation and resumption times. It is more difficult to carry out the analysis of laser cavity stability described in Chap. 7, Sect. 7.4 (planar uranium layers) and in the second section of this chapter, because it is impossible to regard the inhomogeneity distribution in such a system as homogeneous.

Calculation Technique

As in studies [35, 36], the aforementioned laser's cavity stability was analyzed in [34] in a paraxial approximation involving the use of the ray matrix technique. It was assumed that the axis of a system with two-dimensional transverse inhomogeneities is located in a single plane. In this instance, as is generally known (see, e.g., study [29]), the conversion of the projections of light rays in the mutually perpendicular meridional planes can be independently examined within the framework of Gaussian optics. For the two-mirror linear cavity of the laser type under consideration, the circular matrix $\begin{pmatrix} A & B \\ C & D \end{pmatrix}$ is determined as the product of the following matrices: the $M1$ matrix, which describes beam propagation in a homogeneous medium between the first mirror and the boundary of the active volume within which the fission fragment energy is deposited the matrix of beam propagation in the active volume (M_2); the matrix of propagation in a homogeneous medium between the other active volume boundary and the second mirror (M_3); the matrix of beam reflection on the second mirror (M_4); then the M_3, M_2, and M_1 matrices again; and, finally, the M_0 matrix, which describes beam reflection on the first mirror. A cavity is stable if $|m| < 1$, where $m = (A + D)/2$.

For an active volume with a one-dimensional quadratic transverse refractive index distribution,

$$n(y) = n_0 - \frac{1}{2}\alpha y^2, \quad (\alpha > 0), \tag{8.38}$$

the stability parameter takes the form [35, 36]:

$$\begin{aligned} m_y = {} & C_1 \cos 2\gamma - C_2(\cos 2\gamma + 1) - (C_3\gamma^2 + C_4\gamma^{-2})(\cos 2\gamma - 1)- \\ & -(C_5\gamma + C_6\gamma^{-1}) \sin 2\gamma, \end{aligned} \tag{8.39}$$

where

$$C_1 = \left(1 - \frac{2L_1}{n_a R_1}\right)\left(1 - \frac{2L_2}{n_a R_2}\right);$$

$$C_2 = \left(1 - \frac{L_1}{n_a R_1}\right)\frac{L_1}{n_a R_2} + \left(1 - \frac{L_2}{n_a R_2}\right)\frac{L_2}{n_a R_1};$$

$$C_3 = \frac{n_0^2 L_1 L_2}{n_a^2 L_A^2}\left(1 - \frac{L_1}{n_a R_1}\right)\left(1 - \frac{L_2}{n_a R_2}\right);$$

$$C_4 = \frac{L_A^2}{n_0^2 R_1 R_2};$$

$$C_5 = \frac{n_0 L_1}{n_a L_A}\left(1 - \frac{L_1}{n_a R_1}\right)\left(1 - \frac{2L_2}{n_a R_2}\right) + \frac{n_0 L_2}{n_a L_A}\left(1 - \frac{L_2}{n_a R_2}\right)\left(1 - \frac{2L_1}{n_a R_1}\right);$$

$$C_6 = \frac{L_A}{n_0 R_2}\left(1 - \frac{2L_1}{n_a R_1}\right) + \frac{L_A}{n_0 R_1}\left(1 - \frac{2L_2}{n_a R_2}\right);$$

γ is the dimensionless parameter (index gradient strength parameter) introduced in [35, 36],

$$\gamma = L_A\sqrt{\frac{\alpha}{n_0}}. \tag{8.40}$$

Here, L_A is the extent of the metallic uranium layers along the optical axis (the length of the cell's active section); L_1, L_2 are the distances between the first mirror and the cell's active section, and between the second mirror and the cell's active section (the buffer end clearances); R_1, R_2 are the radius of curvature values of the first and second mirrors; and n_a is the medium's refractive index in the region between the active section and the mirrors.

It is not difficult to demonstrate that for a laser in the active medium of which the refractive index distribution takes the form

$$n(x) = n_0 + \frac{1}{2}\beta x^2, \quad (\beta > 0), \tag{8.41}$$

the stability parameter is

$$\begin{aligned} m_x = C_1 \mathrm{ch}2\theta - C_2(\mathrm{ch}2\theta + 1) - \left(C_3\theta^2 + C_4\theta^{-2}\right)(\mathrm{ch}2\theta - 1) + \\ + \left(C_5\theta - C_6\theta^{-1}\right)\mathrm{sh}2\theta, \end{aligned} \tag{8.42}$$

where

$$\theta = L_A\sqrt{\frac{\beta}{n_0}}, \tag{8.43}$$

while the C_{1-6} coefficients are in agreement with the coefficients in equation (8.39).

Experiments [2] aimed at studying optical inhomogeneities in a laser with planar uranium layers (see Chap. 7, Sect. 7.4, and Sect. 8.2 of this chapter) revealed that the refractive index distribution in the $x = \mathrm{const}$ planes is well described by parabola (8.38), while in the $y = \mathrm{const}$ planes, it is fairly well depicted by parabola (8.41). An analysis of the experiments and calculations using the procedure set forth in Sect. 8.2 of this chapter demonstrated that the refractive index distribution in the transverse cross-section of the laser cell's active volume can be described with an accuracy of up to a few percents by the dependence

$$n(x, y, t) = n_0(t) + \frac{1}{2}\beta(t)x^2 - \frac{1}{2}\alpha(t)y^2. \tag{8.44}$$

In a paraxial approximation, geometric optics equations (8.17) take the form [30, 33]

$$\begin{aligned} \frac{d}{dz}\left[n(x, y, z)\frac{dx}{dz}\right] &= \frac{\partial n(x, y, z)}{\partial x}, \\ \frac{d}{dz}\left[n(x, y, z)\frac{dy}{dz}\right] &= \frac{\partial n(x, y, z)}{\partial y}, \end{aligned}$$

For gases, $\Delta n = n - n_0 << 1$, which when taking dependence (8.44) into account, makes it possible to present these equations in the form of two independent equations for the x- and y-components,

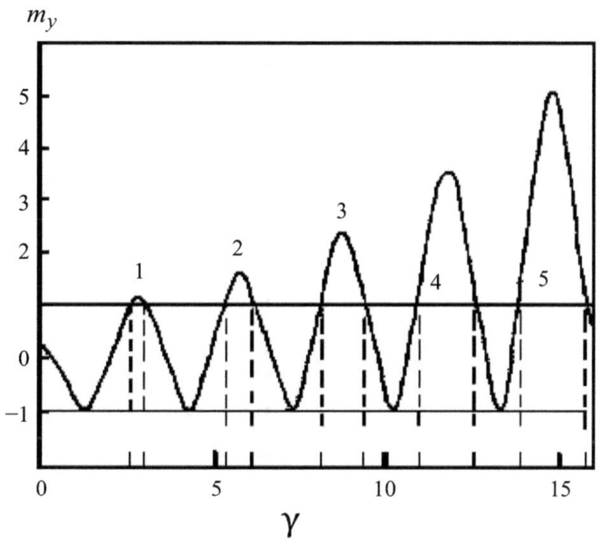

Fig. 8.17 Dependence of the m_y stability parameter upon the γ dimensionless inhomogeneity parameter

$$n_0 \frac{d^2x}{dz^2} = \frac{d\Delta n(x)}{dx},$$

$$n_0 \frac{d^2y}{dz^2} = \frac{d\Delta n(y)}{dy},$$

i.e., to independently examine the passage of the light beams and the fulfillment of the stability criterion in the x, z, and y, z planes. It is obvious that a cavity of this type will be stable if two conditions are simultaneously observed: $|m_x| < 1$ and $|m_y| < 1$.

If the cavity geometry remains unchanged, the stability parameters m_y and m_x (see (8.39) and (8.42)) will then only be dependent upon the corresponding dimensionless inhomogeneity parameters γ and θ, which are determined by equations (8.40) and (8.43). Thus, if the γ values vary over time, then m_y will periodically pass through the $|m_y| = 1$ values, which is explained by the presence of the trigonometric sines and cosines in expression (8.39). In this instance, if the θ values vary over time, then m_x can also intersect the $|m_x| - 1$ values. During the transition of the m_y (or $m_x = 1$) values to $|m_y| = 1$ (or $|m_x| = 1$), the alteration of cavity stability is observed. The γ and θ values at which such a transition occurs are critical values of the dimensionless inhomogeneity parameters. They are designated below as γ_{cr} and θ_{cr}.

Over the course of pumping, the inhomogeneities characterized by the dimensionless γ and θ parameters monotonically increase [2, 23, 35, 36, 45]. The dependence of the m_y stability parameter upon the value of the γ dimensionless inhomogeneity parameter over the variation range of real interest is presented in Fig. 8.17. The corresponding dependence of the m_x stability parameter upon the θ

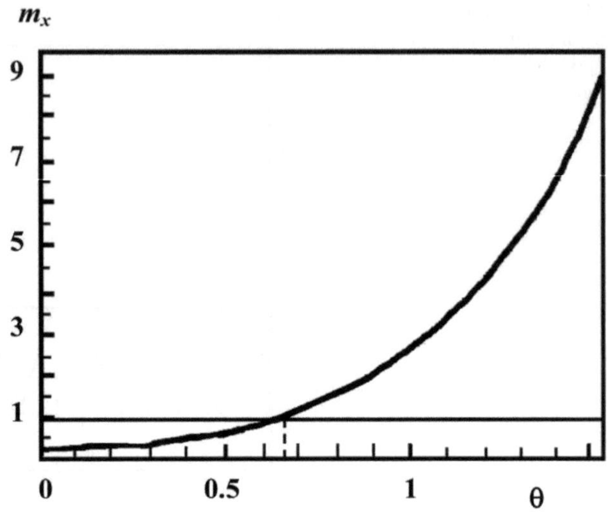

Fig. 8.18 Dependence of the m_x stability parameter upon the θ dimensionless inhomogeneity parameter

dimensionless inhomogeneity parameter is shown in Fig. 8.18. During irradiation time, the m_y parameter periodically intersects the $|m_y| = 1$ value in making the transition from the stable region to the instable one, then again returns to a stable state. The instability region sections in Fig. 8.17 are numbered in order of growth inhomogeneity parameter values: the first-order instability region comes first, then the second-order region, etc.

Changes in the stability of a laser with one-dimensional inhomogeneities of the type described in (8.38) were calculated in [35, 36], which revealed the cavity stability losses and restorations occurring over the course of a pulse that correspond in time to the moments of the total cessation and restoration of lasing observed during the experiments. Over the course of these calculations, the refractive index distribution was approximated by parabola (8.38), which was common over the entire volume; therefore, the loss of stability corresponded to the total cessation of lasing.

It was previously mentioned that during a number of experiments involving planar and cylindrical uranium layers, both fluctuations in laser power and their total cessation were observed. The reason for the fluctuations can be sought in the temporary departures of only a portion of the cavity's volume from the stable state. So, direct calculations demonstrated that during the approximation of inhomogeneities using correlation (8.44) based on their values in the peripheral regions of the transverse cross-section of the cell's active volume, higher α and β coefficients are generally obtained than during an approximation based on the central regions. Apparently, stability is first lost in the peripheral sections, while the central sections remain stable. An artificial technique was used in study [34] to verify this proposition. During the investigation of stability in the y and z planes, the active section of the cavity was evenly split into separate regions along the x axis that fully occupied the gap between the uranium layers, d. An $\alpha_i(t)$ parabola coefficient was selected for each region, i, at a relevant moment in time based on the results of density distribution calculations and it was used to compute the m_{yi} stability parameter. The transverse areas of the regions for which $|m_{yi}| < 1$ were then added. If the result of this addition is assigned to the overall area of the cell's transverse cross-section, we then obtain a characteristic that is dependent upon time, $S_y(t)$, which is found in an approximation of the total absence of a relationship between the individual regions.

We note that during this approach, the stability status of each individual region is examined outside its relationship with the status of the other regions, i.e., as if each region is isolated from the remaining ones. This technique *a priori* precludes the possibility of the origination of a stable (or quasistable) state between different regions, which is far from obvious. The use of this technique can be justified by the fact that during passage through the inhomogeneous media under consideration, each light beam is deflected by a slight angle so that when the transverse dimensions of a region are not too small, the bulk of the beams remain within the confines of this cavity section if its status is stable. Thus, the region Δx_i dimension must be limited from below, because it is apparent that the smaller the transverse dimension, the more numbers of the beams can travel outside the confines of this region to the neighboring ones. As a criterion, it was assumed that a region's size must not be smaller than a beam's displacement after one Cavity pass: $\Delta x_i > L\varphi$, where L is the distance between the mirrors and φ is the beam deflection angle. The deflection angle was estimated from the correlation $\varphi \sim L\,(\partial n \partial x) \sim L\beta_m x_0$, where x_0 is half the dimension of the uranium layer in the direction of the x axis, and β_m is the maximum β coefficient in the region most far removed from the center at the time of pumping pulse termination.

A similar approach was also used when calculating the stability in the x and z planes. The maximum number of splits in all the calculations presented below did not exceed $N = 30$. During the calculations for a cylindrical cell, splitting into coaxial cylindrical regions was carried out. The presence of axial symmetry was assumed.

Once again, we stress that the technique used is artificial and cannot be rigorously substantiated. It a priori rules out the possibility of the origination of a stable state between different regions, which is far from obvious.

Calculation Results

All the calculations of the characteristics of the alteration of cavity stability, $S(t)$, described below were performed using data from experiments for He-Xe and Ar-Xe mixtures (the LUNA-2 M setup with planar uranium layers and the VIR-2 M reactor), as well as Ar-Xe and Kr-Xe mixtures (a cell with a cylindrical uranium layer and the TIBR-1 M reactor). The laser cell characteristics are presented in

Fig. 8.19 Dependences of power deposition (1) and laser power (2) (**a**), as well as stability characteristics (**b** and **c**), upon time in a laser with plane-parallel uranium layer positioning for a mixture Ar:Xe = 2,000:1 at $P_0 = 0.5$ atm and $T_0 = 300$ K

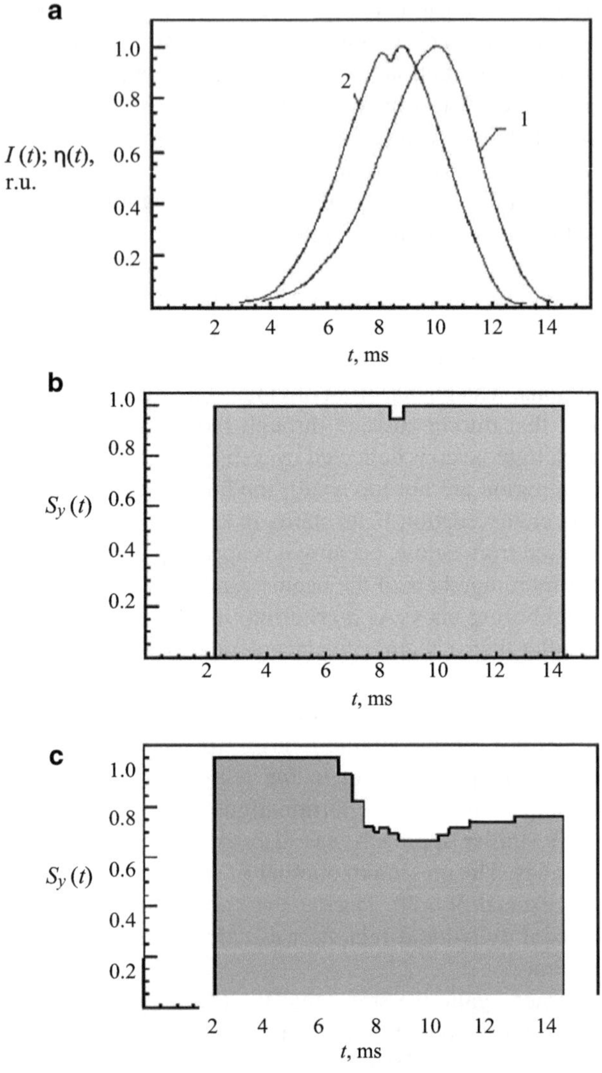

Sect. 7.4 of Chap. 7. Optical inhomogeneity evolution was calculated using the procedure described in the second section of this chapter.

The calculations revealed that, for the cell designs examined under standard pumping conditions, the cavities of lasers with active media that consist of helium-based mixtures do not leave the stable state. For mixtures of this type over the course of an entire pulse, $S(t) = 1$. This result is in full agreement with the experimental data, because the pulse shape therein after the lasing threshold is exceeded virtually duplicates the shape of the pumping pulse. However, when mixtures are used in which the buffer gases have high atomic numbers, as exemplified by Ar or Kr, loss of stability is observed, during which the stability loss

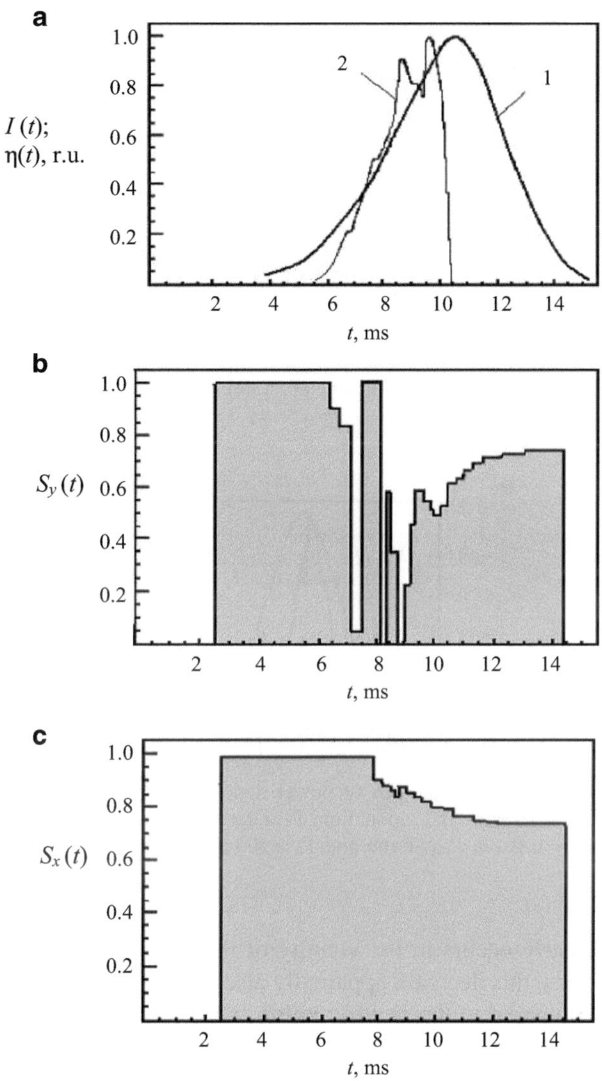

Fig. 8.20 Dependences of power deposition (1) and laser power (2) (**a**), as well as stability characteristics (**b** and **c**), upon time in a laser with plane-parallel uranium layer positioning for a mixture Ar:Xe = 600:1 at $P_0 = 1.5$ atm and $T_0 = 300$ K

effect is more clearly manifested as the mixture pressure is increased. However, if the pressure remains unchanged, this effect is intensified with an increase in the atomic number of the buffer gas.

The experimental dependences of the output laser power of a mixture with a value of Ar:Xe = 2,000:1 at $P_0 = 0.5$ atm in a laser with planar uranium layers (I) and a neutron pumping pulse (η) upon time, normalized to their maximum values, are shown in Fig. 8.19a. The calculated temporal $S_y(t)$ and $S_x(t)$ stability characteristics are shown in Fig. 8.19b. The only small trough on the $S_y(t)$ temporal diagram is in agreement with the trough at the experimental laser pulse peak. The results of the $S_x(t)$ calculations demonstrate that a stability decrease for the x-component

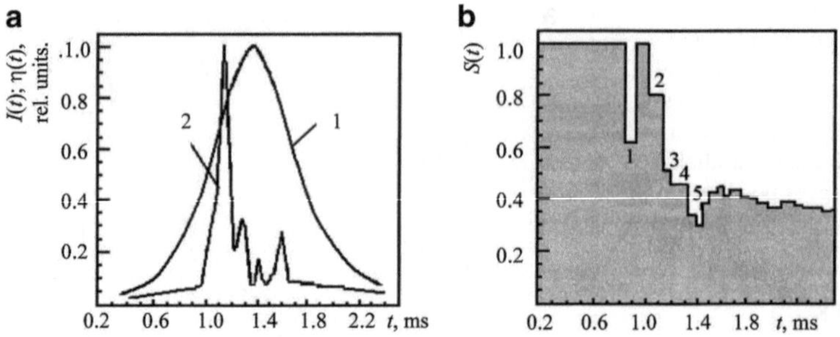

Fig. 8.21 Dependences of power deposition (1) and laser power (2) (**a**), as well as stability characteristics (**b**), upon time in a laser with a cylindrical uranium layer for a mixture Kr: Xe = 200:3 at *P*0 = 1 atm and *T*0 = 300 K

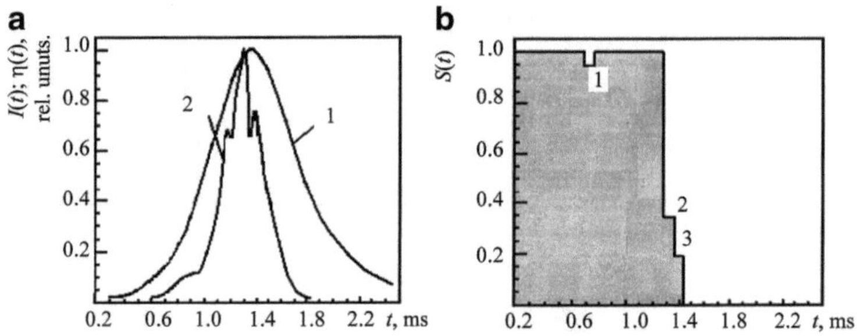

Fig. 8.22 Dependences of power deposition (1) and laser power (2) (**a**), as well as stability characteristics (**b**), upon time in a laser with a cylindrical uranium layer for a mixture Ar: Kr = 100:1 at $P_0 = 1$ atm and $T_0 = 300$ K

clearly occurs in the vicinity of the laser pulse peak. Under experimental conditions, this decrease apparently also leads to the earlier decrease of the laser pulse as compared to the neutron pulse, which creates the outward appearance of a lasing pulse shift relative to the pumping pulse. The slight nonsynchronism between the troughs at the laser peak and on the $S_y(t)$ diagram, as well as between the start of laser pulse decrease and the start of the dip on the $S_x(t)$ diagram, can be explained by the shortcomings of the calculation procedure used, in particular, an escalating error in the calculation of density distributions past the pumping pulse peak. These same shortcomings should apparently also explain the fact that $S_x(t)$ instability develops just as consistently, but not as intensely as the experimentally obtained the behavior of trailing edge of laser pulse.

The similar dependences obtained for a mixture Ar:Xe = 600:1 at $P_0 = 1.5$ atm when using the very same laser cell are presented in Fig. 8.20. Here, as in the previous example, a some correlation is observed between kinks on the laser pulse front and the $S_y(t)$ characteristics. So, the first three laser pulse kinks can be compared to the first three steps on the $S_y(t)$ curves and to the dual dip in the vicinity of the peak—two closely positioned dips at $S_y(t)$ immediately past $t = 8$ ms. Starting at roughly this moment in time, instability develops for $S_x(t)$ and the aggregate effects of both the instabilities lead to the cessation of lasing.

Calculations of $S_y(t)$ and $S_x(t)$ characteristics revealed that the development of a positive gas lens (8.38) in a laser with planar uranium layers leads to cavity stability fluctuations in time, whereas a negative gas lens (8.41), starting at a specific moment in time, produces a consistent stability decrease.

The calculation results for a laser with a cylindrical uranium layer are presented in Figs. 8.21 and 8.22. In this instance, only a positive gas lens originates and cavity stability is described by correlations (8.38)–(8.40), in which the y coordinate must be replaced with the radial coordinate r. A relatively good correlation was observed between the laser pulse and the $S(t)$ characteristics for a mixture Kr:Xe = 200:3 at $P_0 = 1$ atm. The worst correlation between the laser pulse and the calculated $S(t)$ characteristics was obtained for a mixture Ar:Kr = 100:1 at $P_0 = 1$ atm.

References

1. V.Yu. Matyev, *Gazodinamika okolotsentralnykh oblastey v germatichnykh kanalakh LYaN* [Gas Dynamics of the Near-Center Regions in Sealed NPL Channels]. *Tr. 2-oy Mezhdunarodn. konf. Fizika yaderno-vozbuzhdaemoy plazmy i problemy lazerov s yadernoy nakachkoy* [Proceedings of the 2nd International Conference Entitled "Nuclear-Excited Plasma Physics and Problems of Nuclear-Pumped Lasers]. Arzamas-16, 1995, vol. 1, pp. 410–420
2. V.V. Borovkov, B.V. Lazhintsev, S.P. Melnikov, I.N. Mochkayev, V.A. NorArevyan, A.A. Sinyanskiy, G.I. Fedorov, Issledovanie opticheskikh neodnorodnoctey v lazerakh s yadernoy nakachkoy [Investigation of Optical Inhomogeneities in Nuclear-Pumped Lasers]. Izvestiya AN SSSR. Ser. fizicheskaya **54**(10), 2009–2015 (1990)
3. E.P. Magda, S.L. Mukhin, V.A. Kryzhanovskiy, A.V. Bochkov, Opredelenie energovklada v gazovuyu sredu [Determination of Energy Deposition in a Gas Medium]. Pisma v ZhTF [Letters to the JTP (J. Tech. Phy.)] **19**(23), 54–57 (1993)
4. A.M. Voinov, L.Ye. Dovbysh, A.T. Kazakevich, V.N. Krivonosov, S.P. Melnikov, I.V. Podmoshenskiy, A.A. Sinyanskiy, Infrakrasnye lazery s yadernoy nakachkoy na perekhodakh ArI, KrI i XeI [Infrared Nuclear-Pumped Lasers Based on ArI, KrI, and XeI Transitions]. Pisma v ZhTF **5**(7), 422–424, (1979)
5. A.M. Voinov, L.Ye. Dovbysh, V.N. Krivonosov, S.P. Melnikov, I.V. Podmoshenskiy, A.A. Sinyanskiy, Nizkoporogovye lazery s yadernoy nakachkoy na perekhodakh atomarnogo kscenona [Low-Threshold Nuclear-Pumped Lasers Based on Atomic Xenon Transitions]. Doklady AN SSSR **245**(1), 80–83 (1979)
6. A.M. Voinov, L.Ye. Dovbysh, V.N. Krivonosov, S.P. Melnikov, I.V. Podmoshenskiy, A.A. Sinyanskiy, Gely-kriptonovy lazer vysokogo davleniya [High-Pressure Helium-Krypton Laser Excited by Uranium Fission Fragments]. ZhTF [JTP] **52**(7), 1346–1350 (1982)
7. A.M. Voinov, L.Ye. Dovbysh, V.N. Krivonosov, S.P. Melnikov, I.V. Podmoshenskiy, A.A. Sinyanskiy, Infrakrasny lazer s yadernoy nakachkoy na smesi He + Ex and Ar + Xe [Infrared Laser With Nuclear Pumping Based on He + Xe and Ar + Xe Mixtures]. Pisma v ZhTF **7**(16), 1016–1020, (1981)
8. A.M.Voinov, L.Ye. Dovbysh, V.N. Krivonosov, S.P. Melnikov, A.A. Sinyanskiy, A study of nuclear-pumped lasers on rare gas atomic transitions. Proceedings of 17th International Quantum Electronics Conf. Anaheim. USA, 1990. pp. 348–350

9. A.A. Sinyanskiy, *Issledovaniye po sozdaniyu yaderno-lazernykh ustroystv nepreryvnogo deystviya vo VNIIEF* [Research on the Creation of Continuous-Wave Nuclear Lasers at the VNIIEF (All-Russian Research Institute of Experimental Physics)]. Tr. 2-oy Mezhdunarodn. konf. Fizika yaderno-vozbuzhdaemoy plazmy i problemy lazerov s yadernoy nakachkoy [Proceedings of the 2nd International Conference Entitled "Nuclear-Excited Plasma Physics and Problems of Nuclear-Pumped Lasers"]. Arzamas-16, vol. 1, 1995, pp. 16–36
10. M.G. Anuchin, K.F. Grebenkin, V.A. Kryzhanovskiy, E.P. Magda, Effektivnost nakachki gazovykh sred ockolkami deleniya [Efficiency of Gas Medium Pumping by Fission Fragments]. Pisma v ZhTF [Letters to the JTP] **18**(5), 92–95 (1992)
11. E.P. Magda, A.V. Bochkov, V.A. Kryzhanovskiy, S.L. Mukhin, *Issledovanie vliyaniya azota na velichinu izmerennogo impulsa davleniya pri opredelenii energovklada oskolkov deleniya v gely i argon* [Investigation of the Effect of Nitrogen on Measured Pressure Pulse Magnitude When Determining the Energy Deposition of Fission Fragments in Helium and Argon]. Tr. 2-oy Mezhdunarodn. konf. Fizika yaderno-vozbuzhdaemoy plazmy i problemy lazerov s yadernoy nakachkoy [Proceedings of the 2nd International Conference Entitled Nuclear-Excited Plasma Physics and Problems of Nuclear-Pumped Lasers] Arzamas-16, vol. 1, 1995, pp. 452–456
12. A.M. Voinov, S.P. Melnikov, A.A. Sinyanskiy, Kineticheskaya model rekombatsionnykh lazerov na perekhodakh atoma kcenona [Kinetic Model of Recombination Lasers Based on Xenon Atom Transitions]. ZhTP [JTP] **60**(10), 100–113 (1990)
13. A.M. Voinov, S.P. Melnikov, A.A. Sinyanskiy, Kineticheskaya model IK-lazerov vysokogo davleniya na perekhodakh atoma kcenona [Kinetic Model of High-Pressure Recombination IR Lasers Based on Xenon Atom Transitions]. Izvestiya AN SSSR. Ser. fizicheskaya **54**(10), 2040–2044, (1990)
14. A.I. Konak, S.P. Melnikov, A.A. Porkhayev, A.A. Sinyanskiy, Generatsiya na perekhodakh atoma Xe pri vozbuzhdenii smesi Kr-Xe I chistogo ksenona oskolkami deleniya urana [Lasing Based on Xe Atom Transitions During KrXe Mixture and Pure Xenon Excitation by Uranium Fission Fragments]. Kvantovaya Elektronika [Quantum Electronics] **22**(12), 1184–1186 (1995)
15. N.A. Dmitriyev, I.D. Sofronov, B.P. Tikhomirov, Metodika rascheta odnomernykh mnogooblastnykh zadach vysokotemperaturnoy gazovoy dinamiki [Procedure for Calculating One-Dimensional Multiregion Problems in High-Temperature Gas Dynamics]. Voprosy atomnoy nauki i tekniki. Ser. Metodika i programmy chislennogo resheniya zadach matematicheskoy fiziki. **3**(14), 3–8 (1983)
16. N.B. Vargaftik, L.P. Filippov, A.A. Tarzimanov, Ye.Ye. Totskiy, *Teploprovodnost zhidkostey i gazov* [Thermal Conductivity of Liquids and Gases]. *Izd-vo standartov* [Standards Publishing], Moscow, (1978)
17. V.T. Kazazyan, B.A. Litvinenko, L.P. Roginets, I.A. Savushkin, *Fizicheskie osnovy ispolzovaniya kineticheskoy energii oskolkov deleniya radiatsionnoy khimii* [Physical Principles of the Use of Fission Fragment Kinetic Energy in Radiation Chemistry]. *Nauka i tekhnika* [Science and Technology], Minsk, (1972)
18. M.G. Anuchin, K.F. Grebenkin, Ya.Z. Kandiyev, Ye.I. Cherepanova, Raschetnoe issledovanie nakachki gazovoy sredy zaryazhennymi chastitsami – produktami yadernykh reaktsy [Numerical Investigation of Gas Medium Pumping by Charged Particles – Nuclear Reaction Products]. ZhTF **61**(1), 3–8 (1991)
19. V.Yu. Matyev, A.N. Sizov, *Vliyanie neodnorodnostey lazernoy sredy na gazodinamicheskie raschety lazerov s yadernoy nakachkoy* [Effect of Laser Medium Inhomogeneities on Gasdynamic Calculations for Nuclear-Pumped Lasers]. Tr. otraslevoy konf. Fizika yaderno-vozbuzhdaemoy plazmy i problemy lazerov syadernoy nakachkoy [Proceedings of an Specialist Conference Entitled Nuclear-Excited Plasma Physics and Problems of Nuclear-Pumped Lasers]. Obninsk, vol. 2, 1993, pp. 209–218
20. V.Yu. Matyev, Metody rascheta energovklada ionov v neodnorodnykh sredakh [Techniques for Calculating Ion Energy Deposition in Inhomogeneous Media]. Ibid., pp. 79–88

21. V.F. Kolesov, Aperiodicheskie impulsnye reaktory [Aperiodic Pulse Reactors]. *RFYaTs-VNIIEF* [Russian Federal Nuclear Center-VNIIEF], Sarov, 1999
22. A.N. Sizov, Yu.N. Deryugin, Raschet prostranstvennykh neodnorodnostey v tsilindricheskikh gazovykh lazerakh s yadernoy nakachkoy [Calculation of Spatial Inhomogeneities in Cylindrical Gas Nuclear-Pumped Lasers]. *Tez. dokl. VI Vsesoyuz. konf. Optika lazerov* [Abstracts of Papers Presented at the 6th All-Union Conference Entitled Laser Optics]. Leningrad, 1990, p. 132.
23. A.N. Sizov, N. Deryugin Yu, Raschety prostranstvennykh neodnorodnostey v tsilindricheskikh gazovykh lazerakh s nakachkoy oskolkami deleniya [Calculations of Spatial Inhomogeneities in Cylindrical Gas Lasers With Fission Fragment Pumping]. ZhTF [JTP] **62**(9), 107–111 (1992)
24. A.G. Shashkov, N.A. Nesterov, O.A. Kolenchits, Teploprovodnost inertnykh gazov pri shirokom diapazone temperatur [Thermal Conductivity of Inert Gases Over a Wide Range of Temperatures]. Inzhenerno-fizichesky zhurnal **43**(5) 788–795, (1982)
25. N.B. Vargaftik, Yu.D. Vasilevskaya, Teploprovodnost geliya pri temperaturakh (300—6000) K [Thermal Conductivity of Helium at Temperatures of 300–6,000 K]. Inzhenerno-fizichesky zhurnal **42**, 412–417 (1982)
26. S.S. Kutateladze, *Teplootdacha i gidrodinamicheskoe coprotivlenie: Spravochnoe posobie [Heat Transfer and Hydrodynamic Resistance: A Reference Aid]* (Energoatomizdat Publishing, Moscow, 1990)
27. F.Ya Kramerov, Ya. V. Shevelev, *Inzhenernye raschety yadernykh reaktorov* [Engineering Calculations for Nuclear Reactors]. Atomizdat Publishing: Moscow (1964)
28. Ananyev Yu. A. *Opticheskie rezonatory i problema raskhodimosti lazernogo izlucheniya* [Optical Cavities and the Problem of Laser Emission Divergence]. (*Nauka* [Science Publishing], Moscow, 1979)
29. Ye.F. Ishchenko, *Otkrytye opticheskie rezonatory* [Open Optical Cavities]. (*Sovetskoe radio* [Soviet Radio Publishing], Moscow, 1980)
30. D. Marcuse, *Opticheskie volnovody* [Optical Waveguides]. (*Mir* [World Publishing], Moscow, 1974)
31. G.A. Kirillov, S.B. Kormer, G.G. Kochemasov, S.M. Kulikov, V.M. Murugov, V.D. Nikolayev, S.A. Sukharev, V.D. Urlin, Issledovanie raskhodimosti izlucheniya fotodissotsionnogo opticheskogo kvantovogo generatora c neodnorodnoy aktivnoy sredoy [Investigation of the Divergence of the Emission of a Photodissociation Laser With an Inhomogeneous Active Medium]. Kvantovaya elektronika [Quantum Electronics] **2**(4), 666–671 (1975)
32. A.I. Zaretskiy, G.A. Kirillov, S.B. Kormer, S.A. Sukharev, Zavisimost raskhodimosti izlucheniya opticheskhikh kvantovykh generatorov i usiliteley ot opticheskoy neodnorodnostey sredy [Dependence of Laser Emission Divergence Upon Medium Optical Inhomogeneity]. Kvantovaya elektronika [Quantum Electronics] **1**(5), 1185–1190 (1974)
33. M. Born, E. Wolf, *Osnovy optiki* [Principles of Optics]. (Nauka [Science Publishing], Moscow, 1973)
34. Ye.V. Prikhodko, A.N. Sizov, Raschetny analiz ustoychivosti rezonatorov lazerov s yadernoy nakachkoy [Numerical Analysis of the Cavity Stability of Nuclear-Pumped Lasers]. Kvantovaya elektronika **22**(6), 613–618, (1995)
35. D.R. Neal, W.C. Sweatt, J.R. Torczynski, Resonator design with an intracavity time-varying index gradient. Proc. SPIE **965**, 130–141 (1989)
36. D.R. Neal, J.R. Torczynski, W.C. Sweatt, Resonator stability effects in quadratic-duct nuclear-reactor-pumped lasers. Proc. Int. Conf. LASERS'88. McLean, 1989. p. 245–252
37. V.Y. Matyev, V.V. Borovkov, S.P. Melnikov, Opticheskie neodnorodnosti v tsilindricheskikh lazerakh s yadernoy nakachkoy [Optical Inhomogeneities in Cylindrical Nuclear-Pumped Lasers]. Kvantovaya elektronika [Quantum Electronics] **30**(3), 215–220 (2000)
38. J.R. Torczynski, On the motion of a gas experiencing range-dependent volumetric heating. J. Fluid. Mech. **201**, 167–188, (1989)

39. A.V. Gulevich, V.A. Dubovskaya, A.V. Zrodnikov, B.V. Kachanov, *Raschetnoe issledovanie gazodinamicheskikh i teplovykh kharakteristik lazernoaktivnogo elementa* [Numerical Investigation of the Gasdynamic and Thermal Characteristics of a Laser Active Element]. Tr. otraslevoy konf. Fizika yadernovozbuzhdaemoy plazmy i problemy lazerov s yadernoy nakachkoy [Proceedings of an Specialist Conference Entitled Nuclear-Excited Plasma Physics and Problems of Nuclear-Pumped Lasers]. Obninsk, vol. 3, 1993, pp. 77–82
40. B.V. Kachanov, A.V. Gulevich, *O vliyanii neravnomernosti netronnogo potoka i plotnosti gaza na energovklada ockolkov deleniya v lazerno-aktivnom elemennte* [Concerning the Effect of Neutron Flux and Gas Density Inhomogeneity on the Fission Fragment Energy Deposition in a Laser Active Element]. Tr. 2-oy Mezhdunarodn. konf. Fizika yaderno-vozbuzhdaemoy plazmy i problemy lazerov s yadernoy nakachkoy [Proceedings of the 2nd International Conference Entitled "Nuclear-Excited Plasma Physics and Problems of Nuclear-Pumped Lasers"]. Arzamas-16, vol. 1, 1995, pp. 358–363
41. J.R. Torczynski, Heating-induced damping of density perturbations in nuclear-reactor-pumped lasers. Nucl. Sci. Eng. **109**(4), 401–410 (1991)
42. J.R. Torczynski, D.R. Neal, Transient gas motion in nuclear-reactorpumped lasers: computational and experimental results. Nucl. Sci. Eng. **113**(3), 189–206 (1993)
43. J.R. Torczynski, R.J. Gross, G.N. Hays, G.A. Harms, D.R. Neal, D.A. McArthur, W.J. Alford, Fission-fragment energy deposition in argon. Nucl. Sci. Eng. **101**(3), 280–284 (1989)
44. S.P. Melnikov, A.A. Sinyanskiy, Kinetika IK-lazerov s yadernoy nakachkoy na perekhodakh atomov kriptona i argona [Kinetics of IR Nuclear-Pumped Based on Krypton and Argon Atom Transitions]. ZhTP [JTP] **62**(6), 159–169 (1992)
45. J.R. Torczynski, Heat transfer regimes in nuclear-reactor-pumped gas lasers. J. Thermophys. Heat Transfer **5**(3), 318–326, (1991)
46. V.Y. Matyev, V.V. Borovkov, S.P. Melnikov, Opticheskie neodnorodnosti v lazerakh s yadernoy nakachkoy s ploskimi uranovymi sloyami [Optical Inhomogeneities in Nuclear-Pumped Lasers With Planar Uranium Layers]. ZhTP [JTP] **71**(1), 79–85 (2001)
47. A.M. Voinov, A.A. Sinyanskiy, A.G. Vasilenko, O.A. Golubeva, L. Ye. Dovbysh, V.V. Ivanov, M.I. Kuvshinov, B.V. Lazhintsev, A.Ye. Lakhtikov, S.P. Melnikov, A.B. Modenov, V.A. Nor-Arevyan, A.M. Pichugin, A.N. Pokalo, V.V. Porkhayev, Ya.A. Pospelov, A.V. Sinitsyn, I.G. Smirnov, S.L. Turutin, M.V. Khlestkov, *Eksperimentalny kompleks LM-4/BIGR. Poluchenie generatsii v smesi Ar-Xe s dlitelnostyu bolee 1s* [LM-4/BIGR Experimental Complex. Producing Lasing in an Ar-Xe Mixture With a Duration of More Than 1 Sec]. Tr. 2-oy Mezhdunarodn. konf. Fizika yaderno-vozbuzhdaemoy plazmy i problemy lazerov s yadernoy nakachkoy [Proceedings of the 2nd International Conference Entitled Nuclear-Excited Plasma Physics and Problems of Nuclear-Pumped Lasers]. Arzamas-16, vol. 2, 1995, pp. 179–185
48. RF Patent No. 1140668. *Gazovy lazer s yadernoy nakachkoy* [Nuclear-Pumped Gas Laser]. A. N. Sizov. Inventions, No. 12, 1994.
49. S.P. Melnikov, A.A. Sinyanskiy, Kvazinepreryvnaya generatsiya na perekhode 4p-4s atoma argona pri nakachke lazera na smesi Ne-Ar oskolkami deleniya urana [Quasicontinuous Lasing Based on an Argon Atom 4p-4s Transition During Laser Pumping by Uranium Fission Fragments in an He-Ar Mixture]. Kvantovaya elektronika [Quantum Electronics] **18**(12), 1444–1446 (1991)
50. W.J. Alford, G.N. Hays, Measured laser parameters for reactor-pumped He/Ar/Xe and Ar/Xe lasers. J. Appl. Phys. **65**(10), 3760-3766, (1989)
51. M. Ohwa, T.J. Moratz, M.J. Kushner, Excitation mechanisms of the electron-beam-pumped atomic xenon (5d→6p) laser in Ar-Xe mixture. J. Appl. Phys. **66**(11), 5131–5145 (1989)

Chapter 9
Specific Features of NPLs with a Flowing Gas Medium

Investigations of optical inhomogeneities in sealed NPLs have revealed that flowing of the gas medium through the laser cavity is a necessary condition for achieving CW lasing in the stationary and quasi-stationary excitation modes. Gas heating is a no less important factor that requires the implementation of gas flowing. So, for example, at a thermal neutron flux density of $\Phi \approx 10^{14}$ cm$^{-2} \times$ s^{-1}, the average specific power deposition for helium with an average density of $\rho_0 = 3.2 \times 10^{-4}$ g/cm^3 (at initial pressure and temperature of $P_0 = 2$ atm and $T_0 = 300$°K) that fills a cylindrical laser cell with an internal radius of $r_1 = 1$ cm at a metallic uranium active layer thickness of $\delta_U = 2.8 \times 10^{-4}$ cm comes to

$$\bar{\bar{q}}_c = \overline{dq_c/dt} = \frac{2}{r_1}\sigma_f N_U \Phi E_0 \delta_U \varepsilon \sim 10 \quad \text{W/cm}^3.$$

During stationary pumping in the absence of gas replacement, this specific deposition will result in a gas temperature increase to $T > 10^4$°K over a period of 1 s.

One method for implementing continuous gas replacement is longitudinal gas flowing [1] (in the direction of a channel's optical axis), similar to the one used in high-temperature gas-cooled reactors.

9.1 Longitudinal Gas Flowing

The most important parameter that characterizes a gas flow is the Reynolds number

$$Re = \frac{\overline{U}d}{\nu}, \tag{9.1}$$

where d is the typical transverse dimension of the channel; $\overline{U}$ is the velocity of the gas flow averaged over the channel cross-section; and ν is the kinematic viscosity.

S.P. Melnikov et al., *Lasers with Nuclear Pumping*,
DOI 10.1007/978-3-319-08882-2_9

At Reynolds numbers, Re, lower than the critical value, Re_c, a flow is laminar. At $Re > Re_c$, the laminar mode makes the transition to the turbulent mode. Under normal conditions for a flow in a round tube, $Re_c \approx 2{,}300$. For channels with another transverse cross-section shape, it is close to this same value [2, 3].

During a laminar gas flow in a cylindrical channel, the pressure differential along its length, b_L, comes to [4]

$$\Delta P = \frac{8\overline{U}\rho\nu b_L}{r_1^2},$$

while for a channel formed by two parallel planes, the distance between which equals d,

$$\Delta P = \frac{12\rho\nu\overline{U}b_L}{d^2}. \tag{9.2}$$

During a turbulent flow in a channel with any transverse profile, the pressure differential along the channel's length is determined by the expression [5–7]

$$\Delta P = \xi\rho\overline{U}^2\frac{b_L}{2d_e}, \tag{9.3}$$

where ξ is the resistance coefficient. The channel's equivalent diameter, d_e, is expressed by way of the area of the channel's transverse cross-section, S_c, and the total perimeter, Π_c, of the channel surface that the gas washes over,

$$d_e = 4S_c/\Pi_c.$$

The resistance coefficient can be determined based on the Blasius law [6, 7]

$$\xi = 0,3164Re^{-0,25}.$$

Other formulas also exist for determining the ξ parameter [3–5], but they all yield a similar result. An estimate based on formula (9.2) for $\overline{U} \sim 10$ m/s and $b_L \sim 1$ m yields $\Delta P < 3 \times 10^{-4}$ atm. From a similar estimate based on formula (9.3) for $U \sim 100$ m/s, we get $\Delta P < 0.03$ atm. In both cases, the ΔP parameter value is negligible as compared to the optimum pressure value of $P \sim 1$ atm used in NPLs. Consequently, the gas pressure in a laser channel can be roughly regarded as homogeneous.

In this case, the heat balance equation for a laser channel with a stationary gas flow that operates in the steady-state mode takes the forms:

$$\overline{q}_c S_c b_L = \rho_0 U_0 S_c c_p (T_b - T_0),$$

where c_p is the specific heat capacity of the gas at a constant pressure; $\overline{q}_c$ is the specific power deposition averaged over the volume; T_0, T_b are the gas temperatures at the laser channel inlet and outlet; U_0 is the gas velocity at the channel inlet; and ρ_0 is the gas density at the channel inlet. Thus,

$$U_0 = \frac{\overline{q}_c b_L}{\rho_0 c_p (T_b - T_0)}.$$

Because the operating length of the laser channels is $b_L > 1$ m, then at a specific power deposition of $q > 10$ W/cm^3, gas circulation along a channel already requires the use of high gas velocities ($U_0 \sim 100$ m/s) in order to prevent conspicuous active medium overheating (for example, by $\Delta T = T_b - T_0 \leq 100$°K). In the presence of a channel transverse dimension of $d \sim 1$ cm, these velocities correspond to $Re \geq 10^4$. These Re values markedly exceed the Re_c value at which a laminar flow mode makes the transition to a turbulent mode. Because $b_L >> d$ and $Re >> Re_c$, a well-developed turbulent flow is then established in the bulk of the cell volume, which can adversely affect the optical quality of the active medium as a whole.

9.2 Effect of Turbulent Pulses on the Optical Quality of a Medium

Turbulence can lead to perceptible refractive index fluctuations. The danger of wave front distortion and noticeable coherent light beam attenuation arises. But because the effective thermal diffusivity of a turbulent flow considerably exceeds the conventional molecular thermal diffusivity that is inherent in laminar flows and stationary media, the rate of change of temperature, density, and refractive index inhomogeneities increases noticeably.

In the opinion of Schlichting, in a turbulent flow "the details of the pulsating motion superimposed on the main motion are so complex that to describe them theoretically seems hopeless" [2]. A great deal of time has passed since this statement was made. However, despite the longstanding efforts of specialists, the development of reliable mathematical models for the structure of turbulent flows that would make it possible to perform quantitative calculations with acceptable accuracy are far from complete. Nevertheless, based on concepts developed by Kolmogorov [8], certain regularities can be identified that make it possible in a number of cases to qualitatively describe the behavior of the systems under consideration and even to make approximate quantitative estimates [9, 10]. We will examine the structure of a turbulent flow using the model from ref. [8] and assess the potential effect of turbulence on the optical quality of a medium.

Turbulent Flow Structure

According to the Kolmogorov model, during turbulent motion, large vortices are broken down into increasingly smaller ones. Here, the redistribution of the turbulence's kinetic energy occurs, which is transmitted from the large-scale vortices to the small-scale ones. The structure of motion on small scales (at least at distances from solid walls that are greater than the small scales [4]) is locally homogeneous and isotropic. According to ref. [9], the theory of homogeneous turbulence constitutes "an entirely suitable framework for analyzing the problem of scattering" of optical and acoustic waves. In this case, the proposition concerning the homogeneity and even the isotropy of a turbulent field proves to be correct [9].

According to the model [8], a wide range of pulse length and velocity scales exists in a turbulent motion region. For the combination of the length scale, l_0, and the corresponding velocity scale, u_0, at which the Reynolds number takes on the order of unity

$$\frac{u_0 l_0}{\nu} \sim 1, \tag{9.5}$$

the turbulent motion must be damped; i.e., l_0 characterizes turbulent inhomogeneities of minimum dimensions and is called the inner turbulence scale. It is determined by the correlation (see ref. [10] for example)

$$l_0 = \left(\nu^3/w\right)^{1/4}, \tag{9.6}$$

where w is the turbulent motion dissipation rate per unit mass, which transitions to heat at the expense of viscosity. It is fully determined by viscosity and its order of magnitude comes to

$$w \sim \nu u_0^2/l_0^2. \tag{9.7}$$

If the Reynolds number in Eq. (9.1) for a liquid or gas flow is sufficiently large ($Re > Re_c$), the flow loses stability. Here, vortices can be formed that have Λ dimensions up to values on the order of $\Lambda \sim d$, and that have a velocity, the order of magnitude of which may be close to $\overline{U}$. According to the Kolmogorov model, the vortices under consideration are broken down into smaller ones, which are in turn broken down into even smaller ones, all the way down to inhomogeneity dimensions of ~l_0. Thus, between the limiting values of $\overline{U}$, u_0 and Λ, l_0, subintervals of the variation in motion velocities, u, and typical dimensions, l_t, exist. If a w parameter is identified as an average energy dissipation rate throughout the entire gas flow that is dependent upon the ~Λ scale alone for the turbulent flow as a whole, and if the corresponding velocity, ~$\overline{U}$, is used, then for large-scale inhomogeneities, we get [9].

$$w \sim \overline{U}_0^3/\Lambda. \tag{9.8}$$

Over the course of vortex breakdown, energy is transmitted to the smaller-scale inhomogeneities, for which $w \sim u^3/l_t$. Energy dissipation occurs in inhomogeneities with a size on the order of the inner scale [8–10].

From Eqs. (9.5), (9.7), and (9.8), we find

$$l_0 \sim \left(\nu^3\Lambda/\overline{U}^3\right)^{1/4}. \tag{9.9}$$

Consistent with refs. [9, 11], we will estimate the square of the sound velocity in the medium, which as is generally known, is determined by the equation

$$u_s^2 = \left(\frac{\partial P}{\partial \rho}\right)_S$$

(where S is entropy), from the correlation

$$u_s^2 \sim \Delta P/\Delta\rho.$$

As was done in ref. [4, 9], the dynamic pressure for turbulent pulses can be estimated from the correlation

$$\Delta P \sim \rho u^2. \tag{9.10}$$

Thus, for turbulent pulses that are characterized by a velocity of u and a length scale of l_t, we get

$$\Delta\rho/\rho \sim u^2/u_s^2. \tag{9.11}$$

It follows from Eq. (9.11) that the largest density jumps take place in inhomogeneities that are characterized by the highest u velocity values; i.e., in inhomogeneities with a typical Λ scale that have the maximum dimensions.

Combining Eqs. (9.5), (9.7), and (9.8), we get

$$u_0 \sim \left(\nu\overline{U}^3/\Lambda\right)^{1/4}, \tag{9.12}$$

which, after being placed into Eq. (9.11), yields

$$\Delta\rho_0 \sim \rho\left(\frac{\nu}{\Lambda}\right)^{1/2}\frac{\overline{U}_0^{3/2}}{u_s^2}. \tag{9.13}$$

By way of an example, we will examine helium at $T = 300°\text{K}$ and $P_0 = 2$ atm, which corresponds to a density of $\rho \approx 3.2 \times 10^{-4}$ g/cm^3 and a kinematic viscosity of $\nu \approx 0.6$ cm^2/s. Let us suppose that $d = 2$ cm. The sound velocity in helium at this temperature is $u_s \approx 10^5$ cm/s. Assuming that $\Lambda \sim 1$ cm and taking the velocity of gas flow to be $\overline{U} = 10$ m/s, then from Eqs. (9.9), (9.12), and (9.13), we get $l_0 \sim 4 \times 10^{-3}$ cm, $u_0 \sim 10^2$ cm/s, and $\Delta\rho_0 \sim 8 \times 10^{-10}$ g/cm^3. For maximum inhomogeneities with a Λ scale that are characterized by a velocity of $\sim \overline{U}$, according to Eq. (9.11), we then obtain $\Delta\rho_m \sim 3 \times 10^{-8}$ g/cm^3. At $\overline{U} = 100$ m/s for these same parameters, we find $l_0 \sim 7 \times 10^{-4}$ cm, $u_0 \sim 9 \times 10^2$ cm/s, $\Delta\rho_0 \sim 2 \times 10^{-8}$ g/cm^3, and $\Delta\rho_m \sim 3 \times 10^{-6}$ g/cm^3.

The correlation $\tau_t \sim l_t/u$ [9] apparently determines the typical pulse time. Then taking Eq. (9.9) into account for small-scale pulses, we get

$$\tau_{t0} \sim l_0/u_0 \sim \left(\nu\Lambda/\overline{U}^3\right)^{1/2}, \tag{9.14}$$

which, for the parameters we used, yields $\tau_{t0} \sim 3 \times 10^{-5}$ s at $\overline{U} = 10$ m/s and $\tau_{t0} \sim 8 \times 10^{-7}$ s at $\overline{U} = 100$ m/s.

Expression (9.10) was previously used to estimate the pressure jump in inhomogeneities. In ref. [11], the correlation used for this estimate is

$$\Delta P \sim \rho u \overline{U}. \tag{9.15}$$

Without entering into a discussion with the authors of refs. [9, 11] concerning the correctness of a given approach, we will estimate ΔP using both correlation (9.10) and correlation (9.15), because it is important for us to reflect the trend of the decrease in density jumps with a decrease in inhomogeneity scale. We will show that inhomogeneities of the maximum scale must exert a maximum effect on the optical quality of a medium, and that the process of describing the pressure jumps in small-scale inhomogeneities has no effect on the estimation of the density jumps in these inhomogeneities.

Using Eq. (9.15) instead of Eqs. (9.11) and (9.13), we obtain

$$\Delta\rho_0/\rho \sim u_0\overline{U}/u_s^2;\ \Delta\rho_0 \sim \frac{\rho}{u_s^2}\left(\frac{\nu\overline{U}^7}{\Lambda}\right)^{1/4}.$$

Substituting the numerical parameter values we assigned yields $\Delta\rho_0 \sim 5 \times 10^{-9}$ g/cm^3 at $\overline{U} = 10$ m/s and $\Delta\rho_0 \sim 3 \times 10^{-7}$ g/cm^3 at $\overline{U} = 100$ m/s.

Mixing and Temperature Fluctuations in an Inhomogeneously Heated Medium

The reason for density fluctuations in a turbulent flow may be not only the gasdynamic instability of the flow, but also liquid or gas mixing in the presence of the average temperature value's dependence upon coordinates. So, as a result of mixing in a turbulent flow, individual portions of the medium that are characterized by a certain temperature value appear at points where medium particles with different temperature values were heretofore present.

Obukhov [12] and Yaglom [13] examined the structure of a temperature field in such a flow. According to the model they developed, the linear scale of the minimum temperature inhomogeneities is determined by the correlation

$$l_\Theta \sim \left(a^3/w\right)^{1/4}, \tag{9.16}$$

where a is the thermal diffusivity coefficient.

Because the kinematic viscosity coefficient, ν, and thermal diffusivity, a, for gases differ very little from one another (a Prandtl number of $Pr \sim 1$), the minimum linear scales of inhomogeneities, as follows from Eqs. (9.6) and (9.16), are in virtual agreement:

$$l_\Theta/l_0 \approx (a/\nu)^{3/4} \sim 1. \tag{9.17}$$

The mechanism of temperature equalization in sufficiently large volumes of a turbulent flow is due to the combined effects of the medium's turbulent motion and heat conduction [12]. Temperature equalization in large-scale inhomogeneities and intermediate-scale inhomogeneities primarily occurs due to the internal turbulent mixing of the smaller-scale gas particles. The typical time of this equalization is determined by the correlation

$$\tau_\theta \sim l_t/u$$

and corresponds to the lifetime of dynamic l_t-scale inhomogeneities according to the Kolmogorov model. For maximum-scale inhomogeneities

$$\tau_{\theta m} \sim \Lambda/\overline{U}. \tag{9.18}$$

Temperature equalization in the smallest-scale inhomogeneities occurs by way of conventional molecular heat conduction [12]. The typical time required for this equalization can be estimated using the formula

$$\tau_{\theta_0} \sim l_\theta^2/a.$$

From the latter correlation and Eq. (9.14), with allowance for Eq. (9.17), we obtain

$$\tau_{\theta_0}/\tau_0 \sim l_0 u_0/a. \tag{9.19}$$

Because $Pr \sim 1$ for gases (i.e., $a \sim \nu$) then from Eq. (9.19) in concordance with Eq. (9.5), as should be expected, we get $\tau_{\theta 0} \sim \tau_0$.

Thus, the conclusion can be reached that the space–time structure of the turbulent flow temperature inhomogeneities originating as a result of mixing corresponds to the space–time structure of turbulent pulses with a dynamic nature.

Turbulent Pulses in an Optically Active Medium

We will now examine the influence of turbulent pulses on the optical quality of a medium without allowance for the potential effect of turbulent mixing in the presence of an inhomogeneous averaged temperature field. It is obvious that taking this effect into account [10] will not improve the optical characteristics.

The refractive index dependence, *n*, for the gases under consideration is quite accurately imparted by the equation

$$n = 1 + C_n\rho, \tag{9.20}$$

where C_n is the constant for a gas of this type. Using Eq. (9.13) for the refractive index pulses in the smallest-scale inhomogeneities, we obtain

$$\Delta n_0 \sim C_n\rho(\nu/\Lambda)^{1/2}\frac{\overline{U}^{3/2}}{u_s^2}. \tag{9.21}$$

Correlation (9.21) corresponds to definition (9.10) for pressure pulses. However, if dependence (9.15) is used instead of Eq. (9.10), the corresponding expression for estimating the Δn_0 value takes the form

$$\Delta n_0 \sim C_n\rho(\nu/\Lambda)^{1/4}\frac{\overline{U}^{7/4}}{u_s^2}. \tag{9.22}$$

According to Eqs. (9.11) and (9.20), the expression for refractive index pulses in the largest inhomogeneities with a typical scale of Λ can be written as

$$\Delta n_m \sim C_n\rho\overline{U}^2/u_s^2. \tag{9.23}$$

Estimates for helium based on dependence (9.21) using the numerical parameter values previously assigned at $\overline{U} = 10$ m/s yield $\Delta n_{01} \sim 10^{-10}$, while based on dependence (9.22), $\Delta n_{02} \sim 10^{-9}$. According to Eq. (9.23), $\Delta n_m \sim 6 \times 10^{-9}$; at a velocity of $\overline{U} = 100$ m/s, we get $\Delta n_{01} \sim 5 \times 10^{-9}$, $\Delta n_{02} \sim 6 \times 10^{-8}$, and $\Delta n_m \sim 6 \times 10^{-7}$.

Refractive index pulses cause energy dissipation in a coherent light beam passing through a turbulent medium. In an optically active medium, the dissipated energy lost during diffraction on inhomogeneities can be viewed as a coherent beam loss, i.e., as a kind of absorption [14, 15]. In order to calculate the absorption constant (see refs. [15–17] for example), use was made of the results that Sutton [14] obtained under the following assumptions: a turbulent medium with a dimension of L_t is homogeneous and isotropic; the light wave front at the medium inlet is planar; and consideration is given to the effective value averaged over time intervals that exceed the turbulence time scales.

The absorption constant that reflects the effect of turbulent pulses is determined by the formula [14]

$$\alpha_T = 2k_\lambda^2 \langle \Delta n^2 \rangle \Lambda, \tag{9.24}$$

where k_λ is the wave number, and $\langle \Delta n^2 \rangle$ is the mean-square pulse of refractive index.

The α_T coefficient can be estimated by replacing $\langle \Delta n^2 \rangle$ with $(\Delta n_m)^2$. Because, in point of fact $\langle \Delta n^2 \rangle \leq (\Delta n_m)^2$, this replacement can only lead to the overstatement of the absorption coefficient. We assume that $\Lambda \sim 1$ cm. Placing $k_\lambda = 2\pi/\lambda$ into Eq. (9.24) (for certainty's sake, we assume that $\lambda = 1.73$ μm [18, 19]), together with the Δn_m values found above, we obtain $\alpha_T \sim 2 \times 10^{-7}$ cm^{-1} at $\overline{U} = 10$ m/s and $\alpha_T \sim 2 \times 10^{-3}$ cm^{-1} at $\overline{U} = 100$ m/s. For comparison: the gain in NPLs based on rare gas mixtures is $\beta_k \sim 10^{-3}$–10^{-2} cm^{-1} [20].

According to ref. [14], the dependence of coherent beam intensity upon the distance traversed, x, is determined by the law

$$I(x) = I_0 \exp\{(\beta_k - \alpha_T)x\}.$$

Here, if $D/\Lambda \sim 1$ (where D is the transverse dimension of the beam), and

$$\alpha_T L_t < < 1, \tag{9.25}$$

the diffraction pattern in the far-field zone then virtually replicates the diffraction pattern from the inlet following the traversal of a homogeneous medium that does not contain turbulent pulses, and the noticeable attenuation of the laser beam is absent. Thus, if condition (9.25) is satisfied, turbulent pulses in density should not exert a direct influence on the shaping of laser beam power and angular parameters. Only statistically averaged density inhomogeneities identical to those originating in sealed laser cells, or in the presence of laminar flows, will affect these characteristics (see ref. [10] for an example of the determination of this average).

Based on inequality Eq. (9.25), we will estimate the permissible gas flow velocities at which turbulent pulses in density do not necessarily exert a direct influence on the optical quality of a medium. Deliberately reducing the permissible gas-flow velocity values, we will replace $\langle \Delta n^2 \rangle$ with $(\Delta n_m)^2$, then from Eqs. (9.24) and (9.25), we obtain

$$(\Delta n_m)^2 \ll \left(2k_\lambda^2 \Lambda L_t\right)^{-1}. \tag{9.26}$$

Placing Eqs. (9.23) into (9.26), we obtain

$$\overline{U} \ll u_s \left(\frac{\lambda^2}{8\pi^2 \rho^2 C_n^2 \Lambda L_t} \right)^{1/4}. \tag{9.27}$$

We note that in inequality Eq. (9.26), in which the $\overline{U}$ velocity value enters the biquadrate, must be strictly executed; therefore, in order to estimate the permissible $\overline{U}$ values, it is not obligatory to require that the "much less" condition in Eq. (9.27) be satisfied: it is sufficient that its left side be roughly two to three times smaller than its right side. In order to facilitate estimates of the upper limit of the gas-flow velocities at which the effect of turbulent pulses on the optical quality of a medium is small, we will transform Eq. (9.27) into the inequality $\overline{U} < \overline{U}_*$, where

$$\overline{U}_* \approx \frac{1}{3} u_s \left\{ \frac{\lambda^2}{8\pi^2 C_n^2 \rho^2 \Lambda L_t} \right\}^{1/4}. \tag{9.28}$$

During estimates of permissible gas-flow velocity based on formula (9.28) in an amplification regime, the L_t quantity is used (the geometric dimension of the medium along the optical axis, which equals the distance between the amplifier's end windows.) During estimates of gas-flow velocity in the lasing regime, L_t should be replaced with L_p (the effective photon path in the resonator, $L_p = ct_p$, where c is the light velocity in the medium under consideration and t_p is the photon lifetime in the laser cavity). The latter can be determined by means of the formula [21]:

$$t_p = -\frac{2L_A}{c \ln[r_m(1 - R_D)]}. \tag{9.29}$$

Here, L_A is the length of the active amplifying section of the gas medium that fills the cavity (which can be taken to equal the size of the uranium layer in the direction of the optical axis); R_D is the share of diffraction and other losses; and r_m is the reflectivity of the output mirror (the reflectivity of the second mirror is taken to equal unity).

The permissible gas-flow velocities, $\overline{U}_*$, for mixtures based on He ($P_0 = 2$ atm) and Ar ($P_0 = 0.5$ atm) at $L_A = 1$ m, $L_t = 1.4$ m, $\Lambda = 1$ cm, $R_D = 0$, and $r_m = 0.9$ obtained using formulas (9.28) and (9.29) are presented in Table 9.1.

In addition to the direct influence examined, turbulence can have a strong indirect effect. Actually, medium turbulization, as is generally known, leads to an increase in heat conduction process efficiency. Effective turbulent thermal diffusivity coefficient noticeably exceeds conventional thermal diffusivity coefficient in a stationary medium. As a result, the averaged temperature and density profile

Table 9.1 Estimated $\overline{U}_*$ velocity values, in m/s

Gas	Amplification regime	Lasing regime
He	60	30
Ar	20	10

equalization rate increases. When conditions are present that favor the origination of a passive region in a gas within which lasing is absent, this phenomenon can lead to a perceptible increase in the subject region's dimensions as compared to the laminar flow mode.

An approximate estimate of turbulent thermal diffusivity coefficient can be derived from the following considerations. Temperature equalization time in the scales characterized by linear dimensions on the order of the maximum turbulent pulse scale is determined by correlation Eq. (9.18). On the other hand, assuming that the heat exchange process within macroinhomogeneities between individual smaller inhomogeneities has the same mechanism as during conventional molecular heat exchange in a stationary medium (i.e., individual migrating inhomogeneities play the role of molecules during the turbulent mixing process), the temperature equalization time can be estimated from the correlation

$$\tau_{Tm} \sim \Lambda^2 / a_T, \tag{9.30}$$

where a_T is the effective turbulent thermal diffusivity coefficient. Comparing Eqs. (9.18) and (9.30), we obtain

$$a_T \sim \Lambda \overline{U}. \tag{9.31}$$

This expression is commensurate with the well-known correlation $a_T \sim \Lambda \times \Delta u$ [4], where Δu is the typical variation of large-scale pulse velocity at distances on the order of Λ (under specific conditions, it can be comparable to $\overline{U}$).

The average time required for a portion of the gas to traverse a section of a laser channel with a length of x comes to $t \sim x/\overline{U}$. Assuming that the dependence of passive zone development upon gas residence time in the channel comes in accordance with dependence (8.3), then for the size of this zone within the channel part involved in the turbulence at a medium thermal diffusivity coefficient, a_T, that conforms to Eq. (9.31), we get $l \sim \sqrt{\Lambda x}$. Thus, the passive zone will involve almost the entire width of the channel, d, after the gas traverses a turbulent section with a length of $x \sim d^2/\Lambda$, which, assuming that $\Lambda \sim d$, yields $x \sim d << b_L$.

Of course, these are only approximate estimates; however, they illustrate some of the difficulties that would possibly be faced during the development of NPLs with longitudinal gas flowing and that it would require considerable effort to overcome. This situation was one of the reasons for seeking alternative solutions.

9.3 Transverse Gas Flowing

At the VNIIEF, a concept was proposed for a system with transverse gas flowing (relative to the optical axis) [22–24]. This system consists of a set of laser channels with a rectangular cross-section that have planar uranium layers with a width of ~10 cm that are deposited to the inner surface of the channel side walls parallel to the gas flow direction. In order to cool the mixture heated in a channel, a radiator is placed at its outlet. The most convenient and simplest type of radiator consists of a set of thin plates that are positioned parallel to the gas flow and perpendicular to the uranium layers. An identical radiator is placed at the channel inlet which, by virtue of gas heat exchange with the plates, makes it possible to receive the requisite inlet temperature of the working mixture along the entire length of the laser and to equalize its velocity. The laser channel cross-section is shown in Fig. 9.1. The next laser channel can be installed behind the outlet radiator, for which this radiator will also be the inlet radiator, etc. Thus, a chain of laser channels can be built that are merged into a common gas circuit.

Radiators

The shapes, dimensions, and structures of radiators may differ depending upon the operating mode, power density, and location of the laser system relative to the irradiation source. During operation in the pulse and quasi-pulse ($\tau_n \sim 0.1$–10 s) modes, a radiator may take the form of a simple porous structure that functions based on the principle of heat absorption by virtue of intrinsic heat capacity. It is obvious that in the stationary mode, an external heat-transfer agent should cool a radiator. The heat must be removed from this radiator by means of the heat-transfer agent washing over its end surfaces, which constitute a continuation of the outer surfaces of the uranium layer substrates. A radiator should consist of a set of plates

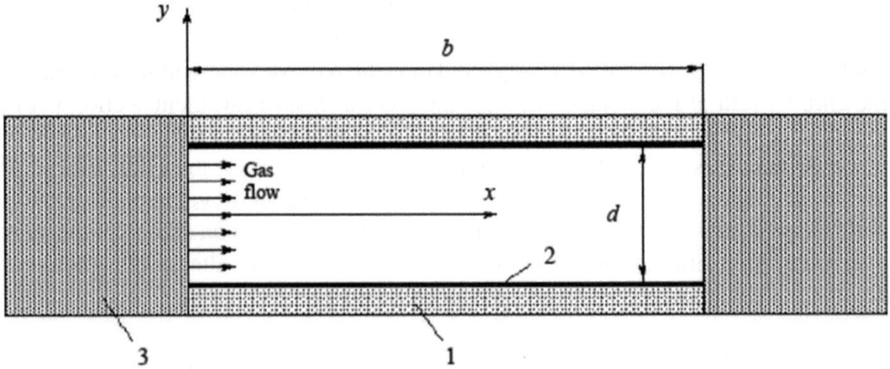

Fig. 9.1 Laser channel cross-section: (*1*) metal substrate; (*2*) uranium layer; (*3*) radiator

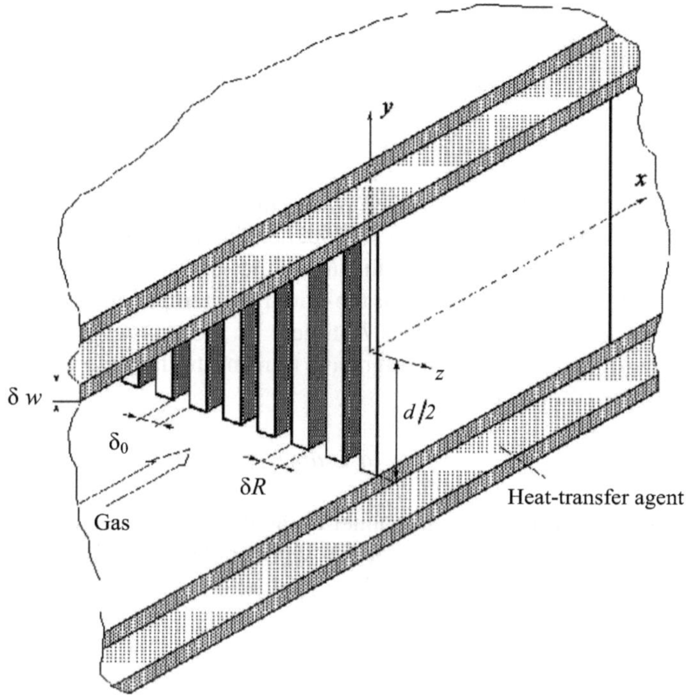

Fig. 9.2 Plate-type radiator directly adjacent to a laser channel and cooled by a liquid heat-transfer agent

(Fig. 9.2) or tubes with microchannels for flowing the cooling heat-transfer agent. The choice of potential engineering solutions is quite broad and will depend upon the conditions under which a specific laser system must operate.

It is not difficult to estimate the efficiency of the radiator under consideration using the results obtained from solving the problem of gas heat exchange with the channel walls. If the wall temperature of a channel as a portion of the gas passes through it can be considered constant, and if the flow itself is laminar, the dependence of the average gas temperature over the channel's transverse cross-section upon the distance traversed, x, then takes the form [25]:

$$T(x) = T_R + (T_0 - T_R)\sum_{i=1}^{\infty} A_i \exp\left\{-\beta_i \frac{ax}{\overline{u}d_e^2}\right\}, \tag{9.32}$$

where a is the thermal diffusivity coefficient of the gas; d_e is the equivalent diameter of the channel; $\overline{u}$ is the average gas velocity over the channel cross-section; T_0 is the gas temperature at the channel inlet; and T_R is the channel wall temperature. The A_i and β_i values are presented in Table 9.2.

When the pressure of a helium-based gas mixture is $P = 2$ atm, $a \approx 1$ cm^2/s. Let us assume that $\overline{u} = 10$ m/s, while the distance between the plates is

Table 9.2 A_i and β_i coefficient values (9.32)

i	1	2	3
A_i	0.819	0.0976	0.0189
β_i	14.62	88.2	212

$\delta_0 = \delta_R = 0.5$ mm, which corresponds to an equivalent diameter of $d_e = 1$ mm (in the radiator channel, $Re = \overline{u}_0 d_e/\nu \sim 150$, i.e., the flow is laminar). An estimate based on formula (9.32) demonstrates that in the presence of a plate length as low as $x = b_R = 3$ cm, the difference between the initial gas temperature and the temperature of the radiator at its outlet drops 70 times.

The amount of heat that fission fragments transmit to the gas in a laser channel over a time frame of t comes to

$$G = \overline{q}_c L_c d b_L t,$$

where L_c is the length of the laser channel in the direction of the optical axis. The amount of heat absorbed by the radiator that is not cooled by the heat-transfer agent during heating from a temperature of T_R to T_0 is

$$G = \omega_R L_c d b_R \rho_R c_R (T_0 - T_R)$$

(we assume that the height of the radiative cooler's plates equals the height of the laser channel, d, in the direction of the y axis). Here, c_R is the specific heat capacity of the radiator's plate material; ρ_R is this material's density; and $\omega_R = \delta_R/(\delta_0 + \delta_R)$ is the radiator's intermittency factor. It follows from these two equations that the radiator's effective operating time is

$$t \sim \frac{\omega_R b_R \rho_R c_R (T_0 - T_R)}{\overline{q}_c b_L}.$$

For radiator made of aluminum plates at $\overline{q}_c = 10$ W/cm^3, $b_R = 3$ cm; $b_L = 6$ cm, and $(T_0 - T_R) = 100$°K, operating time is $t \sim 6$ s. These parameter values proved to be entirely acceptable for the radiators of the LM-4 laser module. The neutron excitation time during its operation in conjunction with a BIGR reactor was ~1.5 s [26].

A plate-type radiator is one alternative for the porous system used to intensify heat exchange processes [27–31]. The equations that describe temperature distribution in the components of a porous system in the stationary mode take the form [27–30]:

$$G c_P \frac{\partial \tau}{\partial x} = h_v (T - \tau), \tag{9.33}$$

$$k\left\{\frac{\partial^2 T}{\partial x^2}+\frac{\partial^2 T}{\partial y^2}\right\}=h_v(T-\tau), \tag{9.34}$$

where G is the specific mass flow of the component being flowed; c_P is this component's specific heat capacity; τ is the gas temperature; h_v is the volumetric inter-pore heat exchange intensity; T is the temperature of a solid porous component; and k is the heat conductivity coefficient of the solid component.

We will demonstrate that the temperature distribution in a plate-type radiator is described by equations similar to Eqs. (9.33) and (9.34).

Gas Cooling in a Radiator. We will select the x axis direction so that it coincides with the gas flow direction (see Figs. 9.1 and 9.2). Under actual conditions, the average gas temperature over the transverse cross-section in the direction of the z axis can change perceptibly at distances of not less than a few dozen radiator plate thickness values; therefore, the dependence of the parameters examined below upon the z coordinate is negligible. We will isolate a gas mass element, m, between the plates:

$$m = \rho(x,y,t)\Delta l(x,y,t)\Delta d(x,y,t)\delta_0 = \rho_0\Delta l_0\Delta d_0\delta_0 = \text{const}, \tag{9.35}$$

where $\rho(x, y, t)$ is the gas density at a point with coordinates of x and y at a moment in time of t; $\Delta l(x, y, t)$ is the length of the gas mass element; $\Delta d(x, y, t)$ is its height; δ_0 is the distance between the radiator's plates; Δl_0 is the length of the element at the radiator inlet; and Δd_0 is its height at the inlet.

For most gases, including rare gases, the enthalpy of a unit mass of gas is determined as

$$w = c_{P1}\tau,$$

where c_{P1} is the heat capacity at a constant pressure; therefore, at a point with coordinates of x and y for a moment in time of t, the enthalpy of the element under consideration is

$$w(x,y,t) = c_{P1}\Delta l(x,y,t)\Delta d(x,y,t)\delta_0\rho(x,y,t)\tau(x,y,t).$$

Accordingly, the increase therein over a time frame of dt is

$$dw = c_{P_1}\Delta l(x,\ y,\ t)\,\Delta d\,(x,\ y,\ t)\,\delta_0\,\rho\,(x,\ y,\ t)\ \frac{d\tau}{dt}\,dt. \tag{9.36}$$

The temperature of the gas mass element under consideration can change as a result of heat transmission to the preceding or subsequent elements, or as a result of heat exchange with the radiator's plates. The effective distance over which heat propagates directly through a gas during a time interval of t is estimated by the correlation

$$l_g \sim \sqrt{a_g t},$$

where l_g is the effective heat propagation distance and a_g is the thermal diffusivity coefficient of the gas. The maximum time of gas residence in a radiator is

$$t_m \sim L/u,$$

where L is the radiator dimension in the x direction and u is gas velocity. Over this time frame, the heat from the element under consideration propagates through the gas a distance of

$$l_g \sim \sqrt{a_g L/u}. \tag{9.37}$$

At optimum pressures ($P \sim 2$ atm for helium and $P \sim 0.5$–0.9 atm for argon), $a_g \leq 1$ cm^2/s, L is on the order of a few centimeters [22], and, $u \geq 10\text{s}^{-1}$. Placing these values into (9.37), we obtain $l_g < 10^{-2}$ cm, which is appreciably smaller than the longitudinal dimension of the gas in the radiator; i.e., longitudinal heat transfer in the gas should not play a noticeable role. However, the heat exchange of the mass element under consideration with the radiator's plates is convective in nature, which, as is generally known, is considerably more efficient than conventional heat conduction.

The pressure drop, ΔP, between the inlet and outlet of a channel formed by two parallel planes can be estimated using the formula [4]

$$\Delta P = 12\eta \bar{u} b_R / \delta_0^2, \tag{9.38}$$

(η is the dynamic viscosity coefficient), according to which $\Delta P \sim 10^{-3}$ atm for the helium cooling example considered above, which is much smaller than its absolute value of $P = 2$ atm. Therefore, the enthalpy decrease ($-dw$) should equal the amount of heat that the mass element under consideration transmits to two adjacent radiator plates. The amount of this heat is determined by the equation

$$dQ = \Delta l(x,\ y,\ t)\Delta d(x,\ y,\ t) q(x,\ y,\ t) dt, \tag{9.39}$$

where q is the bilateral heat flux density at the gas contact boundary with the surfaces of the two radiator plates at a point with coordinates of x and y. During convective heat exchange,

$$q(x,y,t) = \alpha\{\tau(x,y,t) - T_S(x,y,t)\}, \tag{9.40}$$

where α is the heat transfer coefficient in a planar slit and T_S is the radiator's plate surface temperature.

Using a minus sign (as the gas enthalpy decrease) to set Eq. (9.36) equal to expression (9.39) and substituting the result in Eq. (9.40), we obtain

$$\frac{d\tau}{dt} = -\frac{\alpha}{c_{P1}\rho\delta_0}[\tau - T_S]. \tag{9.41}$$

Viewing the gas flow between the plates as the aggregate of the motion of jets with a height of Δd and a width of δ_0, then from the condition for mass flow maintenance, we write

$$u(x, y, t)\rho(x, y, t)\Delta d(x, y, t)\delta_0 = u_0\rho_0\Delta d_0\delta_0,$$

where u is the gas velocity component in the direction of the x axis and u_0 is the gas velocity at the radiator inlet. With allowance for this correlation, Eq. (9.41) takes the form:

$$\frac{1}{u(x,\ y,\ t)} \times \frac{d\tau}{dt} = -\frac{\alpha}{c_{P_1}u_0\rho_0\delta_0} \times \frac{\Delta d}{\Delta d_0}[\tau(x,\ y,\ t) - T_S(x,\ y,\ t)]. \tag{9.42}$$

We present the total derivative of $d\tau/dt$ in the form

$$\frac{d\tau}{dt} = \frac{\partial\tau}{\partial t} + u\frac{\partial\tau}{\partial x} + \nu\frac{\partial\tau}{\partial y}, \tag{9.43}$$

where v is the gas velocity component in the direction of the y axis. An operating mode of practical interest is the one during which the following inequation is satisfied

$$u\frac{\partial\tau}{\partial x} \gg \frac{\partial\tau}{\partial t}. \tag{9.44}$$

Expression (9.44) means that the rate of change in the gas temperature as its location changes noticeably exceeds the rate of change in its temperature at any fixed point of a radiator with the passage of time. Furthermore, due to displacement restraint in the direction of the y axis by the radiator's transverse dimension (see Fig. 9.2), gas jet shifting in this direction is small, so that

$$\left|\frac{\Delta d - \Delta d_0}{\Delta d_0}\right| \ll 1; \ \ v \ll u. \tag{9.45}$$

Taking this into account, Eq. (9.42) is written as

$$\frac{\partial\tau}{\partial x} = -\alpha_0\{\tau(x, y, t) - T_S(x, y, t)\}, \tag{9.46}$$

where α_0 is the thermal interaction coefficient of the gas with the radiator:

$$\alpha_0 = \frac{\alpha}{c_{P_1} \rho_0 u_0 \delta_0}. \tag{9.47}$$

Temperature Field in a Radiator Plate. The heat conduction equation for a plate takes the form:

$$a\left\{\frac{\partial^2 T}{\partial x^2} + \frac{\partial^2 T}{\partial y^2} + \frac{\partial^2 T}{\partial z^2}\right\} = \frac{\partial T}{\partial t}, \tag{9.48}$$

where a is the plate's thermal diffusivity coefficient, which can be regarded as almost constant for the temperature variation range in which we are interested.

We will introduce the average temperature over the transverse cross-section of a planar plate

$$\overline{T}(x, y, t) = \frac{2}{\delta_R} \int\limits_0^{\delta_R/2} T(x, y, z, t)\, dz, \tag{9.49}$$

where δ_R is plate thickness. Integrating Eq. (9.48) for dz from zero to $\delta_R/2$, we obtain

$$a\left\{\frac{\partial^2 \overline{T}}{\partial x^2} + \frac{\partial^2 \overline{T}}{\partial y^2} + \frac{2}{\delta_R} \frac{\partial}{\partial z} T\left(x,\ y,\ \frac{\delta_R}{2},\ t\right)\right\} = \frac{\partial \overline{T}}{\partial t}. \tag{9.50}$$

From the condition of heat fluxes equality at the plate and gas contact boundary, we get

$$k_R \frac{\partial T(x, y, \delta_R/2, t)}{\partial z} = \alpha_1 [\tau\,(x, y, t) - T(x, y, \delta_R/2, t)]. \tag{9.51}$$

Here, k_R is the plate's heat conductivity coefficient and α_1 is the gas's heat transfer coefficient during heat exchange with a single planar surface. The latter is linked to the α coefficient by the correlation

$$\alpha_1 = \alpha/2. \tag{9.52}$$

The plate's maximum temperature differential in the transverse direction can be estimated as

$$\Delta T \sim \frac{\partial T(x, y, d/2, t)}{\partial z} \cdot \frac{\delta_R}{2}.$$

If this differential is much smaller than the difference in the gas temperature and the plate surface temperature

$$\frac{\partial T(x,y,d/2,t)}{\partial z}\cdot\frac{\delta_R}{2} << \tau(x,y,t) - T(x,y,d/2,t), \tag{9.53}$$

the $T(x, y, d/2, t)$ temperature value on the right side of (9.51) can then be replaced with $\overline{T}$ (x, y, t). In this instance, placing Eqs. (9.51) into (9.50), we obtain

$$\left\{\frac{\partial^2\overline{T}(x,y,t)}{\partial x^2}+\frac{\partial^2\overline{T}(x,y,t)}{\partial y^2}\right\}+H\left[\tau(x,y,t)-\overline{T}(x,y,t)\right]=\frac{1}{a}\cdot\frac{\partial\overline{T}(x,y,t)}{\partial t}. \tag{9.54}$$

Here, taking Eq. (9.52) into account

$$H=\frac{\alpha}{k_R\delta_R}. \tag{9.55}$$

It follows from a comparison of Eqs. (9.51) and (9.53) that Eq. (9.54) holds true given the condition that

$$\delta_R \ll \frac{k_R}{\alpha}. \tag{9.56}$$

Accordingly, Eq. (9.46) can also be written in the form

$$\frac{\partial\tau(x,y,t)}{\partial x}=-\alpha_0\left\{\tau(x,y,t)-\overline{T}(x,y,t)\right\}. \tag{9.57}$$

For the stationary case of $\partial\overline{T}/\partial t=0$, Eqs. (9.54) and (9.57) in their form fully agree with Eqs. (9.33) and (9.34), which describe heat exchange in a porous system (for simplicity's sake, we will hereinafter omit the bar over the T, which denotes averaging Eq. (9.49)).

Boundary Conditions. Directly at the inlet end, as is also the case in a porous system [27], heat exchange between the radiator and the oncoming gas flow is virtually absent due to the latter's negligible heat conduction; consequently,

$$\frac{\partial T(0,y)}{\partial x}=0. \tag{9.58}$$

The gas temperature at the inlet ($x=0$) is a function of the y coordinate

$$\tau(0,y) = f(y). \tag{9.59}$$

During gas propagation along a radiator that is cooled by a liquid heat-transfer agent, its temperature can only decrease:

$$\tau(x > 0, y) \leq f_{\max}, \tag{9.60}$$

where $f_{\max}$ is the maximum gas temperature value at the radiator inlet. It is also obvious that the temperature of the radiator plates cannot exceed $f_{\max}$:

$$T(x > 0, y) \leq f_{\max}. \tag{9.61}$$

At $y = 0$, symmetry condition must be satisfied

$$\frac{\partial T(x, 0)}{\partial y} = 0. \tag{9.62}$$

Convective heat exchange with the liquid heat-transfer agent takes place at the $y = d/2 + \delta_w$ boundary:

$$T(d/2 + \delta_w) - T_l = q_l / \alpha_l, \tag{9.63}$$

where δ_w is the thickness of the radiator's substrate; T_l is the temperature of the liquid heat-transfer agent; q_l is the heat flux density at the $y = d/2 + \delta_w$ boundary; and α_l is the heat transfer coefficient at the $y = d/2 + \delta_w$ boundary.

The temperature differential throughout the thickness of the radiator's substrate (the radiator's substrate and its plates may be made from different materials) equals

$$T(d/2) - T(d/2 + \delta_w) = \frac{q_l}{k_w}\delta_w, \tag{9.64}$$

where k_w is the heat conductivity coefficient of the radiator's substrate.

The heat flux at the radiator's plate and substrate contact boundary is

$$q_l = -k_R \frac{\partial T(d/2)}{\partial y}. \tag{9.65}$$

The boundary condition at the radiator's plate contact boundary in the presence of $y = d/2$ follows from Eqs. (9.63)–(9.65)

$$\frac{\partial T(x, d/2)}{\partial y} = h[T_l - T(x, d/2)], \tag{9.66}$$

where

$$h = \frac{1}{k_R \left[\frac{1}{\alpha_l} + \frac{\delta_w}{k_w}\right]}. \tag{9.67}$$

Approximation of an Ideal Radiator. The simplest and apparently the most convenient for elementary estimates and purely qualitative arguments is an "ideal" radiator approximation: gas cooling is examined in a situation when the plate temperature variations of the radiator itself are sufficiently small that their effect can be ignored in Eq. (9.57); i.e., the temperature, T, can be regarded as constant and homogeneous. In the stationary case, when the heat transfer coefficient, α, is not dependent upon time, from Eq. (9.57), we get

$$\tau(x) = T + (\tau_0 - T)\exp\left\{-\int_0^x \alpha_0(x)dx\right\}. \tag{9.68}$$

The gas-flow velocity in the laser channel is $U \sim 10$ m/s and the width of the gap between the radiator's plates, like the thickness of the plates themselves, is $\delta_0 \sim \delta_R \sim 1$ mm. Consequently, the gas velocity in the radiator is $u \sim 20$ m/s. The Reynolds number for a flow of this type is

$$Re = \frac{ud_e}{\nu} \sim 700,$$

where $d_e = 2\delta_0$ is the equivalent diameter of the gap between the plates and ν is the kinematic viscosity of the gas ($\nu \sim 0.6$ cm^2/s). This is appreciably lower than the critical value of $Re_c \approx 2{,}300$, above which the flow is not laminar, but turbulent.

The heat transfer coefficient α is determined by way of the Nusselt number:

$$\alpha = k_g \frac{\text{Nu}}{d_e}, \tag{9.69}$$

where k_g is the heat conductivity coefficient of the gas. In the presence of a laminar flow, the Nu value in the initial part of the flow decreases with distance from the inlet into the slit between the plates to a constant value in the stabilized section [25, 32]. For a planar slit formed by two parallel planes, the distance within which the stabilized heat transfer part begins is determined by the inequality [25]

$$x > ud_e^2/70a_g. \tag{9.70}$$

Besides, if the channel wall temperature is constant, then Nu = 7.5 within the stabilized part. A power law dependence is recommended for the initial part [25]

$$\text{Nu} = 1.85\left(ud_e^2/a_g x\right)^{1/3}. \tag{9.71}$$

A calculation of the $\vartheta = \frac{(\tau - T)}{(\tau_0 - T)}$ parameter (Curve 1) using the formula (9.86) with the recommended Nu(x) dependence [25] is presented in Fig. 9.3. There, too, a similar calculation (Curve 2) using formula (9.32) is offered. The calculations were performed for helium at $P = 2$ atm and $\tau = 400$ K; the gas velocity was $u_0 = 10$ m/s,

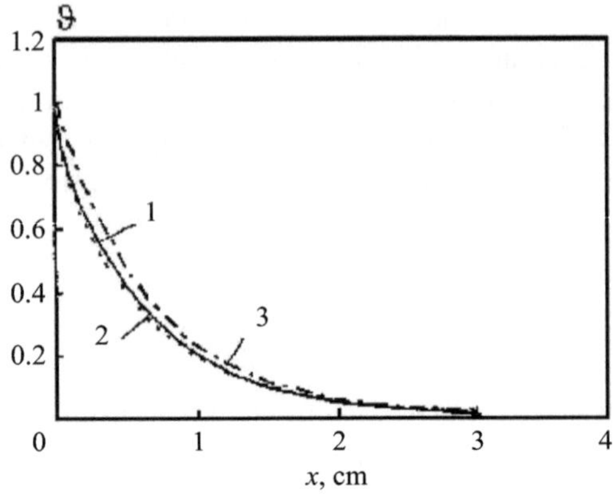

Fig. 9.3 Dependence of the reduced gas temperature upon the distance traversed within a radiator: (*1*) a calculation using formula (9.68) in the presence of a variable Nu value; (*2*) a calculation using formula (9.32); (*3*) a calculation using formula (9.68) at Nu = 7.5

while the distance between the plates and their thickness came to $\delta_0 = \delta_R = 0.5$ mm, which corresponds to an equivalent diameter of $d_e = 1$ mm. For comparison, a similar dependence (Curve 3), obtained using formula (9.68) under the assumption that the Nu number is constant and equals 7.5 over the entire expanse of the radiator, is shown in Fig. 9.3. According to Eq. (9.70), the expanse of the stabilization part is ~2 mm. It is obvious that gas cooling efficiency can be calculated with quite good accuracy using a constant Nusselt number to determine the heat transfer coefficient α.

Real Radiator. We will now examine the stationary problem $\partial\tau/\partial t = \partial T/\partial t = 0$. It is not difficult to reduce equation system (9.54) and (9.57), with boundary conditions (9.58)–(9.62) and (9.66), to a single linear equation in third-order partial derivatives

$$\frac{\partial^3 r}{\partial x^3} + \frac{\partial^3 r}{\partial x^2 \partial y} + \alpha_0 \frac{\partial^2 r}{\partial x^2} + \alpha_0 \frac{\partial^2 r}{\partial y^2} - H\frac{\partial r}{\partial x} = 0 \tag{9.72}$$

and boundary conditions

$$r(0, y) = f(y) - T_l; \tag{9.73}$$

$$r(x > 0, y) \le f_{\max} - T_l; \tag{9.74}$$

$$\frac{\partial r(0, y)}{\partial x} = 0; \tag{9.75}$$

$$\frac{\partial r(x, 0)}{\partial y} = 0; \tag{9.76}$$

$$\frac{\partial r(x, d/2)}{\partial y} + hr(x, d/2) = 0; \tag{9.77}$$

where

$$r(x,y) = \tau(x,y) - T_l. \tag{9.78}$$

The solution to problem (9.72)–(9.78) takes the form:

$$r(x,y) = \sum_{n=1}^{\infty} \frac{C_n}{s_{2n} - s_{1n}} \{s_{2n}\exp(s_{1n}x) - s_{1n}\exp(s_{2n}x)\} \cos\left(\frac{2\beta_n}{d}y\right). \tag{9.79}$$

Here, the range of problem eigenvalues is determined by the transcendental equation

$$tg\,\beta_n = \frac{hd}{2\beta_n}; \quad n = 1, 2, 3, \ldots$$

while the s_{1n} and s_{1n} are the negative roots of the characteristic equation

$$s^3 + \alpha_0 s^2 - (\lambda_n + H)s - \alpha_0 \lambda_n = 0, \tag{9.80}$$

where

$$\lambda_n = \frac{4\beta_n^2}{d^2}.$$

(Generally speaking, Eq. (9.80) has three different real roots. For the actual thermophysical parameter values of the radiator system under consideration, two of them are always negative, while one is positive and increases without limit as the n parameter value increases).

The C_n coefficients take the form:

$$C_n = \frac{\int\limits_0^{d/2} f(y) \cos\left(\frac{2\beta_n}{d}y\right) dy - \frac{dT_l}{2\beta_n} \sin\beta_n}{\|\Phi_n\|^2}.$$

Here, $\|\Phi_n\|^2 = \frac{d}{4} + \frac{d}{8\beta_n} \sin 2\beta_n$ is the square of the norm.

Thus, according to Eq. (9.78), gas temperature distribution takes the form

$$\tau(x,y) = T_l + \sum_{n=1}^{\infty} \frac{C_n}{s_{2n} - s_{1n}} \{s_{2n}\exp(s_{1n}x) - s_{1n}\exp(s_{2n}x)\} \cos\left(\frac{2\beta_n}{d}y\right), \tag{9.81}$$

consequently, the temperature distribution in the radiator's plates is determined by the equation

$$T(x,y) = \tau(x,y) + \frac{1}{\alpha_0}\sum_{n=1}^{\infty} \frac{C_n s_{1n} s_{2n}}{s_{2n} - s_{1n}} \{\exp(s_{1n}x) - \exp(s_{2n}x)\} \cos\left(\frac{2\beta_n}{d} y\right). \quad (9.82)$$

The series in Eqs. (9.81) and (9.82) quickly converge. So, the results of a calculation using ten terms of sum in these equations differ by at least 0.4 % from the results obtained with allowance for 20 terms.

By way of illustration, the first of the ten eigenvalues of $\beta_n = d\sqrt{\lambda_n}/2$ and the s_{in} roots of Eq. (9.80) that correspond to them are presented in Table 9.3. The values cited correspond to helium at $\rho_0 = 4.6 \times 10^{-4}$ g/cm^3 and $P = 2.8$ atm, pumped at a rate of $u_0 = 9$ m/s through an aluminum radiator with plate dimensions of $d \times \delta_R = 2 \times 0.05$ cm. The thickness of the gap between the plates is $\delta_0 = 0.05$ cm; the radiator's substrate (also aluminum) has a thickness of $\delta_w = 0.16$ cm. The heat transfer coefficient of the liquid heat-transfer agent is $\alpha_l = 1.7$ W/cm$^2 \times K$.

The calculations results for the aluminum plates of a radiator with a thickness of $\delta_R = 0.5$ mm are presented in Figs. 9.4, 9.5, 9.6, 9.7, 9.8, 9.9, and 9.10. The width of the slits between the plates is $\delta_0 = 0.5$ mm. It was assumed that the laser channel within which gas mixture excitation by fission fragments occurs has transverse dimensions of $d \times b = 2 \times 6$ cm, that the thickness of the uranium layers is $\delta_U = 2.78 \times 10^{-4}$ cm, that the intensity of uranium nuclear fission is $q_U = 2 \times 10^{16}$ cm$^{-2} \times$ s^{-1}, and that the gas temperature at the laser channel inlet is $T_{g0} = 293$ K. During the calculations, it was assumed that the heat-transfer agent's heat transfer coefficient is $\alpha_l = 1.6$ W/cm^2 K, which corresponds to water at $T_i = 293$ K, pumped at a rate of 5 m/s through an external cooling channel (see Fig. 9.2), the transverse dimension of which comes to 3 cm. The thickness of the radiator's aluminum substrate was taken to equal $\delta_W = 1.6$ mm.

Figures 9.4 and 9.5 reflect the calculation results for the argon temperature distribution within a radiator in the transverse and longitudinal directions following gas transition from a laser channel within which the gas pressure amounted to 0.9 atm, while gas-flow velocity at this channel's inlet was $U_0 = 4.5$ m/s. At the outlet from the laser channel, gas-flow velocity as a result of heating by the fission fragments equals 10.78 m/s, which, in the presence of the δ_R and δ_0 dimensions indicated above, corresponds to an initial gas velocity in the radiator of $u_0 = 21.56$ m/s. The gas temperature distribution at the radiator inlet (the channel outlet), $\tau(0, y) = f(y)$, corresponds to Curve 1 in Fig. 9.4.

The heat transfer coefficient, α, determined by Eq. (9.69), will be dependent upon the gas heat conductivity coefficient, k_g, which is a function of temperature:

$$k_g = k_0\left(\frac{\tau}{\tau_k}\right)^n.$$

For helium, $k_0 = 0.159$ W/m × K, $\tau_k = 320$ K, and $n \approx 0.69$, while for argon, $k_0 = 0.0177$ W/m × K, $\tau_k = 300$ K, and $n \approx 0.71$ [33–35]. Therefore, the heat transfer coefficient is reduced when the gas is cooled as it passes through the radiative cooler, leading to a slight decrease in the thermal interaction coefficient,

Table 9.3 Eigenvalues of β_n and the roots of Eq. (9.80) that correspond to them

n	β_n	s_{1n}	s_{2n}	s_{3n}
1	0.7677	−2.137	−0.336	1.115
2	3.3589	−3.685	−1.184	3.511
3	6.3986	−6.540	−1.310	6.492
4	9.5026	−9.591	−1.336	9.569
5	12.6250	−12.689	−1.346	12.677
6	15.7550	−15.805	−1.350	15.797
7	18.8888	−18.930	−1.353	18.925
8	22.0248	−22.060	−1.354	22.056
9	25.1622	−25.193	−1.355	25.190
10	28.3005	−28.327	−1.356	28.325

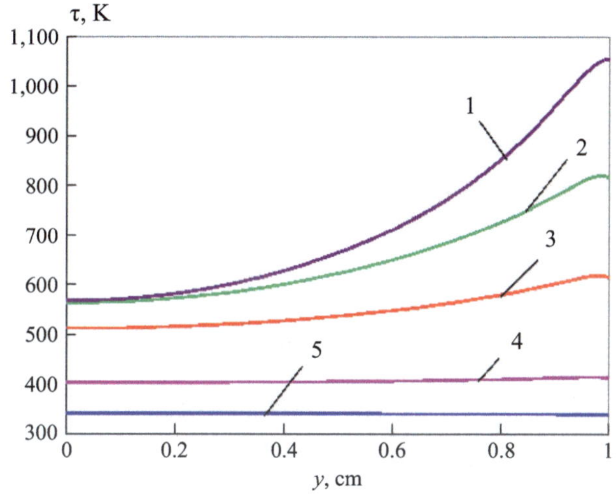

Fig. 9.4 Argon temperature distribution in the transverse direction at $u_0 = 21.56$ m/s ($U_0 = 4.5$ m/s): (*1*) $x = 0$; (*2*) $x = 1$ cm; (*3*) $x = 2$ cm; (*4*) $x = 4$ cm; (*5*) $x = 6$ cm

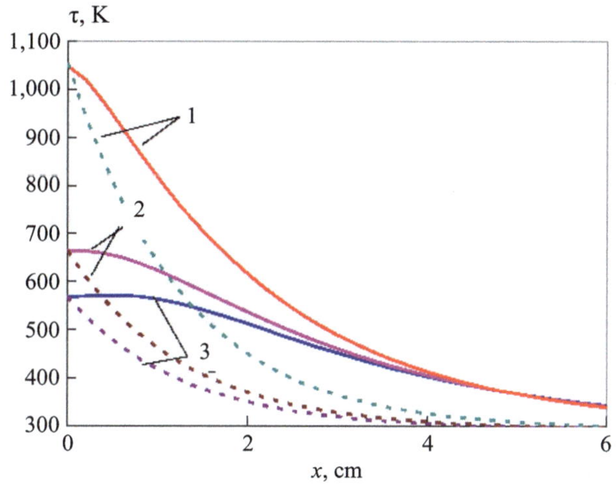

Fig. 9.5 Argon temperature distribution in the longitudinal direction at $u_0 = 21.56$ m/s ($U_0 = 4.5$ m/s): (*1*) $y = 1$ cm; (*2*) $y = 0.5$ cm; (3) $y = 0$; *solid lines* are a real radiator; *broken lines* are an ideal radiator

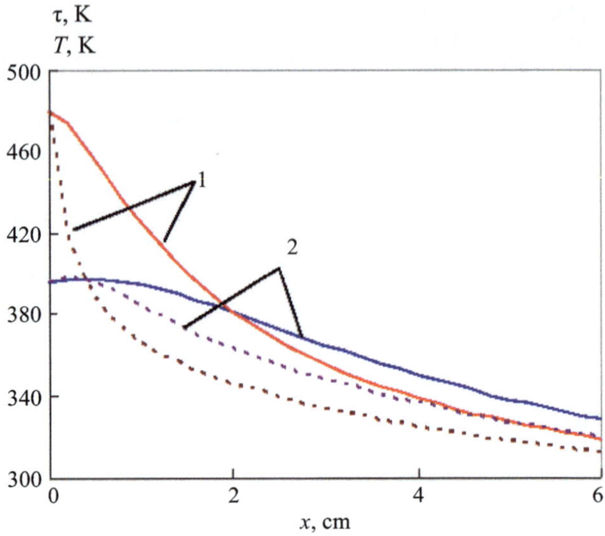

Fig. 9.6 Helium and radiator plate temperature distributions in the longitudinal direction at $u_0 = 12.88$ m/s ($U_0 = 4.5$ m/s): (*1*) y = 1 cm; (2) y = 0; *solid lines* are τ; *broken lines* are T

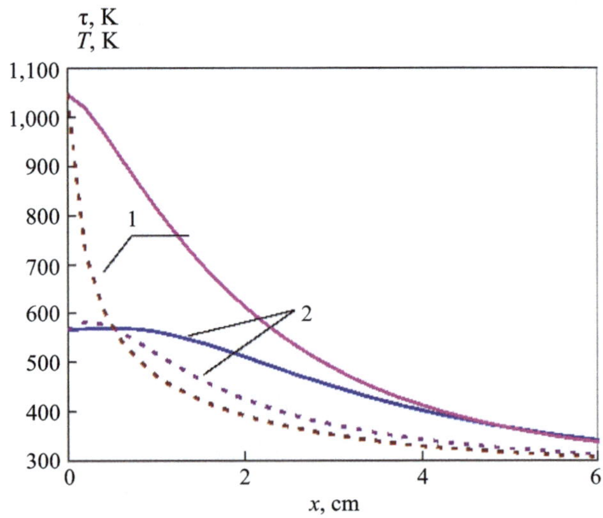

Fig. 9.7 Argon and radiator plate temperature distributions in the longitudinal direction at $u_0 = 21.56$ m/s ($U_0 = 4.5$ m/s): (1) y = 1 cm; (2) y = 0; *solid lines* are τ; *broken lines* are T

α_0, and the H parameter. During the performance of the calculations described above, a heat conductivity coefficient value was used that corresponded to the gas temperature at the radiator inlet, averaged over the transverse cross-section.

Comparative calculations were performed in order to check the extent of the influence of the thermal conductivity coefficient decrease, during which the heat conductivity coefficient was taken for the average gas temperature value (over the transverse cross-section) in a cross-section of $x = 3$ cm. The largest deviations occur in the case of argon; however, even they do not exceed $\beta \approx 30$ % when the relative deviations are calculated using the formula

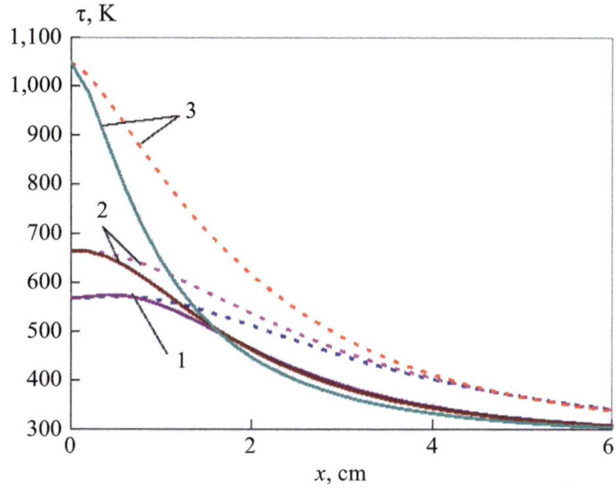

Fig. 9.8 Argon temperature distribution in the longitudinal direction at $U_0 = 4/5$ m/s: (*1*) $y = 0$; (*2*) $y = 0.5$ cm; (*3*) $y = 1$ cm; *solid lines* are $\delta_0 = 0.2$ mm; *broken lines* are $\delta_0 = 0.5$ mm

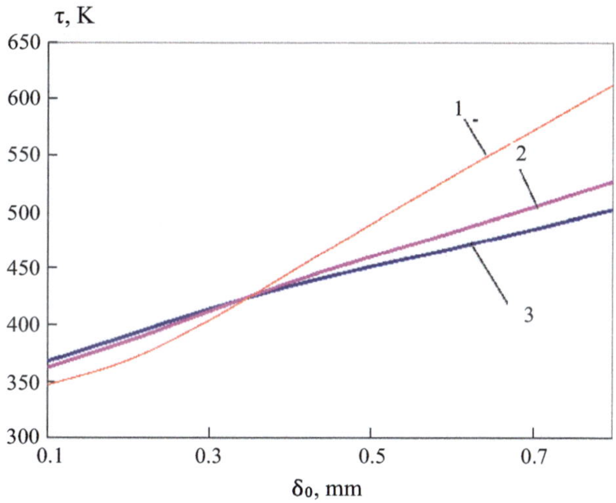

Fig. 9.9 Dependence of argon temperature ($U_0 = 4.5$ m/s) at $x = 3$ cm upon radiator slit width for $\Lambda = 1$ mm: (*1*) $y = 0$; (*2*) $y = 0.5$ cm; (*3*) $y = 1$ cm

$$\beta = \frac{\tau_2 - \tau_1}{\tau_1 - T_l},$$

where τ_1 is the temperature calculation at k_g, determined by means of the average temperature at the inlet, and τ_2 is the temperature calculation at k_g, determined by means of the average temperature in a cross-section of $x = 3$ cm.

The differences between the gas temperature and the radiator plate temperature for helium and argon are shown in Figs. 9.6 and 9.7. As is apparent from these graphs, the differences reach hundreds of degrees at distance of about 1–2 cm from the inlet, which testifies the invalidity of using the local thermal equilibrium model,

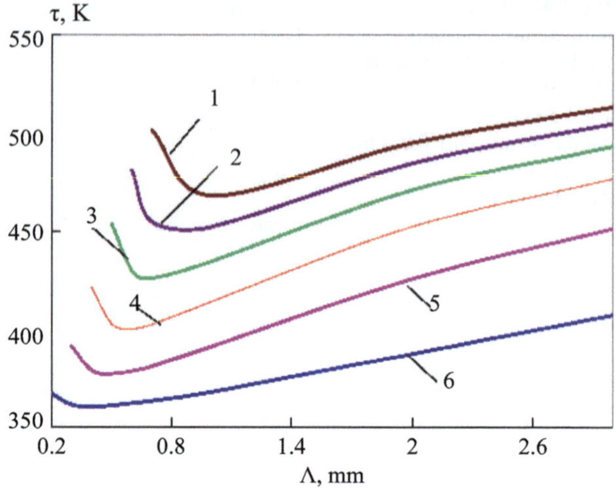

Fig. 9.10 Dependence of argon temperature ($U_0 = 4.5$ m/s) at $x = 3$ cm upon the radiator Λ period for $y = 0$ cm:
$\delta_0 = 0.6$ mm; (*1*)
$\delta_0 = 0.5$ mm; (*2*)
$\delta_0 = 0.4$ mm; (*3*)
$\delta_0 = 0.3$ mm; (*4*)
$\delta_0 = 0.2$ mm; (*5*)
$\delta_0 = 0.1$ mm; (*6*)

$\tau \approx T$, for the radiators under consideration, that is suggested in certain cases involving the calculation of heat exchange in porous systems [27].

The intensity of gas cooling will be substantively dependent upon the thickness of the slit gap between the radiator's plates. The calculation results for longitudinal argon temperature distribution at $\delta_0 = 0.2$ and 0.5 mm ($\delta_R = 0.8$ mm and 0.5 mm, respectively) are shown in Fig. 9.8 for comparison.

Presented in Fig. 9.9 is the dependence of the argon temperature at a distance of $x = 3$ cm from the radiator's inlet upon slit width, δ_0, for the plate lattice spacing, Λ, determined as

$$\Lambda = \delta_0 + \delta_R, \tag{9.83}$$

Figure 9.9 demonstrates that it is possible to select a radiator length and slit width such that the temperature at the outlet will be homogeneous.

The calculated dependence of argon temperature upon the radiator spacing, Λ, at a distance of $x = 3$ cm (for $y = 0$ cm) from the radiator inlet is shown in Fig. 9.10. The dependence presented suggests the possibility of performing radiator optimization on the lattice spacing and the width of the gap between the plates.

Yet another possibility for the enhancement of radiator efficiency consists of expanding its transverse dimension along the y-axis immediately behind the inlet into the radiator itself. Indeed, the gas mass flow in the stationary mode equals $G = \rho_0 u_0 \delta_0 d$, so it follows from Eqs. (9.47) and (9.69) that

$$\alpha_0 = \frac{k_g \mathrm{Nu}}{2 c_P G} \cdot \frac{d}{\delta_0}, \tag{9.84}$$

Here, we took into account that $d_e = 2\delta_0$ for a slit. By leaving the mass flow, G, and the slit width, δ_0, unchanged, but increasing the d dimension after the inlet, we achieve an increase in the thermal interaction coefficient, α_0.

Sequential Circuit of Laser Channels and Radiators. In prospective multichannel laser systems that consist of a gas path made up of a sequential circuit of alternating laser channels and radiators (on the order of 25–50 channels and radiators), the longitudinal dimension of the radiators must be limited. The calculations presented in the preceding subsection revealed that when radiators with a length of $L = 3$ cm (identical to those employed in the LM-4 quasi-pulse laser setup) are used in stationary multichannel laser systems, the gas temperature increase attained in the laser channel, $\tau(0, y) - T_l$, is reduced by a total of two to three times. In this regard, a question arises: What temperature will the gas have after traversing a sequential circuit composed of N laser channels and N radiators?

As the calculation results demonstrated (see above), at distances of $x \geq 3$ cm from the inlet to the radiators, gas temperature distribution in the transverse direction is close to homogeneous; thus, for the sake of simplifying the description, we will assume that the aforementioned gas temperature distribution at the outlet for radiators with a length of $L \geq 3$ cm is scarcely dependent at all on the y coordinate. It follows from the solution to Eqn. (9.81) and the formula for C_n coefficients that the gas temperature at the first radiator's outlet will be

$$\tau_{R1}(L) \approx T_l + \varphi(L) - \mu(L)T_l. \tag{9.85}$$

Here, the following designations are introduced:

$$\varphi(L) \approx \sum_{n=1}^{\infty} \frac{\int\limits_0^{d/2} f(\xi)\cos\left(\frac{2\beta_n}{d}\xi\right) d\xi}{\|\Phi_n\|^2 (s_{2n} - s_{1n})} \cos\left(\frac{2\beta_n}{d} y\right) \{s_{2n}\exp(s_{1n}L) - s_{1n}\exp(s_{2n}L)\};$$

$$\mu(L) \approx \frac{d}{2}\sum_{n=1}^{\infty} \frac{\sin\beta_n}{\beta_n \|\Phi_n\|^2 (s_{2n} - s_{1n})} \cos\left(\frac{2\beta_n}{d} y\right) \{s_{2n}\exp(s_{1n}L) - s_{1n}\exp(s_{2n}L)\}.$$

Let us suppose that the gas temperature at the first laser channel's inlet equals τ_0, while at its outlet (at the first radiator's inlet), $\tau_1(0, y) = f(y)$; i.e., the temperature increment after the laser channel has been traversed comes to

$$\Delta\tau(y) = f(y) - \tau_0.$$

If we assume that all the channels in a common laser-radiator circuit are identical and that they operate under identical conditions, fission fragment energy deposition in the gas within each channel will then be identical, and accordingly, the temperature increments, $\Delta\tau(y)$, will also be roughly identical.

Behind the first radiator, the gas temperature increment relative to the first channel's inlet temperature, τ_0, is

$$\Delta\tau_{R1} = \tau_1 - \tau_0 = T_l + \varphi - \mu T_l - \tau_0.$$

Because the gas temperature at the second channel's inlet equals $\tau_0 + \Delta\tau_{R1}$, while the temperature increment in each channel is identical, the temperature distribution at the second radiator's inlet takes the form

$$f_2(y) = f(y) + \Delta\tau_{R1},$$

then in accordance with the solution to Eqs. (9.81) and (9.85), at the second radiator's outlet, we get

$$\tau_{R_2} = T_l + \varphi - \mu T_l + \mu\Delta\tau_R = \left(1-\mu^2\right)T_l + (1-\mu)\varphi - \mu\tau_0,$$

i.e., the temperature increment at the second radiator's outlet relative to the initial temperature, τ_0, is

$$\Delta\tau_{R2} = (1-\mu^2)T_l + (1+\mu)\varphi - (1+\mu)\tau_0.$$

For the outlet from the Nth radiator, we obtain

$$\Delta\tau_{R_N} = \left(1-\mu^N\right)T_l + (\varphi - \tau_0)\sum_{m=0}^{N}\mu^m.$$

The sum on the right side of Eq. (9.86) is the sum of a decreasing geometric progression with a geometric ratio of μ. Therefore, taking Eq. (9.85) into account, we get

$$\Delta\tau_{R_N} = \frac{\left(1-\mu^N\right)}{1-\mu}\left(\tau_{R_1} - \tau_0\right) = \frac{\left(1-\mu^N\right)}{1-\mu}\Delta\tau_{R_1}. \tag{9.87}$$

The limiting value of $\Delta\tau_{R_N}$ (at $N \to \infty$) is

$$\Delta\tau_{R\infty} = \frac{1}{1-\mu}\Delta\tau_{R_1}. \tag{9.88}$$

In the case of argon for a radiator with a length of $L = 3$ cm at a plate thickness of $\delta_0 = 0.5$ mm and a gap thickness between plates of $\delta_R = 0.5$ mm, $\mu = 0.358$, while for helium, $\mu = 0.348$. The limiting values of the $\varepsilon_\infty = \Delta\tau_{R\infty}/\Delta\tau_{R1}$ ratios for argon and helium equal 1.558 and 1.534, respectively. The $\varepsilon_N = \Delta\tau_{RN}/\Delta\tau_{R1}$ ratios for these gases at $N = 4$ come to $\varepsilon_{4\text{Ar}} = 1.532$ and $\varepsilon_{4\text{He}} = 1.511$, which differ from the limiting value by a total of 1.5 %. Thus, starting with the fifth channel, all the lasers and radiators will operate under virtually identical conditions in the stationary mode.

It is useful to note that at $\delta_0 = 1$ mm and $\delta_R = 1$ mm ($L = 3$ cm), a similar calculation yields $\varepsilon_{\infty \text{Ar}} = 3.497$ and $\varepsilon_{\infty \text{He}} = 2.513$. Here, for $N = 8$, $\varepsilon_{8\text{Ar}} = 3.26$ and $\varepsilon_{8\text{He}} = 2.469$, which differ from the limiting value for argon by 6.8 % and for helium by 1.7 %. Technologically, it is quite a bit simpler to fabricate radiators with a plate thickness of $\delta_0 = 1$ mm and a gap between plates of $\delta_R = 1$ mm; therefore, in a multichannel apparatus with a large number (~50) laser channels sequentially connected in the gas path, it would perhaps be reasonable to give preference to radiators of this type.

Flow Behind a Radiator. The gas flow behind a radiator's outlet was studied on the LUNA-2P setup [23, 24], the gas circuit of which included two laser cells with planar uranium layers (see Fig. 2.13). Three identical heat exchangers (radiators) were located on both sides of the cells: the first was at the first cell's inlet; the second was between the cells; and the third was at the second cell's outlet. The gas circulation system ensured the successive passage of the gas through the radiators and the laser cells. Laser cell active volume was limited by two planar aluminum plates with dimensions of 100×6 cm, which were positioned parallel to one another at a distance of $d = 2$ cm. Thin ^{235}U oxide layers with a thickness 2.8 mg/cm^2 were deposited to the inner surfaces of the plates. The radiators were made from rectangular aluminum plates with a thickness of $\delta = 0.5$ mm and a length along the gas flow of 3 cm. The distance between the plates came to 0.5 mm, while the height of the gas flow in the radiators was 2 cm. A laser channel scheme is presented in Fig. 9.1.

The LUNA-2P setup was irradiated by a pulsed thermal neutral flux from a VIR-2 M reactor that had a half-height pulse duration of ~3 ms. The thermal neutron flux density at the pulse peak, averaged over the laser cell's active length (1 m), was 2.2×10^{15} cm$^{-2} \times$ s^{-1}. The conditions and design of the experiments were the same as in ref. [36], where optical inhomogeneities in NPLs without gas circulation were studied (see Sects. 7.4 and 8.1). Specific energy deposition at the optimum gas pressures ($P_{\text{He}} = 2$ atm and $P_{\text{Ar}} = 0.5$ atm) came to 0.067 J/cm^3.

Optical inhomogeneity measurements in a laser channel with gas circulation were performed using a Mach-Zehnder interferometer at a He–Ne laser wavelength of $\lambda = 0.63$ μm under conditions of a reactor radiation background and slight mechanical fluctuations in the interferometer's mirrors. In order to increase the measurement accuracy to 0.1 λ, a comparative channel was used in the interferometer's optical circuit. The fringe pattern was recorded using a standard SFR-2 M camera with a pulse feed system developed by the authors of refs. [23, 24], which expanded the recording time range to 0.1 s with a resolution no worse than 50 μs. Investigations were performed with helium, argon and air in the presence of a gas flow rate of up to 12 m/s in the laser channel, both at atmospheric pressure and at pressures selected in such a manner that the relative width of the gas gap between the uranium layers, $D = d/R$ (where R is the range of an average uranium fission fragment in a gas of a given type) was close to the optimum value, $D \approx 0.45$ [37] (see Sect. 7.5).

A typical interferogram of the optical inhomogeneities in a laser channel at the time of neutron pulse termination, as well as the results of its processing, are

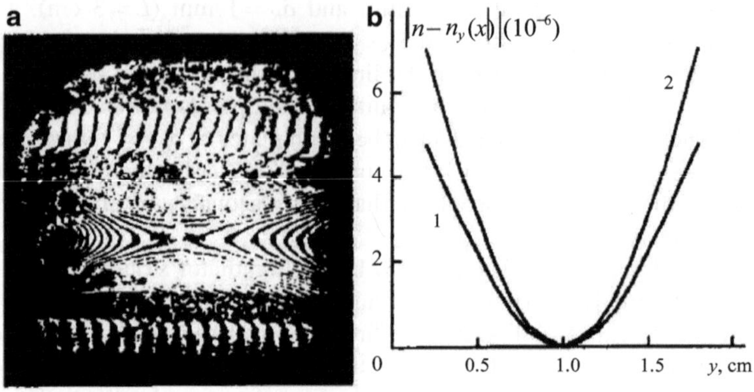

Fig. 9.11 Distribution of an absolute refractive index deviation in the direction of the y-axis from its $n_y(x)$ values in the $y=0$ plane for Ar at $P=0.5$ atm and $\overline{U}=8$ m/s: (*1*) at a distance of $x=1$ cm from the radiator outlet; (*2*) at a distance of $x=3$ cm

presented in Figs. 9.11 and 9.12. The interferogram was divided into three sections: the operating channel is in the center; and the comparative channels are above and below. It was experimentally shown in ref. [36] (see also Sects. 7.4 and 8.1) that when such an excitation mode is used, gas density redistribution in a laser is of the quasi-equilibrium type. Here, the acoustic waves that are usually observed in pulsed lasers are not present in the medium.

In sealed lasers (without gas circulation) during the heating of the active medium, its expansion into the buffer volume occurs, in addition to which a density profile is formed that corresponds to an energy deposition inhomogeneity. The gas circulation mode contributes its own specific features to the nature of the optical

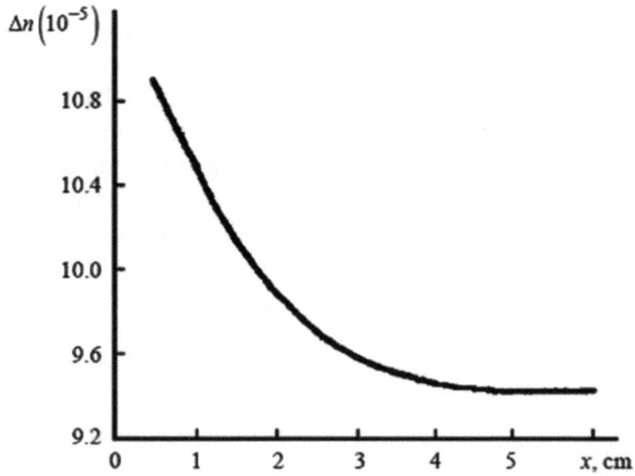

Fig. 9.12 Distribution of refractive index deviations, $\Delta n = n - n_0$, along the x-axis at $y=0$ from its unperturbed value for Ar at $P=0.5$ atm and $\overline{U}=8$ m/s

inhomogeneities that originate. In the cross-section perpendicular to the uranium plate plane (Fig. 9.11), as in the case of a gas at rest, a positive lens originates that has a close-to-parabolic profile. When gas circulation is used, the parabola's steepness increases downstream.

The distribution of refractive index deviations from the initial value, n_0, in the cross-section parallel to the plate plane (along the x-axis in the plane of $y = 0$) is shown in Fig. 9.12. During exposure to an exciting pulse, the gas in the channel at a given velocity is shifted a total of 3 cm from the inlet radiator. Within this part ($0 < x < 3$ cm), the variation of the refractive index along the x-axis is almost linear in nature. The remaining portion of the gas in the region of $x > 3$ cm is present in the channel over the entire course of irradiation; therefore, as it approaches the outlet radiator, the dependence of the refractive index upon x becomes increasingly flat and reaches a constant value roughly ~2 cm before the outlet.

We note that the interferogram in Fig. 9.11 reflects a transient process in gas density redistribution, because the duration of the exciting pulse in these experiments was comparable to the time required for a portion of the gas to pass through the excitation zone.

In refs. [38, 39], calculations demonstrated that gas flow perturbation must occur at the outlet of a plate-type radiator, during which a pair of adjacent symmetrical stable vortex structures will be formed for each rectangular plate. The lateral dimensions of these vortices will decrease downstream and their dimension relative to plate thickness will be proportional to the Reynolds number. If the distance between the plates equals their thickness, the dimension will then be determined by the correlation

$$x_w \approx 0.09 \times \frac{\overline{U}\delta_R^2}{\nu}. \tag{9.89}$$

The circulating motion of the gas in the vortices will lead to convective mixing, which will necessarily bring about the intensification of heat exchange processes as compared to conventional molecular heat conduction.

The laser sounding of an active medium in the gas flowing mode revealed that the fringe pattern near the radiator outlet disappeared due to sounding beam refraction on the radiator's plates. The size of this region (~1 cm) and the large refractive index gradient value (~5×10^{-5} cm^{-1}) confirm the proposition concerning increased thermal diffusivity near a radiator's outlet. The dimension of the region described, x_w, should increase with an increase in the plates' transverse dimension and in gas flow velocity, and will be dependent upon the gas type.

In order to determine the thermal diffusivity associated with the origination of vortex structures at a radiator outlet, investigations were performed during which a nickel-chromium filament with a diameter of 0.2 mm and a length of 25 cm was placed in a laser channel at a distance of 3 mm from a radiator outlet. The filament was heated by an electric current, thereby creating a local thermal perturbation that extended downstream in the steady-state gas flow. The experimental procedure was based on the measurement of the temperature fields created in the cell's volume using an interferometer. Interferograms were recorded using a He-Ne (LG-38) laser

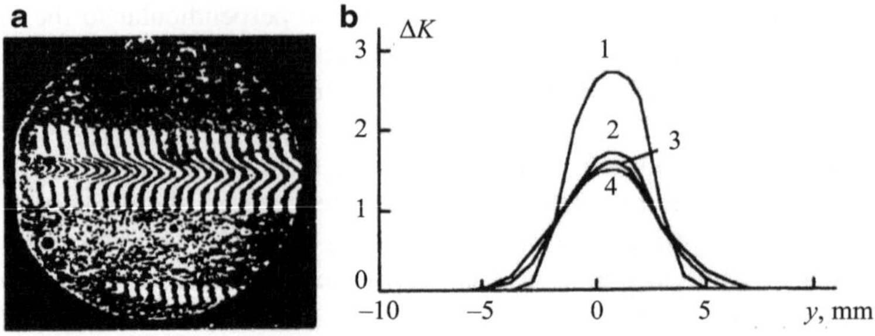

Fig. 9.13 Interference fringe displacement as a function of height during filament heating: (*1*) $x = 1$ cm; (*2*) $x = 3$ cm; (*3*) $x = 4$ cm; (*4*) $x = 5$ cm

radiation and a Michelson interferometer. An SFR-2 M high-speed camera registered the fringe patterns. The nature of heat exchange between the gas flow and the uranium layer substrate was studied separately. To this end, one substrate with a length of 25 cm was heated by $\Delta T = 40\,^{\circ}\mathrm{C}$ using hot water that flowed inside it. The experiments described were carried out without neutron irradiation, i.e., there was no energy release in the uranium layers themselves.

A typical interferogram of the gas density perturbations that originate during filament heating in a helium flow ($\overline{U} = 8$ m/s, $P = 2.2$ atm, and $\delta = 0.5$ mm), together with the results of its processing, are shown in Fig. 9.13. The thermal perturbation had a Gaussian shape and expanded downstream, during which its peak value was reduced. Similar data that correspond to the same gas flow parameters for an experiment involving substrate heating are presented in Fig. 9.14.

The experimental results made it possible to estimate the variation of effective thermal diffusivity coefficient along the gas flow. At the radiator outlet, it exceeds the molecular heat diffusivity coefficient by roughly one order and is reduced twofold at a distance of ~x_w from the radiator outlet. This constitutes confirmation of the existence of vortex structures behind the radiator outlet, the transverse dimension of which is decreased downstream.

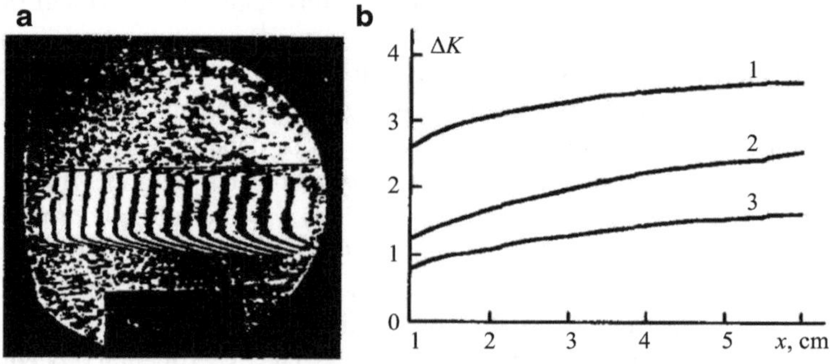

Fig. 9.14 Interference fringe displacement as a function of length during substrate heating: (*1*) $y = -9.5$ mm; (*2*) $y = -8$ mm; (*3*) $y = -6.5$ mm

Based on the experimental results, an attempt was made to verify the hypothesis that the flow behind a radiator is turbulent in nature. The numerical analysis of a flow of this type did not present particular difficulties, because the temperature perturbation distribution behind the heated filament obtained during the experiments had a Gaussian shape. At first glance, this circumstance corresponds to the behavior of the heat wake from a thin linear source in a stationary turbulent flow created by artificially introducing a turbulizing wire grating at a certain distance ahead of the filament [40]. In this instance, thermal perturbation washout with distance from the source occurs in such a way that the maximum value and the half-height width of the perturbation peak are unequivocally linked to the turbulent heat diffusivity coefficient. The numerical analysis results demonstrated that when such allowance is made, the turbulent heat diffusivity coefficients, measured in cm^2/s, are decreased downstream from a few tenths to order-of-unity values at the channel outlet. They will be dependent upon gas velocity and the transverse dimensions of the radiator plates. However, the coefficients obtained from an analysis of the downstream rate of variation in the maximum local thermal perturbation value proved to be several times higher than the similar coefficients found from this perturbation's peak half-width washout rates. This fact suggests that it is impossible to regard the flow behind a radiator as a steady-state turbulent flow, and that the thermal diffusivity coefficient should be determined experimentally.

Experiments with a heated substrate revealed that the gas region near the uranium layers involved in heat exchange with the wall expands downstream (Fig. 9.14). It transverse dimension is roughly proportional to $\sim \sqrt{x}$ (x is the distance from the cell inlet) and is dependent upon gas velocity, density and type. Depending upon the conditions in place, the full volume of the heat exchange zone in the cell may involve 10–30 % of the gas's active volume. In ignoring the viscous boundary layer's influence, it is not difficult to explain this dependence. Here, it can be roughly assumed that the gas velocity is homogeneous throughout the channel cross-section and equals $\overline{U}$. We will isolate a plane with a coordinate of x in the transverse cross-section of the gas. Let us suppose that each such plane up to a moment in time of $t=0$ is situated outside the channel's confines. The time that this plane is located within the channel itself, including the region involved in heat conduction, equals to $x/\overline{U}$. Over this time frame, the thickness of the region involved in near-wall heat exchange in the plane under consideration pursuant to Eq. (8.3) reaches a dimension of

$$y_t(x) \sim \sqrt{\frac{a_{eff}}{\overline{U}}x},$$

where a_{eff} is effective thermal diffusivity near the channel wall.

Thus, the significant role of vortex structures at the radiator outlet during heat conduction processes within a laser cell was demonstrated during the experiments. The most negative aspect consists of the considerable gas layer thickness near the substrate involved in the heat transfer process, as a result of which the lasing region can only occupy a portion of the cell's cross-section. In order to diminish this effect, it is necessary to reduce the size and intensity of the vortex structures in the gas at the radiator outlet by tapering its edges and bringing the temperature of the gas exiting the radiator as close as possible to that of the uranium layer.

Gas Flow Rate Effect on Output Power of Gas-Flowing Lasers

A circuit with continuous transverse gas circulation was used in the LM-4 four-channel laser module operating together with BIGR reactor [41], within which a CW lasing mode was first achieved for NPLs [26] (see Sect. 6.1). The lasing lasted 1.5 s and ended with exciting neutron pulse termination.

The LM-4 apparatus consisted of four laser channels (Fig. 9.15) that were incorporated into a common gas loop and were separated from one another by plate-type radiators. The active length of the channels, determined by the dimension of the uranium layers along the optical axis, was $L_A = 1$ m. The dimension of each channel in the direction perpendicular to the gas flow was $d = 2$ cm, while in the direction parallel to the flow, it was $b = 6$ cm (see Fig. 9.1). The average thickness of the uranium layer was $\delta_U = 5$ mg/cm^2. This layer was covered with a protective aluminum film that had a thickness of $\delta_A = 0.5$ mg/cm^2. The LM-4 apparatus was irradiated by a neutron flux from a BIGR pulse reactor [42] with a duration of ~1.5 s. Thermal neutron flux density in the channel location at the pumping pulse peak reached 3.5×10^{14} cm$^{-2} \times$ s^{-1}. The gas circulation system, based on motion of the hydropneumatic piston with a rectangular cross-section [41], ensured a gas velocity that was homogeneous along the length of the channel. It was switched on roughly ~0.3 s before the neutron pulse start and brought the gas velocity up to its maximum value of 4.5 m/s over this time interval. Gas circulation ceased immediately upon neutron pulse termination. Under such conditions, after the lasing threshold is exceeded by roughly ~2–3 times, the shape of the laser pulse quite closely duplicates that of the pumping pulse.

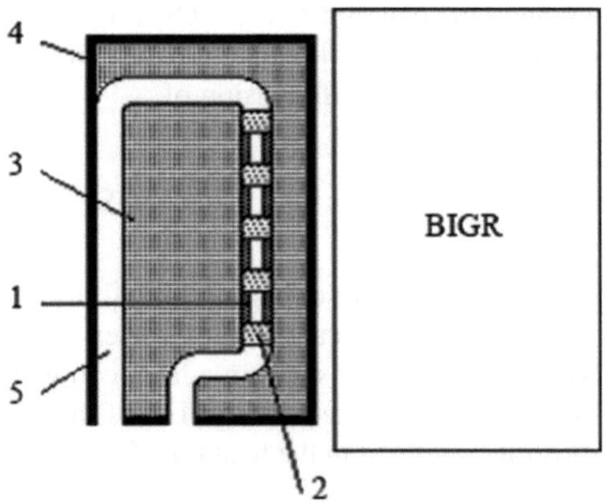

Fig. 9.15 Transverse section of an LM-4 laser module together with a BIGR reactor: (*1*) aluminum substrate with a uranium layer; (*2*) radiator; (*3*) graphite; (*4*) casing; (*5*) gas pipeline

In order to investigate the effect of the gas circulation mode on the shaping of a laser pulse, two experiments were carried out in ref. [26] that involved an Ar–Xe mixture (70:1) at a pressure of 0.35 atm. During the first of these tests, the pumping pulse's leading edge noticeably advanced the start time of gas circulation (delay of circulation start time); during the second test, the circulation system was activated long before the start of the neutron pulse and cut off the gas feed at the pumping pulse's leading edge (premature switching of gas circulation). These experiments demonstrated that lasing does not occur until gas circulation begins, and conversely, that lasing is terminated when circulation stops.

The physical model proposed in ref. [43], which describes the dependence of lasing active volume upon gas mixture composition and density, gas flow velocity, and specific power deposition, provides entirely satisfactory agreement with the experimental time characteristics of laser pulses. According to this model, over the period of time required for a portion of the gas to traverse a laser channel, if gas velocity at the inlet and fission intensity in the uranium layers are negligibly altered,

$$\Delta U_0(t) << U_0, \ \ \Delta\bar{\dot{q}}_c(t) << \bar{\dot{q}}_c, \tag{9.90}$$

then the gas mass flow at each given moment in time can be roughly regarded as constant:

$$\rho_0 U_0(t_0) \ \approx \overline{\rho\,(x,t)} \times \overline{U(x,t)}, \tag{9.91}$$

where $\overline{U(x,t)}$, $\overline{\rho\,(x,t)}$ is the gas velocity and density values in the laser channel in a plane with a coordinate of x at a moment in time of t, averaged over the cross-section; U_0 is the gas velocity at the channel inlet (which is homogeneous throughout the cross-section); t_0 is the moment in time that the portion of gas under consideration enters the channel; and ρ_0 is gas density at the channel inlet (which is kept constant). The average velocity and density values are determined by the equations

$$\begin{gathered}\overline{\rho\,(x,t)} = \frac{2}{d}\int\limits_0^{d/2} \rho\,(x,y,t)\,dy;\\ \overline{U(x,t)} = \frac{2}{d\cdot\overline{\rho(x,t)}}\int\limits_0^{d/2} U(x,y,t)\,\rho\,(x,y,t)\,dy.\end{gathered} \tag{9.92}$$

It is assumed that gas density, velocity, and temperature distribution will not be dependent upon the z coordinate, which is oriented along the system's optical axis. The approach used reduces the description of the gas flow in the channel to a one-dimensional model.

During a gas flow in a plane-parallel channel, the length of which, b, is comparable to its transverse dimension, d, the pressure differential, ΔP, between the channel inlet and outlet should be small ($\Delta P/P << 1$ (see the first section of this chapter)). Thus, from the condition of approximate pressure equality at any given point in the channel, $P(x) \approx$ const, and the state equation for an ideal gas, we get

$$\rho_0 T_0 = \overline{\rho(x,t)} \cdot \overline{T(x,t)}. \quad (9.93)$$

Here, T_0 is the temperature at the channel inlet, and $\overline{T(x,t)}$ is the temperature value in a plane with a coordinate of x at a moment in time of t, averaged over the channel cross-section, which, when small temperature deviations, $\Delta T(x,y,t) << \overline{T(x,t)}$, are present within the confines of the transverse cross-section under consideration, is determined by a formula similar to Eq. (9.92).

It follows from correlations Eqs. (9.91) and (9.93) that

$$\overline{U(x,t)} = \frac{\overline{T(x,t)}}{T_0} U_0(t_0). \quad (9.94)$$

By virtue of the conditions in Eq. (9.90), the following heat balance correlation for a laser channel at each given moment in time should roughly hold true

$$L_A \int_0^x q_S(x,t)dx = L_A d U_0(t_0) \rho_0 c_p \left(\overline{T(x,t)} - T_0\right), \quad (9.95)$$

where q_S is the surface power density of the sources of energy deposition in the gas from both uranium layers. With allowance for Eq. (9.95), correlation Eq. (9.94) takes the form

$$\overline{U(x,t)} = U_0(t_0) + (T_0 \rho_0 c_p d)^{-1} \int_0^x q_S\left(x'\right) dx'.$$

The energy deposition distribution of surface fission fragment sources is quite homogeneous, with the exception of the edge parts at the channel inlet and outlet. Ignoring the edge effects, we can assume that $q_S =$ const. Also, bearing in mind that $\overline{U(x,t)} = dx/dt$, then from the latter equation, we get an equation that reflects, in a one-dimensional model, the relationship between the path traversed by a portion of the gas in the channel and the time over the course of which this path is traversed:

$$dx/dt = U_0(t_0) + Bx,$$

where $B = (T_0 \rho_0 c_p d)^{-1} q_S$.

At the initial condition of $x(t_0) = 0$, the solution to this equation for the time it takes a portion of the gas to traverse a distance of x from the channel inlet, $\tau = t - t_0$, within which the Eq. (9.92) yield the average density and velocity value distribution, takes the form

$$\tau = \frac{1}{B}\ln\left(B\frac{x}{U_0} + 1\right). \tag{9.96}$$

In the special case of sufficiently large pumping velocities,

$$Bx/U_0 < \; < 1, \tag{9.97}$$

we get a simple dependence for the travel time

$$\tau = x/U_0. \tag{9.98}$$

When a laminar gas flow is present, its heat exchange with the laser cell walls occurs by means of conventional molecular heat conduction. If the ȇ τ time value is small enough that the condition $l << d$ is satisfied for the transverse dimension of the near-wall region involved in heat removal, then from Eqs. (8.3) and (9.98) when the condition (9.97) is satisfied, we get

$$l(x) \sim \sqrt{ax/U_0}. \tag{9.99}$$

If it is assumed that the development of the heat removal zone at the channel wall provided gas circulation occurs in accordance with the same regularities as in a cell that contains a stagnant gas, then in the general case, the laser active region volume per unit of laser channel length (along the z-axis) equals

$$V(t) = 2\int_0^b y_A(x)dx, \tag{9.100}$$

where $y_A(x) = d/2 - l(x)$, while the relationship between the x coordinate and travel time, τ, is determined by Eq. (9.96).

It is not difficult to show that in the special case when the $l << d$ and inequality (9.97) are satisfied for the entire range of x, the laser active region volume is determined by the equation

$$V(t) = bd - 2\int_0^b l(x)dx = bd - \frac{4}{3}A_l\sqrt{\frac{ab^3}{u_0}}, \tag{9.101}$$

where A_l is the proportionality factor in correlation (9.99). In the case of sealed (without gas circulation) laser cells, the formula $l(t) = A_l\sqrt{at}$, which corresponds to correlation (9.99) written in the form $l(x) = A_l\sqrt{ax/u_0}$, as a rule satisfactorily describes the $y_A(t)$ dependence obtained from an accurate gasdynamic calculation over an interval of $3d/8 \leq y \leq d/2$ (y varies over limits of 0 to $d/2$). However, for the specific alternative of a planar cell filled with Ar at $P_0 = 0.35$ atm and $T_0 = 293$ K, it allows the calculated dependence of $y_A(t)$ over an interval of $d/4 \leq y \leq d/2$ to be reproduced with an accuracy better than 6 %. Here, $A_l \approx 3$ for sealed cells (see Sect. 8.3), while the A factor can take on values one-and-a-half to two times smaller than for cells with gas circulation.

The model described was used in ref. [43] to calculate the dynamics of the development of a laser region's active volume during experiments involving artificial desynchronization between the start of neutron pulse and gas-circulation system switching. This volume was found from Eq. (9.100) using the results of the calculation of active region size variation with the passage of time in an identical sealed cell. The relationship between the x coordinate in expression (9.100) and travel time, τ, which corresponds to the duration of irradiation in a sealed cell, was determined using Eq. (9.96). The dependences of the neutron flux and gas flow velocity upon time needed for the calculations were assigned based on experimental data.

When the lasing threshold is exceeded, the active volume and the specific power deposition in the gas, which is proportional to the neutron flux, determine the laser output power at each given moment in time. The experimental dependences of neutron flux density, gas velocity, laser output power, as well as the calculated dependence of the product of active region volume times neutron flux density, upon time relative to their maximum values are presented in Fig. 9.16 for an alternative with delay of circulation start time. The results of the calculation of similar dependences for an alternative with premature switching of gas circulation are reflected in Fig. 9.17. The calculated cross-sections of the active volume (the clear area) and the zone involved in heat removal (the dark area) are shown in Fig. 9.18 for several successive moments in time that correspond to Figs. 9.16 and 9.17.

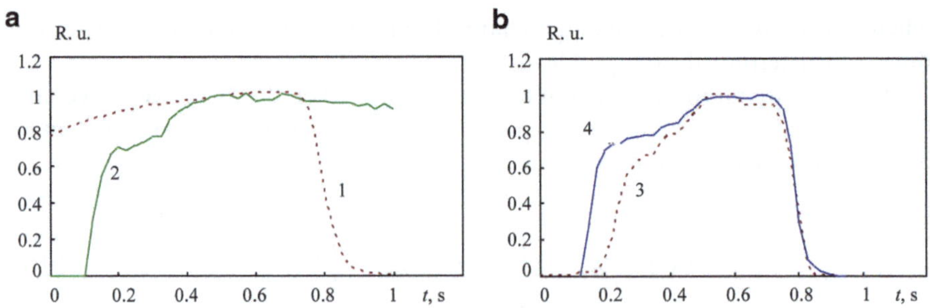

Fig. 9.16 Dependences of neutron flux density (*1*), gas velocity (*2*), laser output (*3*), and the product of gas active volume and neutron flux density (*4*) upon time for alternative with delay of circulation start time

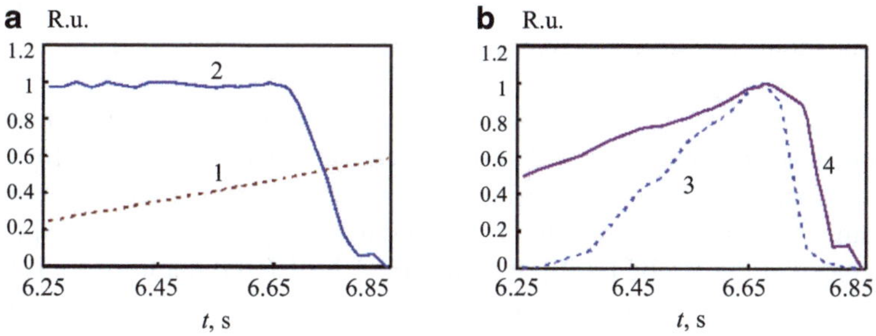

Fig. 9.17 Dependences of neutron flux density (*1*), gas velocity (*2*), lasing output (*3*), and the product of gas active volume and neutron flux density (*4*) upon time for alternative with premature switching of gas circulation

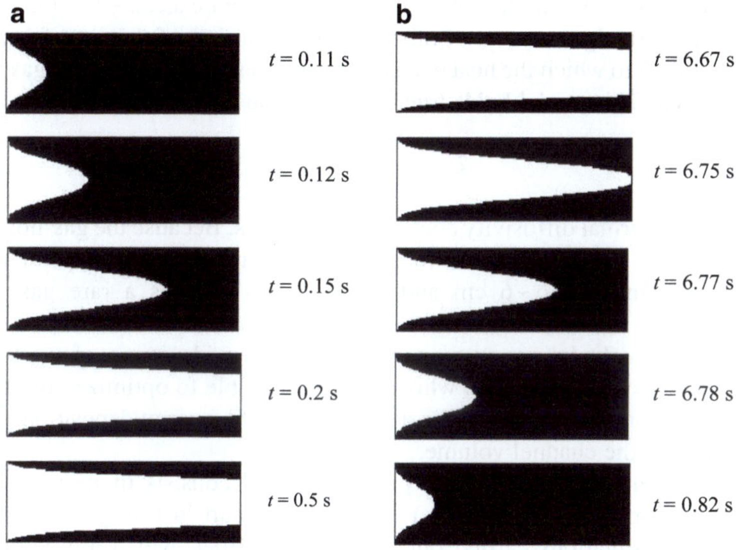

Fig. 9.18 Time dependences of the transverse cross-sections of the laser-channel active volume (*clear area*) and the zone involved in heat removal (*dark area*) for two alternatives: (**a**) delay of circulation start time, and (**b**) premature switching of gas circulation

Entirely satisfactory agreement is observed between the behavior of the experimental dependences of laser output (3) upon time and the curves (4) that reflect the value of the product of the neutron flux and the calculated active region volume values at each given moment in time. The difference in the slopes of the corresponding curves at the leading edge (Fig. 9.17) is explained by the fact that the lasing start has a threshold nature, unlike the process of the formation of a near-wall heat removal zone. Apparently, the medium small-signal gain is minimal at the moment of lasing start and begins to increase with an increase in power deposition. Therefore, the experimental curve of the dependence of laser output upon time runs noticeably steeper. During these experiments, thermal neutron flux density at the lasing threshold was $\Phi \approx 9 \times 10^{13}$ cm$^{-2} \times$ s^{-1}, which corresponds to ≈0.26 r.u. (see Fig. 9.17).

Peculiarities of the Gas Flow in Flowing NPLs. Two-dimensional calculations of spatial inhomogeneities in NPL flowing channels were performed for the first time at the Idaho National Engineering Laboratory [44] using a specially developed gasdynamic program. The results of the calculations performed in ref. [44] were later repeated at the VNIIEF [45]. Because the VNIIEF did not have a gasdynamic program similar to the one used in ref. [44], a simpler calculation procedure was proposed that was based on a series of reasonably simplified physical models [46] (see also the subsection entitled "Calculations of Spatial Inhomogeneities in Flowing NPLs"). The difference in the density distribution calculation results obtained using this procedure and those of the similar calculations performed using the two-dimensional program [44], which makes strict allowance for gasdynamic and heat exchange processes in the entire gas volume, do not exceed ~2 % [45, 46].

One of the principal aspects of this procedure consists of the use of a physical model according to which the heat exchange processes in the bulk of the gas volume can be ignored. This model holds true, provided that

$$U_0 > \ > ab/d^2,$$

where a is the thermal diffusivity coefficient of the gas. Because the gas-flow rate in the NPLs under consideration comes to $U_0 \sim 10$ m/s, then at typical channel transverse dimensions, $b \sim 6$ cm and $d \sim 2$ cm, as well as a rare gas thermal diffusivity of $a \sim 0.2$–0.6 cm^2/s, the condition written above is virtually ensured.

In ref. [46], calculations were performed for a wide range of working gas pressure and velocity variations, which made it possible to optimize both the total energy deposition in the gas and the degree of specific energy deposition inhomogeneity through the channel volume.

A weak facet of the calculation procedure used consists of its passive zone model, i.e., the near-wall region that does not take part in lasing. Numerical and theoretical investigations carried out during 1979–1984, which are described in detail in Chap. 8, revealed that the temperature profile has a clearly pronounced maximum and the density profile has a clearly pronounced minimum in sealed NPLs at times of $t \leq 0.1$ s. The latter is explained by the fact that in direct proximity to the laser channel side wall to which a uranium layer is deposited, a gas zone is formed from which the heat is intensely removed to the wall. This is called the passive (dead) zone.

It is also obvious that the same passive zone formation effect must take place in NPL flowing channels (where, as is generally known, specific energy release increases from the center to the side surfaces), the wall temperature of which is lower than the gas flow temperature. Accordingly, a proposition was advanced in ref. [43] that the development of a heat removal zone at the channel wall during gas flowing occurs qualitatively under the same laws as in a cell without gas flowing, i.e., the existence of a passive zone is proposed, the boundary of which is determined by the density minimum and the temperature maximum. The relationship between temperature and density is unequivocally determined by a state equation. Because the pressure in the laser channel's transverse cross-section is almost homogeneous ($P \approx$ const), then in order to fully describe both the density and the temperature profile, it is sufficient to examine the dependence upon the coordinate of one of these parameters.

However, in an actual situation involving a gas flow, a boundary layer with a thickness of l_ν is formed near the wall, within which velocity increases from zero at the wall itself to a value equaling that in the flow core. Over the course of temperature equalization between the wall and the heated gas, hydrodynamic effects and heat conduction effects have a strong influence on one another [2].

For a compressible gas, temperature and velocity distribution near a planar wall in the presence of stationary motion is described by an equation system [2] that includes heat conduction, motion, and continuity equations

$$c_p\rho(x,y)\left\{u(x,y)\frac{\partial T(x,y)}{\partial x}+v(x,y)\frac{\partial T(x,y)}{\partial y}\right\} = \frac{\partial}{\partial y}k(x,y)\frac{\partial T(x,y)}{\partial y}+w(x,y); \tag{9.102}$$

$$\rho(x,y)\left\{u(x,y)\frac{\partial u(x,y)}{\partial x}+v(x,y)\frac{\partial u(x,y)}{\partial y}\right\} = -\frac{\partial P}{\partial x}+\frac{\partial}{\partial y}\eta(x,y)\frac{\partial u(x,y)}{\partial y}; \tag{9.103}$$

$$\frac{\partial[\rho(x,y)u(x,y)]}{\partial x}+\frac{\partial[\rho(x,y)v(x,y)]}{\partial y}=0, \tag{9.104}$$

as well as a state equation

$$T = P/(\gamma - 1)c_v\rho. \tag{9.105}$$

Here, k is the heat conductivity coefficient; u is the longitudinal velocity component; v is the transverse velocity component; w is the specific power of the internal energy sources; and η is the dynamic viscosity coefficient.

The heat conductivity and dynamic viscosity coefficients of rare gases have a power dependence upon temperature [34]:

$$k = k_0T^n;\ \eta = \eta_0T^n. \tag{9.106}$$

For these gases, $n \approx 0.7$ [34]. Even within the viscous and thermal boundary layers (as well as the passive zone, if it has not been eliminated), under conditions of actual powerful flowing NPLs, the absolute temperature values vary by roughly two times both in the longitudinal and the transverse directions. According to the relation (9.106), temperature differentials of this type will lead to variations of

~60 % in the heat conductivity and dynamic viscosity coefficients. Therefore, in order to find the dependence of viscous boundary layer thickness and the velocity distribution within its confines under actual NPL operating regimes, upon gas mixture flowing conditions and parameters, it is necessary to solve the system of essentially nonlinear differential equations in partial derivatives (9.102)–(9.106). Here, in addition to the boundary condition for velocity at the wall itself, it is correct to give the boundary condition far from the wall not at infinity ($y=\infty$), as is frequently done in order to simplify certain problems, but rather in the plane of symmetry, i.e., at a finite distance of $y=d/2$. The same thing also pertains to temperature. At the wall itself, a Derichlet boundary condition, $T(x,\,0)=f_1(x)$, or a Neumann boundary condition, $\partial T(x,\,0)/\partial y=f_2(x)$, is given for temperature.

Thus, the solutions for the effective thickness of a viscous boundary layer, l_ν, and the velocity distribution therein, as well as for the width of the passive zone, l_a, and the temperature distribution therein, are functionals of the type

$$F\left[x, y, d, T_0, U_0, w(x, y), \frac{\partial T(x, 0)}{\partial y}\right],$$

where T_0 is the temperature value at the channel inlet (in these functionals, depending upon the conditions assigned at the wall, the temperature $T(x,0)$ may be present instead of the derivative $\partial T(x,0)/\partial y$. In addition, for simplicity's sake, the dependence upon thermophysical parameters, especially upon the Prandtl number, is not reflected here). In point of fact, the problem is even more complex, because the distribution of specific sources in the gas will be dependent not only upon the fission intensity of the uranium nuclei in the uranium layers deposited to the channels walls, but also upon velocity and temperature distribution. A problem of this type for actual NPL modes can only be solved by means of numerical techniques involving the use of two-dimensional gasdynamic programs.

Of course, it may also be useful to a certain extent to obtain analytical solutions by introducing various simplifying assumptions, for example, of the $T(x, y)-T_0<<T_0$, $\nu=\text{const}$, and $w=\text{const}$ types, etc. This helps mark the paths for optimizing energy depositions in the gas by means of gas pressure and velocity variations, an even to estimate share of passive zone volume (however small). But it must be said that any viscous boundary layer and passive zone models not based on the results of rigorous two-dimensional numerical calculations or direct experiments for NPLs operating at real power depositions cannot be regarded as correct.

From equation system (9.102)–(9.106), it is not difficult to obtain an integral equation for the passive zone. Actually, from continuity Eq. (9.104) and state Eq. (9.105), we get

$$u\frac{\partial T}{\partial x}+v\frac{\partial T}{\partial y}=T\left\{\frac{\partial u}{\partial x}+\frac{\partial v}{\partial y}\right\}. \quad (9.107)$$

Placing this result into Eq. (9.102) and integrating from zero to l_a with allowance for the fact that $v(x\,,0)=0$ and that $\rho T=\rho_0 T_0$ at $P=$const, we obtain

$$c_p\rho_0 T_0\left\{\int_0^{l_a}\frac{\partial u(x,y)}{\partial x}dy+v(x,l_a)\right\}=\int_0^{l_a}w(x,y)dy-k(x,0)\frac{\partial T(x,0)}{\partial y}. \quad (9.108)$$

Possibility of Passive Zone Elimination

We will now examine the conditions under which a passive zone may be eliminated. Let us suppose that $l_a=0$; it then follows from Eq. (9.108) that $\partial T(x\,,0)/\partial y=0$, i.e., there is no heat flux at the wall. On the other hand, setting the heat flux at the wall to equal zero, then from Eq. (9.108), we get

$$\frac{\partial u(x,y)}{\partial x}+\frac{\partial v(x,y)}{\partial y}=\frac{w(x,y)}{c_p\rho_0 T_0}.$$

The latter equation means that when there is no heat flux at the wall, a passive zone can only exist if the specific heat source power distribution profile is similar to the sum of profiles of the derivative for x from the longitudinal velocity component and the derivative for y from the transverse component. It is obvious that the simultaneous existence of such a coincidence over the entire extent of a flow in an NPL channel under normal operating conditions is unrealistic.

Thus, a necessary and sufficient condition for passive zone elimination consists of the heat flux at the gas and channel wall boundary equaling zero. Moreover, there will be no passive zone if the heat flux is directed from the wall to the gas ($\partial T(x,0)/\partial y>0$). This result makes it possible to estimate the channel wall heating needed for passive zone elimination. We will then perform an upper estimate. To this end, we will assume that the gas within a channel part from the inlet with a length of x over a time interval with a duration of t is in a state of rest. Here, its density is homogeneous and equals the gas density at the inlet, ρ_0, while the specific energy deposition in the entire volume under consideration coincides with that in an infinitely expanded planar channel. In this model, the specific energy deposition near the channel wall can only be overstated, because in an actual instance when there is no passive zone, $\partial\rho/\partial y\leq 0$ at any y value. Consequently, specific energy deposition near the wall in an actual situation will also be lower than in the model under consideration. Furthermore, in a real case, gas density steadily decreases in the direction of x, dropping roughly half in size near outlet (see refs. [44, 46]), which also leads to a decrease in specific energy deposition.

During gas pumping through any given cross-section of $x = \text{const}$ at any given instant in time, a new portion of gas flows in that is heated to a lesser extent than in the case when this gas remains motionless. The gas velocity in each cross-section increases from the wall to the center from zero to $U(x)$. Therefore, the transverse temperature gradients when pumping is absent do not at least exceed the absolute value of the similar gradients when gas pumping is present. Consequently, the heat flux from the wall to the center in an unpumped model does not exceed the corresponding fluxes in a flowing system. Thus, in the nonflowing model under consideration, the wall temperature can only be overstated as compared to a flowing system.

A comparison of an actual flowing system and a model nonflowing system in a plane at a distance of x from the inlet can be made over a time interval of $t = x/\overline{U}(x)$, where $\overline{U}(x)$ is the average gas velocity over the transverse cross-section. It steadily increases with an increase in the distance from the inlet [44, 46]. However, the t time value can be determined in a simpler manner:

$$t = x/U_0, \tag{9.109}$$

knowingly overstating it in this instance, and accordingly, overstating the warm-up temperature.

So, ignoring longitudinal heat conduction, we get the following boundary value problem for the gas temperature distribution in a nonflowing model:

$$\frac{\partial}{\partial y} k(y,t) \frac{\partial T(y,t)}{\partial y} + w(y) = c_p \rho_0 \frac{\partial T(y,t)}{\partial t}; \tag{9.110}$$

$$\partial T(0,t)/\partial y = 0; \tag{9.111}$$

$$\partial T(d/2,t)/\partial y = 0; \tag{9.112}$$

$$T(y,0) = T_0. \tag{9.113}$$

The boundary condition in Eq. (9.111) reflects the fact established above that a necessary and sufficient condition for the absence of a passive zone consists of the heat flux at the gas and solid wall boundary equaling zero.

Making the upper-bound estimate even more rigorous, we will now assume that the heat conductivity coefficient is equal to its minimum value everywhere, which corresponds to a value at the inlet of $k(x, y, t) = k_0$, i.e., we ignore its dependence upon temperature in Eq. (9.106), thereby understating heat removal from the wall to the center of the channel. In this instance, Eq. (9.110) takes the form:

$$a \frac{\partial^2 T}{\partial y^2} + f(y) = \frac{\partial T}{\partial t}, \tag{9.114}$$

where

$$f = w/c_p\rho_0; \quad a = k_0/c_p\rho_0. \tag{9.115}$$

The solution of the problem (9.111)–(9.114) takes the form:

$$T(y,t) = T_0 + \frac{2x}{dU_0}\int_0^{d/2} f(y)dy + \\ + \frac{d}{a\pi^2}\sum_{n=1}^{\infty}\frac{1}{n^2}\left\{1 - \exp\left[-\frac{4a\pi^2n^2x}{d^2U_0}\right]\right\}\cos\left(\frac{2\pi n}{d}y\right)\int_0^{d/2} f(y)\cos\left(\frac{2\pi n}{d}y\right)dy. \tag{9.116}$$

The upper temperature values of the channel wall at its contact surface with the gas that are needed for passive zone elimination were calculated using Eq. (9.116). The temperature values obtained for uranium fission intensity in the active layers, $q = 2 \times 10^{16}$ cm$^{-3} \times$ s^{-1}, at an active layer thickness of $\delta_U = 2.78 \times 10^{-4}$ cm, as well as channel transverse dimensions of $b = 6$ cm and $d = 2$ cm, are presented in Fig. 9.19.

The specific power deposition in a gas was calculated using the results obtained in ref. [49]. The gas temperature at the channel inlet was given as equaling $T_0 = 293$ K.

By way of illustrating the depth of heat exchange zone penetration into the gas, the dependences of the relative temperature increments, $\theta = (T(x, y) - T_0)/(T(x,0) - T_0)$, and specific power deposition, $\omega = w(x, y)/w(x,0)$, upon the transverse coordinate, calculated at a point of $x = b = 6$ cm for helium at a velocity of $U_0 = 4.5$ m/s, are presented in Fig. 9.20.

In order to achieve the stationary mode, it is necessary to ensure wall cooling through the use of a liquid heat-transfer agent, such as water, for heat exchange.

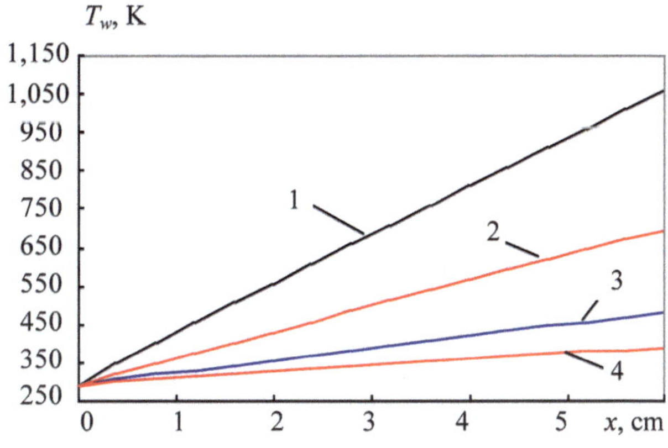

Fig. 9.19 Dependence of the upper channel wall temperature boundary upon the longitudinal coordinate: (*1*) argon, $P_0 = 0.9$ atm and $U_0 = 4.5$ m/s; (*2*) argon, $P_0 = 0.76$ atm and $U_0 = 9$ m/s; (*3*) helium, $P_0 = 2.78$ atm and $U_0 = 4.5$ m/s; (*4*) helium, $P_0 = 2.6$ atm and $U_0 = 9$ m/s

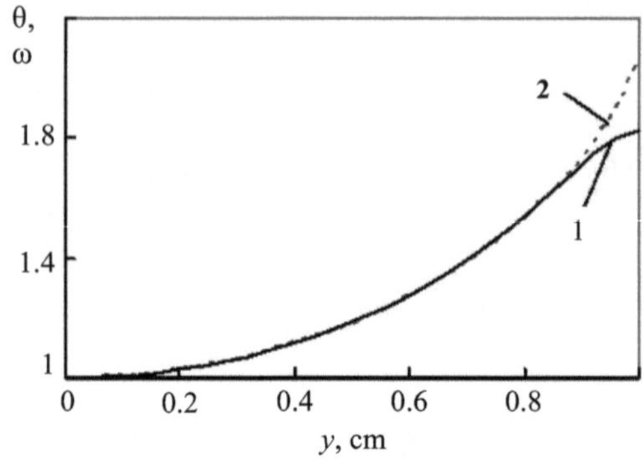

Fig. 9.20 Relative temperature increments in helium (*1*) and relative specific power deposition at the channel outlet at $U_0 = 4.5$ m/s (*2*)

At $P = 1$ atm, the boiling point of water is $T_\kappa = 100\,°\mathrm{C}$. Water boiling in the external cooling channel can lead not only to the noticeable worsening of heat exchange, but also to vibrations and acoustic perturbations on the outer cooled surfaces of the laser channels that occur both during the formation and during the detachment of steam bubbles. The latter must inevitably lead to the transmission of the aforementioned perturbations through the channel wall to the laser gas. Consequently, small-scale density inhomogeneities will occur, the result of which will be a deterioration of laser characteristics, or perhaps, simply lasing failure.

For a planar wall in the stationary mode, the temperature differential between its limiting surfaces equals

$$\Delta T = Q\delta_w / k_w, \tag{9.117}$$

where Q is the density of the heat flux through the wall; δ_w is the wall thickness; and k_w is the wall heat conductivity coefficient. A uranium layer is deposited to the wall surface that adjoins the gas (see Fig. 9.1). The fission fragments carry a portion of the energy released in the uranium layer, which equals ε, into the gas. The remaining portion of the energy $(1 - \varepsilon)$ in the form of heat is released within the layer itself. If the requirement of passive zone elimination, $\partial T(x,0)/\partial y = 0$, is satisfied—i.e., the heat flux from the wall (or more precisely, from the uranium layer) into the gas equals zero—the excess portion of the heat $(1 - \varepsilon)$ must then flow through the wall to the heat-transfer agent. In this instance, the heat flux density will be

$$Q = (1 - \varepsilon) E_0 q \delta_U, \tag{9.118}$$

where E_0 is the energy released in the single fission.

As in the preceding calculations, let us assume that $q = 2 \times 10^{16}\ \mathrm{cm}^{-3} \times \mathrm{s}^{-1}$. When the argon velocity at the inlet is $U_0 = 4.5$ m/s, then based on the previously cited upper-bound estimates, the wall temperature at the contact boundary with the gas will equal $T_w(b) = 1{,}055$ K. In order to avoid heavy water boiling near the outer wall surface, the temperature differential at the wall must exceed $\Delta T = T_w(b) - T_\text{к} = 682$ K. Pursuant to Eqs.(9.117) and (9.118)

$$\delta_w = \frac{k_w \Delta T}{(1-\varepsilon)E_0 q \delta_\text{U}}$$

at $\varepsilon \approx 0.1$ and for the thickness of a wall made from zirconium ($k_w = 0.21$ W/cm × K), we get $\delta_w = 1$ cm. Similarly, for $U_0 = 9$ m/s ($T_w(b) = 696$ K), we obtain $\delta_w \approx 0.5$ cm.

Cooling with an External Heat-Transfer Agent. In the stationary operating mode, if all the energy released in the laser channel is removed exclusively through the gas by means of its pumping, then at a thermal neutron flux density of $\Phi \sim 10^{14} - 10^{15}$ cm^{-2} s^{-1}, even during transverse pumping, the requisite velocity, as follows from a balance equation of the (9.4) type, must be $U_0 \sim 100$ m/s. However, as previously stated, heat removal through the outer surface of the laser channel substrate can be accomplished concurrently with gas pumping using the heat-transfer agent that washes around this substrate. But in the quasi-stationary mode, an appreciable portion of the heat released in the uranium layer can be absorbed in the substrate itself, as is the case in the LM-4 apparatus [41], or in the material mass directly adjacent to the substrate.

If the directions of movement of the external heat-transfer agent and the gas in the laser channel coincide, it is not difficult to establish the correlation between the geometric and thermophysical parameters of a system, during which the projection of the refractive index gradient in the y direction perpendicular to the uranium layer's plane is $(\nabla n)_y \le 0$ in the entire gas volume, i.e., there is no near-wall passive zone. This can take place provided that $(\nabla T)y \ge 0$ everywhere, including near the uranium layer surface. This means that the heat can only flow from the substrate to the gas. In order to implement this condition, it is essential that the heat fraction carried out of the channel by the gas is not less than ε.

The total amount of energy carried away by the gas per unit of time through the cross-section of a laser channel with a coordinate of x per unit of channel length in the z direction equals

$$Q_1 = dU_0 \rho_0 c_p \left(\overline{T(x)} - T_0\right). \tag{9.119}$$

It is obvious that the realization of the condition $(\nabla T)_y \ge 0$ must satisfy the inequality

$$Q_1 \ge \int_0^x q_S(x)dx, \tag{9.120}$$

where q_S is the power surface density of energy deposition in the gas from both of the uranium layers that is directly due to the escaping fission fragments,

$$q_S = 2E_0 N_U \langle \sigma_f \Phi \rangle \delta_U \varepsilon, \tag{9.121}$$

where the angle brackets denote averaging over the neutron spectrum.

The total amount of energy per unit length in the z direction that the heat-transfer agent carries away per unit of time through the cross-section with a coordinate of x for each of the two external channels intended for the heat-transfer agent equals

$$Q_2 = d_l U_{l_0} \rho_{l_0} c_{pl} \left(\overline{T_l(x,t)} - T_0 \right), \tag{9.122}$$

Here, the l index indicates that the corresponding parameters are related to the external heat-transfer agent.

The amount of energy carried away per unit of time from a unit of uranium layer surface through the substrate to the external heat-transfer agent that is required in order to implement the condition $(\nabla T)_y \geq 0$ must not exceed

$$q_2 = \mu_f E_0 N_U \langle \sigma_f \Phi \rangle \delta_U (1 - \varepsilon), \tag{9.123}$$

i.e.,

$$Q_2 \leq \int_0^x q_2(x) dx. \tag{9.124}$$

Here, μ_f is the ratio of the thermal energy released directly into the uranium layer during the fission of a single uranium nucleus to the total kinetic energy of the fragments, E_0.

Combining Eqs. (9.119)–(9.124) and assuming in this instance that the uranium layer and the fission density within the confines of the channel's expanse along the direction of the x-axis are homogeneous, we obtain

$$\frac{d}{d_l} \frac{U_0}{U_{l_0}} \frac{\rho_0 c_p}{\rho_{l_0} c_{pl}} \frac{\left(\overline{T(x)} - T_0 \right)}{\left(\overline{T_l(x)} - T_0 \right)} \geq \frac{2}{\mu_f} \left[\int_0^x \varepsilon(x) dx \Big/ \left(x - \int_0^x \varepsilon(x) dx \right) \right]. \tag{9.125}$$

Here, allowance is made for the fact that the gas density, $\overline{\rho(x)}$, can be perceptibly altered downstream; therefore, ε can be dependent upon x.

When an oncoming or transverse gas and heat-transfer agent flow is present, the situation becomes considerably more complex. So, for example, if the heat flux at the gas and uranium layer contact boundary equals zero during an oncoming flow in a given transverse cross-section plane of the system under consideration, the heat flux density vector on different sides of the that plane will have an opposite direction.

Passive Zone Transverse Dimension. As previously indicated, it is not possible to obtain a precise solution for the width of a passive zone in a gas-flowing channel under actual pumping conditions. Nonetheless, even an approximate estimative description of its behavior would be of interest, which can be obtained using a series of simplifying propositions. As far as order of magnitude, the average travel time of a gas particle in a channel comes to $\tau \sim b/U_0$. If δ is flow line displacement in the transverse direction, the transverse velocity component is then $v \sim \delta/\tau$. We will assume that flow line displacement within the confines of a passive zone in the presence of the flows originating in NPLs with gas pumping is much smaller than this zone's thickness ($\delta << l_a$), i.e., $v << l_a U_0/b$. The transverse temperature gradient in the zone under consideration is estimated as T/l_a; therefore, as far as order of magnitude, the second addend in the braces on the left side of Eq. (9.102) comes to $v(\partial T/\partial y) << U_0 T/b$. For the first addend in these same braces, we get $u(\partial T/\partial x) \sim U_0 T/b$. Thus, the second addend in the braces can be ignored. In ref. [46] (see the ensuing subsection), the quasi-Lagrangian coordinates associated with the flow lines in the active zone, x and y_0, are introduced. They are expressed through Eulerian coordinates by the correlation $dy_0 U_0 \rho_0 = U(x, y_0)\rho(x, y_0)dy(x, y_0)$, where U is the gas velocity in a transverse cross-section with a coordinate of x for a flow line that has a Eulerian coordinate of $y = y_0$ at the channel inlet. By analogy, introducing a simpler relationship between the Eulerian x, y coordinates and the new x, y' coordinates,

$$dy(x,y') = \rho_0 dy'/\rho(x,y'), \tag{9.126}$$

and taking into account the neglect of the second addend on the left side of Eq. (9.102), we obtain

$$\frac{\partial}{\partial y'}\left\{\rho\left(x,y'\right)k\left(x,y'\right)\frac{\partial T\left(x,y'\right)}{\partial y'}\right\} = \rho_0^2 c_p u\left(x,y'\right)\frac{\partial T\left(x,y'\right)}{\partial x} - \rho_0^2\frac{w\left(x,y'\right)}{\rho\left(x,y'\right)}. \tag{9.127}$$

We will now examine the behavior of a gas near a planar wall, assuming for simplicity's sake that the channel's opposite wall stands an infinitely great distance away. The conditions on approaching the wall's leading edge are:

$$T(0,y') = T_0; u(0,y') = U_0;\ \rho(0,y') = \rho_0. \tag{9.128}$$

The boundary conditions in the gas itself at the passive zone boundary, where the temperature reaches a peak, are:

$$\partial T(x,\ l_a)/\partial y' = 0. \tag{9.129}$$

On the wall itself, we will assume that

$$T(x,\ 0) = T_0. \tag{9.130}$$

Let us suppose that source power, W, per unit mass is constant:

$$W = w(x, y')/\rho(x, y') = \text{const}, \tag{9.131}$$

while the gas velocity outside the confines of the viscous boundary layer is constant and equals U_0:

$$u(x, y' \geq l_\nu) = U_0. \tag{9.132}$$

According to the model being described, because the heat exchange processes in the bulk of the gas volume—i.e., outside the confines of the passive zone—can be ignored, the temperature of each flow line before it intersects that passive zone will then equal

$$T\left(x, y'\right) = T_0 + \int_0^x \frac{w(x, y')}{c_p u(x, y')\rho(x, y')} dx = T_0 + \frac{W}{c_p}\int_0^x \frac{dx}{u(x, y')}. \tag{9.133}$$

We will determine the thickness of the viscous boundary layer near the surface of a planar plate using the correlation [2]

$$l_\nu = R\sqrt{\nu x/U_0}, \tag{9.134}$$

where R is a numerical coefficient, and ν is the kinematic viscosity. According to one of the viscous boundary layer models most frequently used in practice, the velocity profile therein is presented in the form of a linear dependence [2]:

$$u(x, y) = \begin{cases} \dfrac{U_0}{l_\nu(x)} y, & 0 \leq y \leq l_\nu(x); \\ U_0, & y \geq l_\nu(x). \end{cases} \tag{9.135}$$

In this instance, pursuant to the impulse theorem, $R = 3.46$ [2].

We will now examine the successive intersection of the viscous boundary layer and the thermal passive zone under consideration by a given flow line (Fig. 9.21). We will avail ourselves of a model according to which the zone in question and the viscous layer have a clearly pronounced boundary. Let us assume that l_a is the coordinate of the boundary of the zone under consideration in a cross-section with a coordinate of x_a. If $l_\nu > l_a$, then a liquid jet with a transverse coordinate of y' all the way up to the intersection with the viscous layer's boundary will have a velocity of U_0. Pursuant to Eq. (9.134), starting with a longitudinal coordinate of $x_\nu = y'^2 U_0/R^2\nu$, the jet flows within a region where the velocity, $u(x, y')$, is determined by the linear dependence of Eq. (9.135) upon y'. For a similar flow line at

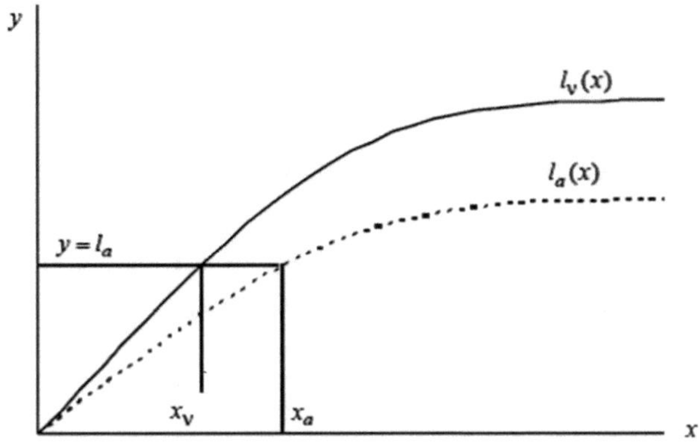

Fig. 9.21 Viscous boundary layer and passive zone near a planar wall

$\rho = \rho_0$, Eq. (9.133) can be presented as (for simplicity's sake, we will omit the prime marks associated with the y coordinate below):

$$T(x,y) = T_0 + \frac{W}{c_p U_0}\left\{\frac{U_0}{R^2 \nu} y^2 + \int\limits_{y^2 U_0/R^2 \nu}^{x} \frac{l_\nu(x)}{y} dx\right\}. \tag{9.136}$$

We will employ the hypothesis of the similarity of viscous and thermal boundary layers, extending it to the passive zone as well. A comparison of calculations and experiments [43] confirm the latter in an entirely satisfactory manner. Let us suppose that

$$l_a(x) \;=\; A_0 l_\nu(x). \tag{9.137}$$

From expression (9.136) for $y = l_a$ under the condition $l_a \leq l_\nu$, we get

$$T(x, l_a) = T_0 + \frac{W}{A_1 c_p U_0} x;\;\; A_1 = \frac{A_0}{A_0^3 - A_0^2 + 1}. \tag{9.138}$$

We will seek a solution to the problem using the following separation of variables for the transverse derivative from temperature:

$$\partial T(x,y)/\partial y \;=\; f(x)\varphi(x,y). \tag{9.139}$$

Pursuant to the boundary condition in Eq. (9.129), the φ function becomes zero at the passive zone boundary. At the wall itself, when $y = 0$, it can be assumed that $\varphi = 1$. We will present $\varphi(x, y)$ as a function of a single combined variable, $Y = y/l_a(x)$; i.e., $\varphi(x, y) = \varphi(Y)$. Then, pursuant to Eq. (9.139)

$$T(x,\ l_a) = T_0 + f(x) l_a(x) \int_0^1 \varphi(Y) dY. \tag{9.140}$$

From Eqs. (9.138) and (9.140), we get

$$f(x) = \frac{W}{A_1 c_p U_0 \Phi} \cdot \frac{x}{l_a(x)}, \quad \Phi = \int_0^1 \varphi(Y) dY,$$

whence, according to Eq. (9.139),

$$T(x, Y) = T_0 + \frac{Wx}{A_1 c_p U_0 \Phi} \int_0^Y \varphi(Y) dY, \quad \frac{\partial T(x, Y)}{\partial Y} = \frac{W \varphi(Y) x}{A_1 c_p U_0 \Phi},$$

$$\frac{\partial T(x, Y)}{\partial x} = \frac{W}{A_1 c_p U_0 \Phi} \left\{ \int_0^Y \varphi(Y) dY - \frac{x}{l_a(x)} Y \varphi(Y) \frac{dl_a(x)}{dx} \right\}.$$

Equation (9.127) is written in the new coordinates in the following manner:

$$\frac{\partial}{\partial Y} \left\{ \rho(x, Y) k(x, Y) \frac{\partial T(x, Y)}{\partial Y} \right\} = \rho_0^2 c_p l_a^2(x) u(x, Y) \frac{\partial T(x, Y)}{\partial x} - \rho_0^2 W l_a^2(x).$$

Integrating it from zero to unity with allowance for the conditions $\varphi(Y = 0) = 1$ and $\varphi(Y = 1) = 0$, as well as the dependence $u(x, Y) = A_0 U_0 Y$, for the velocity within the confines of the passive zone, we get

$$-\frac{ax}{A_1 U_0 \Phi l_a(x)} = \frac{l_a^2(x)}{A_1 \Phi l_v(x)} \left\{ \int_0^1 Y dY \int_0^Y \varphi(\theta) d\theta - \frac{x}{l_a(x)} \frac{dl_a(x)}{dx} \int_0^1 Y^2 dY \right\} - l_a(x),$$

where $a = k(x,0)/c_p \rho_0$. We introduce

$$\Psi = \int_0^1 Y^2 \varphi(Y) dY$$

then using Eqs. (9.134), (9.137), and (9.138) for A_1 , we obtain an equation for the A_0 coefficient:

$$R^2Pr\left\{\frac{1}{2}\Phi - \Psi\right\}A_0^6 - R^2Pr\left\{\frac{1}{2}\Phi - \Psi\right\}A_0^5 - \left\{\left[\frac{1}{2}\Phi + \Psi\right]R^2Pr - 1\right\}A_0^3$$
$$- A_0^2 + 1 = 0, \tag{9.141}$$

where $Pr = \nu/a$ is the Prandtl number.

Analysis demonstrates that at any uniform dependence, $\varphi(Y)$, that satisfies the conditions $\varphi(Y=0)=1$ and $\varphi(Y=1)=0$, Eq. (9.141) has two pairs of complex conjugate roots and two real positive roots, the first of which is smaller than unity, while the second is greater than unity. The first of the real roots, $A_0 < 1$, has physical significance by virtue of the assumption made that $l_a < l_\nu$. However, if we proceed under the assumption that $l_a > l_\nu$, then operating under this same premise, we obtain $A_0 < 1$ for $Pr \geq 0.3$ ($Pr > 0.6$ for rare gases), i.e., $l_a < l_\nu$, which contradicts the basic assumption.

An analysis performed for the Prandtl numbers that correspond to rare gases revealed that the A_0 coefficient may be found within limits of $0.5 \leq A_0 \leq 0.67$. Thus, according to the correlations used, Eqs. (9.134) and (9.137), the coefficient $A = A_0 \times R \times \sqrt{Pr}$ in a dependence of

$$l_a = A\sqrt{ax/U}. \tag{9.142}$$

falls within limits of $1.4 < A < 1.9$. Therefore, a conclusion can be reached: unlike sealed lasers, the size of a passive zone in gas-flowing lasers is dependent upon the Prandtl number, i.e., it is determined not only by heat conduction, but also by viscosity, and is roughly one-and-a-half to two times smaller than its value for sealed lasers without gas circulation.

Calculations of Spatial Inhomogeneities in NPLs with Gas Circulation

During the design of a particular laser system, the need may arise for the preliminary estimation and optimization of the spatial distribution of gas density in the laser channel's active region. The behavior of a gas in systems where the density deviations are comparable to, or exceed, their average values over the channel's volume can only be correctly calculated using two-dimensional numerical programs similar to that of the Idaho National Engineering Laboratory [44]. In ref. [45], a simplified technique was proposed for the direct numerical calculation of spatial inhomogeneities in the channel of a gas-flowing NPL. The only example of a similar calculation available at the time of publication performed using a simplified technique for determining the specific energy deposition of fission fragments in a channel of finite width with a plane-parallel uranium layer arrangement was also presented in ref. [45] by way of an illustration. A comparison demonstrated that the difference in the results of density distribution calculations using the procedure described in ref. [45] and the calculations performed using the two-dimensional gasdynamic program described in ref. [44] does not exceed ~2 %.

The fundamentals of the procedure described below [46] consist of the following. As a gas flows through a plane-parallel channel, the length of which, b, is comparable to its transverse dimension, d, the pressure in the laser channel's transverse cross-section is almost homogeneous (see the first section of this chapter). The typical gas temperature equalization time in the channel comes to $\tau_T \sim d^2/a$, where a is the thermal diffusivity coefficient of the gas. For the gas mixtures used in NPLs, $a \leq 1$ cm^2/s; hence, it follows that, at $d \sim 1$ cm, the temperature equalization time in the channel is $\tau_T \geq 1$ s. The time during which a portion of the gas traverses the channel in the x direction is determined by the obvious correlation $\tau \sim b/U$. At gas-flow velocities of $U > 1$ m/s, the average time that the portion of gas is present in the channel comes to $\tau < 0.1$ s. We will now apply the technique under consideration for the two-dimensional calculation of the gas flow in a channel to the gas flow velocities at which the time that the portion of gas is present in the channel, τ, is much shorter than the typical heat exchange time, τ_T. In this instance, the heat exchange processes in the bulk of the gas volume can be ignored during approximate calculations. The only exception consists of the thin near-wall passive zone layers, the method for considering the effect of which will be discussed next.

In the first approximation of the proposed technique, the flow lines, and accordingly, the gas jets separated along each line are assumed to be rectilinear, while the spatial distribution of fragment energy deposition is calculated under the assumption that the gas density distribution is homogeneous throughout the channel volume. The dependence of gas temperature upon the x coordinate is then calculated for each flow line, by means of which, using a state equation and proceeding on the basis of the condition that the gas pressure is almost homogeneous throughout the entire channel, the spatial gas density distribution is calculated. Thereafter, pursuant to the results obtained in refs. [44, 50], according to which the gas's longitudinal velocity profile, $U_n(x)$, is transversely homogeneous (with the exception of the narrow near-wall viscous boundary layer), and based on the requirements of mass flow constancy both within the channel as a whole and in each of the jets separated, the gas velocity and the variation in the transverse dimensions of the jets as a function of the x coordinate are calculated. The latter makes it possible to determine the curvature of the flow lines, and using them to correct the spatial distribution of gas density. During the next step, the specific energy deposition distributions of the fragments are calculated based on the density distribution obtained during the previous step, then the procedure for finding the density distribution is repeated using the arrangement described above, etc.

When determining the basic computational correlations of the stationary problem, it was assumed that the gas's density, velocity, and temperature distributions, as well as the specific intensity of uranium nuclear fission, are not dependent upon the z coordinate directed along the system's optical axis (in Fig. 9.1, the z axis is perpendicular to the figure's plane and coincides with the direction of the laser's optical axis), i.e., a purely two-dimensional problem was examined. The gas flow in the planar channel was represented by superimposing the independent jets, the camber and thickness of which can vary with the distance traversed, x. The effect of the viscosity within the jets and the friction between adjacent jets was ignored. Each jet was tagged with an Eulerian coordinate at the inlet, y_0, and its width was assumed to be so small that the density, velocity, and specific energy deposition within the confines of any of its transverse cross-sections could be regarded as homogeneous. Heat exchange between the jets was ignored. Then in the active region outside the confines of the passive zone by virtue of pressure homogeneity for a given jet, from the energy balance equation, we get

$$U(x,y_0)\rho(x,y_0)T(x,y_0)c_p\delta y(x,y_0) - U_0\rho_0T_0c_p\delta y_0 = \int\limits_0^x w(x,y_0)\delta y(x,y_0)dx,$$

where $T(x, y_0)$ is the gas temperature in the jet under consideration at a point with a coordinate of x; $U(x, y_0)$ is the gas velocity in the jet under consideration at a point with a coordinate of x; U_0 is the gas velocity at the channel inlet; $w(x, y_0)$ is the specific power deposition at a point with a coordinate of x for a jet marked with a coordinate of y_0; $\delta y(x, y_0)$ is the thickness of the jet under consideration at a point with a coordinate of x; δy_0 is the thickness of the jet under consideration at the channel inlet; $\rho(x, y_0)$ is the gas density in the jet at a point with a coordinate of x; and ρ_0 is the gas density at the channel inlet.

From the condition of mass flow constancy in each jet, we get

$$U(x,y_0)\rho(x,y_0)\delta y(x,y_0) = U_0\rho_0\delta y_0, \tag{9.143}$$

then, according to the previous equation,

$$T(x,y_0) = T_0 + \frac{1}{c_p}\int\limits_0^x \frac{w(x,y_0)}{U(x,y_0)\rho(x,y_0)}dx. \tag{9.144}$$

In the main body of the flow (outside the confines of the viscous boundary layers), the gas's longitudinal velocity profile, as previously mentioned, is transversely homogeneous ($U(x, y_0) = U_n(x)$) and Eq. (9.144) is simplified

$$T(x,y_0) = T_0 + \frac{1}{c_pU_n(x)}\int\limits_0^x \frac{w(x,y_0)}{\rho(x,y_0)}dx.$$

For rare gases, the relationship between density, temperature, and pressure is described with good accuracy by the ideal gas state equation

$$P = c_v(\gamma - 1)\rho T.$$

Thus, for the gas density in each jet, we get

$$\rho(x, y_0) = \frac{P}{c_v(\gamma - 1)} \cdot \frac{1}{\frac{1}{c_p}\int\limits_0^x \frac{w(x, y_0)}{U(x, y_0)\rho(x, y_0)} dx}.$$

According to Eq. (9.143), the relationship between the y_0 coordinate and the y Eulerian coordinate takes the form

$$y(x, y_0) = \rho_0 \frac{U_0}{U_n(x)} \int\limits_0^{y_0} \frac{dy_0}{\rho(x, y_0)}.$$

The dependence of the near-wall passive zone's transverse dimension upon time is described by an equation of the (9.142) type. Direct calculations (see Chap. 8) revealed that the proportionality factor for sealed lasers is $A \approx 3$ and is slightly dependent upon cell dimensions, neutron pulse shape and intensity, and uranium layer thickness, as well as upon the gas mixture's thermophysical properties. A passive zone is also formed in the channels of NPLs with gas pumping, the wall temperature of which is lower than the gas flow temperature. Its size is a function of many parameters that reflect the geometry and physical conditions of a flow, especially including viscosity and heat conduction. In the general case, it is not possible to derive an explicit analytical expression for the dependence of passive zone size upon the longitudinal x coordinate. However, since the time required for a portion of the gas to traverse a distance of x in the main section of the channel (outside the viscous layer) is

$$t = \int\limits_0^x \frac{dx}{U_n(x)}, \tag{9.145}$$

it is then proposed that the transverse dimension of the near-wall passive zone at this distance be roughly determined using the correlation

$$l_a(x) = A \left[a \int\limits_0^x \frac{dx}{U_n(x)} \right]^{1/2}. \tag{9.146}$$

As was shown in the preceding section (also see ref. [51]), even in the simplest case of $U_n(x) = U_0$ and $w/\rho = \text{const}$, the A factor is dependent upon the Prandtl number, $\text{Pr} = \nu/a$, and may lie within limits of $1.4 < A < 1.9$.

During the calculation procedure used in ref. [46], temperature distribution in the passive zone was described by the parabolic approximation

$$T_p(x,y) = \alpha_0(x) + \alpha_1(x)y + \alpha_2(x)y^2.$$

The α_i coefficients are determined from the conditions at the passive zone boundary and at the channel wall. So, for example, when a constant wall temperature is present, the following conditions are used:

$$T_p(x,l_a) = T(x,l_a);\ \partial T_p(x,l_a)/\partial y = 0;\ T_p(x,0) = T_0.$$

If the velocity in the flow core is also not dependent upon the longitudinal x coordinate (i.e., $U_n = U_0$) then the thickness of the viscous layer is determined by the correlation [2]

$$l_\nu(x) = R\sqrt{\nu\frac{x}{U_0}}. \tag{9.147}$$

The situation is more complicated when U_n is a function of the longitudinal x coordinate. Extant literature source contains no direct recommendations on how to proceed in this instance. Therefore, since the time required for a portion of the gas to traverse a distance of x within the channel outside the viscous layer is determined by Eq. (9.145), it is proposed that the transverse dimension of the viscous layer at this distance be approximately determined using the correlation

$$l_\nu(x) = R\left[\nu\int_0^x \frac{dx}{U_n(x)}\right]^{1/2}. \tag{9.148}$$

The following approximation was used to describe velocity distribution in a viscous layer at $y_0 \le l_\nu(x)$

$$u(x,y) = U_n\left(\frac{3}{2}\frac{y}{l_\nu(x)} - \frac{1}{2}\left[\frac{y}{l_\nu(x)}\right]^3\right).$$

In this case, according to the impulse theorem, $R = 4.64$ [2]. Such a distribution duplicates the true velocity distribution [2] near a planar surface washed around by an incompressible fluid that has a constant viscosity and a constant velocity outside the confines of the viscous layer with an error of less than 1 %.

Thus, an artificial technique was used to describe the width of the passive zone and the viscous layer, as well as temperature and velocity distribution within their confines. Nonetheless, at sufficiently high velocities, U_0, the dimensions of these structures are small; therefore, it was anticipated that the proposed approach would not lead to noticeable errors during the calculation of density, temperature, and velocity distributions in the active zone.

The spatial distribution of the specific power deposition due to fission fragment slowing-down was calculated using the procedure described in ref. [49].

During the third iteration step, the gas density distribution differs from the distribution obtained at the end of the first step by not more than 1 %, which is indicative of the good precision of the iterative scheme used.

A comparison demonstrated that the difference between the density distribution calculation results obtained using the procedure described in ref. [46] and similar calculations [44] performed using a two-dimensional gasdynamic program like the one employed during the calculations in ref. [45] based on a simplified procedure for determining the specific energy deposition of fission fragments, does not exceed 2 %. In addition, it was learned that the variation of the A factor value from 1.4 to 3 only results in perceptible density changes during calculations of the same problem alternative in the passive region; for the active region, these changes do not exceed ~1 %. At a gas-flow velocity of $U_0 = 4.5$ m/s, A factor variability leads to relative fluctuations in the active region volume of ~10 %. These fluctuations diminish with an increase in gas-flow velocity; so, at $U_0 = 9$ m/s, they come to ~6 %. Nonetheless, at all A values, the nature of the behavior of all the dependences cited in this work is preserved.

The channel studied during the calculations, which has dimensions of $d = 2$ cm and $b = 6$ cm, has open layers (without a protective film) that consist of metallic uranium with a thickness of $\delta_U = 2.78$ μm (a normalized thickness of $D_1 = 0.5$). It was assumed that uranium fission intensity in the active layers comes to $q = 2 \times 10^{16}$ cm$^{-3} \times$ s^{-1}. The spatial distributions of density and specific energy deposition in active volume of the gas were calculated for a wide range of helium and argon pressure and velocity variations. It was shown that in the central regions of the gas volume (at distances from the channel inlet and outlet on the order of the third fission fragment range in the gas), the dependence of these parameters upon the longitudinal coordinate parallel to the gas flow direction is almost linear in nature, while their dependence upon the transverse coordinate is close to parabolic. The results cited below correspond to a value of $A = 3$ (apparently the least probable value [51]), which corresponds to the smallest active region volume.

An important parameter that characterizes a real laser channel with a finite dimension of b along the gas flow direction is the efficiency, ε, of total energy deposition in a laser-active region not involved in near-wall passive zones. It is defined as the ratio of the total fission-fragment energy deposition in the active region solely to the total fragment energy released in the uranium layers. It is useful to compare this parameter to the efficiency of energy deposition, ε_0, in an unperturbed gas in a channel without gas pumping ($U_0 = 0$) that has a homogeneous density distribution (there are no near-wall passive zones). Below, the former will be called the real energy deposition efficiency, while the latter will be called the ideal energy deposition efficiency of an unperturbed gas.

In ref. [37] (also see Sect. 7.5), an energy deposition optimization parameter was introduced for NPLs without gas pumping, excited by fairly short neutron pulses with a duration in the millisecond range,

$$\omega_0 = D_1 \varepsilon_0 \frac{w_0}{\langle w \rangle},$$

where $D_1 = \delta_U/R_1$ is the normalized uranium layer thickness; R_1 is the range of an average fission fragment in this layer; w_0 is the specific power deposition in the center of the laser channel; and $\langle w \rangle$ is the average value of the specific power deposition over the channel cross-section.

In laser channels with gas pumping that operate in the stationary mode, a near-wall passive zone can reach appreciable dimensions at the channel outlet in the presence of relatively low gas-flow velocities. In addition, gas density steadily decreases in the longitudinal direction. For these reasons, real energy deposition efficiency, ε, can differ conspicuously from unperturbed channel energy deposition efficiency, ε_0; thus, a more rigorous statement of the optimization parameter should be introduced:

$$\omega_1 = D_1 \varepsilon \frac{\langle w(0) \rangle_x}{\langle w(x, y) \rangle}$$

or

$$\omega_2 = D_1 \varepsilon \frac{w(b/2, 0)}{\langle w(x, y) \rangle},$$

where

$$\langle w(0) \rangle_x = \frac{1}{b} \int_0^b w(x, 0)\, dx,$$

$$\langle w(x, y) \rangle = \frac{1}{bd} \int_0^b \int_{-d/2}^{d/2} w(x, y)\, dx\, dy,$$

while $w(b/2, 0)$ is the specific energy deposition value at a point that corresponds to the geometric center of the channel.

The ω_1 and ω_2 parameters are distinguished from ω_0 by the fact that the calculated real energy deposition efficiency, ε, is used therein instead of unperturbed channel energy deposition efficiency, ε_0. In the first of these, the fraction characterizes the ratio of the specific energy deposition averaged over the channel's entire central plane, $y = 0$, to the specific energy deposition averaged over the channel's cross-section. In the second, it characterizes the ratio of specific energy deposition in the center of the channel to the specific energy deposition averaged over the channel's cross-section.

The dependences of the ω_1 and ω_2 optimization parameters upon argon pressure in the channel at $U_0 = 4.5$ m/s are shown in Fig. 9.22. Here, too, a similar

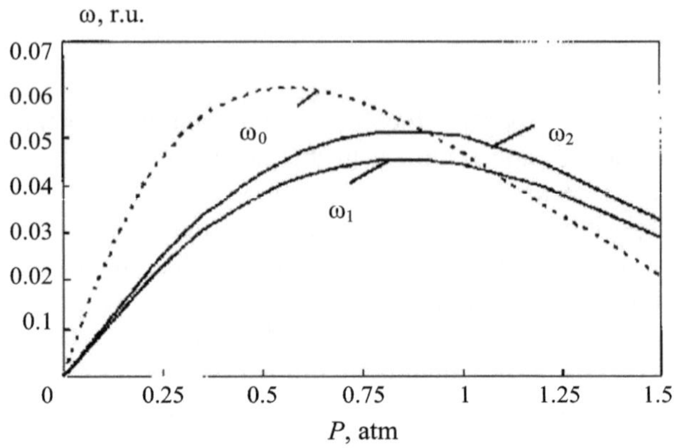

Fig. 9.22 Dependence of the optimization parameters for argon upon pressure at $U_0 = 4.5$ m/s

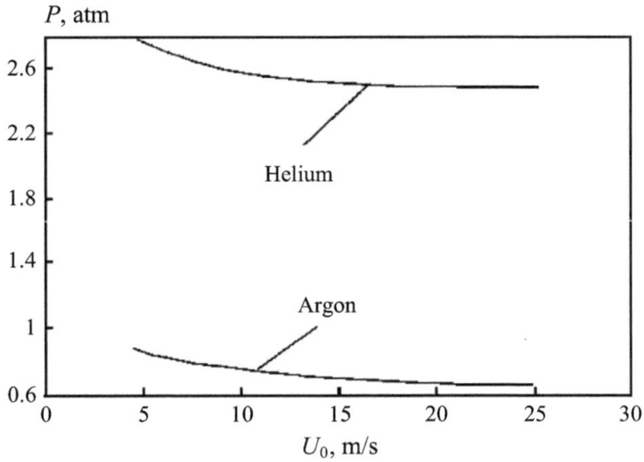

Fig. 9.23 Dependence of optimum gas pressure upon gas-flow velocity

dependence for the ideal parameter, ω_0 (at $U_0 = 0$), is shown. The calculated dependences of the optimum (as far as energy deposition) argon and helium pressures upon gas-flow velocity are presented in Fig. 9.23. We note that the optimum argon pressure in a channel with gas pumping over the velocity interval studied, U_0, may exceed the optimum pressure for a sealed unpumped laser operating in a comparatively short (≤ 10 ms) pulse lasing mode ($P \approx 0.6$ atm) by 1.5 times (up to $P = 0.9$ atm), whereas the optimum helium pressure in a channel with gas pumping differs little (not more than 15 %) from the optimum pressure for a sealed unpumped laser. The spatial distributions of argon density at an optimum pressure of $P = 0.9$ atm and a velocity of $U_0 = 4.5$ m/s are presented in Figs. 9.24 and 9.25 for $T_0 = 293$°K.

The calculations made it possible to find out a quite important fact. At the optimum pressure precisely, the distribution of specific energy deposition in the central plane, $y = 0$, and in the adjacent planes, $y = \text{const}$, is symmetrized relative to the central cross-section, $x = b/2$ (see Fig. 9.26b), equalizing in a considerable

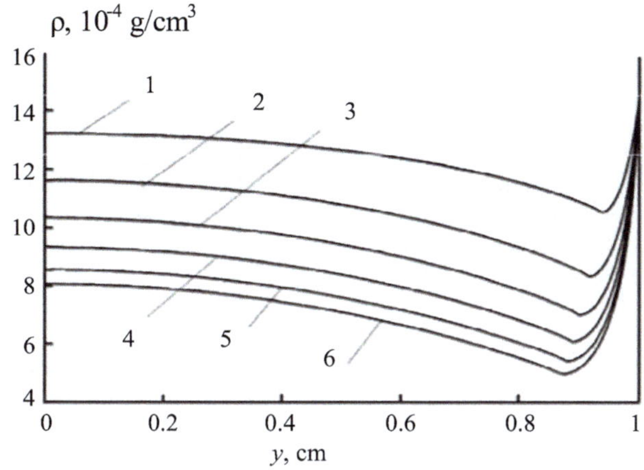

Fig. 9.24 Gas density distribution in a laser channel in the transverse direction at the optimum argon pressure ($U_0 = 4.5$ m/s): (*1*) $x = 1$ cm; (*2*) $x = 2$ cm; (*3*) $x = 3$ cm; (*4*) $x = 4$ cm; (*5*) $x = 5$ cm; (*6*) $x = 6$ cm

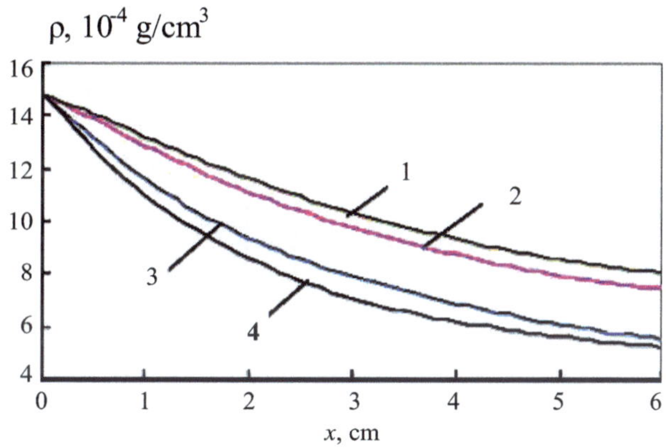

Fig. 9.25 Gas density distribution in a laser channel in the longitudinal direction at the optimum argon pressure ($U_0 = 4.5$ m/s): (*1*) $y = 0$; (*2*) $y = 0.4$ cm; (*3*) $y = 0.8$ cm; (*4*) $y = 0.9$ m

portion of the gas volume. At pressures lower that the optimum value in a gas region that stands away from the inlet and outlet sections by a distance on the order of one-third of the fragment range in the gas, specific energy deposition steadily decreases with an increase in x for any $y = \text{const}$ plane (Fig. 9.26a). At pressures higher than the optimum value, a reverse pattern emerges (Fig. 9.26c), with the exception of the near-wall regions, where the decrease in energy deposition with an increase in x is retained at all pressure values.

It follows from flow calculations in real NPL channels that the dependence of the main flow's velocity upon the coordinate is close to linear (this fact is in full agreement with the results of the calculations performed in ref. [44]):

$$U_n(x) \approx U_0 + Bx. \tag{9.149}$$

In this case, according to Eq. (9.146), we get:

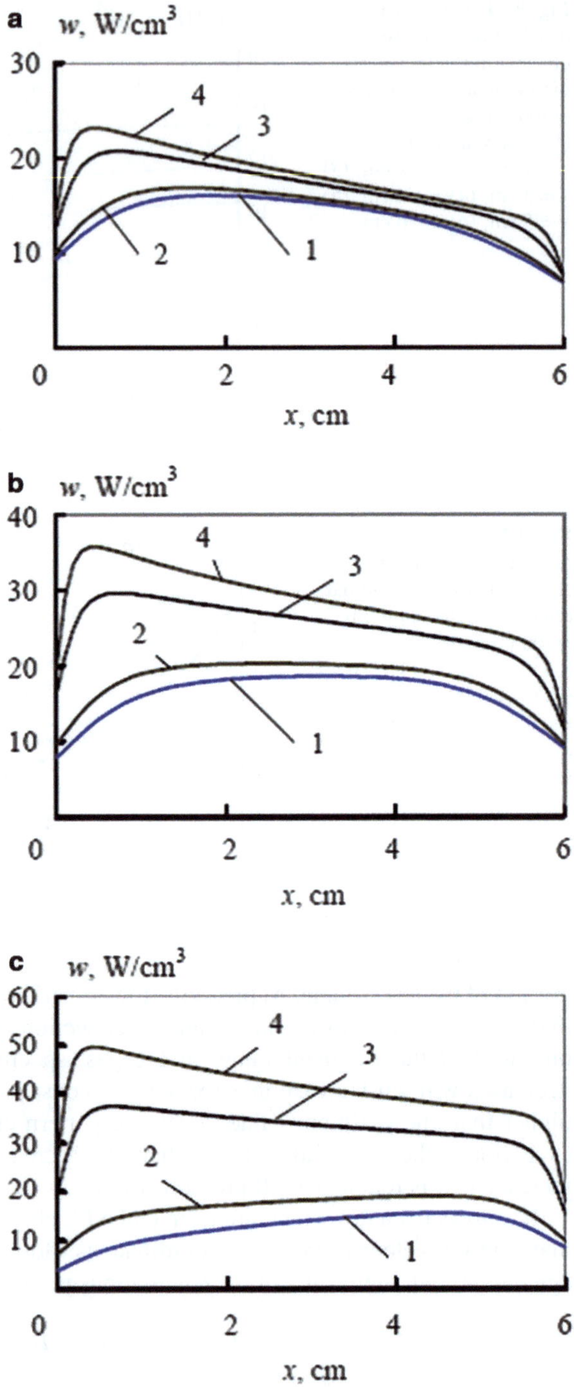

Fig. 9.26 Distribution of specific energy deposition in a laser channel in the longitudinal direction at an argon pressure of ($U_0 = 4.5$ m/s): $P = 0.5$ atm; (**a**) $P = 0.9$ atm; (**b**) $P = 1.4$ atm (**c**); (*1*) $y = 0$; (*2*) $y = 0.4$ cm; (*3*) $y = 0.8$ cm; (*4*) $y = 0.9$ cm

$$l_a(x) \approx A\sqrt{\frac{a}{B}\ln\left(1+B\frac{x}{U_0}\right)}, \tag{9.150}$$

It follows from a comparison of the Eq. $l_a(t) \approx A\sqrt{at}$ and relation (9.150) that the time required for a portion of the gas to traverse a distance of x from the channel inlet comes to

$$t \approx \frac{1}{B}\ln\left(1+B\frac{x}{U_0}\right).$$

The latter is in agreement with the results obtained in ref. [43], in which

$$B = \frac{2E_0\varepsilon_1\delta_U q}{T_0\rho_0 c_p d},$$

where E_0 is the kinetic energy of the uranium fission fragments, and ε_1 is a certain effective value of the share of energy of the fission fragments that transmit to the gas. An analysis of the calculation results demonstrated that ε_1 is slightly smaller than the ideal energy deposition efficiency, ε_0; however, the deviation of ε_1 from ε_0 at all pressure and velocity values does not exceed 20 %. In the presence of sufficiently high velocities and weak fission intensities, when $Bx/U_0 << 1$, we get a simple dependence

$$l_a(x) = A\sqrt{a\frac{x}{U_0}}.$$

Radiation Intensity Distribution in Gas-Flowing Laser

The distribution of the radiation intensity, shape, and area of a light spot on the cavity mirror surface of a sealed laser with plane-parallel arranged uranium layers was studied in ref. [52]. At the moment in time that lasing originates, the radiative zone is concentrated near the laser cell axis, then it begins to enlarge and its shape becomes close to rectangular at the lasing pulse peak. Radiation intensity distribution is symmetrical relative to the planes of symmetry, $y = \text{const}$ and $x = \text{const}$, that passes through the optical axis. The lasing volume reaches ~60 % of the laser cell's volume.

This behavior is determined by the dynamics of the development of gas density inhomogeneities and the spatial redistribution of specific energy deposition in the laser's sealed cell over the course of the pumping pulse. It is natural that the dynamics for a laser with transverse gas pumping have a different, more complex nature. Consequently, the pattern of radiation intensity distribution throughout the laser channel cross-section should change as well, which was also experimentally proven in ref. [53].

During experimentation, an Ne–Ar–Xe mixture (300:100:1) at a pressure of 1 atm was used, which was flowed through the laser channel of an LM-4 module at a velocity of ~7.2 m/s. The average thickness of the metallic uranium layers (90 % enriched by ^{235}U) was $\delta_U = 2.67$ μm. The uranium layers were deposited to aluminum substrates with a thickness of 4 mm and were covered with a thin (0.5 μm) aluminum film for the purpose of preventing ^{235}U atom ejection into the gas [54, 55]. Just as during the experiments described in ref. [52], which involved a sealed channel, the distance between the layers was 2 cm and the channel dimension along the gas flow direction was 6 cm. The length of the laser channel along the optical axis direction was 1 m. A high-reflectivity spherical dielectric mirror with a radius of curvature of 20 m and a semi-transparent planar dielectric mirror were used as the cavity mirrors. The distance between the mirrors was 1.4 m.

Excitation was accomplished by means of neutrons from the core of a BIGR reactor. Diagrams of the gas flow direction, as well as the mutual position of the laser module and the BIGR reactor, are presented in Figs. 9.1 and 9.15. The duration of the reactor's exciting quasi-pulse was ~1.5 s. The average specific power deposition in the gas over the length of the laser channel at the exiting pulse peak reached ~6.2 W/cm^3. The lasing threshold came to ~35 % of the maximum neutron flux. The maximum laser output power reached 12 W.

A charge-coupled camera operating in the infrared region of the spectrum recorded an image of the laser spot. An image of the laser beam's transverse cross-section obtained from a single laser module channel was formed on the sensing element of a camera with a pixel size of ~0.1 mm. During neutron pumping, the camera operated in the burst recording mode and captured images of the laser spot at an interval of 30 ms. Over the course of the pulse, 25 frames were produced that depicted the laser beam's development in chronological order.

FSA-G1 sensors with a photoresistor based on lead sulfide and IMO-2N optical radiation power meters were used to determine laser output power.

The experimental dependences of neutron flux density and laser output power upon time are shown in Fig. 9.27. Lasing begins at the moment in time that the

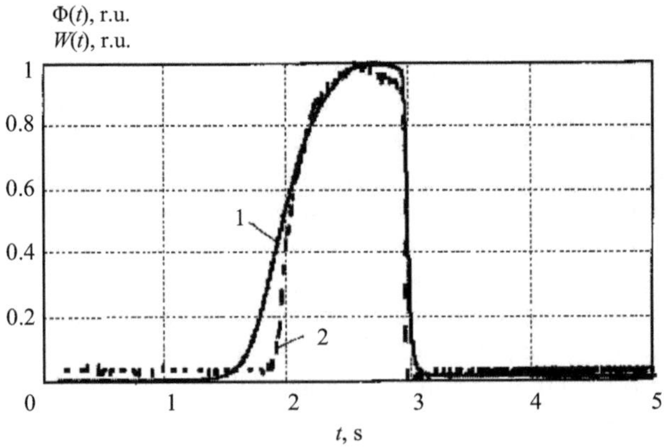

Fig. 9.27 Time dependencies of neutron flux density (*1*) and laser output (*2*)

threshold is reached and thereafter its dependence upon time almost fully duplicates the similar dependence of neutron flux density.

Radiation intensity distributions at the surface of output mirror are presented in Fig. 9.28 for a number of successive moments in time.

The relative distribution of laser radiation intensity in the perpendicular and parallel directions relative to the gas flow's motion is shown in Fig. 9.29.

At a velocity of $U \sim 7$ m/s and a channel dimension along the gas flow direction of $b = 6$ cm, the typical time that a portion of the gas resides in the channel is $\tau_0 \sim b/U \sim 10^{-2}$ s, which is considerably shorter than the neutron pulse's duration. This fact makes it possible to apply the procedure described in ref. [46] for calculating the gasdynamic characteristics of the stationary excitation mode to the calculation of the spatial distributions of gas density and specific power deposition in the laser channel at each given moment in time.

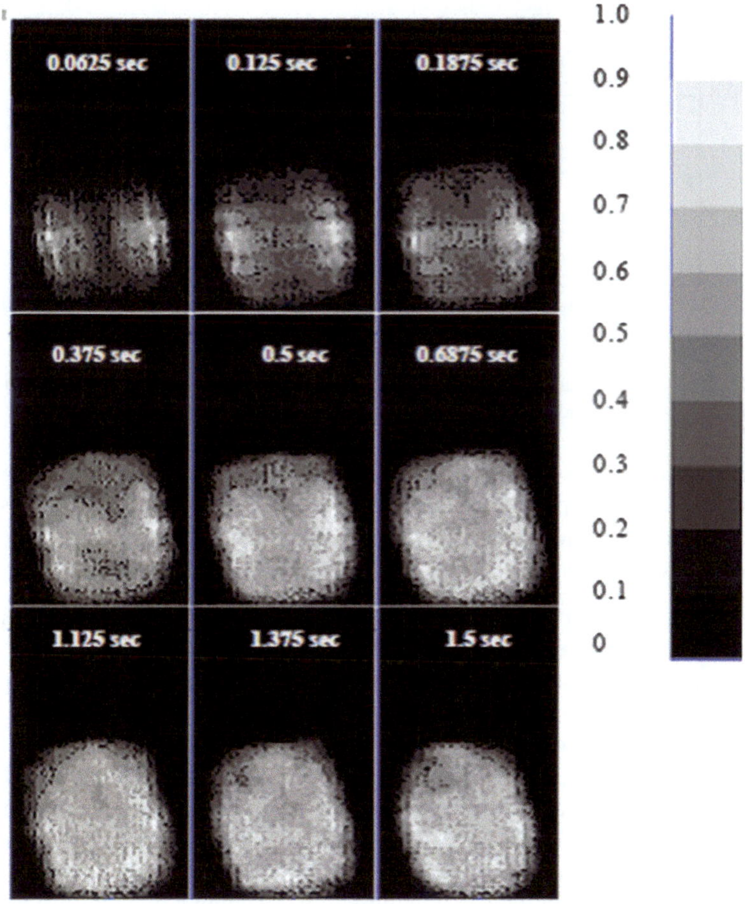

Fig. 9.28 Distribution of laser radiation intensity (in r.u.) at the surface of output mirror at different moments in time

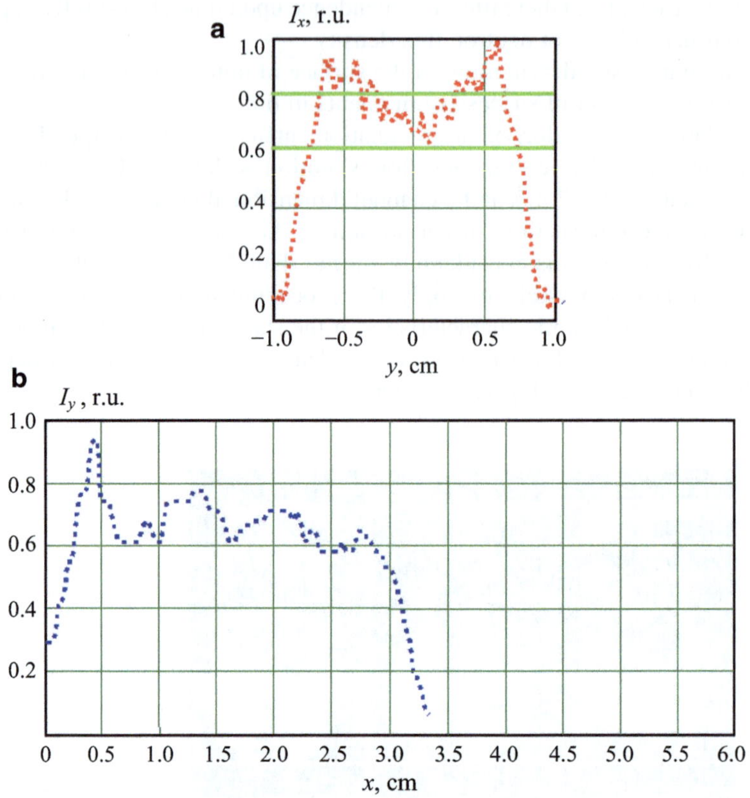

Fig. 9.29 Relative laser radiation intensity distribution in the transverse (a) and longitudinal (**b**) directions at $x = 1.5$ cm and $y = 0$, respectively

The cumulative results of the experiments and the calculations (Fig. 9.30) performed using the procedure described in ref. [46] under conditions of an excitation and a geometry that fully corresponds to the experimental conditions categorically evidences that:

- In flowing nuclear-pumped lasers, lasing is “spawned" in two regions that are symmetrical relative to the longitudinal plane of symmetry and that closely border the laser channel gas inlet, within which the specific power deposition and gas density approach the maximum values;
- In direct proximity to the uranium layer surface where power deposition reaches the maximum value, two narrow regions exist that do not take part in lasing, the presence of which can be explained by the origination of passive zones; and
- As the neutron flux density and subsequently power deposition increases at each given point in the gas volume, both of the symmetrical lasing zones merge into one, which expands in the gas flow direction, reaching a maximum size (~60 % of the laser channel’s volume) at a moment in time that corresponds to the exciting pulse peak.

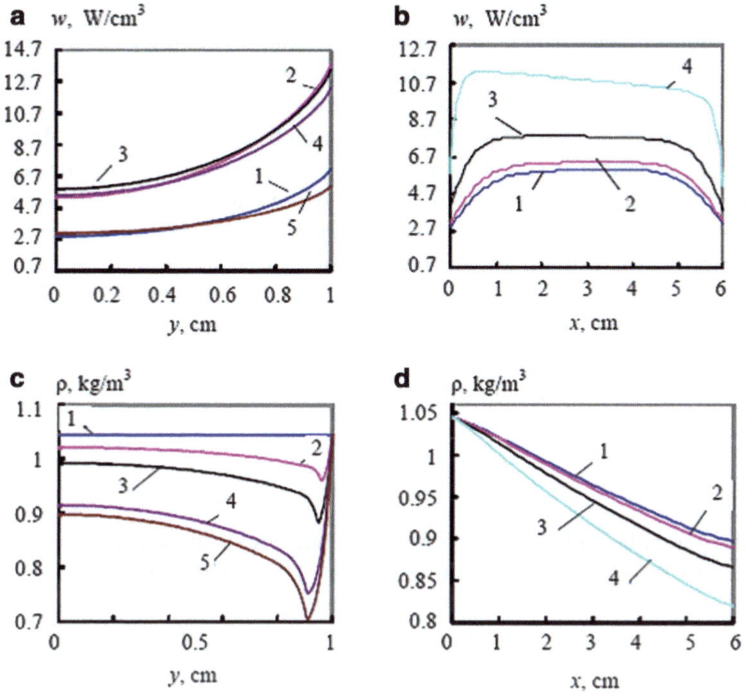

Fig. 9.30 Spatial distribution of specific power deposition (**a** and **b**) and gas density (**c** and **d**) in the transverse (**a** and **c**: *(1)* $x = 0$; *(2)* $x = 1$ cm; *(3)* $x = 2$ cm; *(4)* $x = 5$ cm; *(5)* $x = 6$ cm) and longitudinal (**b** and **d**: *(1)* $y = 0$; *(2)* $y = 0.3$ cm; *(3)* $y = 0.6$ cm; *(4)* $y = 0.9$ cm) directions at the moment in time that the exciting pulse peak is reached

Permissible Gas Overheating and Cavity Stability

In the $y = \text{const}$ planes, gas density, and accordingly, the refractive index in the bulk of a flowing laser's gas volume decrease in an almost linear manner with an increase in x; only within the fairly short channel inlet and outlet parts near to edge of a uranium layer and further outside its confines the gas density not dependent upon the x coordinate. The latter results from a noticeable decrease in power deposition in direct proximity to the edges of the uranium layers and outside their confines. In any $x = \text{const}$ plane, refractive index distribution (with the exception of a near-wall passive zone) is satisfactorily described by a negative parabola. The parabola's curvature increases in a linear manner with the x distance value. It reaches a maximum value near the edge of a uranium layer, $x = b$, and subsequently remains unchanged.

A conventional two-mirror cavity was used in the LM-4 apparatus [26, 41] that consisted of planar and spherical (with a radius of curvature of 20 m) mirrors. However, the two-dimensional stable-unstable cavity proposed in refs. [22, 24] is more convenient for a similar NPL with gas pumping. It should be unstable in the longitudinal direction in the $y = \text{const}$ planes and stable in the transverse direction in the $x = \text{const}$ planes. The latter can be achieved, for example, using mirrors with a predetermined radius of curvature in these planes.

The distinctive features noted above for gas density distribution in gas flowing NPLs do not make it possible to obtain a solution to the equation describing a light beam propagation that is suitable for studying cavity stability using the ray matrix technique in a channel's entire volume, as was done for a sealed NPL in refs. [56, 57]. However, an obviously simplified approach can be used to derive stationary mode criterion for estimating the permissible gas heating from an analysis of stability in the $x = \text{const}$ plane. So, the infringement of the stability condition in the space immediately past the right edge of a uranium layer is examined in ref. [58] (the gas is pumped in the direction of the x axis). In this region, the refractive index distribution has a parabolic shape

$$n(y) = n_0(b) - \frac{1}{2}\alpha(b)y^2, \tag{9.151}$$

where the parabola coefficient, α, is maximal and equals the value reached in direct proximity to the edge of the uranium layer. This coefficient increases with an increase in the power deposition in the gas, and accordingly, its overheating. If it exceeds the value at which the cavity becomes unstable in the $x = \text{const}$ plane in the $x \geq b$ region, it is then natural to assume that the instability in the transverse direction relative to x will be shifted into the $x < b$ region with a further increase in the power deposition.

As previously demonstrated, with the exception of the thin near-wall passive zone layers, heat exchange in the bulk of the gas volume can be ignored. Because the pressure in the channel is almost homogeneous, density and temperature distribution is determined by the distribution of specific fission fragment power deposition. The fission fragment energy losses per unit path are proportional to the medium's density [59]. We will ignore the edge effects, which have a comparatively minor influence on integral energy deposition in the gas as it passes through the channel when the dimension of the uranium layers is sufficient. Therefore, along each flow line that a particular gas particle traverses, the power deposition per unit mass of gas, f, can be roughly regarded as constant. This approach is equivalent to the one used in refs. [57, 60] to represent the distribution of specific power deposition in the form $F(r) = F_0(r)\rho(r)/\rho_0$, where $F_0(r)$ is the specific power deposition at a point, r, in the presence of a homogeneous unperturbed gas density distribution, ρ_0. For an infinitely expanded planar geometry that is characterized by dependence upon a single y coordinate alone, this equality is strict due to the absence of edge effects [61].

For the gas velocities at which inequality (9.97) holds true, the temperature increment of a particular gas particle of unit mass after it has traversed a distance of x, according to the approximations used, equals

$$T(y,x) - T_0 = \frac{f(y)}{c_p} \times \frac{x}{U_0}.$$

Because the gas pressure differential in the channel is small, $(\Delta P/P) << 1$, then pursuant to the ideal gas state equation, $\rho(x, y) = T_0\rho_0/T(x, y)$. Thus,

$$\rho(x,y) = \rho_0\left[1 + \frac{f(y)x}{c_pU_0T_0}\right]^{-1}. \tag{9.152}$$

The surface power density of the fission fragments from the uranium layers into the gas can be expressed by way of the specific power deposition of the gas averaged over the transverse cross-section, $q_S = \rho_0\bar{f}d$. Thus, condition (9.97), in the presence of which correlation (9.98), used to derive Eq. (9.152), holds true, is equivalent to the condition $(fx/c_pU_0T_0) << 1$. Therefore, expanding (9.152) into a series and selecting the first-order terms, we obtain

$$\rho(x,y) = \rho_0\left[1 - \frac{f(y)x}{c_pU_0T_0}\right]. \tag{9.153}$$

With a plane-parallel uranium layer arrangement, the specific power deposition in a unit mass of gas in the coordinate system shown in Fig. 9.1, ignoring the edge effects, is determined by the equations [61]

$$\begin{aligned} f(y) &= \frac{1}{\rho_0}E_0q\frac{R_1}{R_0}I(Y); \\ I(Y) &= G\left(\frac{D_0}{2} - Y\right) + G\left(\frac{D_0}{2}Y\right) - G\left(\frac{D_0}{2} - Y + D_1\right) \\ &\quad - G\left(\frac{D_0}{2} + Y + D_1\right), \end{aligned} \tag{9.154}$$

where $Y = y/R_0$; R_0, R_1 are the average range of a fission fragment in the gas and in the uranium layer material, respectively; $D_0 = d/R_0$ is the normalized gas channel width; and $D_1 = \delta/R_1$ is the normalized uranium layer thickness, while the G function is determined by the correlation

$$G(\xi) = \frac{1}{2}(1 - \xi^2 + 2\xi\ln\xi). \tag{9.155}$$

Following the appropriate transformations and the expansion of the addends containing the logarithms into a series for the small parameter $2Y/D_0$, with an accuracy up to terms of the second-order of smallness for central regions located near the channel's plane of symmetry, we obtain

$$I(Y) = A_n + \frac{4D_1}{D_0(D_0 + 2D_1)} Y^2, \tag{9.156}$$

where

$$A_n = \left(\frac{D_0}{2}\right)^2 + \left(\frac{D_0 + 2D_1}{2}\right)^2 + D_0 \ln\frac{D_0}{2} - (D_0 + 2D_1)\ln\frac{D_0 + 2D_1}{2}.$$

From the heat balance condition in the stationary circulation mode

$$\int_0^x q_S(x)dx = U_0 c_p \rho_0 d\left[\overline{T(x)} - T_0\right],$$

in the approximation $q_S(x) = \text{const}$ for overheating averaged throughout the channel's cross-section, we get

$$\overline{\Delta T} = \overline{T(b)} - T_0 = \frac{q_S b}{c_p \rho_0 d u_0}. \tag{9.157}$$

We will express the surface power density of the fission fragments from the uranium layers into the gas, q_S, in terms of the total energy of the uranium nuclear fission fragments, E_0, the intensity of uranium fissions per unit volume, q, uranium layer thickness, δ_U, and energy deposition efficiency, ε (the share of fission fragment energy transmitted to the gas):

$$q_S = 2E_0 q \delta_U \varepsilon. \tag{9.158}$$

Determining the refractive index based on its unperturbed value at the inlet, n_{00},

$$n(x, y) = 1 + (n_{00} - 1)\frac{\rho(x, y)}{\rho_0},$$

then from Eqs. (9.153), (9.154), and (9.156) with allowance for the relationship of Eqs. (9.157) and (9.158) between q and $\overline{\Delta T}$ at $x = b$, we obtain Eq. (9.151), in which

$$\alpha = \frac{4(n_{00} - 1)}{(D_0 + 2D_1)R_0^2 \varepsilon T_0} \overline{\Delta T}; n_0 = n_{00} - \frac{(n_0 - 1)AD_0}{2D_1 \varepsilon T_0} \overline{\Delta T}. \tag{9.159}$$

The dimensionless γ parameter determined by formula (8.40) was introduced in ref. [56] (see Chap. 8.5). Under a given spatial distribution of specific sources, it increases with an increase in energy deposition. The value of this parameter at which the infringement of the stability condition $|m| < 1$ begins is called the critical value. The γ_{cr} parameter reaches its highest values in the case of a cavity made up of

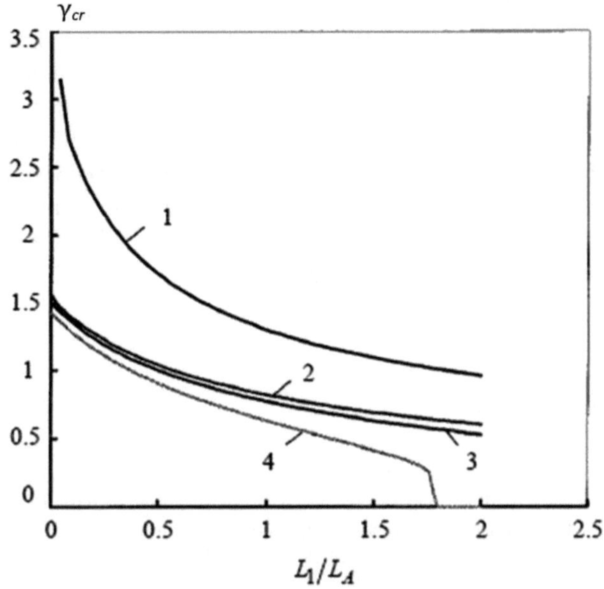

Fig. 9.31 Dependence of the critical stability parameter upon the relative clearance between the mirrors and the active volume at $r_1=\infty$: (*1*) $r_2=\infty$; (*2*) $r_2=20$ m; (*3*) $r_2=10$ m; (*4*) $r_2=5$ m

planar mirrors (with radii of curvature of $r_1=r_2=\infty$), as well as a cavity with codirectional mirrors of identical curvature ($r_1=-r_2$). In the case of other mirror curvature combinations, the γ_{cr} value does not exceed 1.5. The dependence of γ_{cr} upon the relative length of the clearance between a mirror and the active volume boundary, L_1/L_A, calculated using the technique described in ref. [56], is presented in Fig. 9.31 for several radius of curvature values, r_1, of one of the mirrors. It was assumed that $L_1=L_2$, while the radius of curvature of the second mirror was $r_2=\infty$.

If it is assumed that the stability condition should not be infringed in the $x=\text{const}$ planes within the gas regions of $x\geq b$, where energy deposition reaches the maximum value, then from the requirement $\gamma<\gamma_{cr}$, as well as Eqs. (8.40) and (9.159), we obtain

$$\overline{\Delta T} < \frac{n_0(D_0+2D_1)R_0^2\varepsilon T_0\gamma_{\text{кр}}^2}{4(n_{00}-1)L_A^2}. \tag{9.160}$$

During medium overheating, $\overline{\Delta T}<10^4$ K $(n_0-1)<<1$; therefore, in Eq. (9.160), n_0 can be replaced with unity. Thus, according to the criterion under consideration, permissible gas overheating through values of D_0, D_1, R_0, ε, and L_A is determined by gas mixture density, the geometric dimensions of the active gas volume, and uranium layer thickness, while through γ_{cr}, it is determined by resonator mirror curvature and the distances from the mirrors to the active volume.

The L_1 and L_2 distance values are selected in such a way that the uranium nuclear fission fragments do not reach the cavity mirrors. In actual designs, they usually come to $L_1 = L_2 \approx 10$ cm; thus, at $L_A \sim 1$ m, it can be roughly assumed for estimates that $\gamma_{cr} \leq 1.5$. For the laser channel of the LM-4 apparatus ($d = 2$ cm, $\delta_U = 2.7 \times 10^{-4}$ cm, and $L_A = 1$ m) during the pumping of a gas mixture, the basic component of which is helium ($P_0 = 2$ atm) at $T_0 = 300$ K, an estimate using formula (9.160) yields $\overline{\Delta T} < 1{,}900$ K, while for an argon-based mixture at $P_0 = 0.5$ atm and $T_0 = 300$ K, $\overline{\Delta T} < 860$ K. According to Eq. (9.160), a twofold increase in active length, L_A, while the other parameters remain unchanged, as exemplified by an argon-based mixture, will result in the stability limitation becoming more rigorous ($\overline{\Delta T} < 230$ K.)

References

1. F.V. Bunkin, V.I. Derzhiyev, A.V. Karelin, A.L. Nefedov, V.I. Subbotin, V.V. Kharitonov, K.R. Chikin, S.I. Yakovlenko, *Statsionarnyy reaktor-lazer s poverkhnostnoy nakachkoy* [Stationary Laser Reactor with Surface Pumping], No. 321. *Preprint IOF AN SSSR* [Preprint of the Institute of General Physics (IGP) of the AS USSR (Academy of Sciences, Union of Soviet Socialist Republics, Moscow, 1985)]
2. H. Schlichting, *Teoriya pogranichnogo sloya* [Boundary Layer Theory]. *Inostr. lit-ra* [Foreign Literature Publishing House, Moscow, 1956]
3. L.N. Polyanin, M.Kh. Ibragimov, G.I. Sabelev, *Teploobmen v yadernykh reaktorakh* [Heat Exchange in Nuclear Reactors]. *Energoizdat* [Scientific and Technical Publishing House of Power-Engineering Literature, Moscow, 1982]
4. L.D. Landau, Y.M. Lifshits, *Mekhanika sploshnykh sred* [Continuum Mechanics]. *Gostekhteorizdat* [State Publishing House of Technical and Theoretical Literature, Moscow, 1954]
5. F.Y. Kramerov, Y.V. Shevelev, *Inzhenernyye raschety yadernykh reaktorov* [Engineering Calculations of Nuclear Reactors]. *Atomizdat* [Publishing House of the State Committee on the Use of Atomic Energy, Moscow, 1964]
6. I.I. Novikov, K.D. Voskresenskiy, *Prikladnaya termodinamika i teploperedacha* [Applied Thermodynamics and Heat Transfer]. *Gosatomizdat* [State Publishing House of Literature in the Area of Nuclear Science and Technology, Moscow, 1961]
7. Y.F. Ratnikov, S.D. Tetelbaum, *Gazy kak teplonositeli i rabochiye tela yadernykh energeticheskikh ustanovok* [Gases as Heat Carriers and Working Media of Nuclear Power Installations]. [*Atomizdat*] (Publishing House of the State Committee on the Use of Atomic Energy, Moscow, 1978)
8. A.N. Kolmogorov, Lokalnaya struktura turbulentnosti v neszhimayemoy zhidkosti pri ochen bolshikh chislakh Reynoldsa [Local structure of turbulence in an incompressible fluid at very high Reynolds numbers]. Doklady AN SSSR **30**(4), 299–303 (1941)
9. G.U. Lipman, Vzlet i padeniye idey v turbulentnosti [Rise and fall of ideas in turbulence]. Uspekhi fizicheskikh nauk **143**(4), 641–656 (1984)
10. V.I. Tatarskiy, *Rasprostraneniye voln v turbulentnoy atmosfere [Wave Propagation in a Turbulent Atmosphere]* (Nauka [Science] Publishing House, Moscow, 1967)
11. D.I. Aveliani, S.S. Kutateladze, Vzaimodeystviye sveta s turbulentnym potokom zhidkosti [Light interaction with a turbulent fluid flow]. Prikladnaya mekhanika i tekhnicheskaya fizika (4), 115–123 (1973)
12. A.M. Obukhov, Struktura temperaturnogo polya v turbulentnom potoke [Structure of a temperature field in a turbulent flow]. Izvestiya AN SSSR. Ser. geogr. i geofiz. **13**(1), 58–69 (1949)

13. A.M. Yaglom, O lokalnoy strukture polya temperatur v turbulentnom potoke [On the local structure of a temperature field in a turbulent flow]. Doklady AN SSSR **69**(6), 743–746 (1949)
14. G.W. Sutton, Effect of turbulent fluctuations in an optically active fluid medium. AIAA J. **7**(9), 1737–1743 (1969)
15. G. Brederlow, E. Fill, K. Witte, *Moshchnyy yodnyy lazer* [High-Power Iodine Laser]. *Energoatomizdat* [Publishing House for Power Engineering and Nuclear Science and Technology, Moscow, 1985]
16. M.G. Ktalkherman, V.M. Malkov, Aerooptika soplovykh blokov gazodinamicheskikh lazerov [Aerooptics of the nozzle clusters of gasdynamic lasers]. Prikladnaya mekhanika i tekhnicheskaya fizika (6), 20–28 (1993)
17. I.A. Alekseyev, A.K. Zinchenko, *Rezonatory moshcnykh tekhnologicheskikh lazerov: Obzor OA-35* [High-power industrial laser cavities: review OA-35]. *NIIEFA* [Electrophysical Apparatus Research Institute, Leningrad, 1980]
18. S.P. Melnikov, A.A. Sinyanskiy, *O predelnom KPD [koeffitsient poleznogo deystviya] gazovykh lazerov s yadernoy nakachkoy* [On the ultimate efficiency of nuclear-pumped gas lasers]. *Tr. otraslevoy konf. "Fizika yaderno-vozbuzhdayemoy plazmy i problemy lazerov s yadernoy nakachkoy"* [Proceedings of Specialist Conference "Nuclear-Excited Plasma and Problems of Nuclear-Pumped Lasers"]. Obninsk, vol. 2, pp. 133–148 (S.P. Melnikov, A.A. Sinyanskii, Ultimate efficiency of nuclear-pumped gas lasers//*Laser Part. Beams.* **11** (4), 645–654 (1993)
19. A.A. Sinyanskiy, *Issledovaniye po sozdaniyu yaderno-lazernykh ustroystv nepreryvnogo deystviya vo VNIIEF* [Research on the Creation of Continuous-Wave Nuclear Lasers at the VNIIEF (Russian National Experimental Physics Research Institute)]. *Tr. 2-y Mezhdunarod. konf. "Fizika yaderno-vozbuzhdayemoy plazmy i problemy lazerov s yadernoy nakachkoy"* [Proceedings of the 2nd International Conference "Nuclear-Excited Plasma and Problems of Nuclear-Pumped Lasers"], vol. 1 (Arzamas-16, 1995), pp. 16–36
20. A.M. Voinov, S.P. Melnikov, A.A. Sinyanskiy, Kineticheskaya model rekombinatsionnykh lazerov na perekhodakh atom ksenona [Kinetic model of recombination lasers based on xenon atom transitions]. ZhTF [J. Tech. Phys.] **60**(10), 100–113 (1990)
21. K. Smith, R. Thomson, Numerical modeling of gas lasers. *Mir* [World]. (Publishing House, Moscow, 1981)
22. V.V. Borovkov, B.V. Lazhintsev, V.A. Nor-Arevyan, A.N. Sizov, A.A. Sinyanskiy, *Patent RF* [Russian Federation Patent] No. 3207271/25 (1984). Gas laser. *Izobreteniya* [Inventions], No. 5, (1996)
23. I.I. Borovkov, B.V. Lazhintsev, V.A. Nor-Arevyan, A.N. Sizov, A.A. Sinyanskiy, G.F. Fedorov, *Osobennosti gazodinamiki prokachnykh LYaN i problemy formirovaniya kachestva svetovogo puchka* [Gasdynamic features of flowing NPLs and problems in the definition of light beam quality]. *Tr. 2-y Mezhdunarod. konf. "Fizika yaderno-vozbuzhdayemoy plazmy i problemy lazerov s yadernoy nakachkoy"* [Proceedings of the 2nd International Conference "Nuclear-Excited Plasma and Problems of Nuclear-Pumped Lasers"], vol. 1 (Arzamas-16, 1995), pp. 399–406)
24. V.V. Borovkov, B.V. Lazhintsev, V.A. Nor-Arevyan, A.N. Sizov, A.A. Sinyanskiy, G.F. Fedorov, Osobennosti gazodinamiki prokachnykh lazerov, vozbuzhdayemykh oskolkami deleniya yader urana [Gasdynamic features of flowing lasers excited by uranium fission fragments]. Kvantovaya elektronika [Quant. Electron.] **22**(12), 1187–1191 (1995)
25. S.S. Kutateladze, *Osnovy teorii teploobmena* [Principles of the heat exchange theory]. *Atomizdat* [Publishing House of the State Committee on the Use of Atomic Energy, Moscow, 1979]
26. A.M. Voinov, A.A. Sinyanskiy, A.G. Vasilenko, O.A. Golubeva, L.Y. Dovbysh, V.V. Ivanov, M.I. Kuvshinov, B.V. Lazhintsev, A.Y. Lakhtikov, S.P. Melnikov, A.B. Modenov, V.A. Nor-Arevyan, A.M. Pichugin, A.N. Pokalo, V.V. Porkhayev, Y.A. Pospelov, A.V. Sinitsyn, I.G. Smirnov, S.L. Turutin, M.V. Khlestkov, *Eksperimentalnyy kompleks LM-4/BIGR. Polucheniye generatsii v smesi Ar-Xe s dlitelnostyu boleye 1s* [LM-4/BIGR

Experimental Facility. Achievement of lasing in an Ar-Xe mixture with a duration of more than 1 sec]. *Tr. 2-y Mezhdunarod. konf. "Fizika yaderno-vozbuzhdayemoy plazmy i problemy lazerov s yadernoy nakachkoy"* [Proceedings of the 2nd International Conference "Nuclear-Excited Plasma and Problems of Nuclear-Pumped Lasers"], vol. 2 (Arzamas-16, 1995), pp. 179–185

27. V.A. Mayorov, V.M. Polyayev, L.L. Vasilyev, A.I. Kiselev, Intensifikatsiya konvektivnogo teploobmena v kanalakh s poristym vysokoteploprovodnym zapolnitelem. I. Teploobmen pri lokalnom teplovom ravnovesii vnutri pronitsayemoy matritsy [Intensification of convective heat exchange in channels containing a porous highly conductive filler. I. Heat exchange during local heat equilibrium within a permeable matrix]. Inzhenerno-fizicheskiy zhurnal **47**(1), 13–24 (1984)
28. V.A. Mayorov, V.M. Polyayev, L.L. Vasilyev, V.N. Bogdanov. Intensifikatsiya konvektivnogo teploobmena v kanalakh s poristym vysokoteploprovodnym zapolnitelem. II. Forsirovannyy rezhim teploobmena [Intensification of Convective Heat Exchange in Channels Containing a Porous Highly Conductive Filler. II. Accelerated Heat Exchange Mode]. Ibid., No. 2, pp. 199–205
29. V.A. Mayorov, Granichnyye usloviya dlya sistemy intensivnogo transpiratsionnogo okhlazhdeniya [Boundary Conditions for an Intense Transpiration Cooling System]. Ibid., No. 4, pp. 587–594
30. A.V. Kurpatenkov, V.M. Polyayev, A.L. Sintsov, Otsenka effektivnosti protochnogo poristogo okhlazhdeniya (granichnoye usloviye tretyego roda) [Estimation of the efficiency of direct-flow porous cooling (boundary condition of the third kind]. Inzhenerno-fizicheskiy zhurnal **53**(3), 361–365 (1987)
31. A.V. Kurpatenkov, V.M. Polyayev, A.L. Sintsov, O matematicheskikh modelyakh protsessov poristogo okhlazhdeniya [On Mathematical Models of Porous Cooling Processes]. Ibid., No. 2, pp. 243–249
32. B.S. Petukhov, L.G. Genin, S.A. Kovalev, *Teploobmen v yadernykh energeticheskikh ustanovkakh* [Heat Exchange in Nuclear Power Installations]. *Energoatomizdat* [Publishing House for Power Engineering and Nuclear Science and Technology, Moscow, 1986]
33. N.B. Vargaftik, D. Vasilevskaya Yu, Teploprovodnost geliya pri temperaturakh 300–6000°K [Helium heat conduction at temperatures of 300–6,000°K]. Inzhenerno-fizicheskiy zhurnal **42**(3), 412–417 (1982)
34. A.G. Shashkov, N.A. Nesterov, O.A. Kolenchits, Teploprovodnost inertnykh gazov v shirokom diapazone temperatur [Rare gas heat conduction over a wide temperature range]. Inzhenerno-fizicheskiy zhurnal **43**(5), 788–795 (1982)
35. N.B. Vargaftik, *Spravochnik po teplofizicheskim svoystvam gazov i zhidkostey* [Handbook on the Thermophysical Properties of Gases and Liquids]. *Nauka* [Science Publishing House, Moscow, 1972]
36. V.V. Borovkov, B.V. Lazhintsev, S.P. Melnikov, I.N. Mochkayev, V.A. Nor-Arevyan, A.A. Sinyanskiy, G.I. Fedorov, Issledovaniye opticheskikh neodnorodnostey v lazerakh s yadernoy nakachkoy [Investigation of optical inhomogeneities in nuclear-pumped lasers]. Izvestiya AN SSSR. Ser. fizicheskaya **54**(10), 2009–2015 (1990)
37. A.N. Sizov, O svyazi prostranstvennykh neodnorodnostey nakachki s moshchnostyu generatsii lazerov, vozbuzhdayemykh oskolkami deleniya urana [On the relationship between pumping spatial inhomogeneities and the lasing output of lasers excited by uranium fission fragments]. Pisma v ZhTF [Lett. J. Tech. Phys.] **20**(9), 64–67 (1994)
38. F.S. Milos, A. Acrivos, Steady flow past sudden expansions at large Reynolds number. Part I: boundary layer solutions. Phys. Fluids. **29**(5), 1353–1359 (1986)
39. F.S. Milos, A. Acrivos, J. Kim, Steady flow past sudden expansions at large Reynolds number. Part II: Navier-Stokes solutions for the cascade expansion. Phys. Fluids **30**(1), 7–18 (1987)
40. M.S. Anand, S.B. Pope, *Diffusion behind a line source in grid turbulence*. In: Turbulent Shear Flow 4. (Spinger-Verlag, 1985), pp. 46–61

41. V.N. Bogdanov, A.G. Vasilenko, V.V. Zherebtsov, V.V. Ivanov, M.I. Kuvshinov, A.Y. Lakhtikov, V.S. Mayornikov, A.B. Modenov, A.V. Nikulin, V.A. Nor-Arevyan, A.N. Pokalo, A.A. Sinyanskiy, I.G. Smirnov, S.L. Turutin, M.V. Khlestkov, V.V. Shirokov, *Eksperimentalnyy kompleks LM-4/BIGR. Ustroystvo i printsip raboty* [LM-4/BIGR Experimental Facility. Arrangement and Operating Principle]. *Tr. 2-y Mezhdunarod. konf. "Fizika yaderno-vozbuzhdayemoy plazmy i problemy lazerov s yadernoy nakachkoy"* [Proceedings of the 2nd International Conference "Nuclear-Excited Plasma and Problems of Nuclear-Pumped Lasers"], vol. 2 (Arzamas-16, 1995), pp. 172–178
42. V.F. Kolesov. *Aperiodicheskiye impulsnyye reaktory* [Aperiodic Pulse Reactors]. *RFYaTs-VNIIEF* [Russian Federal Nuclear Center at the Russian National Experimental Physics Research Institute, Sarov, 1999]
43. A.N. Sizov, V.V. Porkhayev, Vliyaniye skorosti gazovogo potoka na moshchnost generatsii protochnykh lazerov, vozbuzhdayemykh oskolkami deleniya yader urana [Effect of gas flow velocity on the lasing output of flowing lasers excited by uranium nuclear fission fragments]. Kvantovaya Elektronika [Quant. Electron.] **23**(6), 510–514 (1996)
44. L.M. Montierth, W.A. Neuman, D.W. Nigg, B.J. Merrill, Energy deposition in direct nuclear pumped optical cavities. J. Appl. Phys. **69**(10), 6776–6788 (1991)
45. V.Y. Matyev, A.N. Sizov, *Profil plotnosti v slaboprogrevayemykh prokachnykh kanalakh LYaN* [Density profile in the slightly heated flowing channels of NPLs]. *Tr. 2-y Mezhdunarod. konf. "Fizika yaderno-vozbuzhdayemoy plazmy i problemy lazerov s yadernoy nakachkoy"* [Proceedings of the 2nd International Conference "Nuclear-Excited Plasma and Problems of Nuclear-Pumped Lasers"], vol. 1 (Arzamas-16, 1995), pp. 407–409
46. A.N. Sizov, L.P. Babich, I.M. Kutsyk, Y.V. Prikhodko. *Raschet prostranstvennykh neodnorodnostey v prokachnykh LYaN* [Calculation of spatial inhomogeneities in flowing NPLs]. *Tr. 3-y Mezhdunarod. konf. "Problemy lazerov s yadernoy nakachkoy i impulsnyye reaktory"* [Proceedings of the 3rd International Conference "Problems of Nuclear-Pumped Lasers and Pulse Reactors"], (Snezhinsk, 2003), pp. 298–306
47. A.N. Sizov, Y.N. Deryugin, *Raschet prostranstvennykh neodnorodnostey v tsilindricheskikh gazovykh lazerakh s yadernoy nakachkoy* [Calculation of spatial inhomogeneities in cylindrical nuclear-pumped gas lasers]. *Tez. dokl. VI Vsesoyuz. konf. "Optika lazerov"* [Abstracts of Papers Presented at the 6th All-Union Conference "Laser Optics"], (Leningrad, 1990), p. 132
48. A.N. Sizov, Y.N. Deryugin, Raschety prostranstvennykh neodnorodnostey v tsilindricheskikh gazovykh lazerakh s nakachkoy oskolkami deleniya [Calculations of spatial inhomogeneities in cylindrical fission-fragment pumped gas lasers]. ZhTF [J. Tech. Phys.] **62**(9), 107–111 (1992)
49. V.Y. Matyev, *Metody rascheta energovklada ionov v neodnorodnykh sredakh* [Techniques for calculating ion energy deposition in inhomogeneous media]. *Tr. otraslevoy konf. "Fizika yaderno-vozbuzhdayemoy plazmy i problemy lazerov s yadernoy nakachkoy"* [Proceedings of the Specialist Conference "Nuclear-Excited Plasma and Problems of Nuclear-Pumped Lasers"], vol. 2 (Obninsk, 1993), pp. 79–88
50. V.Y. Matyev, *Dvumernyy profil skorosti techeniya gaza v kanalakh LYaN* [Two-dimensional gas flow velocity profile in the channels of NPLs]. *Tr. 2-y Mezhdunarod. konf. "Fizika yaderno-vozbuzhdayemoy plazmy i problemy lazerov s yadernoy nakachkoy"* [Proceedings of the 2nd International Conference "Nuclear-Excited Plasma and Problems of Nuclear-Pumped Lasers"], vol. 1 (Arzamas-16, 1995), pp. 430–442
51. A.N. Sizov, *Osobennosti techeniy gaza v prokachnykh LYaN* [Distinctive features of gas flows in flowing NPLs]. *Tr. 3-y Mezhdunarod. konf. "Problemy lazerov s yadernoy nakachkoy i impulsnyye reaktory"* [Proceedings of the 3rd International Conference "Problems of Nuclear-Pumped Lasers and Pulse Reactors"]. (Snezhinsk, 2003), pp. 263–272
52. S.P. Melnikov, I.N. Mochkayev, A.A. Sinyanskiy, *Raspredeleniye izlucheniya lazera s yadernoy nakachkoy na perekhode atoma neona v blizhney zone* [Radiation distribution of a nuclear-pumped laser based on neon atom transition in a near-field region]. *Tr. otraslevoy konf. "Fizika yaderno-vozbuzhdayemoy plazmy i problemy lazerov s yadernoy nakachkoy"*

[Proceedings of an Specialist Conference "Nuclear-Excited Plasma and Problems of Nuclear-Pumped Lasers"], vol. 2 (Obninsk, 1993), pp. 233–239
53. A.N. Korzenev, M. Limar Yu, A.N. Sizov, A.A. Sinyanskiy, Raspredeleniye intensivnosti protochnogo gazovogo lazera s yadernoy nakachkoy [Intensity distribution of a flowing gas nuclear-pump laser]. Kvantovaya elektronika [Quant. Electron.] **35**(9), 795–798 (2005)
54. L.L. Kazakov, *Rezultaty resursnykh ispytaniy energovydelyayushchikh elementov dlya LYaN* [Results of life cycle tests of energy-releasing elements for NPLs]. *Tr. otraslevoy konf. "Fizika yaderno-vozbuzhdayemoy plazmy i problemy lazerov s yadernoy nakachkoy"* [Proceedings of the Specialist Conference "Nuclear-Excited Plasma and Problems of Nuclear-Pumped Lasers"], vol. 3 (Obninsk, 1993), pp. 41–49
55. L.L. Kazakov, *Strukturnyye i fazovyye izmeneniya v energovydelyayushchikh elementakh pri dlitelnom obluchenii* [Structural and phase changes in energy-releasing elements during prolonged irradiation]. *Tr. 2-y Mezhdunarod. konf. "Fizika yaderno-vozbuzhdayemoy plazmy i problemy lazerov s yadernoy nakachkoy"* [Proceedings of the 2nd International Conference "Nuclear-Excited Plasma and Problems of Nuclear-Pumped Lasers"], vol. 2 (Arzamas-16, 1995), pp. 60–68
56. D.R. Neal, W.C. Sweatt, J.R. Torczynski, Resonator design with an intracavity time-varying index gradient. Proc. SPIE. **965**, 130–141 (1989)
57. Y.V. Prikhodko, A.N. Sizov, Raschetnyy analiz ustoychivosti rezonatorov lazerov s yadernoy nakachkoy [Numerical analysis of nuclear-pumped laser cavity stability]. *Kvantovaya electronika* [Quant. Electron.]. **22**(6), 613–618 (1995)
58. Y.V. Prikhodko, A.N. Sizov, Dopustimyy peregrev gazovoy sredy v prokachnom lazere, vozbuzhdayemom oskolkami deleniya yader urana [Permissible overheating of the gas medium in a flowing laser excited by uranium fission fragments]. Kvantovaya electronika [Quant. Electron.]. **28**(3), 207–211 (1999)
59. V.T. Kazazyan, B.A. Litvinenko, L.P. Roginets, I.A. Savushkin, *Fizicheskiye osnovy ispolzovaniya kineticheskoy energii oskolkov deleniya v radiatsionnoy khimii* [Physical principles of the use of fission-fragment kinetic energy in radiochemistry]. *Nauka i tekhnika* [Science and Technology] (Publishing House, Minsk, 1972)
60. V.V. Borovkov, G.V. Vlokh, B.V. Lazhintsev, V.A. Nor-Arevyan, A.N. Sizov, A.A. Sinyanskiy, G.E. Filippov, Effektivnost energovklada v lazerakh s nakachkoy oskolkami deleniya urana [Energy deposition efficiency in uranium fission-fragment-pumped lasers]. Kvantovaya electronika [Quant. Electron.] **22**(3), 219–224 (1995)
61. V.Y. Matyev, A.N. Sizov, *Vliyaniye neodnorodnostey lazernoy sredy na gazodinimicheskiye raschety lazerov s yadernoy nakachkoy* [Effect of laser medium inhomogeneities on gasdynamic calculations of nuclear-pumped lasers]. *Tr. otraslevoy konf. "Fizika yaderno-vozbuzhdayemoy plazmy i problemy lazerov s yadernoy nakachkoy"* [Proceedings of the Specialist Conference "Nuclear-Excited Plasma and Problems of Nuclear-Pumped Lasers"], vol. 2 (Obninsk, 1993), pp. 209–218

Chapter 10
Design Concepts for Stationary Reactor Lasers

10.1 Main Types of Stationary Reactor Lasers

Chapter 6 gives a brief overview of designs for nuclear laser setups with a pulse duration of ~10 ms (OKUYAN and LIRA setups). Their operation is based on the "initiating reactor–subcritical laser block" scheme. The literature has repeatedly suggested different design schemes for stationary nuclear laser facilities or reactor-laser (RL) based on gas laser media. The term RL was apparently first proposed in article [1], where the potential of removing energy in the form of laser radiation from the reactor core during the relaxation stage was noted in principle. The most interesting version of the RL is a device where, as a consequence of spatial overlapping of the nuclear fuel and laser medium, a significant portion of the nuclear energy is converted into laser radiation bypassing the intermediate thermal energy stage.

The idea of creating a stationary nuclear-laser facility based on using a gas medium was expressed in 1964 before the appearance of the first nuclear-pumped lasers (NPLs). In this study [2], a reactor core was examined that consisted of 1,500 aluminum tubes surrounded by moderators. These tubes had layers of ^{235}U deposited on their inner surfaces and were filled with a He-Ne mixture at ~10 Torr of pressure. There are some schemes of stationary RLs that are differentiated by the nuclear fuel used for reactor core operation and pumping of the laser medium (Fig. 10.1): (a) a gas compound UF_6, uranium vapor, or uranium aerosol particles; (b) thin layers of uranium or its compounds on metal substrates.

Interest arose in UF_6, because it is practically the only gas-phase compound of uranium at low temperatures, and the energy deposition efficiency for fission fragments in the "UF_6 + laser gas" mixture approaches 100 %. In addition, a series experiments [3, 4] were carried out that show the potential for creating gas-core nuclear reactors with comparatively low critical masses of UF_6 (~6 kg).

Thus, there is currently no doubt of the potential for creating gas-core reactors based on UF_6, although these devices are rather complicated and dangerous.

S.P. Melnikov et al., *Lasers with Nuclear Pumping*,
DOI 10.1007/978-3-319-08882-2_10

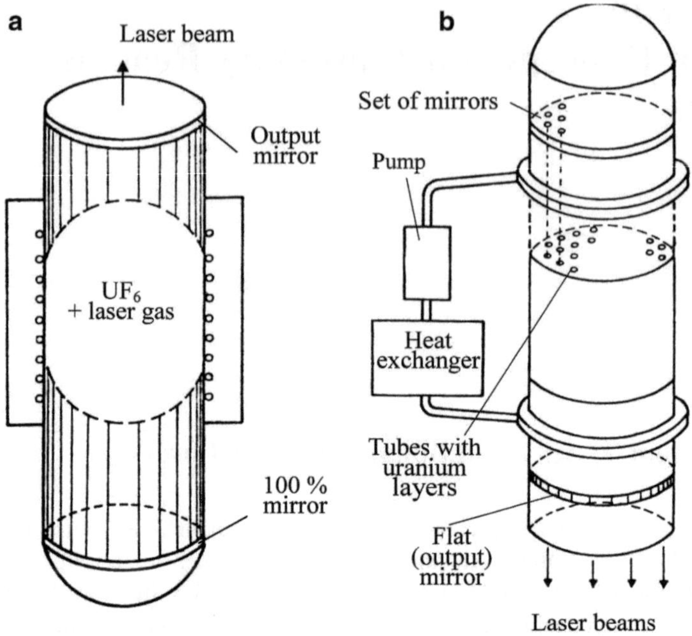

Fig. 10.1 Schematics of the main variants for a stationary RL [6]: (**a**) a gas-core reactor whose core is a mixture of uranium fuel (UF_6, uranium aerosol, or vapor) and laser gas; (**b**) an RL using thin-layer uranium fuel

However, when choosing laser gas media together with UF_6, a number of difficulties arise, which are caused by the chemical aggressiveness of UF_6 and its radiolysis products, high losses in the active medium due to absorption of the laser radiation by UF_6 molecules, and the high quenching rate of the excited atoms and molecules in collisions with UF_6 molecules.

As laser media for gas-core RL based on UF_6, these mixtures have been suggested: UF_6-CO_2-N_2-He [6], UF_6-He, UF_6-TlF-F_2 [7], UF_6-He-Xe [8], UF_6-Ar-Xe [9], etc. However, there are currently no laser media using UF_6, for which the feasibility of pumping by nuclear radiation would have been established experimentally. Experimental [10] and theoretical [11] studies have shown that, at the present time, the most effective laser media with nuclear pumping at the infrared transition of atoms $M =$ Xe, Kr, and Ar may not operate at UF_6 concentrations higher than 1 % due to a reduction in the ionized atom M^+ concentrations and the high rate of collisional quenching of the excited atoms M^* by the UF_6 molecules. One paper [9] concludes that designs of gas-core RLs using UF_6 are apparently unrealistic because it is difficult to provide the conditions for laser operation with sufficiently high concentrations of UF_6, which are necessary to attain reactor criticality. The authors of study [9] came to the same conclusion when analyzing the possibility for creating an RL using aerosol fuel with uranium particles several micrometers in diameter.

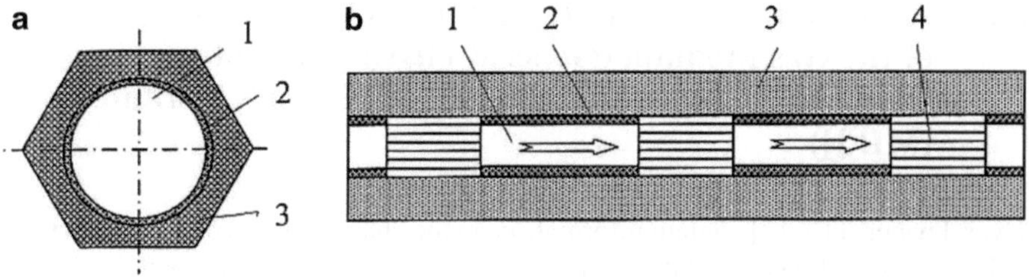

Fig. 10.2 Cross-sections of laser cells for RL [15]: (**a**) cylindrical cell; (**b**) rectangular cell with transverse gas flow (*1*) laser channels; (*2*) uranium layers; (*3*) neutron moderators; (*4*) radiators; the direction of the gas circulation is indicated with *arrows*

The design of an RL using uranium vapors was not examined in detail. In the authors' opinion [12], the main difficulty for creating such facilities is the necessity to maintain the active medium at a high temperature (~4,000 K). This substantially complicates the construction of the device and severely limits the choice of laser medium.

At present, the most realistic designs are heterogeneous RLs using thin-film uranium fuel. The core of this RL is, in essence, a specific quantity of laser cells with uranium layers appropriately placed in a neutron moderator matrix. Calculations show that, with the appropriate selection of components, this system has a sufficient neutron multiplication factor when only uranium layers inside laser cells are used, that is, the conditions for the operation of the RL are provided without utilizing additional fuel (uranium). The number of laser cells may vary from a hundred to several thousand. The total amount of uranium may vary from 5.7 to 40–70 kg, and characteristic linear dimensions are 2.5 m.

Figure 10.2 shows the two main alternatives for laser cells: (a) a cylindrical cell capable of longitudinal gas circulation relative to the axis of the cell; (b) a rectangular cell with transverse gas circulation relative to the axis of the cell (in the direction indicated by the arrows). Longitudinal gas circulation is possible with low power deposition and slow ($\leq$10 m/s) gas velocities. At high power, transverse circulation with the energy transfer to internal radiators is more effective (Fig. 10.2b). Some characteristics of laser cells with transverse gas flow are examined in Sect. 9.3 of Chap. 9. This laser cell design was proposed at the All-Union Scientific Research Institute for Experimental Physics [VNIIEF] [13, 14] and verified experimentally on the LM4/BIGR complex (see Chap. 6, Sect. 6.1).

Stationary RLs using thin-layer uranium fuel have been considered in Russia [15–18] and abroad [2, 6, 19]. The main characteristics of these systems and the issues that arise with their design are considered below in examples of the RLs that are discussed in studies [15–18].

10.2 A Reactor Laser with Longitudinal Circulation of the Gas Medium (General Physics Institute [IOFAN]-Moscow Engineering and Physics Institute [MIFI])

Several papers [16, 17] contain information on the characteristics of a stationary RL with longitudinal gas circulation relative to the axes of the laser channels. Calculations for this design were made by employees of the IOFAN and MIFI.

A schematic of one RL channel (cell) is shown in Fig. 10.3. In this case, an RL is considered with a cylindrical core consisting of neutron moderators and a penetrated system of parallel channels, which are filled with a laser gas medium. A critical mass of uranium is accumulated from the uranium layers deposited on the internal surfaces of the channels. According to one study [16], the optimal diameter of the laser channel, which is determined by the path length of the fission fragments in the gas medium and the profile of the energy deposition function along the channel radius, should be approximately 30–40 % of the path length of the fission fragments in gas. For example, the path length of fragments for helium at room temperature and 3 atm is 3.4 cm (see Table 7.1), and the optimal diameter of the channel according to [16] is 10–14 mm. At a reactor core temperature of ~1,000 K, the fragment path length and the optimal channel diameter for helium at 3 atm are equal to 11 cm and 30–45 mm, respectively.

Neutron calculations of the critical RL parameters as functions of the volume concentration of the moderators in the core, the diameter of laser channels, and the thickness of the uranium layer were performed in a diffuse multigroup approximation. BeO, Be_2O, and graphite were examined as neutron moderators. As shown by the calculations, each moderator has a certain minimum of critical R and H (radius and height of the cylindrical core), which is a function of the volume concentration of the moderators in the core. With the laser channel diameter at 20 mm, a core with a BeO moderator has the minimal dimensions $R = H = 80$ cm at an 80 % volume concentration for the moderators (the volume concentration of the gas medium, therefore, is 20 %). The base version for further examination was chosen as a core with a BeO moderator surrounded by side and end neutron reflectors made of 15-cm

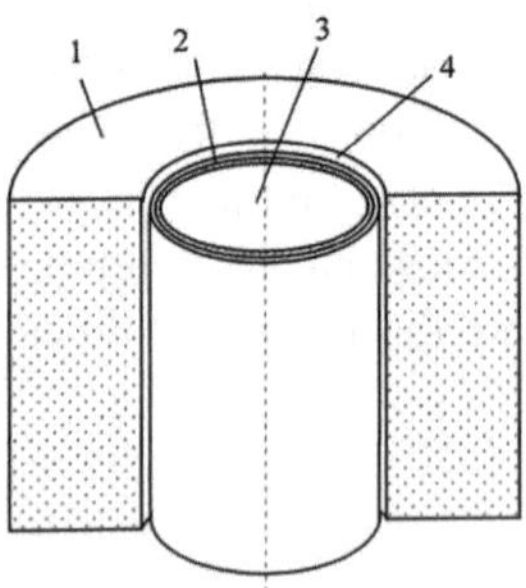

Fig. 10.3 Schematic of RL channels [16]: (*1*) neutron moderator; (*2*) the metal tube of a laser channel with a layer of fuel; (*3*) laser gas medium; (*4*) channel with coolant

Table 10.1 Characteristics of the base version RL [16]

Neutron moderator	BeO
Thickness of the UO_2 layer with 60 % ^{235}U enrichment, μm	2.5
Core diameter, m	3
Core height, m	1.5
Volume concentration of the gas medium, %	65
Diameter of the laser channels, mm	40
Number of laser channels	3,656
Critical mass of UO_2, kg	17.3
Mass of BeO, kg	1.9×10^4

Table 10.2 Characteristics of the RL using a He-Xe medium [16]

Thermal power, MW	150
Gas density, cm^{-3}	3×10^{19}
Gas pressure, atm	4
Specific power deposition in the laser medium, W/cm^3	3
Efficiency of energy transfer from the fission fragments to the laser medium, %	16
Laser power efficiency, %	1
Gas velocity, m/s	150
Gas flow, kg/s	132
Power consumed by flowing gas through the core, MW	1.4
Gas temperature at the core input, K	1,080
Gas temperature at the core output, K	1,300
Maximum temperature of the uranium layer, K	1,657

thick beryllium. The results of neutron calculations for this variant of the RL are given in Table 10.1.

The output performance of the base version RL using He-Xe as the laser medium is given in Table 10.2. To operate this RL in stationary mode, gas flowing must be provided along the closed path at a sufficiently high velocity ~150 m/s. This requires significant power consumption.

As the authors note [16], this variant of the RL is not definitive, because choosing the optimal design for the RL is a difficult problem with a large number of constraints for different parameters, many of which are known to a poor degree of accuracy. It should be pointed out here that at a gas velocity of $U = 150$ m/s, the Reynolds number, $Re = U/d\nu \sim 10^5$, significantly exceeds the critical $Re_{cr} = 2{,}300$, therefore the gas flow will be substantially turbulent. This is a serious obstacle to realizing the laser operation (see also Chap. 9, Sects. 9.1 and 9.2).

10.3 Conceptual Designs for Reactor Lasers (VNIIEF)

On the basis of calculations and experimental studies, engineering and technical examinations, and experience in the construction of domestic reactors and the operation of research nuclear reactors and laser experimental systems, the VNIIEF associates made preliminary assessments of the main energy, nuclear and physical, technical, and operational parameters for alternative RL designs operating in the time span from a fraction of a second to stationary mode.

All possible alternatives for the RL can be relatively divided into two types. The first type of RL has transverse gas flow and external cooling of the laser channels using a coolant (for example, heavy water D_2O, which simultaneously plays the part of the neutron moderator) with subsequent withdrawal of the heat beyond the boundaries of the core. This type of RL can operate in stationary mode as well as in protracted pulse mode $\geq$5-s long. The second type is a heat-capacity RL. The heat released during operation of this reactor is accumulated in the core moderator. As a result, the temperature of the moderator and laser-active gas grow with time. Therefore, the second-type of RL can only operate in pulsed mode. If laser-active gas is not circulated in these RLs during the active pulse, the laser elements may be represented as extended cylinders (Fig. 10.2a) or channels formed with flat plates (Fig. 10.2b). In the latter, radiators do not have to be added to the design. The pulse duration of these RLs should be $\tau \leq 0.1$ s (see Chap. 8). After each pulse, forced cooling may be carried out by flowing the same gas using external pumps and blowers. For second-type of RL, the pulse durations $0.1 \leq \tau \leq 100$ s may be obtained. In this case, the optimal RL design has rectangular laser elements with flat uranium layers. They should be joined with the radiators as shown in Fig. 10.2b and combined into one or several gas circuits, along which, the gas laser medium is circulated during the pulse.

Some RL versions have been examined in studies [15, 18]: RLs with transverse gas flow and the removal of heat energy beyond the boundaries of the core, and an RL with stored heat in the reactor core (heat-capacity mode). The main characteristics of these RLs are presented in Table 10.3, and their design and operational features are described below.

Table 10.3 Main characteristics of two types of RLs [15, 18]

	RL with transverse gas flow	Heat-capacity RL
Type of design	Rectangular, modular	Cylindrical, monoblock (sections)
Operation mode	Stationary (up to ~30 h)	Brief startups (up to ~100 s)
Overall dimensions, m	4.5 × 4.5 × 4.5	2.4 × 2.4
Mass of the facility, t	100	15
Number of channels	1,000	300
Uranium load, kg	70	15
Thermal power, MW	800	65
Laser medium	He(Ne, Ar)-Xe	He(Ne, Ar)-Xe
Laser wavelength, μm	1.73; 2.03; 2.65	1.73; 2.03; 2.65
Laser output power, kW	1,000	200

RLs with Transverse Gas Flow

To support stationary operation of a NPL with transverse gas flow through the laser channels, the gas flow velocity must be ≥100 m/s. Chapter 9 shows that, at these circulation speeds, the gas flow is turbulent and gas density fluctuations emerge, which negatively affect the optical quality of the medium and, consequently, on the characteristics of the laser radiation. The calculations show that the gas velocity at which turbulent density fluctuations do not have a direct influence on the laser parameters should be <30 m/s for helium-based laser media and <10 m/s for argon-based.

Velocities ~10 m/s may be realized with transverse circulation (see Chap. 9, Sect. 9.3). In this case, the RL core consists of separate laser channels with transverse (with respect to the optical axis) circulation of the gas medium [13, 14]. A single laser channel with a rectangular profile contains a ~10-cm wide flat layer of uranium deposited to the internal surface of the side walls parallel to the direction of gas flow. A cross section of this channel is shown in Fig. 10.2b (see also Figs. 9.1 and 9.2). To cool the mixture heated in the channel, a radiator is placed at its output. The next laser channel can be placed beyond the radiator output, which serves as its input and so on. Thus, a chain of laser channels can be built that is combined in a single gas loop. The gasdynamic and thermophysical characteristics of the laser channel, their relationship with laser parameters, and also the optimal operation modes for this channel in the case of transverse gas flow are examined in detail in Chap. 9.

The RL core should contain a sizable (~1,000) number of laser cells. Due to design, manufacturing, and operational considerations, this system is conveniently implemented in sections consisting of functionally complete modules.

A possible schematic for the RL is given in Fig. 10.4. The device does not have a special enclosure. This function is fulfilled by the reactor matrix together with a hard support layer made of a solid neutron moderator (4) and biological shielding (5). The reactor matrix may be produced with graphite blocks and zirconium tubes containing nuclear reactivity regulators (2) of the reactor (boron carbide rods) and devices for monitoring nuclear-physical parameters (3). The matrix contains replaceable laser modules (1) forming, in total, the RL core.

Heat-Capacity RLs

A simpler variant of an RL with minimal critical mass parameters does not use coolant and excess heat is absorbed inside the core. In this case, the operating time of the RL is limited by the permissible temperature of the core.

An RL with energy release in the core ≤10 MW/m^3 has a simple construction (Fig. 10.5) if the laser cell (1) is carried out according to the schematic in Fig. 10.2a. Beryllium (or beryllium oxide) is used as the neutron moderator (2) as it is the

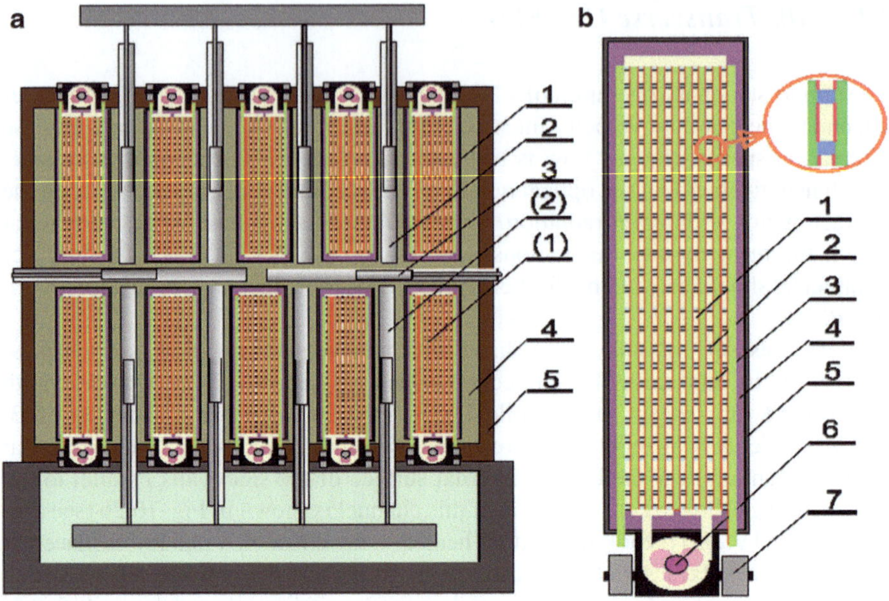

Fig. 10.4 RL with transverse gas flow [15, 18]: (**a**) RL core (*1*) core module; (*2*) control devices; (*3*) monitoring devices; (*4*) reflector; (*5*) biological shielding; (**b**) RL module (*1*) laser channel; (*2*) radiators; (*3*) moderator-coolant; (*4*) solid moderator; (*5*) enclosure; (*6*) gas flowing system; (*7*) traverser

substrate for the uranium layer. With the output laser power at ~200 kW, the RL core contains almost 300 cells that are 2-m long (the core is cylindrical with a 2-m diameter). The central channel contains reactor control devices (5). The inside of the core may be supplied with auxiliary mechanisms and gas-intake devices. In addition, the cavity of the central channel may have an initiating small-scale pulsed reactor intended to accelerate output of the device to the given power level and to simply operations with control devices at this output. A neutron reflector (3) is used (the beryllium thickness ~100 mm) to vary the neutron flux round core volume. The entire structure is placed in a cylindrical metal enclosure (4) with optical elements (6). To reduce the flowing velocity of the laser medium, the core may be formed in sections in the axial direction as shown in Fig. 10.5.

The energy release in one startup (that is the operation time at the given power) of the core is limited by the thermal capacity of the core. If the permissible rise in the core temperature is accepted to be at most 500 °C, then stationary operation is ~100 s (with other power levels, this time changes accordingly). After this time, downtime of approximately 1 h is needed for forced cooling of the core. The startup frequency may be ~5 min with forced cooling of the core.

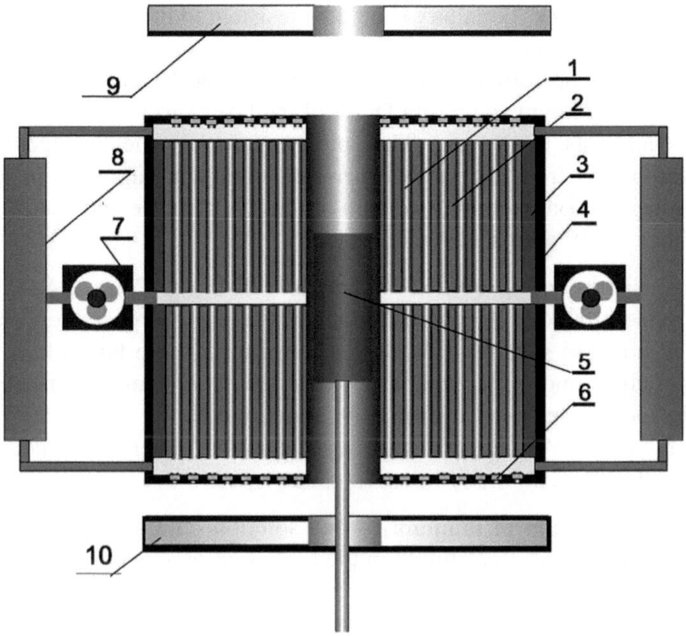

Fig. 10.5 Schematic of a heat-capacity RL [15, 18]: (*1*) laser channels; (*2*) neutron moderator; (*3*) neutron reflector; (*4*) enclosure; (*5*) reactivity control devices; (*6*) internal optical elements; (*7*) gas flowing system for RL cooling after a pulse; (*8*) heat exchangers; (*9*, *10*) outside optical elements

10.4 Thin-Film Uranium Fuel

One of the main problems that arise when creating NPLs and RLs is the development of uranium layers that are stable relative to different influencing factors. These layers serve a dual function in RLs: they serve as reactor fuel and provide fission fragments to excite the laser medium.

Experience with NPLs using different types of uranium layers makes it possible to define their main performance requirements.

1. The uranium layer must be deposited on substrate made of a material that absorbs neutrons poorly, for example, aluminum or zirconium. The optimum thickness of the layer is about half the path length of the fission fragments in the layer. The density, thickness, and ^{235}U concentration of the layer should be as uniform as possible.
2. The layer should be mechanically durable and thermally stable and have high adhesive bond to the substrate. Localized bond decrease between the layer and substrate disrupts heat removal and leads to the mechanical deterioration of the layer. This effect may arise with long-term storage of the layer and as a result of thermocyclic tension.

3. The "substrate-layer" boundary should be resistant to the thermal and radiation-stimulated diffusion of the layer material into the substrate.
4. To increase the operational lifetime of the NPL and RL, layers must be developed with a minimum sputtering factor of the layer surface during output of the fission fragments to the gas. The sputtering factor (the average number of dispersed atoms from the surface per one emergent fragment) in a complex manner depends on the microstructure of the layer surface and changes within a wide range 1–10^4 atoms/fragment [20, 21].

The first experiments at VNIIEF on reactor VIR-2 (see Chap. 2, Sect. 2.3) used uranium oxide-protoxide layers manufactured using chemical deposition from solutions of uranium salts onto aluminum substrates. These layers have a loosely coupled structure and poor durability. This led to contamination of the laser channel, including the cavity mirror, with uranium dust.

To study NPLs, VNIIEF associates used uranium layers produced using the following technologies:

- Electric precipitation of uranium from a solution onto an aluminum substrate (see, for example, [22]), resulting in the formation of a layer that is a mixture of uranium oxide (UO_2) and uranium protoxide (U_3O_8).
- Deposition of uranium from gas-phase compound onto a heated aluminum substrate (see, for example, [23]) with the formation of UO_2.
- Magnetron ion-beam sputtering in a vacuum [24] resulting in the formation of metallic layers (α-uranium) or intermetallic compounds (UAl_3 and UAl_4). To protect the metallic uranium from corrosion and reduce the sputtering factor, the surface of the metallic uranium layer may be covered by a thin aluminum film with a thickness of about 0.5 μm.

The main characteristics of layers produced using the above two technologies are given in Table 10.4. Table 10.5 shows the comparative efficiency of different layers with ^{235}U thicknesses of about 4.5 mg/cm^2 (the efficiency of the layer was determined as the ratio of the power output by the fragments into the gas to the power released in the layer).

Table 10.4 Characteristics of the layers containing uranium [25]

Characteristics of the layer	Production technology: Deposition from gas-phase compound (VNIIEF)	Production technology: Magnetron sputtering (NIIAR [Institute for Atomic Reactors, Dimitrovgrad])
Phase composition	UO_2 (density ~10 g/cm^3)	α-uranium (density ~19 g/cm^3)
Adhesion	Medium	High (pull strength of about 100 atm)
Structure of the layer	Coarse-grain (grain-size 2–5 μm)	Fine-grain (grain-size <1 μm)
Surface appearance	Dull	Reflective
Linear non-uniformity of the layer thickness	<0.3 %/cm	<1 %/cm

Table 10.5 The relative efficiency of different layers [25]

Composition of uranium-containing layer	α-uranium	UO_2	U_2O_3	UAl_3	UAl_4	U + 0.5 μm Al
Efficiency of the layer, arb. units	1.0	0.74	0.79	0.76	0.70	0.80

Each technology has its advantages. For example, magnetron sputtering yields metallic layers (α-uranium) with maximum efficiency. However, to prevent corrosion of these layers, layer protection, which reduces efficiency, is necessary. Magnetron sputtering allows layers to be deposited in the intermetallic phase of uranium compounds (UAl_3 and UAl_4), which have exceptionally high strength properties, but lower efficiency.

The adhesive strength of the uranium oxide layers is somewhat lower than that of the metallic layers. However, not one destruction has been noted in almost 20 years. Over this period, the VIR-2M/LUNA-2M experimental setup (see Chap. 2, Sect. 2.4) generated about 2,000 pulses with a total neutron fluence of ~10^{16}/cm^2. Moreover, no deterioration in NPL characteristics was observed. Tests [25] of layers made of metallic α-uranium showed that the adhesion of the layers does not diminish when the layer is kept for 10 years and after 150 heating cycles from 20 to 300 °C with a 5 min cycle time.

Oxide layers have a grain texture that is coarser than metallic layers. This is an advantage from the point of view of the suppression effect of sputtering. Metallic layers were studied by VNIIEF associates on the VIR-2M pulse reactor using mass spectrometry, electron microscopy and α- and γ-spectrometry. These investigations established the following: sputtering of the layer mainly occurs by inelastic energy transfer from the fragment to the target atoms [20] and, for oxide layers, the sputtering factor is 10^3 atoms per fragment [25]. This factor may be reduced by about tenfold using special thermal processing [26]. Deposition of thin films (for example, 0.5 μm thick Al) to the surface of the uranium layer reduces the sputtering factor to several units.

Investigations of radiation-induced diffusion have shown [25] that this effect arises at a neutron fluence of about 2×10^{18}/cm^2. In this case, the uranium layer "descends" into the substrate to a depth of about 0.5 μm. This reduces its efficiency by 7 %.

Life tests of the metallic layers, which were performed on the RPT-6 reactor (NIIAR, Dimitrovgrad) up to a neutron fluence of ~10^{20}/cm^2, have shown that the uranium layers on the whole retain their initial texture to a 2×10^{19}/cm^2 fluence of thermal neutrons [27]; the operational life of metal layers with a protective aluminum layer exceeds 0.1 % with respect to burn-up and, for intermetallic layers, 0.4 %. With this, the main characteristics of the layer remained highly stable. Potential methods for increasing the operational life of layers up to 1–2 % with respect to burn-up have been examined in study [28].

The problems of operational life and durability of uranium layers, discussed in this section do not belong among the principal physical problems determining the potential for creating RLs. However, they broadly impact the appearance of the RL and its possible applications.

10.5 Shaping of Laser Radiation

The schematic for shaping laser radiation to a significant degree is determined by the purpose of the RL. Practice frequently requires that light beams be integrated from individual cells of the RL to a united light field. All methods of radiation adding can be divided into two major groups: incoherent and coherent. The first are primarily used where the quality parameters of the laser beam do not play a leading role, but the following are required: reservation of radiation sources; obtaining more output power at a given wavelength, which technically cannot be obtained on a single laser; forming a spatially limited optical field (beam) with different wavelengths, etc. Coherent addition of beam radiation in multi-channel laser facility repeatedly increases the intensity and brightness of the laser radiation and reduces its divergence. This is important for transportation of the radiation over a large distance. Coherent addition of N light beams, each of which has a beam divergence close to the diffraction limit, increases beam intensity in the far-field by N^2 times. Coherent, serial addition of light beams, as a rule, requires quick-acting adaptive optics that correct the phase of radiation in all laser channels. Implementation of this operation mode for RL is a complicated task.

Schemes for the Convergence of Light Beams Using Linear Optical Methods

A scheme for shaping an incoherent radiation field may contain a set of laser cavities with subsequent periscopic addition of radiation into a piecewise continuous (mosaic) field. The field is then telescoped and focused onto the target.

The radiation of N laser channels may be added into a single (incoherent) mosaic beam using periscopes (Fig. 10.6). If divergence of laser beam from one channel is equal to ~5×10^{-4} rad, after telescoping a united light beam up to the dimensions of the transverse section of the RL, the radiation divergence may be $\leq 5 \times 10^{-5}$ rad.

For systems with a large number of identical laser channels it is expedient to use serial, parallel, or mixed addition of radiation from individual channels to increase the intensity of the laser radiation. Figure 10.7 gives the principal schematics of this addition for three channels. As shown by the calculations, the total length of a "multipart" cavity with a high Q-factor may be 30–50 m (~10 channels). With parallel addition, the number of channels are limited by the losses in the semi-

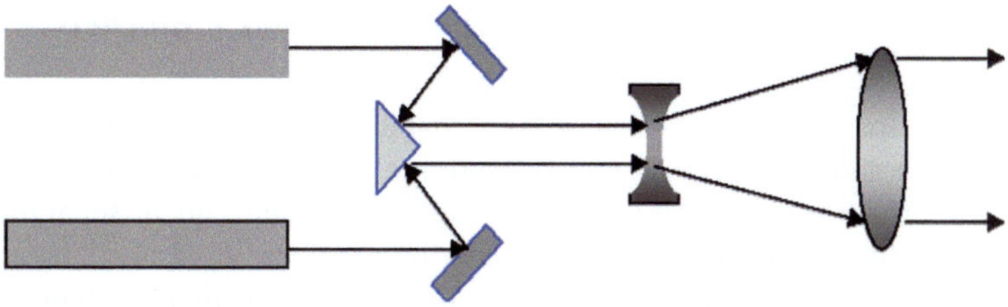

Fig. 10.6 Shaping of a mosaic radiation field

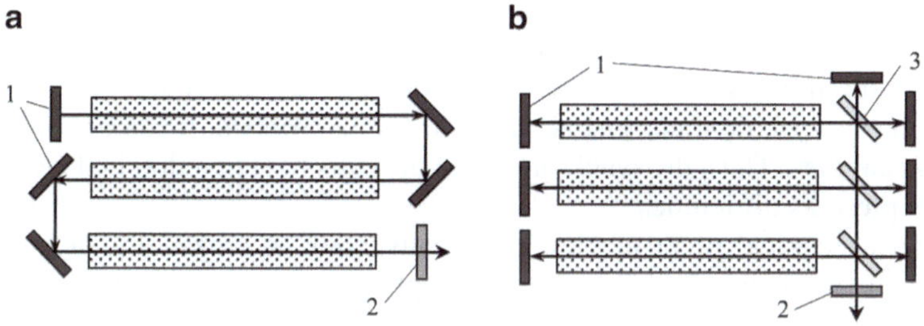

Fig. 10.7 Schematics of serial (**a**) and parallel (**b**) addition of radiation from laser channels: (*1*) high-reflectivity mirrors; (*2*) output mirrors; (*3*) translucent plates

transparent plates and also may be more than 10; with mixed (serial-parallel), respectively, more than 100 channels. In this case, the laser radiation is withdrawn from the RL by several (from 2–4 to 20–30) beams with a laser power of ~100–200 kW each. These beams may either be added by external systems (including coherent), or each is directed to its target independently.

VNIIEF researchers are currently studying methods of "direct" addition (parallel and serial) in different laser setups (LUNA-2M, LM-4, and LM-8). The first results from these studies are given in studies [29–32]. Direct addition methods are preferable when system size increases. In this case, difficulties arise with the application of methods using a common mirror. On the other hand, direct addition methods introduce increased demands for optical quality (as to coefficients of refraction and transmittance) and for optical uniformity in the active laser medium.

With nuclear pumping, startup of the individual lasers of a multi-channel system occurs in a random way. Use of the direct addition method permits the first laser that starts lasing to operate in "master oscillator" mode. Thus, it is possible to achieve complete phasing of all lasers in a set. In this case, the radiation from all channels will have identical polarization, phase, and frequency.

Coherent Addition of the Radiation of RL Laser Channels

The potential for coherent addition of laser channels is both interesting and important. Coherent addition methods known from the literature are based on the injection of the radiation of one channel into another and subsequent separation of a specific field distribution using various selection methods.

For sets of lasers with periodic arrangements of elements, there is an opportunity for coherent phasing due to the effect of the self-reproduction of the periodic structure of the monochromatic field at some distance (Talbot optical coupling) [33]. Individual lasers may also be coupled on a common nonlinear cell in the laser cavity (mirror based on stimulated Brillouin scattering) [34].

Hereafter, as we examine a large number of optically coupled lasers, we will assume that, during independent lasing, each laser operates in a single transverse mode. If the dimensions of the active medium of an individual laser grow, the selection of a specific transverse mode turns into a problem. It can be solved using unstable cavities. Thus, the synchronization of lasers having individual independent cavities presents difficulties.

From a technical point of view, it is convenient to create sets of lasers with periodically arranged elements. Optical coupling in such sets can be implemented using external mirrors deflecting a small portion of the radiation into neighboring lasers. It is sufficient to simply use a flat mirror as the coupling element which is placed at some distance from the aperture of the set of lasers. Radiation reflected from the mirror leads to optical coupling between elements due to diffraction "spreading." However, the efficiency of this coupling remains rather low. To increase the portion of radiation coming from one laser to another, the mirror must be moved aside. Due to diffraction, the radiation spreads along the whole aperture, and losses increase in the coupling channel. With a periodic arrangement of the set of lasers, this consideration is incorrect, because it does not take into account the reproduction effect of the periodic structure of the monochromatic field at a specific distance (Talbot diffraction coupling).

If the mirror is placed at half the reproduction distance of the periodic structure (Fig. 10.8), the image of the face plane of the channel, if it is emitted in phase, is projected onto the very same face plane. At the same time, because there is a contribution from the radiation of many channels to the image of each spot, coupling between individual lasers is sufficiently high. Without in-phase radiation of individual channels, the diffraction image breaks up, and losses increase sharply. Thus, this effect leads to selection of a phase-synchronization mode. It has been shown in study [33] that the distance z at which the mirror should be placed is equal to half of the "Talbot" distance $Z_T = \beta a^2/\lambda$, where a is the lattice pitch and β is a numerical coefficient that is a function of the geometry of the lattice.

As exemplified by a multi-beam waveguide CO_2 laser with periodically arranged elements, there was experimental confirmation [35, 36] of the potential for effective phase synchronization by Talbot diffraction coupling. It was shown that the graph of the laser power as a function of z can be divided into three parts. The first part relates

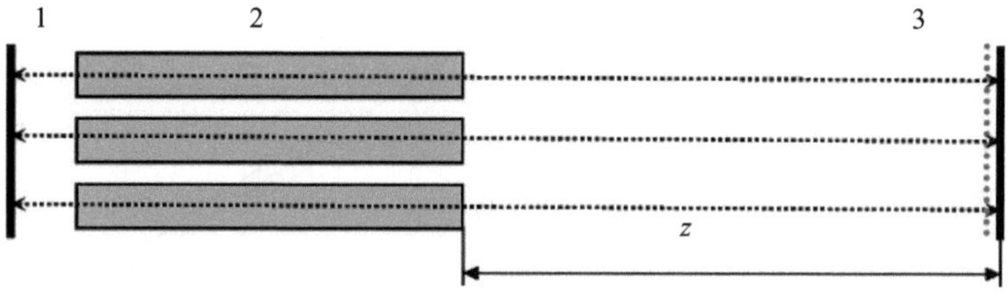

Fig. 10.8 Optical coupling of laser channels on a flat mirror

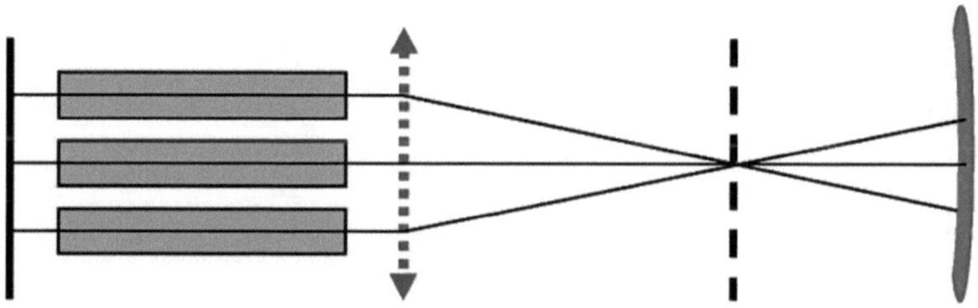

Fig. 10.9 Use of a spatial intra-cavity filter schematics

to independent lasing. In this case, the field distribution in the far-field region is determined by the aperture of a single channel, and the laser power is at a maximum. The second part represents a region with a partially coherent mode, with spots corresponding to interference orders are visible in the far-field and the total power drops by approximately twofold. In the third part, the radiation of each channel is highly coherent. However, the power is ~20 % the power of independent lasing.

A method that uses a spatial intra-cavity filter (Fig. 10.9) does not add up to optical coupling between lasers. The concept of a spatial filter consists of the selection of a specific configuration of the generated field that has minimal losses in the filter. If this configuration corresponds to an in-phase mode, it can be selected [36]. It was experimentally confirmed that with the use of an intra-cavity filter, there is absolutely no incoherent lasing, and complete synchronization is possible even with rough positioning of the filter. The total laser power in the principal maximum is 12 % with respect to the power of independent lasing. To enhance synchronization, it is recommended that the lengths of the laser cavities be increased and varied between themselves.

The first experiments [37] studying the coherent addition of NPL channels were performed by VNIITF researchers on reactor EBR-L. These experiments used the spatial filter method to realize an in-phase operation mode for a NPL package consisting of three parallel 60-cm long tubes with an internal diameter of 11 mm. The parameters of the laser medium, including the laser wavelength, are not given in study [37]. Diaphragms 1–2 mm in diameter were used as the spatial filters. The experiments showed that the use of a laser cavity including a telescope with a spatial filter reduces the laser radiation divergence due to phase synchronization of individual channels.

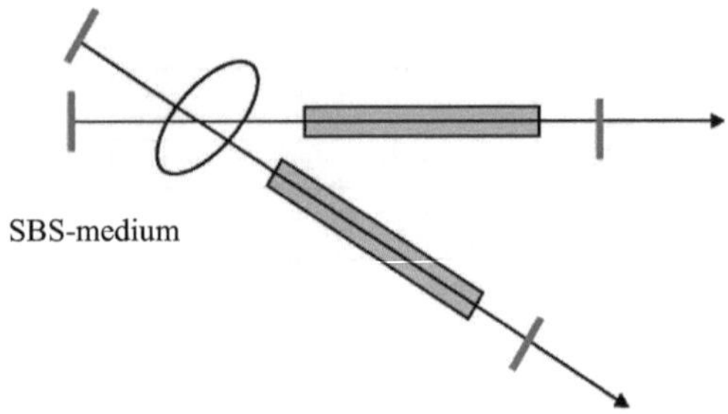

Fig. 10.10 Optical coupling in a nonlinear cell

Table 10.6 Gain increments and critical energy values [34]

Medium	g, cm/GW	E_{cr}, J	gE_{cr}, cm × ns
$TiCl_4$ (liquid)	14.2	1.1	16
CCl_4 (liquid)	3.8	11	40
GLS-1 (neodymium glass)	1.2	20	24
Xe (gas at 50 atm)	90	5	480

Individual lasers may also be coupled by redistributing the beams on a common nonlinear cell in the laser cavity. SBS mirror can be used as this cell The nonlinear cell based on stimulated Brillouin scattering (SBS). This method is especially useful for NPLs that make it possible to obtain the significant laser power necessary to reach the SBS threshold. The schematic of a laser with SBS mirror is simple (Fig. 10.10). The laser radiation is somehow focused in the cell with a transparent medium. When a certain threshold power is exceeded, the radiation due to SBS is reflected back and returns to the laser cavity. If the radiation of two or more lasers is focused on one nonlinear cell, they will be synchronized due to redistribution of the energy.

To maintain a steady SBS process, the threshold radiation power W_{thr}, which corresponds to the threshold energy $E_{thr} \approx W_{thr} \times \tau$ (where τ is the pulse duration) and is inversely proportional to the gain increment of the medium g, should enter the nonlinear medium. The energy released in the medium leads to the development of processes that compete with SBS. Therefore, when exceeding some critical energy E_{cr}, the effectiveness of SBS falls sharply. Thus, the determining criterion for the choice of a SBS medium is that the critical energy E_{cr} exceeds the threshold energy, that is, the product gE_{cr}, which only depends on the properties of the medium. Values for the gain increment and critical energy of different media are given in Table 10.6 [34]. It follows from the table that high-purity xenon is preferred as the SBS medium.

The experiments [34] performed by VNIIEF researchers studied the effect of SBS within the range 0.69–4.2 μm. They demonstrated the potential to:

- Compensate for aberrations and increase the radiation brightness.
- Phase the radiation of a large number of laser channels (up to 1,000).
- Obtain high-frequency (≥1 MHz) pulsed-periodic lasing.

In pulsed power NPLs, the pulse length is 10^{-3}–1 s. This significantly increases the time for pressure equalization in the focal region. This means that the efficiency of SBS may be influenced by a change in refraction coefficient due to electrostriction and heating of the medium. Experiments on xenon (at 46 atm) with a wavelength $\lambda = 1.315$ μm gave values determined by electrostriction: $W_{cr} = 3.2 \times 10^5$ W and $E_{cr} = 5$ J.

An experimentally obtained a laser beam divergence for a single laser channel of ~0.5×10^{-3} rad, close to that of a diffraction one. Linear methods (serial and parallel addition, telescoping, etc.) make it possible to obtain laser beam divergences of up to ~10^{-5} rad. Use of coherent addition methods brings the beam divergence up to values no worse than 10^{-6} rad.

10.6 Possible Uses for Reactor Lasers

The advantages of RLs over other types of laser systems are provided by high specific and absolute power capacities, which are practically unattainable in other types of lasers, a wide range of radiation wavelengths (from the near ultraviolet region to the infrared), the absence of intermediate steps in energy conversion, and flexibility of control. The consequences of this are a relatively simple construction, compactness and reliability, the system's high self-containment, operability for 20 years, the potential to develop mobile systems with minimal mass and dimensions at a given energy, especially in range of high laser powers. Multiple repeat startups do not require additional fuel, laser materials, or other materials.

Without chemically aggressive media and a large amount of uranium, the RL is safe for storage, use, and the environment. The disadvantage connected with the presence of nuclear radiations during RL operation is sufficiently neutralized by biological protection methods.

Proposals for potential uses of NPLs and RLs already started to appear in the first publications, which discussed the potential for the direct transformation of nuclear energy into a laser radiation. A review [12] gives a list of published works and notes on potential uses for NPLs and RLs. Some noted are:

1. Long distance communication.
2. Energy transfer (obtaining energy for space probes or satellites from generating stations placed on the earth or moon).
3. Photochemistry (for synthesis of chemical compounds on an industrial scale).
4. Fog dispersion (resulting from the evaporation of drops of liquid).
5. Illumination of surfaces (illumination of cities is possible with a stationary orbit).
6. High temperature heating of materials at a distance.

Issues concerning the use of NPLs and RLs in stationary as well as pulsed and pulsed-periodic modes are discussed in almost all reviews (see Chap. 1) and in more detail in specific studies. The following variants have been proposed for using NPLs and RLs:

- To supply energy to satellites and other space objects and stations on the dark side of the moon [19, 38, 39].
- To clean space of "debris" [19, 38, 40, 41].
- Laser drivers to transfer space devices from one orbit to another [19, 42] and delivering loads to a near-earth orbit [19, 43, 44].
- Reactor laser drivers for inertial confinement fusion systems [45–47].
- Heating and processing the surfaces of different materials (welding and cutting, changing surface properties, and deposition of thin coatings) [19, 38].
- Technological processes destined to the production of nanoparticles [19, 38].
- Development of new types of neutron detectors [48, 49], which can be used, for example, in systems for rapid nuclear reactor protection systems [50]. In this case, the detector is a NPL, which is a threshold system actuated at a specific level of neutron flux.

The area of use of an RL depends, naturally, on its characteristics, primarily its power, wavelength, and laser beam divergence, and also on the mode and duration of its operation. At the present time, the most realistic options for RLs with thin films of uranium fuel have an overall efficiency (η_{rl}) with respect to the nuclear energy released in the core that does not exceed 0.4 %: $\eta_{rl} = \varepsilon \times \eta_l$, where $\varepsilon \leq 20$ % is the share of the total energy absorbed in the gas laser medium, and $\eta_l \leq 2$ % is the laser efficiency (ratio of output laser power to power deposition in laser medium). To increase η_{rl} and expand the area of application for RLs, it is necessary to search for more effective laser media in different spectral ranges and for RL designs that have more effective transfer of nuclear energy to the laser media.

References

1. L.A. Gudzenko, S.I. Yakovlenko, Atomnyy reaktor-lazer [An atomic reactor laser]. Kratkiye soobshcheniya po fizike [Brief. Inform. Phys.] (2), 14–15 (1974)
2. L.O. Herwig, Concepts for direct conversion of stored nuclear energy to laser beam power. Trans. Am. Nucl. Soc. **7**(1), 131 (1964)
3. V.A. Dmitriyevskiy, Ye.M. Voinov, S.D. Tetelbaum, Primeneniye geksaftorida urana v yadernykh energeticheskikh ustanovykh [Application of Uranium Hexafluoride in Nuclear Energy Devices]. Atomnaya energiya [Atom. Energ.] **29**(4), 251–255 (1970)
4. K. Thom, R.T. Schneider, Gaseous fuel reactor research. IEEE Trans. Plasma. Sci. **PS-5**(4), 259–272 (1977)
5. K. Thom, F.C. Schwenk, Gaseous-fuel reactor systems for aerospace applications. J. Energy. **1** (5), 267–276 (1977)
6. R.T. Schneider, F. Hohl, Nuclear-pumped lasers. Adv. Nucl. Sci. Technol. **16**, 123–287 (1984)
7. L.I. Gudzenko, S.I. Yakovlenko, Plazmennyye lazery [Plasma lasers]. M: Atomizdat (1978)
8. I.S. Lakoba, Perebiraya varianty sredy reaktora-lazera [Sorting through the options for laser-reactor media]. Preprint FIAN [Preprint PIAS [Physical Institute of the Academy of Science]], No. 131, Moskva (1982)
9 F.V. Bunkin, V.I. Derzhiyev, V.I. Subbotin, V.V. Troyanskiy, V.V. Kharitonov, K.R. Chikin, S.I. Yakovlenko, Problemy gazofaznogo reaktora-lazera [Issues with the gas-phase reactor laser]. Preprint FIAN [Preprint PIAS], No. 199, Moskva (1984)

10. R.J. DeYoung, Y.J. Shiu, M.D. Williams, Fission-Fragment Nuclear Lasing of Ar(He)-Xe. Appl. Phys. Lett. **37**(8), 679–681 (1980)
11. H.A. Hassan, Analysis of the UF_6-Xe direct nuclear-pumped lasers. AIAA J. **18**(12), 1490–1494 (1980)
12. K. Thom, R.T. Schneider, Nuclear pumped gas lasers. AIAA J. **10**(4), 400–406 (1972)
13. V.V. Borovkov, B.V. Lazhintsev, V.A. Nor-Arevyan, A.N. Sizov, A.A. Sinyanskiy, Gazovyy lazer [A gas laser], Patent RF No. 3207271/25 (1984); (Izobreteniya, 1996), No. 5
14. V.V. Borovkov, B.V. Lazhintsev, V.A. Nor-Arevyan, A.N. Sizov, A.A Sinyanskiy, G.I. Fedorov, Osobennosti gazodinamiki prokachnykh lazerov, vozbuzhdayemykh oskolkami deleniya urana [The features of the gas dynamics of flowing-gas lasers excited with uranium fission fragments]. Kvantovaya elektronika [Quant. Electron.] **22**(12), 1187–1191 (1995)
15. A.A. Sinyanskii, V.N. Krivonosov, Powerful nuclear-optical converters for applications in scientific investigations and national economy. Proc. SPIE **3574**, 440–448 (1998)
16. F.V. Bunkin, V.I. Derzhiyev, A.V. Karelin, A.L. Nefedov, V.I. Subbotin, V.V. Kharitonov, K.R. Chikin, S.I. Yakovlenko, Statsionarnyy reaktor-lazer s poverkhnostnoy nakachkoy [A stationary reactor laser with surface pumping]. Preprint IOF AN SSSR [Preprint IGP [Institute of General Physics] AS USSR], No. 321, Moskva (1985)
17. A.V. Karelin, A.A. Sinyanskiy, S.I. Yakovlenko, Lazery s yadernoy nakachkoy i fizicheskiye problemy sozdaniya reaktora-lazera [Lasers with nuclear pumping and physical issues related to creating a reactor laser]. Kvantovaya elektronika [Quant. Electron.] **24**(5), 387–414 (1997)
18. A.M. Voinov, S.V. Vorontsov, V.N. Krivonosov, S.P. Melnikov, A.A. Pikulev, V.T. Punin, A.N. Sizov, A.A. Sinyanskiy, S.L. Turutin, Raboty, vypolnennyye vo VNIIEF, po issledovaniyu voszmozhnosti sozdaniya reaktora-lazera [Studies performed at VNIIEF, on investigating the possibility for creating a reactor laser]. Voprosy atomnoy nauki i tekhniki. [Questions of atomic science and technology] Ser. Fizika yadernykh reaktorov [Series on the physics of nuclear reactors] (2), 19–28 (2007)
19. J.R. Felty, R.J. Lipinski, D.A. McArthur, P.S. Pickard, DOE reactor-pumped laser program. Proc. SPIE **2121**, 2–9 (1994)
20. I.A. Baranov, Yu.V. Martynenko, S.O. Tsepelevich, Yu.N. Yavlinskiy, Neyprugoye raspyleniye tverdykh tel ionami [Inelastic disintegration of hard bodies by ions]. Uspekhi fizicheskikh nauk **156**(3), 477–511 (1988)
21. I.A. Baranov, V.V. Obnorskiy, Raspyleniye materialov oskolkami deleniya [Atomization of materials by fission fragments]. Voprosy atomnoy nauki i tekhniki. [Questions of atomic science and technology] Ser. Fizika radiatsionnykh povrezhdeniy i reaktornoye materialovedeniye [Series on the physics of radiation damage and reactor material science] **5**(19), 37–51 (1981)
22. G.I. Khlebnikov, Yu.T. Cheburkov, Polucheniye tolstykh sloyev toriya, urana, neptuniya, plutoniya, ameritsiya [Obtaining thick layers of thorium, uranium, neptunium, plutonium, and americium. Atomnaya energiya [Atom. Energ.], **9**(5), 406–408 (1960)
23. V.Ya. Mishin, S.M. Solovyev, A.S. Voronin, P.P. Chinenov, V.I. Kapitonov, Ye.M. Rubtsov, P.S. Soloshenkov, V.K. Isupov, Yu.Z. Prokopchuk, Letuchiya β-diketonaty aktinoidov i nekotoryye vozmozhnosti ikh prakticheskogo ispolsovaniya [Volatile â β-Diketonates of Actinoids and Some Possibilities for their Use]. Radiokhimiya [Radiochemistry] **27**(3), 354–361 (1985)
24. B.S. Danilin, Primeneniye nizkotemperaturnoy plazmy dlya naneseniya tonkikh plenok [Using low temperature plasma to deposit thin films]. M: Energoatomizdat (1989)
25. G.V. Vlokh, A.A. Sinyanskiy, G.E. Filippov, L.L. Kazakov, N.S. Kosulin, V.N. Cherevatyuk, *Plenochnyye energovydelyayushchiye elementy dlya lazerov s yadernoy nakachkoy*. [Thin-film energy releasing components for nuclear-pumped lasers]. Trudy 2-y mezhdunarodnoy konf. [Proceedings of the 2nd International Conference] "Fizika yaderno-vozbuzhdayemoy plazmy i problemy lazerov s yadernoy nakachkoy" [Physics of nuclear-excited plasma and problems of nuclear-pumped lasers], Arzamas-16, 1995, vol. 2, pp. 47–54

26. G.V. Vlokh, A.M. Korochkin, N.G. Krylov, I.N. Sidenko, G.E. Filippov, *Vliyaniye predvaritelnoy termoobrabotki uranovykh sloyev na effektivnost ikh raspyleniya oskolkami deleniya* [The impact of the preliminary heat treatment of uranium layers on the efficiency of their atomization by fission fragments]. Trudy otraslevoy konf. [Proceedings of the Specialist Conference] "Fizika yaderno-vozbuzhdayemoy plazmy i problemy lazerov s yadernoy nakachkoy" [The physics of nuclear-excited plasma and problems of nuclear-pumped lasers], Obninsk, 1993, vol. 3, pp. 91–94
27. L.L. Kazakov, N.S. Kosulin, V.N. Cherevatyuk, *Rezultaty resursnykh ispytaniy energovydelyayishchikh elementov dlya lazerov s yadernoy nakachkoy* [Test results for the energy releasing components of lasers with nuclear pumping]. Trudy otraslevoy konf. [Proceedings of the Specialist Conference] "Fizika yadernovozbuzhdayemoy plazmy i problem lazerov s yadernoy nakachkoy" [The physics of nuclear-excited plasma and problems of nuclear-pumped lasers], Obninsk, 1993, vol. 3, pp. 41–49
28. L.L. Kazakov, N.S. Kosulin, V. N. Cherevatyuk, *Strukturnyye i fazovyye izmeneniya v energovydelyayishchikh elementakh pri dlitelnom obluchenii* [Structural and phase changes in energy releasing elements with long-term irradiation]. Trudy 2-y mezhdunarodnoy konf. [Proceedings of the 2nd International Conference] "Fizika yaderno-vozbuzhdayemoy plazmy i problemy lazerov s yadernoy nakachkoy" [The physics of nuclear-excited plasma and problems of nuclear-pumped lasers], Arzamas-16, 1995, vol. 2, pp. 60–68
29. A.A. Sinyanskiy, S.V. Patyanin, A.A. Pikulev, P.V. Sosnin, S.L. Turutin, V.M. Tsvetkov, *Formirovaniye opticheskogo izlucheniya mnogokanalnykh lazerov s yadernoy nakachkoy.* [Shaping the optical radiation of multichannel lasers with nuclear pumping]. Trudy 4-y mezhdunarodnoy konf. [Proceedings of the 4th International Conference] "Fizika lazerov s yadernoy nakachkoy i impulsnyye reaktory" [The physics of nuclear-pumped lasers and pulsed reactors], Obninsk, 2009, vol. 1 pp. 196–205
30. A.A. Pikulev, A.A. Abramov, The numerical investigation efficiency of parallel and successive laser cells composition. Proc. SPIE **6263**, 186–196 (2006)
31. A.A. Pikulev, V.M. Tsvetkov, P.V. Sosnin, A.A. Sinyanskiy, Posledovatelnoye slozheniye dvukh lazernykh kanalov pri vozbuzhdenii smesey He-Ar-Xe (2.03 mkm) i Ar-Xe (1.73 mkm) oskolkami deleniya urana [Serial addition of two laser channels at excitation of He-Ar-Xe (2.03 μm) and Ar-Xe (1.73 μm) mixtures by Uranium fission fragments]. Kvantovaya electronica [Quant. Electron.] **39**(6), 494–500 (2009)
32. A.A. Pikulev, V.M. Tsvetkov, P.V. Sosnin, A.A. Sinyanskiy, *Rabota lazera s yadernoy nakachkoy v rezhime parallelnogo slozheniya dvukh lazernykh kanalov* [Operating of nuclear-pumped laser at the parallel addition of two laser channels]. Trudy 4-y mezhdunarodnoy konf. [Proceedings of the Fourth International Conference] "Fizika lazerov s yadernoy nakachkoy i impulsnyye reaktory" [The physics of nuclear-pumped lasers and pulsed reactors], Obninsk, 2009, vol. 1, pp. 213–218
33. A.A. Golubentsev, V.V. Likhanskiy, A.P. Napartovich, Teoriya fazovoy sinkhronizatsii nabora lazerov [The theory of phase synchronization for laser sets]. ZhETF [J. Exp. Theor. Phys.] **93**(4), 1199–1209 (1987)
34. Yu.V. Dolgopolov, A.M. Dudov, V.A. Yeroshenko, L.I. Zykov, G.G. Kochemasov, S.M. Kulikov, V.N. Novikov, S.A. Sukharev, A.F. Shkapa, *Lazery c obrashcheniyem volnovogo fronta v rezhime svobodnoy generatsii.* [Lazers with phase conjugation operating in free-running lasing mode]. Trudy 2-y mezhdunarodnoy konf. [Proceedings of the 2nd International Conference] "Fizika yadernovozbuzhdayemoy plazmy i problemy lazerov s yadernoy nakachkoy" [The physics of nuclear-excited plasma and problems of nuclear-pumped lasers], Arzamas-16, 1995, vol. 2, pp. 229–245
35. V.V. Antyukhov, A.F. Glova, O.R. Kachurin, F.V. Lebedev, V.V. Likhanskiy, A.P. Napartovich, V.D. Pismennyy, Effektivnaya fazovaya sinkhronizatsiya nabora OKG [Effective phase synchronization of a set of optical quantum generators]. Pisma v ZhETF [Lett. J. Exp. Theor. Phys.] **44**(2), 63–65 (1986)

36. V.V. Antyukhov, A.A. Golubentsev, Ye.V. Danshchikov, F.V. Lebedev, V.V. Likhanskiy, A.P. Napartovich, O.R. Kachurin, *Generatsiya opticheski svyazannykh lazerov* [Lasing of optically coupled lasers]. Izvestiya AN SSSR. Ser. fizicheskaya [Proceedings of the Academy of Science of the USSR], 1990, vol. 54, No. 10, pp. 2002–2008
37. A.V. Bochkov, E.P. Magda, G.S. Sofiyenko, V.V. Mironenko, V.M. Murzin, *Issledovaniye sinfaznogo rezhima raboty paketa LYaN* [Investigation of an in-phase operation mode for a NPL package]. Trudy 3-yey mezhdunarodnoy konf. [Proceedings of the 3rd International Conference] "Problemy lazerov s yadernoy nakachkoy i impulsnyye reaktory" [Problems of nuclear-pumped lasers and pulsed reactors], Snezhinsk, 2003, pp. 244–246
38. R.J. Lipinski, D.A. McArthur, *Application for Reactor-Pumped Lasers*. Trudy 2-y mezhdunarodnoy konf. [Proceedings of the 2nd International Conference] "Fizika yaderno-vozbuzhdayemoy plazmy i problemy lazerov s yadernoy nakachkoy" [The physics of nuclear-excited plasma and problems of nuclear-pumped lasers], Arzamas-16, 1995, vol. 1, pp. 44–51
39. Ye.A. Pashin, A.V. Gulevich, P.P. Dyachenko, O.F. Kukharchuk, A.P. Barzilov, *O vozmozhnosti peredachi lazernogo izlucheniya v kosmos ot nazemnogo reaktora-lasera*. [On the potential to transmit laser radiation into space from a earth-based reactor-laser]. Izvestiya vuzov. Yadernaya yenergetika [Proceedings of Higher Education Institutions. Nuclear Energy], 1997, No. 4, pp. 31–35
40. P.I. Bystrov, L.A. Gorshkov, N.I. Zelenshchikov, A.A. Maslennikov, V.V. Pukhov, Yu.P. Semenov, V.V. Sinyavskiy, Yu.A. Sobolev, *O vozmozhnosti ispolzovaniya termoemis-sionnogo reaktora-preobrazovatelya s yadernoy nakachkoy lazernoy sistemy primenitelno k ochistke kosmosa ot musora* [On the possibility of using a thermal emission reactor-converter with nuclear pumping by a laser system relating to the removing debris from space]. Trudy otraslevoy konf. [Proceedings of the Specialist Conference]. "Fizika yadernovozbuzhdayemoy plazmy i problemy lazerov s yadernoy nakachkoy" [The physics of nuclear-excited plasma and problems of nuclear-pumped lasers], Obninsk, 1993, vol. 2, pp. 257–261
41. C.R. Phipps, M.M. Michaelis, *Clearing near-earth space debris in 4 years using a 20-kW, 530-nm repetitively pulsed laser*. Trudy 2-y mezhdunarodnoy konf. [Proceedings of the 2nd International Conference] "Fizika yadernovozbuzhdayemoy plazmy i problemy lazerov s yadernoy nakachkoy" [The physics of nuclear-excited plasma and problems of nuclear-pumped lasers]. Arzamas-16, 1995, vol. 2, pp. 252–259
42. C.R. Phipps, M.M. Michaelis, *Space propulsion concept using high energy, pulsed laser ablation*. Trudy otraslevoy konf. [Proceedings of the Specialist Conference] "Fizika yaderno-vozbuzhdayemoy plazmy i problem lazerov s yadernoy nakachkoy" [The physics of nuclear-excited plasma and problems of nuclear-pumped lasers], Obninsk, 1993, vol. 3, pp. 196–205
43. P.P. Dyachenko, *Nuclear-laser power engineering as a prospective direction in nuclear energy utilization*. Proceedings of the 8th International Conference on Emerging Nuclear Energy Systems, Obninsk, 1996, vol. 1, pp. 296–303
44. E.P. Magda, *Lazerno-reaktornyy drayver dlya vyvedeniya poleznykh gruzov na okolozemnuyu orbity* [A laser reactor driver for ascent of useful loads in near-earth orbit]. Trudy 3-y mezhdunarodnoy konf. [Proceedings of the 3rd International Conference] "Problemy lazerov s yadernoy nakachkoy i impulsnyye reaktory" [Problems of nuclear-pumped lasers and pulsed reactors], Snezhinsk, 2003, pp. 34–40
45. A.V. Zrodnikov, *Perspektivy primeneniya lazerov s yadernoy nakachkoy v nauke, tekhnike i tekhnologii* [The prospects for applying nuclear-pumped lasers in science, technology, and manufacturing]. Trudy otraslevoy konf. [Proceedings of the Specialist Conference] "Fizika yaderno-vozbuzhdayemoy plazmy i problemy lazerov s yadernoy nakachkoy" [The physics of nuclear-excited plasma and problems of nuclear-pumped lasers], Obninsk, 1992, vol. 1, pp. 122–143
46. D.E. Beller, J.M. Jacobson, G.H. Miley, M. Petra, Y. Shaban, *Parametric design study of a nuclear-pumped-laser-driven inertial confinement fusion power plant*. Trudy otraslevoy konf. [Proceedings of the Specialist Conference] "Fizika yaderno-vozbuzhdayemoy plazmy i

problemy lazerov s yadernoy nakachkoy" [The physics of nuclear-excited plasma and problems of nuclear-pumped lasers], Obninsk, 1993, vol. 3, pp. 121–135

47. P.P. Dyachenko, A.V. Zrodnikov, Yu.A. Prokhorov, V.Ya. Pupko, S.S. Satin, V.I. Subbotin, N.G. Basov, V.B. Rozanov, Yu.V. Senatskiy, L.P. Feoktistov, A.A. Mak, S.B. Papernyy, V.A. Serebryakov, V.Ye. Yashin, Gibridnaya yaderno-termoyadernaya elektrostantsiya [Hybrid nuclear-thermonuclear power plant]. Preprint FEI [Preprint IPP], No. 2327, Obninsk (1993)
48. A. Okumura, S. Soramoto, H. Hayashida, H. Nakamura, M.A. Nakazawa, Proposal of a new in-core neutron monitor using nuclear pumped laser. Nucl. Instr. Meth. Phys. Res. **A306**(3), 530–535 (1991)
49. V.V. Gorbunov, L.Ye. Dovbysh, S.P. Melnikov, A.V. Sinitsyn, A.A. Sinyanskiy, V.M. Tsvetkov, *Neutronnyy detektor na osnove yaderno-opticheskogo preobrazovatelya* [A neutron detector based on nuclear optical conversion]. Trudy RFYaTs-VNIIEF [Proceedings RF NC [Nuclear Center]-VNIIEF], 2007, Issue 11, pp. 168–179
50. M.I. Kuvshinov, I.G. Smirnov, A.M. Voinov, L.Ye. Dovbysh, *Pryamodeystvuyushchaya bystraya avariynaya zashchita reaktora s datchikom neytronov na osnove yaderno-opticheskogo preobrazovatelya* [Direct-acting rapid reactor protection system with a neutron sensor based on nuclear optical conversion]. Voprosy atomnoy nauki i tekhniki. [Questions of Atomic Science and Technology] Ser. Fizika yadernykh reaktorov [Series on the Physics of Nuclear Reactors], 2000, No. 2/3, pp. 117–119

Chapter 11
Studies of Condensed-Media NPLs

The most progress in searching for active media for NPLs, studying their characteristics, and developing various nuclear-laser devices was achieved when high-pressure gas media were used. However, there have been repeated attempts in Russia and the United States to pump different condensed media with nuclear radiation and carry out feasibility studies on developing powerful nuclear-laser devices based on these media. The active material in solid-state and liquid lasers is a dielectric in a condensed phase. As compared to gases, it is possible to create a higher density of active particles in condensed media, and, as consequence, to obtain more specific output energy of laser radiation.

When uranium, for example, is introduced into a condensed media it is possible to achieve essentially uniform excitation of the active volume at close to 100 % deposition efficiency (due to a small path length for charged nuclear particles in a condensed medium). Therefore in the initial stage of NPL investigations at the end of the 1960s and the beginning of the 1970s, when known gas lasers operated only at low pressures, the emphasis was on condensed media. At that time, these studies were carried out at VNIIEF [1–3] in Russia and in the laboratories of the North American Rockwell Corporation, General Atomics, etc. in the United States (see the literature cited in publications [4–6]). Later, other laboratories became involved in the investigation of condensed-media NPLs. A significant number of studies were fulfilled at FEI [Physics and Power Engineering Institute] (Russia) and the University of Missouri-Columbia (USA).

Two pumping alternatives were considered for excitation of condensed-media NPLs: immediate (direct) pumping using nuclear radiation when nuclear reactions occur inside the laser medium; and pumping with help of optical radiation from intermediate devices—nuclear-optical transformers or convertors (nuclear-excited plasma and solid-state or liquid scintillation media). In the first case, the pumping region and the region where population inversion is created coincide, in the second, they are spatially separated.

S.P. Melnikov et al., *Lasers with Nuclear Pumping*,
DOI 10.1007/978-3-319-08882-2_11

11.1 Condensed-Media Lasers with Direct Nuclear Pumping

NPLs with direct pumping absorb the kinetic energy of nuclear particles (neutrons, γ-quanta, fission fragments) and convert it into laser radiation in the laser medium itself. Especially interesting are laser media with fissile elements uniformly distributed in material (for example, ^{235}U), because, in this case, higher specific energy depositions are possible. The main obstacle on the road to creating condensed-media NPLs is their radiation damage: the formation of displaced atoms and color centers in solid-state lasers (see, for example, [7]); radiolysis [8] and the formation of gas-vapor bubbles [9] on the tracks of nuclear particles in liquid lasers.

Solid-State Laser Media

To our knowledge, there have been no experiments on the direct pumping of solid-state laser media with nuclear radiation of pulsed reactors, although one study [2], for example, was devoted to investigation of the luminescent characteristics of some laser-active materials (ruby, glass with neodymium, $CaWO_4:Nd^{3+}$, $Y_3Al_5O_{12}:Nd^{3+}$, etc.) and the radiation resistance of ruby and neodymium glass lasers under the effect of the *n*,γ-ray radiation of pulsed reactors. Some materials had rather high energy conversion efficiencies η_λ, where η_λ is the ratio of the specific output of luminescent radiation to the specific energy deposition. Thus, with high-energy electron beam excitation the following values of η_λ were obtained: $\eta_\lambda \approx 10$ % ($\lambda \approx 611$ nm) for Y_2O_3 doped with europium [10], and $\eta_\lambda = 4$ % ($\lambda = 900$–1,100 nm) for the crystal $Y_3Al_5O_{12}:Nd^{3+}$ [11]. The value of η_λ depends on the type of charged particle. When excited by heavy charged particles (α-particles and fission fragments), the crystal $Y_3Al_5O_{12}:Nd^{3+}$ has lower values of η_λ: $\eta_\lambda = 2$ % in the range 900–1,100 nm, and $\eta_\lambda = 0.28$ % for the more intense laser line 1,063 nm [12]. We note that in another study [11], when the crystal $Y_3Al_5O_{12}:Nd^{3+}$ was excited by an electron beam, laser action or, at least, superluminescence was obtained.

Yet another study [13] examined the possibility of lasing when neodymium-doped, uranium-containing media is irradiated by the pulsed neutron flux of the BARS-6 reactor [14]. The $Y_3Al_5O_{12}$ crystal and GLS2 silicate and phosphate (GLS22, GLS27, and LGS40) glasses were selected as the dielectric matrices. With ^{235}U concentrations in these media at 5×10^{19} cm^{-3}, the specific energy deposition was 31 J/cm^3 (the neutron pulse duration was 140 μs). We note that the possibility of introducing uranium into these media and its influence on laser characteristics have not yet been verified in practice. In the authors' opinion [13], under these conditions and using direct pumping with fission fragments, a laser effect is possible for active media based on $Y_3Al_5O_{12}$ and GLS27 and LGS40 glasses. Of the effects arising during a pumping pulse and resulting to additional losses, the calculations only took into account light radiation scattering on the fission fragment tracks. In fact, a substantially larger contribution to the total detrimental loss factor may be caused by radiation-induced (neutrons, γ-ray emission, and fission fragments) light absorption due to formation of color centers (see Chap. 2, Sect. 1). Some media (for example, $Y_3Al_5O_{12}$ without ^{235}U impurities) completely lose their transparency on the 1.06 μm line at the maximum of the reactor pulse when the dose rate is about 2×10^5 Gr/s. Therefore, conclusions of study [13] concerning the direct pumping of solid-state neodymium media with nuclear radiation in experiments on pulsed reactors are excessively optimistic.

Liquid Laser Media Based on Organometallic Compounds

Liquid laser media have a number of advantages as compared to solid-state media: there are no technical difficulties when making lasers with large amounts of active media, the liquids have sufficiently high volumetric optical uniformity (providing there are no temperature gradients), and the laser medium may be cooled and cleaned using flowing mode of operation.

It appears that the first experiments verifying the possibility of laser action in a liquid medium directly pumped with nuclear radiation were performed in study [15]. These experiments used an active medium based on europium chelate: a solution of $Eu^{3+}(TTA)_3TBPh$ in toluene. To excite the medium using fission fragments, the ^{235}U isotope in the form of the $UO_2(NO_3)_2$ compound was introduced into the solution. Irradiation of the solution with a neutron flux from the KEWB pulsed reactor [16] with a 1.2 ms pulse duration resulted in a rather high value for $\eta_\lambda \approx 4\ \%$ ($\lambda = 615$ nm). We note that electron beam excitation of a series of europium chelate compounds resulted in an even higher value for $\eta_\lambda \leq 10\ \%$ [17].

The experiments [15] attempted to obtain a laser effect, for which a container with the solution was placed inside the laser cavity. However, there was no lasing since the gain of the active medium, 3.3×10^{-2} cm^{-1}, was lower than the detrimental loss factor.

After the article [15] was published, VNIIEF [3] began studies directed to searching for liquid active media for NPLs. X-ray excitation with a 10 ns pulse duration and ~10^{-4} Gr dose was used for preliminary study of the specific light output and the $^5D_0$ level lifetime of the Eu^{3+} ion as a function of the solvent type (Table 11.1). The specific light output could be increased by cooling the medium. For example, with an approximately 100 °C reduction in temperature, the light output of $Eu^{3+}(BA)_4HP$ in acetone increased by ten times, and that of Eu^{3+}-(BTFA)^{4}HP in acetonitrile, by 1.5 times.

Next, the luminescent characteristics of europium chelate solutions were studied by irradiating them with the n,γ-ray radiation of the VIR-2 pulsed reactor [16] with a pulse duration of ~4 ms as a function of the specific energy deposition and the europium concentration. The maximum dose of γ-radiation was $D_\gamma = 3.4 \times 10^3$ Gr, and the neutron fluence was $F = 3.5 \times 10^{14}$ cm^{-2}. Evaluations have shown that

Table 11.1 The specific light output and $^5D_0$ level lifetime of the ion Eu^{3+} for the $Eu^{3+}(TTA)_3 2P$ compound in different solvents [3]

Solvent	Lifetime (ms)	Specific light output, rel. units	Solvent	Lifetime (ms)	Specific light output, rel. units
Ditolyl methane	0.65	100	Acetonitrile	0.77	12
α-Methylnapthalene	0.52	97	Methyl methacrylate	0.62	7.1
Benzol	0.75	86	Ethyl acetate	0.60	5.9
Toluene	0.62	74	Dimethyl formamide	0.62	5.4
Xylene	0.65	73	Acetylacetone	0.51	4.7
Acetophenol	0.45	67	Styrene	0.53	4.2
Acetone	0.50	44	Ethanol	0.51	3.4
Cyclohexane	0.37	39	Diethyl ether	0.50	3.0
Chloroform	0.72	24	Ethyl cellosol [sic]	0.37	2.3
Tetrachloride carbon	0.62	14			

Note: $T = 300$K; the europium concentration is 0.005 g/cm^3

about 70 % of the specific energy deposition was provided by recoil protons that arise at elastic scattering of fast neutrons on hydrogen nuclei. Some results from these studies are given in Fig. 11.1 and Table 11.2. For reference, Table 11.2 also gives the conversion efficiency for x-ray excitation of europium chelates.

The luminescence spectrum and lifetime of the metastable level $^5D_0$ of the Eu^{3+} ion did not change with the introduction of $^{235}UO_2(NO_3)_2$ in concentrations up to 0.05 g/cm^3 into the europium chelate solutions. However, in this case, the conversion efficiency had a sharp decrease to 0.01 %. This decrease was explained by optical losses in the medium due to the formation of gas-vapor bubbles [9] on the tracks of fission fragments and radiolysis of the europium compounds. Experiments on the VIR-2 reactor studying the transparency of the aqueous solution $^{235}UO_2SO_4$ (the uranium concentration was about 0.05 g/cm^3) have shown that, the transmission coefficient of the solution at $\lambda = 1.06$ μm slowly decreases during the reactor pulse, and with a specific energy deposition of about 30 J/cm^3, the solution boils up. To decrease the influence of gas-vapor bubbles, the pressure can be increased in the liquid, or, for example, a laser cell without free space [9] can be used.

Verification of the laser action possibility when europium chelate solutions are excited with pulsed n,γ-ray radiation was carried out on the TIBR reactor at $D_\gamma = 2 \times 10^3$ Gr; $F = 4.7 \times 10^{14}$ cm^{-2} and a pulse duration of about 1.5 ms [1, 3]. A cylindrical quartz laser cell, which was 30-cm long and had a 6 mm internal diameter, was filled with a solution of $Eu^{3+}(BTFA)_4HDPhH$ in acetone ($C_{Eu} = 0.02$ g/cm^3) and placed in the horizontal through port of the reactor (Fig. 11.2). Laser cavity was formed by two mirrors which were pressed on the ends of the cavity. These mirrors represented Ag layers on surfaces of radiation-resistant

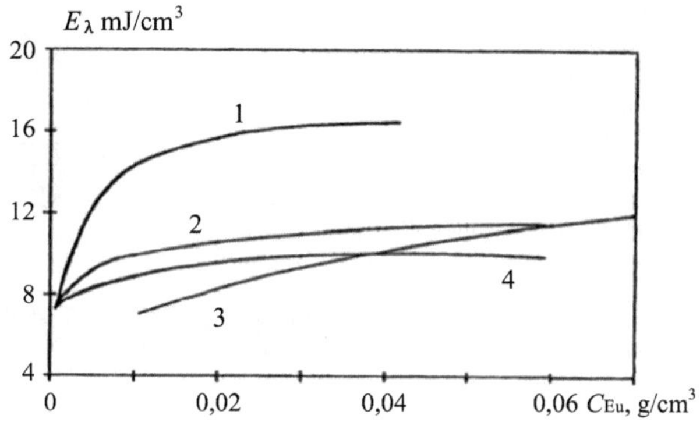

Fig. 11.1 The dependence of light output E_λ on the concentration of Eu for $Eu^{3+}(BTFA)_4HP$ solutions in benzol (1), acetone (2), and acetonitrile (3), and also $Eu^{3+}(BTFA)_4HDPhH$ in acetone (4) [3]

Table 11.2 The light output and conversion efficiency of europium chelates [3]

Solution	E_λ, 10^{-2} J/cm^3	$\eta_\lambda^{n\gamma}$ (%)	η_λ^{r} (%)
$Eu^{3+}(TTA)_3 2P$ (toluene)	3.2	0.20	1.0
$Eu^{3+}(BTFA)_4HP$ (benzol)	1.6	0.18	0.52
$Eu^{3+}(BTFA)_4HP$ (acetone)	1.1	0.13	0.42
$Eu^{3+}(BTFA)_4HP$ (acetonitrile)	0.7	0.08	0.23
$Eu^{3+}(BTFA)_4HDPhH$ (benzol)	1.1	0.13	0.42

Note: E_λ is light output; $\eta_\lambda^{n\gamma}$ at irradiation with the n-, γ-pulse of the VIR-2 reactor; η_λ^{r} at irradiation with X-ray

quartz (one of the mirrors had a 1.7 % transmission). The solution was cooled to 200 K with liquid nitrogen vapor. Signals from photodetectors, when a laser cavity was used, exceeded the level of the luminescent radiation by several times. This testifies to the presence of superluminescence. The absence of lasing in these experiments was apparently related to the use of low-quality mirrors.

Inorganic Liquid Laser Media

The first studies of the luminescent properties of inorganic liquids doped with europium and neodymium and excited with uranium fission fragments were apparently conducted at VNIIEF [3]. Experiments on the VIR-2 pulsed reactor with uranium fission fragment (specific energy deposition up to 25 J/cm^3) excitation of a $POCl_3$-$SnCl_4$-$^{235}UO_2^{2+}$:Nd^{3+}(Eu^{3+}) solution with a uranyl concentration of ~2×10^{19} cm^{-3} obtained low conversion efficiencies $\eta_\lambda \leq 0.1$ % [3, 18]. The conversion efficiency for perchlorates with europium under analogous conditions was significantly lower.

The spectral-luminescent characteristics of laser media based on inorganic liquids with excitation by nuclear particles under various conditions were studied in more detail in studies at the FEI. The main results of these studies are given in

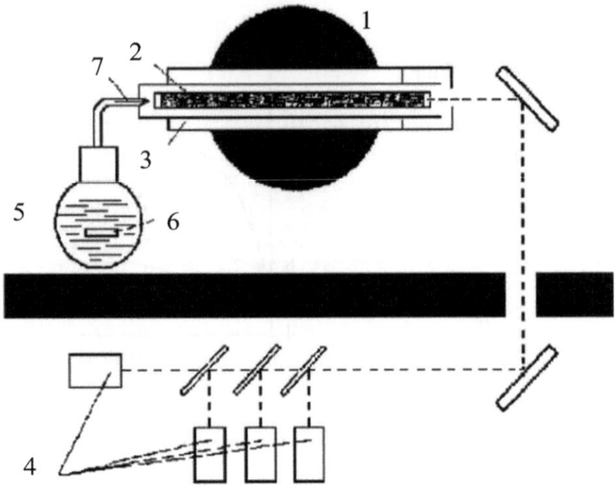

Fig. 11.2 A schematic of experiments on direct nuclear pumping of a liquid europium chelate medium on the TIBR reactor [1]: (1) core of the TIBR reactor; (2) a laser cell with a europium chelate solution; (3) thermostat; (4) photodetectors; (5) Dewar's flask; (6) electric heating unit; (7) nitrogen vapor

reviews [19, 20]. The following inorganic liquids were examined as test objects: $POCl_3$-$SnCl_4$:Nd^{3+}(Eu^{3+},Tb^{3+}) [21–27]; $ZnCl_2$-$GaCl_3$:Nd^{3+}(Tb^{3+}, Mn^{2+}) [22]; $SOCl_2$-$GaCl_3$:Nd^{3+} [22, 25]; SO_2Cl_2-$GaCl_3$:Nd^{3+} [25, 28]; and D_2O:Eu^{3+}(Tb^{3+}) [22, 29], doped with Nd^{3+}, Eu^{3+}, Tb^{3+}, Mn^{2+}.

The main effort was directed toward the search for laser-active liquid media containing ^{235}U. For this purpose, technology was developed to introduce uranium into laser media in the form of uranyl UO_2^{2+} in concentrations $\leq 10^{20}$ cm^{-3}, and the interactions between uranyl and ions of rare-earth elements were studied [30, 31]. The laser media were excited with the products of nuclear reactions taking place upon the interaction between the nuclei and neutrons. The neutron sources were either the accelerator KG-0.3 with a neutron energy of 14 MeV [22, 23, 25, 28, 29] operating in pulsed and continuous modes or the BARS-6 pulsed reactor [26, 27]. In the first case, during continuous operation of the accelerator, the specific pump power of the laser media was ~5×10^{-7} W/cm^3 (for liquids based on D_2O, it is approximately five times higher), and in the second case, while using a medium containing uranium, the specific energy deposition reached 30 J/cm^3 at a pulse duration of 140 μs.

The studies of liquid laser media for NPLs carried out at FEI obtained information on the luminescence spectra, conversion efficiency, and lifetimes of the upper laser level of media based on inorganic liquids as a function of activator concentrations, and also their characteristics were compared using different methods and excitation modes. Some experimental information on the conversion efficiencies and lifetimes of upper laser levels for different media is given in Fig. 11.3 and Table 11.3.

The data cited testify that the conversion efficiency depends on the type of solvent and activator, the concentration of the activator, and also the presence of uranyl in the solution. The maximum value, $\eta_\lambda = 1.4$ %, was obtained for the solution $SOCl_2$-$GaCl_3$:Nd^{3+}. However, with the addition of the uranyl UO_2^{2+} to

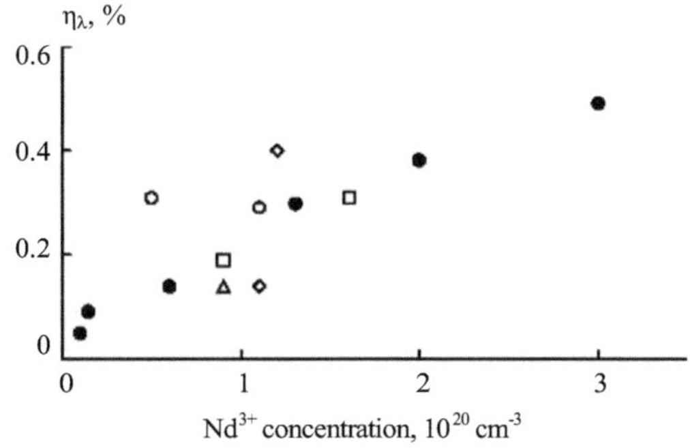

Fig. 11.3 The dependence of the conversion efficiency (η_λ) on the concentration of Nd^{3+} for the inorganic liquid $POCl_3$-$SnCl_4$-O_2^{2+}: Nd^{3+} in the absence of uranyl (filled circle), and also with a uranyl concentration of 1×10^{19} (*open circle*), 2×10^{19} (*open square*), 3×10^{19} (*open triangle*) and 5×10^{19} (*open diamond*) cm^{-3} [21, 22]

this solution, processes begin that lead to uranyl transition from a 6-valent state to a 4-valent state, making the laser characteristics of the medium drastically worse [20, 25]. Therefore, for subsequent experiments, the laser liquid $POCl_3$-$SnCl_4$-UO_2^{2+}:Nd^{3+} was selected and studied in the most detail.

The possibility for reaching the lasing threshold when pumping laser media with nuclear radiation not only depends on the spectral and luminescent characteristics of the medium, but also on its radiation resistance, that is, the ability to retain optical uniformity and transparency at the lasing wavelength under the radiation influence. As was already noted, the main causes that could lead to the appearance of detrimental losses in a liquid laser medium excited with heavy charged particles are the formation of gas-vapor bubbles (light scattering) [9, 32] and radiolysis on the tracks of the fission fragments, and also thermal optical disturbances (thermal lens and optical wedge) caused by non-uniformity of the energy deposition over the active laser volume [18].

Apparently, these specific circumstances were the reasons for the absence of laser action during attempts to pump the liquid laser medium with uranium fission fragments in experiments on the BARS-6 pulsed reactor [20, 33]. Figure 11.4 shows a schematic of experiments, which used the laser liquid $POCl_3$-$SnCl_4$-$^{235}UO_2^{2+}$: Nd^{3+} with equal neodymium and uranyl concentrations of 3×10^{20} cm^{-3}. A 34-cm long liquid-filled laser cell surrounded by a polyethylene high-speed fast neutron moderator was placed alongside one of the BASRS-6 reactor cores or between two cores (Fig. 11.4 shows the second variant). The specific energy deposition to the laser medium over a 140 μs pulse is 32 J/cm^3. The laser cavity was formed by two mirrors with reflectivities of 100 and 65 %.

Semiconductor Lasers

There is no information in the literature about semiconductor lasers excited by nuclear radiation although the possibility for pumping them, for example, with reactor γ-ray radiation, is discussed in article [35]. To pump semiconductor lasers,

Table 11.3 The conversion efficiency (η_λ) and lifetime (τ_u) of upper laser levels for laser media based on inorganic liquids with excitation by the products of nuclear reactions

Composition of the medium	Ion activator	Ion-activator concentration, 10^{20} cm^3	Uranium concentration, 10^{20} cm^3	η_λ (%)	τ_u (μs)	Reference
$POCl_3$-$SnCl_4$	Nd^{3+}	1.1	–	0.29	220	[22]
$POCl_3$-$SnCl_4$	Nd^{3+}	1.1	0.1	0.29	230	[22]
$POCl_3$-$SnCl_4$	Nd^{3+}	1.1	0.5	0.14	120	[22]
$POCl_3$-$SnCl_4$	Nd^{3+}	1.3	–	0.30	210	[22]
$POCl_3$-$SnCl_4$	Nd^{3+}	1.3	0.1	0.23	170	[22]
$POCl_3$-$SnCl_4$	Nd^{3+}	3.0	–	0.49	250	[22]
$POCl_3$-$SnCl_4$	Nd^{3+}	2.4	0.36	0.56	230	[26, 27]
$POCl_3$-$SnCl_4$	Eu^{3+}	2.0	–	0.52	4,000	[25]
$POCl_3$-$SnCl_4$	Tb^{3+}	1.0	–	0.22	2,000	[25]
$ZnCl_2$-$GaCl_3$	Nd^{3+}	0.6	–	0.09	130	[22]
$ZnCl_2$-$GaCl_3$	Nd^{3+}	0.6	1.2	0.06	100	[22]
$ZnCl_2$-$GaCl_3$	Nd^{3+}	2.4	0.3	0.27	190	[22]
$ZnCl_2$-$GaCl_3$	Tb^{3+}	0.6	–	–	1,700	[22]
$ZnCl_2$-$GaCl_3$	Mn^{2+}	4.0	–	–	4,700	[22]
$SOCl_2$-$GaCl_3$	Nd^{3+}	0.6	0.3	0.12	260	[22]
$SOCl_2$-$GaCl_3$	Nd^{3+}	10	–	1.4	280	[25]
SO_2Cl_2-$GaCl_3$	Nd^{3+}	0.72	–	0.23	250	[28]
SO_2Cl_2-$GaCl_3$	Nd^{3+}	0.42	0.4	0.09	240	[28]
$D_2O:Eu^{3+}$	Eu^{3+}	2.5–5.0	0–3	~0.05	2,000–2,500	[29]

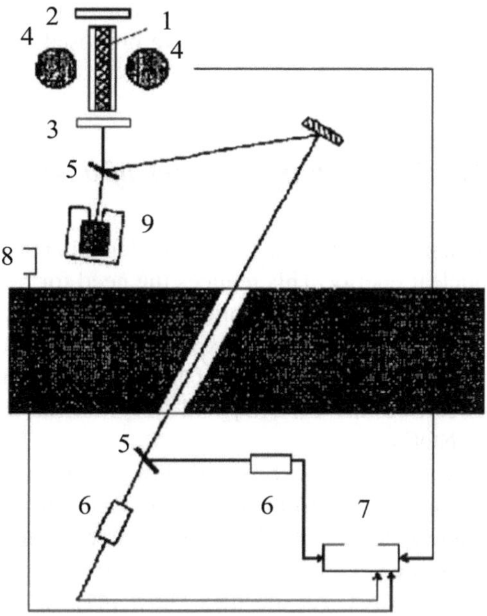

Fig. 11.4 A schematic of experiments on pumping laser liquid with uranium fission fragments using the BARS-6 reactor [33, 34]: (1) a container that contains the laser medium that is placed in a moderator; (2) 100 % mirror in the cavity; (3) semi-transparent mirror in the resonant cavity; (4) two reactor cores; (5) beam splitters; (6) FD-24 photodiodes; (7) multichannel pulse analyzer with a personal computer; (8) neutron detector; (9) an IMO-2N device for measuring the energy of the laser pulses

electron beams are another type of ionizing radiation that is widely used. Numerous studies in this area have produced semiconductor lasers based on CdS, InSb, InAs, GaAs, etc. that radiate within a broad spectral range, 0.33–30 μm, and have very different operation modes: continuous-wave mode with excitation by a focused electron beam, a picosecond pulsed mode with a repetition frequency in the several tens of gigahertz, and a pulse-periodic mode with an average power in the tens of watts and intrinsic efficiency of ~10 % [36].

In addition to electron beams, semiconductor lasers were pumped using the bremsstrahlung of pulsed electron accelerators. In experiment [37], by pumping monocrystal CdS ($\lambda \approx 500$ nm) with the bremsstrahlung of a pulsed electron accelerator, 5 kW lasing was achieved with a pulse duration of about 20 ns.

11.2 Pumping of Condensed Laser Medium Using Nuclear-Optical Converters

When using a nuclear-optical converter (NOC) to pump NPLs, immediate excitation of the laser medium is done using a well-known method: optical pumping. In this case, the kinetic energy of the nuclear radiation is initially converted by the NOC into light (luminescent) emission. (Sometimes NOCs are called "nuclear" lamps). The possibility of using NOCs for laser pumping was apparently examined for the first time in a U.S. patent [38] where various crystal and plastic scintillators and liquid and gas luminescent media were proposed as NOCs, and dielectric materials doped with neodymium and chromium and other condensed and gas laser media known at that time were proposed as laser elements.

The Main Characteristics of a NOC

When using NOCs, the laser element may be removed from the source of the nuclear energy. This reduces the need for radiation resistance of the laser medium. For example, some publications [39, 40] report on experiments using the TRIGA pulsed reactor. In these experiments, light radiation from the NOC was transmitted using 5-m long bundles of light-guide fibers beyond the biological shielding and directed to the $Y_3Al_5O_{12}$:Nd^{3+} crystal. Gas as well as condensed media can be used as NOCs.

Gas NOCs may be divided by emission spectrum into two types: sources with a continuous spectrum close to the spectrum of black-body radiation and sources with a line spectrum consisting of separate luminescent lines or bands. In the first case, the source of the light radiation is nuclear-excited plasma heated to sufficiently high temperatures. As the data show [41, 42], when gas media based on rare gases (krypton, xenon, etc.) containing the isotopes ^{3}He, ^{10}B, or ^{235}U at atmospheric pressure are irradiated with neutron fluxes from pulsed reactors with specific energy depositions up to ~10 J/cm^3, the media temperature can reach 10^4K. In this case, the spectra of the luminescence closely correspond to the emission spectrum of an absolutely black body. Intensity lines belonging to the atom transitions of rare gases in the 800–1,000 nm spectral range are sometimes superimposed on these spectra.

With specific energy depositions $\leq$1 J/cm^3, the main contribution to the emission spectra at $\lambda < 1$ μm is made by luminescent radiation. Naturally, in this case, gas media with maximum conversion efficiencies should be used for NOCs. In the UV spectral range, these media are mixtures based on the excimer molecules of the rare gases R_2* (R = Xe, Kr, Ar) and the halides of rare gases RX* (X = F, Cl, Br, I) [40, 43–47]. With the use of excimer molecules, there is no radiation reabsorption because when a photon is emitted, the excimer molecules transit to a lower dissociable state or a weakly linked ground state.

A simple expression was used to evaluate the maximum values for the conversion efficiency η_λ (Table 11.4): $\eta_\lambda{}^{max} = h\nu/w_a$, where $h\nu$ is the photon energy of a luminescent transition, and w_a is the average energy consumed for the formation of one primary active particle (an ion or excited atom of a buffer gas). During subsequent relaxation processes, this active particle can produce maximum of one excited excimer molecule. Using Platzman's formula (see Chap. 4, Sect. 2), a link may be obtained between the w_a and ionization potential of the buffer gas V_i: $w_a \approx 1.2V_i$. It follows from Table 11.4 that the conversion efficiency for some excimer molecules can reach 50 %. The bottom line of the table gives experimental data for some excimer molecules at excitation of the gas media by proton [48] and electron [49] beams, and also nuclear radiations [50, 51].

Table 11.4 The emission wavelength and conversion efficiency of several excimer molecules in the UV spectrum

Excimer molecule	Ar_2*	Kr_2*	Xe_2*	KrCl*	KrF*	XeBr*	XeCl*	XeF*
λ, (nm)	129	147	172	222	248	282	308	351
Buffer gas	Ar	Kr	Xe	Ar	Ar	Ar	Ar	Ar
Maximum calculated conversion efficiency, (%)	51	50	50	30	26	23	21	19
Experimental data on conversion efficiency, (%)	29 [48] 30 [49]	12 [48]	30 [49] 68 [50]	–	23 [50]	~10 [51]	~10 [51]	14 [50]

For effective pumping of active laser elements based on the condensed media, their absorption spectra must correspond to the emission spectra of the NOC. The most intense absorption bands of ruby and neodymium lasers are located in the visible and near IR spectra. These laser media can be pumped using NOCs based on excimer alkaline molecules [40, 44–46] or rare gases [40, 46, 52, 53]. According to estimates [40, 45, 46], the maximum conversion efficiency for alkaline molecules Li_2* ($\lambda_{max}=458$ nm), Na_2* ($\lambda_{max}=436$ nm), K_2* ($\lambda_{max}=575$ nm), Rb_2* ($\lambda_{max}=601$, 603, 606 nm), and Cs_2* ($\lambda_{max}=703$, 713, 718 nm) may reach 40–50 %.

Experiments [40, 52, 53] have shown that in NOCs based on mixtures of rare gases, almost all of the most intense lines are located in the near-IR range of the spectrum, 800–1,000 nm, and belong to the transitions $(n+1)p-(n+1)s$ of the Xe, Kr, and Ar atoms ($n=5,4,3$ with respect to Xe, Kr, Ar). Table 11.5 gives the results of a study of the luminescent characteristics of He, Ne, Ar, Kr, Xe, and their binary mixtures in the spectral region 740–1,100 nm when excited by uranium fission fragments [53]. The VIR-2 M pulsed reactor with a pulse duration of approximately 3 ms was used as the neutron source [16]. For the most intense lines, 912.3 and 965.8 nm (ArI) in the He-Ar mixture, 892.9 nm (KrI) in the He-Kr mixture, and 980.0 nm (XeI) in the Ar-Xe mixture, the conversion efficiency was 0.1–0.15 %. With excitation of ^{3}He-Ar(Kr,Xe) mixtures with the products of the nuclear reaction $^3He(n,p)^3H$ ($q=20$ W/cm^3), similar values $\eta_\lambda \le 0.5$ % were obtained for the spectral range 740–840 nm [52].

The active substance for NOCs based on condensed media is phosphor as liquid or solid. The characteristics of some liquid luminescent media are considered in the previous section. Of the other liquid media for NOCs, one study [54] proposes a saline solution of the isotope ^{245}Cm in heavy water. The luminescence of this solution when excited by α-particles occurs in the 560–620 nm region at a comparatively low $\eta_\lambda \sim 0.1$ %.

In VNIIEF, some characteristics of alkali halide crystals and luminescent plastics were studied in experiment [2] at the VIR-2 and at the BIR-2 pulsed reactors [16]. The CsI(Tl), CsI(Na), and NaI(Tl) crystals were considered in the most detail. Earlier these crystals were mainly used at low excitation levels in the nuclear particle counting mode. The VNIIEF experiments showed that the luminescence spectra of these crystals change insignificantly at the absorbed doses of the

Table 11.5 The luminescent characteristics of rare gases and their binary mixtures with excitation by uranium fission fragments

Gas medium	Pressure (atm)	q (W/cm^3)	w_t (mW/cm^3)	η_λ (%)
He	1.8	60	85	0.15
Ne	0.64	34	94	0.28
Ar	0.45	47	85	0.18
Kr	0.32	46	120	0.26
Xe	0.23	38	120	0.32
He-Ne (0.56 % Ne)	1.8	43	51	0.12
He-Ar (0.56 % Ar)	1.8	46	180	0.39
He-Kr (0.56 % Kr)	1.8	46	260	0.57
He-Xe (0.56 % Xe)	1.8	40	230	0.57
Ne-Ar (1.7 % Ar)	0.64	34	260	0.76
Ne-Kr (1.7 % Kr)	0.64	37	210	0.57
Ne-Xe (1.7 % Xe)	0.64	34	240	0.71
Ar-Kr (2.2 % Kr)	0.45	47	85	0.18
Ar-Xe (2.2 % Xe)	0.45	44	260	0.59
Kr-Xe (3.1 % Xe)	0.32	52	150	0.29

Note: q is the specific deposition power; w_t and η_λ is the specific power of the luminescence and the conversion efficiency within the range 740–1,100 nm

γ-radiation $D_\gamma \leq 5 \times 10^3$ Gr and the neutron fluences $F \leq 5 \times 10^{14}$ cm^{-2} (Fig. 11.5). Other important parameters for the CsI(Tl), CsI(Na), and NaI(Tl) crystals were studied as a function of the activator concentration and the dose rate of γ-radiation: the specific output of luminescent radiation, and the conversion efficiency. It is possible to conclude from the results that the light output, for example, of the CsI (Tl) crystal increases nearly linearly with the growth in the absorbed dose up to 4.5×10^3 Gr. At this maximum dose, the specific light output is 1.1 J/cm^3, and the conversion efficiency is about 6 %.

The Application of NOCs for Laser Pumping

For condensed laser media, pumping with the help of NOCs was more successful than direct pumping of these media with nuclear radiation, as was considered in the previous section. There are two experimental studies which show that neodymium lasers may be pumped using NOCs based on xenon plasma [41] and the phosphor CsI(Tl) [55]. This method was also used successfully in experiments on the TRIGA reactor, in which a photo-dissociation iodine laser was pumped with the luminescent emission of XeBr* molecules (see Chap. 3, Sect. 6).

Experiments that pumped solid-state lasers with NOC emission were carried out on pulsed reactors [40, 41, 55].

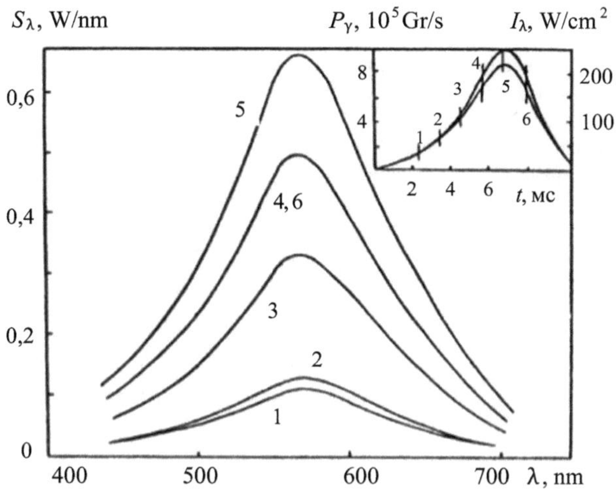

Fig. 11.5 The spectra of the luminescent crystal CsI (0.5 % Tl) at different points of the pulsed irradiation on the VIR-2 reactor at $D_\gamma = 3.5 \times 10^3$ Gr and $\Phi = 4 \times 10^{14}$ cm^{-2}. The insert shows the time dependence of the dose rate P_γ (*upper curve*) and the luminescent power density I_γ (*lower curve*) and indicates the moment in time (1, 2, 3, 4, 5, 6) when the spectra were recorded [2]

A series of experiments [40] on the TRIGA reactor used gas NOCs based on ^{3}He-Kr(Ar,Xe) mixtures with excitation by the products of the nuclear reaction ^{3}He $(n,p)^3$H. The spectral-luminescent characteristics of these mixtures are given in two works [40, 52]. The specific power deposition at the maximum of a 12-ms reactor pulse was about 20 W/cm^3. A schematic of the experiment is shown in Fig. 11.6. The light radiation from the NOC, which was placed near the reactor core, was transmitted to the active laser element $Y_3Al_5O_{12}$:Nd^{3+}, 3 mm in diameter and 62.5-mm long, using a bundle of ~2,000 quartz light-guide fibers that were 5-m long. To pump the volume of the laser element uniformly, the end faces of the fibers were uniformly distributed about its surface (Fig. 11.7). Laser action was not observed in these experiments. In study [40] absence of laser action is explained by the low specific power deposition in the laser medium due to loses during transportation of the light radiation and poor correlation between the plasma luminescence spectrum and the absorption spectrum of the $Y_3Al_5O_{12}$:Nd^{3+} crystal.

Experiments performed in VNIIEF on the VIR-2 M reactor were more successful. As part of these experiments, fiber neodymium lasers were pumped with the light radiation of xenon plasma [41]. Prior studies have shown that the spectrum of the plasma radiation that appears when xenon at 1 atm is excited with uranium fission fragments (the specific energy deposition is ~5 J/cm^3) is close to the emission spectrum of an absolutely black body at temperatures $\leq$6,000 K [41].

In experiments on the VIR-2 M reactor, the laser active element was placed inside the NOC between two plates with uranium layers (Fig. 11.8). The 10 $\times$ 6 cm plates were placed parallel to each other 2 cm apart. Their internal surfaces have a deposited layer of uranium oxide-protoxide. The thickness of the metallic ^{235}U layer was about 5 mg/cm^2. A fiber laser 1–4-m long with a fiber diameter of 40–50 μm was wound around a frame and placed symmetrically between the

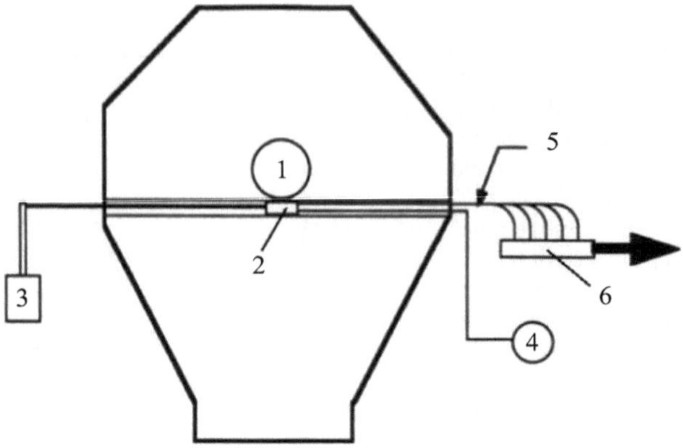

Fig. 11.6 A schematic of experiments using the TRIGA reactor on pumping the solid-state laser $Y_3Al_5O_{12}:Nd^{3+}$ with the optical radiation of a gas NOC [40]: (1) core reactor; (2) NOC; (3) vacuum system; (4) device for filling the NOC with gas; (5) bundle of optic fibers; (6) $Y_3Al_5O_{12}:Nd^{3+}$ crystal

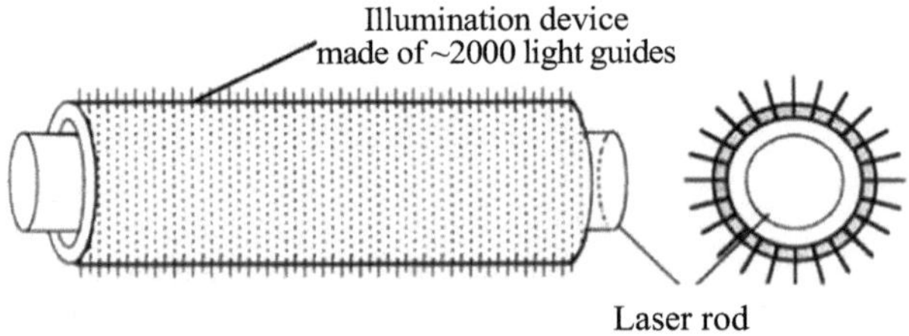

Fig. 11.7 The relative position of the illumination device and laser rod [40]

uranium layers. Irradiation of the NOC was done at the bottom of the central reactor channel. This series of experiments obtained laser action when using fiber neodymium lasers based on silicate and phosphate glasses. The results of one experiment on a neodymium laser based on phosphate glass with the NOC filled with xenon at 0.5 atm (plasma temperature 4,700 K) are shown in Fig. 11.9. The maximum energy of the laser radiation over a single pulse was 2.5 J.

Besides using gas NOCs to pump neodymium lasers, in VNIIEF also used NOCs based on solid-state scintillator. A brief overview [55] of these experiments reported that lasing was obtained when pumping a laser element based on neodymium glass with radioluminescent radiation from the CsI(Tl) crystal. The laser element and CsI(Tl) crystal were irradiated in the central channel of the BR-1 pulsed reactor [16]. The maximum laser power was equal to 4 kW at a laser pulse of approximately 60 μs duration.

In conclusion, a recently published study [56] should be noted where pumping a laser based on the crystal $YAP:Tm^{3+}$ ($\lambda = 1.85$–2.0 μm) was fulfilled using gas NPLs at the transitions of Ar ($\lambda = 1.79$ μm) and Xe ($\lambda = 1.73$ μm) atoms as a NOC. The active region of the laser crystal was 5-mm long with a volume of about 4 mm^3. From the oscilloscope traces in Fig. 11.10, it can be supposed that the experiments were performed on the EBR-L pulsed reactor [16]. When a xenon NPL ($\lambda = 1.73$ μm) was used as the source for pumping, the threshold specific power deposition was about 35 kW/cm^3, and the conversion efficiency of the absorbed optical radiation into laser radiation was 4 %. Such low laser efficiency is rather

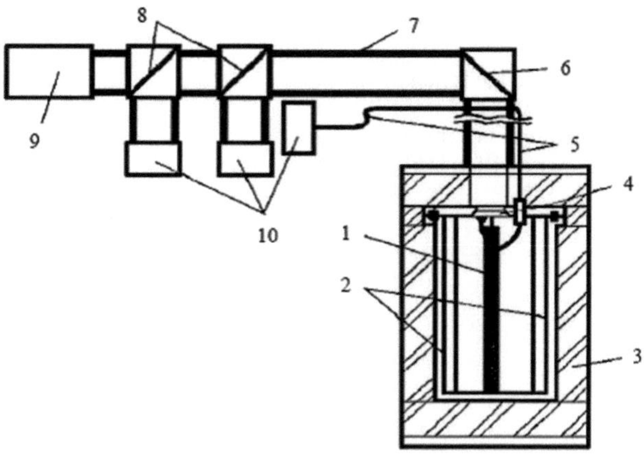

Fig. 11.8 A schematic of a laser experiment on the VIR-2M reactor [41]: (1) neodymium fiber laser; (2) flat surfaces with uranium layers; (3) fast neutron moderator; (4) optical connecter; (5) quartz light-guide fiber; (6) rotary mirror; (7) aluminum pipe; (8) quartz plates; (9) laser pulse energy meter; (10) photodetectors

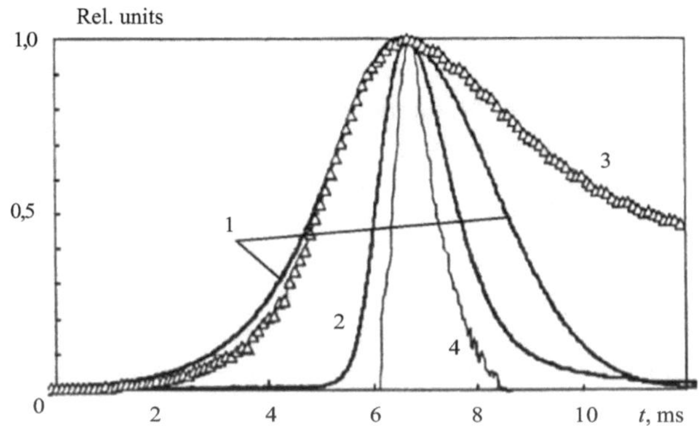

Fig. 11.9 Oscilloscope traces of the following pulses: reactor pulse (1), light radiation of the NOC (2), pressure in the NOC (3) and laser pulse (4) for a fiber laser based on neodymium phosphate glass with an NOC filled with xenon at a pressure of 0.5 atm [41]

unexpected since prior executed studies [56] showed that the conversion efficiency of laser radiation on lines 1.73 and 1.79 μm into luminescent radiation was equal to 54 and 76 %, respectively.

Other Applications for NOCs

Laser pumping is not the only application for NOCs. For example, they were considered as a possible means for initialization chemical processes and for photo-dissociation of water or carbon dioxide. In this case, the full efficiency for the conversion of nuclear energy into the end products of these reactions may reach 10 % when a NOC based on Xe_2* ($\lambda_{max} = 172$ nm) molecules is used [43, 44].

One more possible application of the NOC is the conversion of light emission into electricity. In this case, the efficiency for the conversion of nuclear energy into electrical energy is 30–40 % using dielectric transformers with a wide band gap

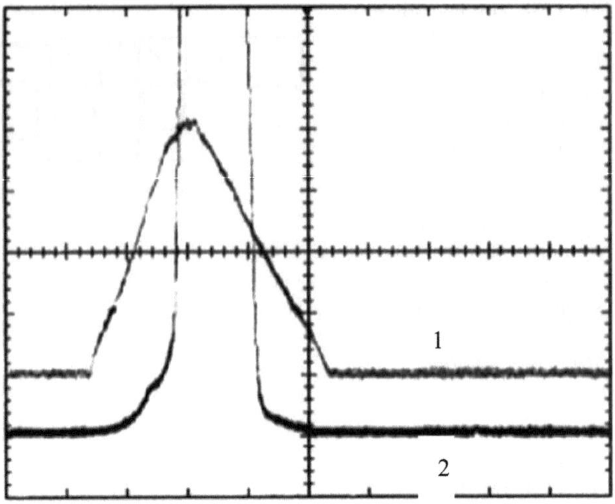

Fig. 11.10 Oscilloscope traces of the xenon NPL pulse (1) and the lasing pulse of YAP:Tm^{3+} crystal (2). The division value is 100 μs per division [56]

(for example, A1N, diamond or corundum) and the appropriate choice for the NOC emission spectrum [43].

Another variant for photoelectric conversion of NOC radiation has been proposed [57]. In this case, the study examines the potential for using radioactive waste as a source of energy for pumping NOCs with subsequent conversion of the optical radiation into electrical energy using photoelectric transformers. The main source of ionizing radiations from the spent fuel of nuclear reactors is the γ-ray ^{137}Cs (the half-life is 10.7 years, the energy of γ-quanta is 662 keV). As an active medium for NOCs, an Ar-N_2 mixture has been proposed that primarily radiates in the ranges $\lambda = 350–410$ and 750–1,050 nm on the vibrational transitions of the N_2 molecule. The output characteristics were evaluated for a device consisting of 200 radioactive waste containers, 0.5 m in diameter and 1-m high, placed in three tiers. The expected specific electrical power is about 1 W/kg, and the full electrical power of this device with a 40 m radius (200 containers) may equal to 1 MW.

NOC radiation may also be used for photosynthesis of biological products, for example, microalgae chlorella. For large-scale photosynthesis, a paper [58] proposes to develop a NOC whose central part is a solution reactor-breeder on thermal neutrons with uranium-thorium fuel, and the cover consists of a luminescent material. In this case, biomass production may be about 1,000 g/(m^2/day). This significantly exceeds the analogous value for solar illumination $\leq$150 g/(m^2/day).

And, finally, based on the NOC, it is possible to create neutron detectors for operating control of pulsed as well as stationary nuclear reactors. In these detectors, a luminescent emission that has an intensity linearly related to the neutron flux density is withdrawn beyond the biological reactor shielding using light-guide fibers and recorded using photodetectors [59, 60]. Neutron detectors based on NOCs have a series of advantages as compared to ionization chambers: the lack of a power supply; a low sensitivity to γ-ray emission; the potential to produce a small NOC for intrareactor measurements; etc.

11.3 Some Schemes for Nuclear Laser Devices Using Condensed Media

Thus, NPLs using condensed media have for the present given rather modest results: pumping has only been done on glass neodymium lasers with optical radiations from NOCs. Moreover, there is no detailed information on the characteristics of these lasers. Despite this, the literature has proposed some schemes for powerful nuclear-laser devices based on condensed laser media. We will review some proposals without discussing issues related to feasibility.

Colloidal Nuclear Reactor → Optical Thermal Emission → Liquid Laser Medium

An article [61] reviews one of the first designs for a nuclear laser device using condensed laser media. In a simplified form, this device consists of two coaxial cylinders, wherein the liquid laser medium $POCl_3$-$ZrCl_4$:Nd^{3+} is placed in the inner cylinder with a transparent cover, and the reactor core occupies the space between the two cylindrical covers. A colloidal reactor core operating in stationary mode consists of a suspension of 0.5–5 μm uranium carbide particles (total mass of 4.5 kg) in a vortex flow of helium. The power density that will be released inside the reactor core, according to estimates, is 20 kW/cm^3. It is supposed that the laser media is pumped with the optical emission of fuel particles, which are heated to a temperature of 3,000K. The geometric dimensions and the energy characteristics of this device are not given in the work [61].

Reactor with an Aerosol Core → Luminescent Radiation → Solid-State Laser Medium

Two studies [40, 62] considered a nuclear-laser device that operates in pulsed-periodic mode with a 1 Hz frequency and a pulse duration of about 1 ms (Fig. 11.11). The schematic (Fig. 11.12) shows the photon path from its origin in the NOC to its absorption in the laser material. The gas medium is excited in the NOC by fission fragments escaping from the fissile material: uranium aerosol particles of several microns in diameter. Luminescent photons reflected by particles from the fissile material and NOC walls pass through the window and are then transmitted to the active laser medium using a light-conducting tube. The laser medium is manufactured as a plate and is a so-called "active mirror" (see, for example, [63]). A design analysis [40] of different configurations for the "container-tube" shows that the extraction efficiency of the light radiation from the NOC container and its transport to the laser medium may reach 70–80 %.

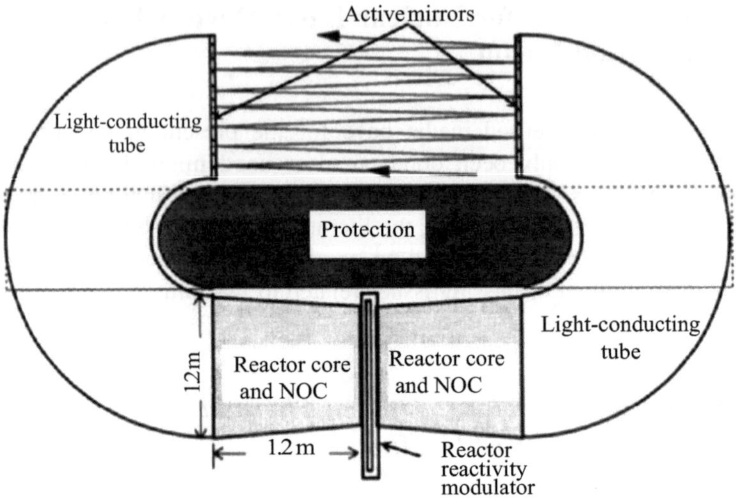

Fig. 11.11 Schematic of a nuclear laser device [40, 62]

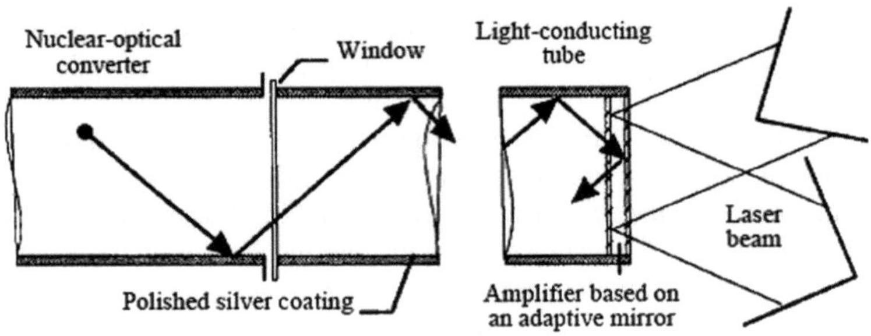

Fig. 11.12 Schematic of the transmission of luminescent radiation from the NOC container to the laser medium [40]

The reactor core consists of two parts which contain a mixture of aerosol uranium fuel (2–5 μm UO_2 particles) and a luminescent gas medium. Between the parts of the core, there are three disks that contain ^{235}U, ^{10}B, and Be and rotate at different frequencies. These disks, which are modulators of the core reactivity, are used to realize the pulsed operation mode of the reactor. We note that a similar method for obtaining periodic bursts of fissions using external reactivity modulation is used in IBR-type pulsed reactors [64]. The parameters of the proposed reactor in studies [40, 62] are: core volume 2.3 m^3; uranium load 0.5 g/cm^3; pulse repetition frequency 1 Hz; average power 400 kW (170 W/cm^3); and the peak neutron flux density 2×10^{16} cm^{-2} s^{-1}.

Using light-conducting tubes, the luminescent radiation from the two parts of the reactor core is withdrawn beyond the biological shielding. Two flat mosaic blocks of laser elements are installed at the ends of the light-conducting tubes. Each of these blocks consists of tens of "active mirrors" manufactured from 0.5 cm thick GSGG:Cr^{3+}:Nd^{3+} crystals up to 20 cm in diameter. The suggested luminescent media is a gas mixture based on excimer alkali molecules with an assumed conversion efficiency of 30 %. Under these conditions, the estimated peak power of the laser radiation is 12 MW (average power is 12 kW) with a 3 % total conversion efficiency of the nuclear energy released in the reactor core into laser radiation.

Reactor (Uranium Layers or Uranium Hexafluoride) → Optical Thermal Emission → Solid-State Laser Medium

The investigations performed at VNIIEF on different types of NOCs and the neodymium lasers excited by them permit consideration of a series of schemes for pulsed reactor lasers (RLs) using neodymium active elements with laser radiation energy $\leq$10 MJ and intended for studies of some problems of inertial confinement fusion [41, 42].

One of the variants is a modular design. The modules are placed in the reactor matrix, which contains nuclear fuel, reactor control units, a neutron moderator and reflector, etc. Inside the module, which has an 80-cm diameter, are about 100 neodymium rods 45 mm in diameter and 1,200-cm long (Fig. 11.13a). The space between the rods, which are arranged as a hexagonal grate with 80 mm steps, is filled with a UF_6-Xe (4:1) mixture at 2 atm.

The laser rods are pumped with the optical radiation of the plasma (temperature $\leq 10^4$ K), which forms under the impact of neutron pulses ~3 ms duration with a fluence of ~3×10^{14} cm^{-2}. The total energy of the laser radiation of a single module is about 400 kJ at ~2.5 % efficiency for conversion of the released nuclear energy into laser radiation.

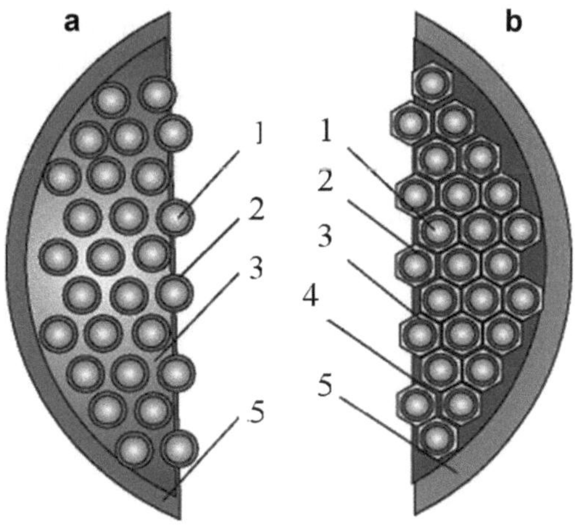

Fig. 11.13 A schematic of a laser module with uranium hexafluoride (**a**) and uranium layers (**b**) [42]: (1) laser elements; (2) filters that absorb UV-emission; (3) NOC; (4) NOC body with a uranium layer; (5) the module body

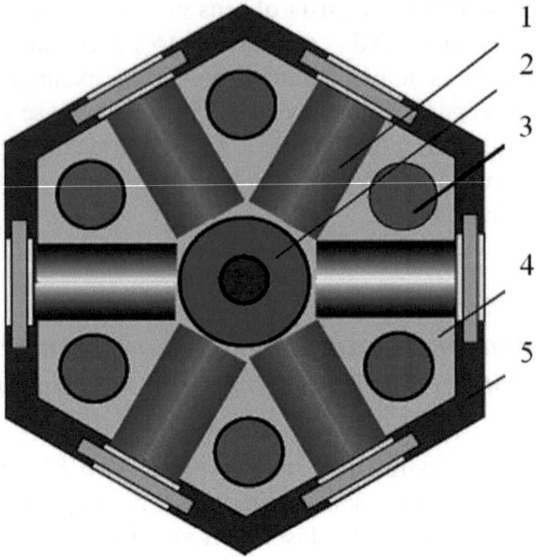

Fig. 11.14 Schematic of a pulsed RL [42]: (1) laser module; (2) "start-up" reactor; (3) peripheral reactor module-multiplicator; (4) neutron moderator; (5) reactor body

Another possible variant in the module design uses uranium fission fragments that escape from the uranium layers to excite the xenon plasma (Fig. 11.13b). The uranium layers are deposited to the internal surface of the NOC body (hexagonal in cross-section), which is filled with xenon at ~1 atm. There are 60–70 laser elements at the same module dimensions, and the energy of the laser radiation is no more than 250 kJ. If this variant of the module design is used, the efficiency will be approximately ten times lower due to the lower transmission efficiency of the fission fragment energy to the gas medium. This requires an increase in the total energy release in the module.

It should be noted that the modules considered have a significant neutron multiplication factor, and, as follows, contribute to the total reactivity of the reactor system. This allows loading of additional nuclear fuel into the reactor to decrease.

A RL based on these modules with an output laser energy of ~10 MJ can be proposed, for example, in an 18 module design with the modules placed in six channels and three tiers (Fig. 11.14) [42]. In the central part, there is a "start-up" reactor with control units. All of the device elements are placed in a matrix-neutron moderator and comprise a single nuclear-physical system. The pulse duration of the nuclear fission and laser radiation in the free-running mode may be 1–10 ms. The typical liner dimensions of such a facility are 3 m; the total mass is no more than 100 t.

"Start-Up" Reactor → Subcritical Multiplicator Block with a Laser Medium

Unlike the schemes presented previously, which use the NOC principle, some studies [65–68] consider the possibility of creating a powerful pulsed-periodic RL with direct conversion of the kinetic energy of the fission fragments into laser radiation. It is proposed that this RL may be used in inertial confinement fusion facilities as a final amplifier for the laser driver [65, 66].

A schematic of a pulsed-periodic RL based on the separation of the reactor and subcritical laser block is shown in Fig. 11.15. The laser block is a cylindrical longitudinal cavity containing the core of the IBR-type "start-up" pulsed reactor

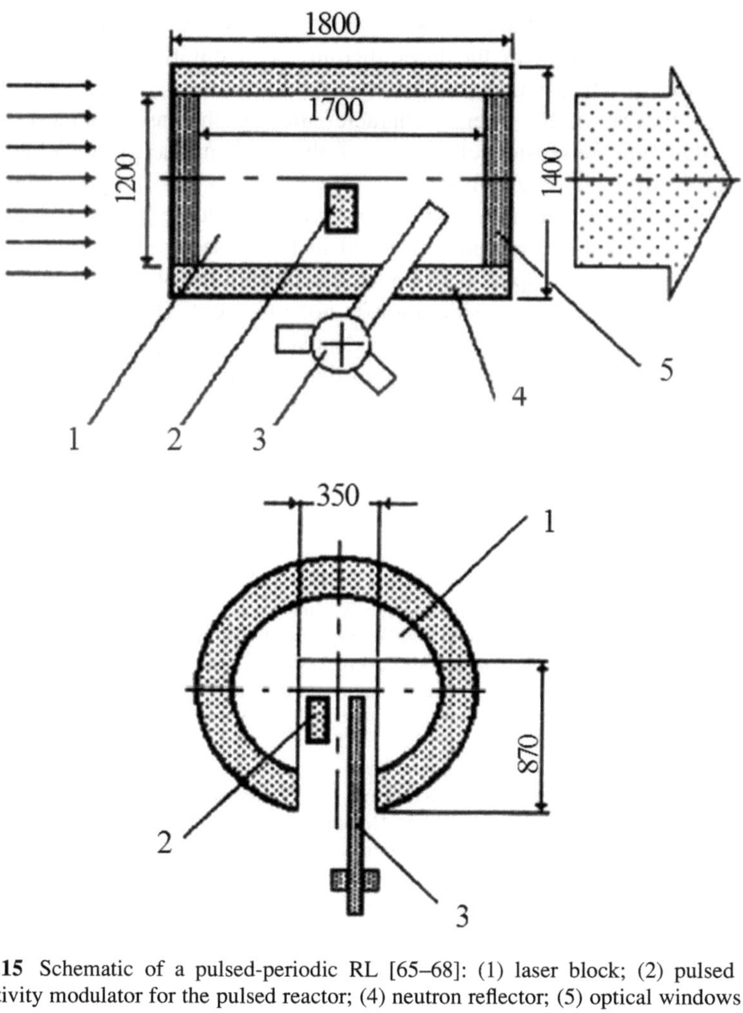

Fig. 11.15 Schematic of a pulsed-periodic RL [65–68]: (1) laser block; (2) pulsed reactor; (3) reactivity modulator for the pulsed reactor; (4) neutron reflector; (5) optical windows

Table 11.6 Parameters of the pulsed-periodic RL [65]

Parameter	Value
Energy of the "start-up" pulsed reactor, MJ	2
The effective neutron multiplication factor in the laser block	0.9
Energy released in the laser block, MJ	8
Volume of the laser block, m^3	1.0
The conversion efficiency of nuclear energy into laser emission, %	2.0
Output energy of the laser beam, MJ	0.16

[64] with a reactivity modulator. It is filled with a laser medium containing fissile material and includes elements of the neutron moderator. Neutron calculations [67, 68] have shown that the system operating at a frequency $\leq$5 Hz with about 8 MJ of energy release in the laser block over one pulse can provide ~1 kW/cm^3 of specific power deposition in the laser medium at a duration of 1–5 ms. Some parameters for this RL are given in Table 11.6.

Upon evaluation of the energy characteristics of the laser radiation, it was supposed that the conversion efficiency of the nuclear energy absorbed in the laser medium into the laser emission is $\eta_l = 2.0$ %. It was proposed in studies [65, 66] that the laser medium could use inorganic aprotonic liquids (for example, $POCl_3$-$ZrCl_4$:Nd^{3+}) with uranium dissolved in them (see the first section of this chapter). In the authors' opinion [65, 66], the possibility of also using gas laser media in the design under consideration is not ruled out. However, in this case, at least the problem of effective energy deposition to the gas medium must be solved, and a laser medium must be found that not only has a high η_l, but a long upper laser level lifetime (~1 ms).

References

1. A.A. Sinyanskiy, Issledovaniya po cozdaniyu yaderno-lasernykh ustroystv nepreryvnogo deystviya vo VNIIEF [Investigations of the creation of continuous nuclear laser devices at VNIIEF]. Trudy 2-y mezhdunarodnoy konf. [Proceedings of the Second International Conference]. "Fizika yaderno-vozbuzhdayemoy plazmy i problemy lazerov s yadernoy nakachkoy" [The Physics of Nuclear-Excited Plasma and Problems of Nuclear-Pumped Lasers]. Arzamas-16, 1995, vol. 1, pp. 16–36
2. A.M. Voinov, L.Ye. Dovbysh, V.N. Krivonosov, S.P. Melnikov, A.A. Sinyanskiy, Metodicheskoye obespecheniye issledovaniy po radiatsionnoy stoykosti opticheskikh elementov lazerov s yadernoy nakachkoy [Methodological support for investigations of the radiation resistance of the optical elements of lasers with nuclear pumping]. Trudy 2-y mezhdunarodnoy konf. [Proceedings of the Second International Conference]. "Fizika yaderno-vozbuzhdayemoy plazmy i problemy lazerov s yadernoy nakachkoy" [The Physics of Nuclear-Excited Plasma and Problems of Nuclear-Pumped Lasers]. Arzamas-16, 1995, vol. 1, pp. 52–69

3. T.A. Babicheva, A.M. Voinov, L.Ye. Dovbysh, L.M. Pavlovskaya, A.A. Sinyanskiy, Lazery s yadernoy nakachkoy na zhidkikh aktivnykh sredakh. [Lasers with nuclear pumping on liquid active media]. Trudy otraslevoy konf. [Proceedings of the Specialist Conference] "Fizika yaderno-vozbuzhdayemoy plazmy i problemy lazerov s yadernoy nakachkoy" [The Physics of Nuclear-Excited Plasma and Problems of Nuclear-Pumped Lasers]. Obninsk, 1993, vol. 3, pp. 146–155
4. K. Thom, R.T. Schneider, Nuclear pumped gas lasers. AIAA J. **10**(4), 400–406 (1972)
5. N.W. Jalufka, Direct nuclear-pumped lasers. NASA Technical Paper 2091 (1983)
6. V.E. Derr, G.T. McNice, P.M. Rushworth, Application of nuclear radiation to the pumping of lasers, in *Radioisotopes for Space. Part 2: Systems and Applications*, (Plenum Press, 1966), pp. 309–346
7. S.M. Brekhovskikh, Yu.N. Viktorova, L.M. Landa, Radiatsionnyye effekty v steklakh [The radiation effects in glasses]. M: Energoizdat (1982)
8. A.K. Pikayev, Sovremenaya padiatsionnaya khimiya. Radioliz gazov i zhidkostey. [Modern radiation chemistry. Radiolysis of gases and liquids]. M: Nauka (1986)
9. A.N. Sizov, Rasprostraneniye sveta cherez zhidkost, vozbuzhdayemuyu oskolkami deleniya [Light propagation through liquid excited with fission fragments]. Trudy 2-y mezhdunarodnoy konf. [Proceedings of the Second International Conference] "Fizika yaderno-vozbuzhdayemoy plazmy i problemy lazerov s yadernoy nakachkoy" [The Physics of Nuclear-Excited Plasma and Problems of Nuclear-Pumped Lasers]. Arzamas-16, 1995, vol. 1, pp. 397–398
10. R.A. Buchanan, K.A. Wickersheim, Rare earth phosphors and scintillators. IEEE Trans. Nucl. Sci. **NS-15**(3), 95–101 (1968)
11. Y.K. Voronko, E.L. Nolle, V.V. Osiko, M.I. Timoshechkin, Indutsirovannoye izlucheniye $Y_3Al_5O_{12}$:$(Nd)^{3+}$ pri vozbuzhdenii elektronnym puchkom [Induced emission of $Y_3Al_5O_{12}$: $(Nd)^{3+}$ with excitation by an electron beam]. Pisma v ZhETF [Lett. J. Exp. Theor. Phys.] **13**(3), 125–128 (1971)
12. Ye.A. Seregina, P.P. Dyachenko, V.V. Kalinin, O.D. Shevchuk, Lyuminestsentnyye svoystva kristalla ittriy-alyuminiyevogo granata, aktivirovannogo neodimom, pri vozbuzhdenii α-chastitsami i oskolkami deleniye [The luminescent properties of yttrium aluminum garnet, doped with neodymium, with excitation by α-particles and fission fragments]. Zhurnal prikladnoy spektroskopii [J. Appl. Spectros.] **54**(5), 788–792 (1991)
13. A.A. Seregin, P.P. Dyachenko, V.I. Lapidus, Ye.A. Seregina, Fizicheskiye aspekty sozdaniya tverdotelnogo lazera s yadernoy nakachkoy [Physical aspects of the design of a solid-state laser with nuclear pumping]. Kvantovaya elektronika [Quant. Electron.] **26**(2), 98–102 (1999)
14. A.V. Gulevich, P.P. Dyachenko, A.V. Zrodnikov, O.F. Kukharchuk, Svyazannyye reaktornyye sistemy impulsnogo deystviya [Coupled pulsed reactor systems]. M: Energoatomizdat (2003)
15. E. Matovich, In pursuit of a pulsed homogeneous nuclear laser. IEEE J. Quant. Electron. **QE-4** (5), 379 (1968)
16. V.F. Kolesov, Aperiodicheskiye impulsnyye reaktory [Aperiodic pulsed reactors]. Sarov: RFYaTs-VnIIEF (1999). Also see Kolesov V.F. Aperiodic pulsed reactors: monograph in two volumes. (RFNC-VNIIEF, Sarov, 2006)
17. B.A. Knyazev, V.M. Moralev, Ye.P. Fokin, Lyuminestsentsiya kompleksov evropiya v rastvorakh pri vozbuzhdenii elektronnym puchkom [The luminescence of europium compounds in solutions with excitation by an electron beam]. Optika i spektroskopiya [Optics Spectros.] **40**(1), 93–98 (1976)
18. B.V. Alekhin, L.Ye. Dovbysh, B.V. Lazhintsev, V.A. Nor-Arevyan, Lazernyye sredy na neorganicheskikh zhidkostyakh: predelnyye energeticheskiye kharakteristiki i vozmozhnosti primeneniya v moshchnykh lazerakh s yadernoy nakachkay [Laser media based on inorganic liquids: maximum energy characteristics and the potential for use in high-energy lasers with nuclear pumping]. Trudy 2-y mezhdunarodnoy konf. [Proceedings of the Second International Conference] "Fizika yaderno-vozbuzhdayemoy plazmy i problemy lazerov s yadernoy nakachkoy" [The Physics of Nuclear-Excited Plasma and Problems of Nuclear-Pumped Lasers]. Arzamas-16, 1995, vol. 2, pp. 208–212

19. Ye.A. Seregina, Problemy poiska lazerno-aktivnykh sred dlya preobrazovaniya energii deleniya atomnykh yader v kogerentnoye elektromagnitnoye izlucheniye [Problems in the Search for Laser Active Media for the Conversion of the Fission Energy of Atomic Nuclei into Coherent Electromagnetic Emission]. Khimicheskaya fizika [Chem. Phys.] **15**(8), 23–27 (1996)
20. Ye.A. Seregina, Poisk i issledovaniye sred dlya zhidkostnykh lazerov s yadernoy nakachkoy [Media for liquid lasers with nuclear pumping: search and investigation]. Obzor rabot, vypolnennykh v GNTs RF FEI [A Review of Work Performed at SSC RF IIP]. Trudy 3-y mezhdunarodnoy konf. [Proceedings of the Third International Conference]. "Problemy lazerov s yadernoy nakachkoy i impulsnyye reaktory" ["Problems of Nuclear-Pumped Lasers and Pulsed Reactors"], Snezhinsk, 2003, pp. 129–133
21. Ye.A. Sergina, P.P. Dyachenko, V.V. Kalinin, O.D. Shevchuk, O.N. Gilyarov, Yu.I. Krasilov, B.N. Kulikovskiy, T.L. Novoderezhkina, Spektralnolyuminestsentnyye svoystva lazerno-aktivnykh zhidkostey na osnove oksidakhlorida fosfora pri vozbuzhdenii produktami yadernykh reaktsii [The spectral-luminescent properties of laser active liquids based on oxide chloride phosphorous with excitation by the products of nuclear reactions]. Neorganicheskiya materialy [Inorg. Mater.] **28**(1), 162–169 (1992)
22. P.P. Dyachenko, V.V. Kalinin, Ye.A. Seregina, G.V. Tikhonov, O.D. Shevchuk, O.N. Gilyarov, Yu.I. Krasilov, B.N. Kulikovskiy, T.L. Novoderezhkina, I.M. Batyayev, Yu. A. Kabatskiy, Issledovaniye spektralno-lyuminestsentnykh svoystv neorganicheskikh lazernykh zhidkostey, aktivirovannykh redkozemelnymi i aktinidnymi elementami, pri vozbuzhdenii produktami yadernykh reaktsiy [An investigation of the spectral-luminescent properties of inorganic laser liquids doped with rare earth and actinide elements with excitation by the products of nuclear reactions]. Trudy otraslevoy konf. [Proceedings of the Specialist Conference] "Fizika yaderno-vozbuzhdayemoy plazmy i problemy lazerov s yadernoy nakachkoy" ["The Physics of Nuclear-Excited Plasma and Problems of Nuclear-Pumped Lasers"], Obninsk, 1993, vol. 2, pp. 156–166
23. P.P. Dyachenko, V.V. Kalinin, Ye.A. Seregina, G.V. Tikhonov, O.D. Shevchuk, O.N. Gilyarov, Yu.I. Krasilov, B.N. Kulikovskiy, T.L. Novoderezhkina, Lazer na osnovye neorganicheskoy zhidkosti, legirovannoy neodimom i uranilom [A laser based on an inorganic liquid doped with neodymium and uranyl]. Trudy otraslevoy konf. [Proceedings of the Specialist Conference]"Fizika yadernovozbuzhdayemoy plazmy i problem lazerov s yadernoy nakachkoy" [The Physics of Nuclear-Excited Plasma and Problems of Nuclear-Pumped Lasers], Obninsk, 1993, vol. 3, pp. 190–195
24. A.F. Dobrovolskiy, P.P. Dyachenko, V.V. Kalinin, Ye.A. Seregina, G.V. Tikhonov, Problemy sinteza i generatsionnyye svoystva uransoderzhashchey neorganicheskoy zhidkosti na osnove oksikhlorida fosfora, aktivirovannogo neodimon [Issues of the synthesis and laser properties of an uranium-containing inorganic liquid based on phosphorous oxychloride doped with neodymium]. Trudy 2-y mezhdunarodnoy konf [Proceedings of the Second International Conference] "Fizika yaderno-vozbuzhdayemoy plazmy i problemy lazerov s yadernoy nakachkoy" [The Physics of Nuclear-Excited Plasma and Problems of Nuclear-Pumped Lasers]. Arzamas-16, 1995, vol. 1, pp. 227–233
25. Ye.A. Seregina, P.P. D'yachenko, G.V. Tikhonov, Application of an accelerator for subthreshold diagnostics of active liquid media for direct nuclear pumped lasers. Nucl. Instr. Meth. Phys. Res. B **161–163**, 1022–1026 (2000)
26. V.I. Lapidus, A.F. Dobrovolskiy, P.P. Dyachenko, Ye.A. Seregina, G.V. Tikhonov, Effektivnost nakachki i vremya zhizni urovnya $^4F_{3/2}$ iona neodima pri vozbuzhdenii zhidkosti $POCl_3$-$SnCl_4$-$UO_2{}^{2+}$-Nd^{3+} oskolkami deleniya urana [The pumping efficiency and level lifetime of the $^4F_{3/2}$ ion of neodymium with excitation of the liquid $POCl_3$-$SnCl_4$-$UO_2{}^{2+}$-Nd^{3+} by uranium fission fragments]. Trudy 3-y mezhdunarodnoy konf. [Proceedings of the Third International Conference]. "Problemy lazerov s yadernoy nakachkoy i impulsnyye reaktory" [Problems of Nuclear-Pumped Lasers and Pulsed Reactors], Snezhinsk, 2003, pp. 76–81
27. V.I. Lapidus, A.F. Dobrovolskiy, P.P. Dyachenko, Ye.A. Seregina, G.V. Tikhonov, Effektivnost nakachki i vremya zhizni urovnya $^4F_{3/2}$ iona neodima v zhidkoy lazerno-aktivnoy srede $POCl_3$-$SnCl_4$-$UO_2{}^{2+}$:Nd^{3+} pri impulsnom obluchenii oskolkami deleniya na reaktore

BARS-6 [The pumping efficiency and level lifetime of the $^4F_{3/2}$ ion of neodymium in the liquid laser active medium $POCl_3$-$SnCl_4$-UO_2^{2+}-Nd^{3+} with irradiation by fission fragments on the BARS-6 reactor]. Kvantovaya elektronika [Quant. Electron.] **33**(6), 507–510 (2003)
28. A.F. Dobrovolskiy, P.P. Dyachenko, V.V. Kalinin, Ye.A. Seregina, Podporogovaya diagnostika i osobennosti uransoderzhashchey lazerno-aktivnoy zhidkosti na osnove sulfurila [Subthreshold diagnostics and the features of uranium-containing laser active liquid based on sulfuryl]. Trudy 2-y mezhdunarodnoy konf [Proceedings of the Second International Conference] "Fizika yaderno-vozbuzhdayemoy plazmy i problemy lazerov s yadernoy nakachkoy" [The Physics of Nuclear-Excited Plasma and Problems of Nuclear-Pumped Lasers]. Arzamas-16, 1995, vol. 1, pp. 142–153
29. P.P. Dyachenko, V.V. Kalinin, Ye.A. Seregina, A.A. Seregin, G.V. Tikhonov, Rassmotreniye primeneniya zhidkosti D_2O:Eu^{3+}: UO_2^{2+} v kachestve aktivnoy sregy dlya lazerov s yadernoy nakachkoy [An examination of the use of liquid D_2O:Eu^{3+}: UO_2^{2+} as an active medium for lasers with nuclear pumping]. Trudy 2-y mezhdunarodnoy konf [Proceedings of the Second International Conference] "Fizika yaderno-vozbuzhdayemoy plazmy i problemy lazerov s yadernoy nakachkoy" [The Physics of Nuclear-Excited Plasma and Problems of Nuclear-Pumped Lasers]. Arzamas-16, 1995, vol. 1, pp. 219–226
30. T.L. Novoderezhkina, Ye.A. Seregina, A.F. Borina, B.N. Kulikovskiy, Vzaimodeystviye i perenos energii vozbuzhdeniya mezhdu uranom i RZE v smeshannom aprotonnom rastvoritele $POCl_3$-$SnCl_4$ [The interaction and transfer of excitation energy between uranium and rare earth elements in the mixed aprotonic solvent $POCl_3$-$SnCl_4$]. Zhurnal neorganicheskoy khimii [J. Inorg. Chem.] **43**(2), 314–319 (1998)
31. Ye.A. Seregina, A.F. Borina, T.L. Novoderezhkina, B.N. Kulikovskiy, Perenos energii vozbuzhdeniya i kharakteristika poliyadernykh kompleksov v sisteme $POCl_3$-$SnCl_4$-UO_2^{2+}-Nd^{3+}(Er^{3+}) [The transfer of excitation energy and the characteristics of polynuclear complexes in the system $POCl_3$-$SnCl_4$-UO_2^{2+}-Nd^{3+}(Er^{3+})]. Zhurnal neorganicheskoy khimii [J. Inorg. Chem.] **44**(7), 1201–1206 (1998)
32. Ye.A. Seregina, A.F. Dobrovolskiy, V.V. Kalinin, A.A. Seregin, G.V. Tikhonov, Prokhozhdeniye opticheskogo izlucheniya cherez zhidkost, vozbuzhdennuyu oskolkami deleniya [The propagation of optical emission through liquid excited by fission fragments]. Khimia vysokikh energiy [Chem. High Energ.] **33**(2), 127–131 (1999)
33. Ye.A. Seregina, A.F. Dobrovolsky, P.P. Dyachenko, A.A. Seregin, G.V. Tikhonov, Liquid laser with nuclear pumping. Proc. 9th Int. Conf. on Emerging Nuclear Energy Systems, Tel-Aviv, 1998, vol. 2, pp. 824–831
34. A.A. Seregin, A.F. Dobrovolskiy, P.P. Dyachenko, Ye.A. Seregina, Model zhidkostnogo lazera s yadernoy nakachkoy [A model of a liquid laser with nuclear pumping]. Kvantovaya elektronika [Quant. Electron.] **27**(2), 127–130 (1999)
35. A.G. Molchanov, Yu.M. Popov, Vozbuzhdeniye poluprovodnikovykh lazerov rentgenovskim i γ-izlucheniyem [Excitation of semiconductor lasers with X-rays and γ-rays]. Fizika tverdogo tela [Solid-State Phys.] **11**(7), 1965–1967 (1969)
36. O.V. Bogdankevich, Poluprovodnikovyye lazery s nakachkoy elektronnym puchkom [Semiconductor lasers with electron beam pumping]. Kvantovaya elektronika [Quant. Electron.] **21**(12), 1113–1136 (1994)
37. M.D. Tarasov, V.A. Balakin, S.I. Lobov, N.G. Pavlovskaya, A.M. Sidoruk, V.A. Tsukerman, Lazernoye izlucheniye kristallov sulfida kadmiya pri vozbuzhdenii impulsnym tormoznym rentgenovskim izlucheniyem [The laser emission of cadmium sulphide crystals with excitation by pulse bremsstrahlung X-ray emission]. Doklady AN SSSR [Proc. Acad. Sci. USSR], **241**(5), 1058–1060 (1978)
38. K. Held, L.R. Solon, Phosphor transducer means for pumping coherent light generators by nuclear energy sources. (1969), U.S. Patent 3,470,490 (filled November 12, 1963)
39. G.H. Miley, Dual-media nuclear-pumped lasers. Proc. Int. Symp. "Short-Wavelength Lasers and Their Applications," Osaka, 1988, pp. 14–25
40. F.P. Boody, Nuclear-driven fluorescence pumped solid-state lasers. Ph.D. Dissertation, University of Missouri-Columbia, 1991

41. A.M. Voinov, L.Ye. Dovbysh, V.N. Krivonosov, V.A. Krotov, B.V. Lazhintsev, L.C. Mkhitaryan, V.A. Nor-Arevyan, A.N. Sizov, A.V. Sinitsin, A.A. Sinyanskii, Glass 1-MJ laser for laser fusion with nuclear-optical pumping on the pulse reactor. Proc. 8th Int. Conf. on Emerging Nuclear Energy Systems, Obninsk, 1996, vol 1, pp. 310–321
42. A.M. Voinov, L.Ye. Dovbysh, V.N. Krivonosov, A.N. Sizov, A.V. Sinitsyn, A.A. Sinyanskiy, Impulsnyye reaktory-lazery s yaderno-opticheskoy nakachkoy. [Pulsed reactor lasers with nuclear optical pumping]. Voprosy atomnoy nauki i tekhniki. [Questions of Atomic Science and Technology] Ser. Fizika yadernykh reaktorov [Series on the Physics of Nuclear Reactors], 2000, No. 2/3, pp. 16–18
43. M.A. Prelas, F.P. Boody, G.H. Miley, J.F. Kunze, Nuclear driven flashlamps. Laser Part. Beams. **6**, Part 1, 25–62 (1988)
44. M.A. Prelas, Lasers with combined nuclear pumping. Trudy otraslevoy konf. [Proceedings of the Specialist Conference] "Fizika yaderno-vozbuzhdayemoy plazmy i problemy lazerov s yadernoy nakachkoy" [The Physics of Nuclear-Excited Plasma and Problems of Nuclear-Pumped Lasers], Obninsk, 1992, vol. 1, pp. 75–100
45. F.P. Boody, M.A. Prelas, Efficient visible nuclear-driven fluorescer lamps. Trudy otraslevoy konf. [Proceedings of the Specialist Conference] "Fizika yadernovozbuzhdayemoy plazmy i problemy lazerov s yadernoy nakachkoy" [The Physics of Nuclear-Excited Plasma and Problems of Nuclear-Pumped Lasers], Obninsk, 1992, vol. 1, pp. 161–165
46. M.A. Prelas, Lin Li-Te, J.T. Bahns, Progress in nuclear-driven flashlamps. Int. Embedded Topical Meeting on Physics, Safety, and Applications of Pulse Reactors, Washington, 1994, pp. 155–162
47. J.W. Wilson, A. Shapiro, Nuclear-induced excimer fluorescence. J. Appl. Phys. **51**(5), 2387–2393 (1980)
48. T.E. Stewart, G.S. Hurst, T.E. Bortner, J.E. Parks, F.W. Martin, H.L. Weidner, Proton excitation of continuous emission in the noble gases. J. Opt. Soc. Am. **60**(10), 1290–1297 (1970)
49. J. Wiser, D.E. Murnick, A. Ulrich, H.A. Huggins, A. Liddle, W.L. Brown, Vacuum ultraviolet rare gas excimer light source. Rev. Sci. Instrum. **68**(3), 1360–1364 (1997)
50. R.A. Walters, J.D. Cox, R.T. Schneider, J. Hagefstration, Generation, measurement, and utilization of xenon excimer radiation produced by nuclear reaction products. Trans. Am. Nucl. Soc. **34**(1), 808–810 (1980)
51. A.A. Mavlyutov, A.I. Miskevich, B.S. Salamakha, Spektralnyye kharakteristiki eksimernykh aktivnykh sred pri yadernoy nakachke [The spectral characteristics of excimer active media during nuclear pumping]. Trudy 2-y mezhdunarodnoy konf [Proceedings of the Second International Conference] "Fizika yadernovozbuzhdayemoy plazmy i problemy lazerov s yadernoy nakachkoy" [The Physics of Nuclear-Excited Plasma and Problems of Nuclear-Pumped Lasers]. Arzamas-16, 1995, vol. 1, pp. 162–171
52. F.P. Boody, M.A. Prelas, Absolutely calibrated spectra of nuclear-driven rare gases, 400–950 nm. Trudy otraslevoy konf. [Proceedings of the Specialist Conference]. "Fizika yaderno-vozbuzhdayemoy plazmy i problemy lazerov s yadernoy nakachkoy" [The Physics of Nuclear-Excited Plasma and Problems of Nuclear-Pumped Lasers]. Obninsk, 1993, vol. 2, pp. 149–155
53. A.A. Abramov, V.V. Gorbunov, S.P. Melnikov, A.K. Mukhamatullin, A.A. Pikulev, A.V. Sinitsyn, A.A. Sinyanskii, V.M. Tsvetkov, Luminescence of nuclear-induced rare-gas plasmas in near infrared spectral range. Proc. SPIE **6263**, 279–296 (2006)
54. A.A. Seregin, P.P. Dyachenko, Ye.A. Seregina, Model zhidkostnogo lazera s yadernoy-opticheskoy nakachkoy [A model of a liquid laser with nuclear optical pumping]. Kvantovaya elektronika [Quant. Electron.] **33**(6), 503–506 (2003)
55. A.M. Voinov, L.Ye. Dovbysh, V.N. Krivonosov, S.P. Melnikov, A.A. Sinyanskiy, Lazery s nakachkoy opticheskim izlucheniyem konvertora na osnove radiolyuminestsentnogo i plazmennogo istochnikov sveta. [Lasers with nuclear pumping by optical emission of a converter based on radioluminescent and plasma light sources] Trudy otraslevoy konf. [Proceedings of the Specialist Conference] "Fizika yaderno-vozbuzhdayemoy plazmy i problemy lazerov s yadernoy nakachkoy" [The Physics of Nuclear-Excited Plasma and Problems of Nuclear-Pumped Lasers], Obninsk, 1993, vol. 3, pp. 224–226

56. A.V. Bochkov, A.V. Zagidulin, E.P. Magda, V.V. Mironenko, G.S. Sofiyenko, A.S. Podymako, Generatsiya na perekhode $^3F_4$-$^3H_6$ iona Tm^{3+} pri pryamom vozbuzhdenii sostoyaniya $^3F_4$ izluchenim gazovogo lazera s yadernoy nakachkoy [Lasing at the transition of the $^3F_4$-$^3H_6$ ion Tm^{3+} with direct excitation of the $^3F_4$ state by gas laser emission with nuclear pumping]. Pisma v ZhTF [Lett. J. Theor. Phys.] **32**(19), 42–46 (2006)
57. A.V. Karelin, R.V. Shirokov, Utilizatsiya radioaktivnykh otkhodov na osnove yaderno-opticheskikh preobrazovateley [The use of radioactive wastes as a basis for nuclear-optical transformation] Tezisy dokladov 4-y mezhdunarodnoy konf. [Thesis of the Fourth International Conference] "Fizika lazerov s yadernoy nakachkoy i impulsnyye reaktory" [The Physics Nuclear-Pumped Lasers and Pulsed Reactors], 2007, pp. 143–144
58. A.M. Voinov, S.V. Vorontsov, L.Ye. Dovbysh, R.I. Ilkayev, V.Ye. Marshalkin, V.T. Punin, A.V. Rudnev, Laguna vysokykh tekhnologiy [High tech lagoon.] Voprosy atomnoy nauki i tekhniki. [Questions of Atomic Science and Technology] Ser. Fizika yadernykh reaktorov [Series on the Physics of Nuclear Reactors], 2000, No. 2/3, pp. 91–94
59. A.A. Sinyanskii, S.P. Melnikov, L.Ye. Dovbysh, G.L. Johnson, Nuclear-optical converters for neutron detection. Proc. 4th Am. Nucl. Soc. Int. Topical Meeting on Nuclear Plant Instrumentation, Controls and Human-Machine Interface Technologies, Columbus, OH, 2004
60. V.V. Gorbunov, L.Ye. Dovbysh, S.P. Melnikov, A.V. Sinitsyn, A.A. Sinyanskiy, V.M. Tsvetkov, Neutronnyy detektor na osnove yaderno-opticheskogo preobrazovatelya [A neutron detector based on nuclear optical conversion] Trudy RFYaTs-VNIIEF [Proceedingsof the RF NC [Nuclear Center]-VNIIEF], 2007, Issue 11, pp. 169–179
61. J.D. Cox, R.T. Schneider, Y.S. Tang, Nuclear pumped lasers utilizing radiant transfer. Laser Interact. Relat. Plasma Phenom. **6**, 73–85 (1984)
62. F.P. Boody, M.A. Prelas, Design of a large-scale nuclear-driven fluorescer pumped solid-state laser system. Trudy otraslevoy konf. [Proceedings of the Specialist Conference] "Fizika yaderno-vozbuzhdayemoy plazmy i problemy lazerov s yadernoy nakachkoy" [The Physics of Nuclear-Excited Plasma and Problems of Nuclear-Pumped Lasers], Obninsk, 1993, vol. 3, pp. 95–97
63. J.A. Abate, L. Lund, D. Brown, S. Jacobs, S. Refermat, J. Kelly, M. Gavin, J. Waldbilling, O. Lewis, Active mirror: a large-aperture medium-repetition rate Nd:glass amplifier. Appl. Optics **20**(2), 351–361 (1981)
64. Ye.P. Shabalin, Impulsnyye reaktory na bystrykh neytronakh [Pulsed reactors based on fast neutrons]. M: Atomizdat (1976)
65. P.P. Dyachenko, A.V. Zrodnikov, V.N. Kononov, V.Ya. Pupko, V.V. Korobkin, A.M. Prokhorov, Kombinirovannaya termoyadernaya ustanovka s reaktornolasernym drayverom dlya promyshlennogo polucheniya elektroenergii [A combined thermonuclear device with a laser-reactor driver for industrial electrical energy]. Preprint FEI [Preprint IPP], No. 2097, Obninsk (1990)
66. P.P. Dyachenko, A.V. Zrodnikov, Yu.A. Prokhorov, V.Ya. Pupko, S.S. Satin, V.I. Subbotin, N.G. Basov, V.B. Rozanov, Yu.V. Senatskiy, L.P. Feoktistov, A.A. Mak, S.B. Papernyy, V.A. Serebryakov, V.Ye. Yashin, Gibridnay ayaderno-termoyadernaya elektrostantsiya [Hybrid nuclear-thermonuclear power plant]. Preprint FEI [Preprint IPP], No. 2327, Obninsk (1993)
67. A.P. Barzilov, A.V. Gulevich, P.P. Dyachenko, A.V. Zrodnikov, V.N. Kononov, O.F. Kukharchuk, V.Ya. Pupko, V.V. Korobkin, Lazernaya sistema impulsno-periodicheskogo deystviya s nakachkoy ot yadernogo reaktora tipa IBR2. [A pulsed-periodic laser system with pumping from a IBR-2 type nuclear reactor]. Trudy 2-y mezhdunarodnoy konf [Proceedings of the Second International Conference] "Fizika yaderno-vozbuzhdayemoy plazmy i problemy lazerov s yadernoy nakachkoy" [The Physics of Nuclear-Excited Plasma and Problems of Nuclear-Pumped Lasers], Arzamas-16, 1995, vol. 2, pp. 186–194
68. A.P. Barzilov, A.V. Gulevich, P.P. Dyachenko, O.F. Kukharchuk, E.A. Pashin, S.S. Satin, Investigation of a pulse periodic nuclear pumped laser system. Proc. 9th Int. Conf. on Emerging Nuclear Energy Systems, 1998, Tel-Aviv, vol. 2, pp. 897–904

Chapter 12
Gas Lasers Excited by Radiation from Nuclear Explosions

In previous chapters, we have reviewed pumping methods for gas-medium and condensed-medium lasers using nuclear reactors as neutron sources. We also reviewed the characteristics of these lasers and the potential to use them in different nuclear-laser devices. The highest specific power depositions for gas media ($q \leq 5$ kW/cm^3) can be obtained if a laser cell that has a uranium layer or is filled with ^{3}He is placed inside the core of a pulsed reactor at the minimal possible pulse length for pulsed reactors, ~100 μs. However, these specific power depositions are sometimes not sufficient to reach the lasing threshold of certain active media. Precisely for this reason, attempts to achieve lasing with help of the neutron fluxes of pulsed reactors to pump excimer or chemical lasers were unsuccessful (see Chap. 3, Sect. 6).

If electron or ion beams with a ~10 ns pulse duration are used, significantly higher specific power depositions can be obtained for gas media: $\leq 10^7$ W/cm^3 × atm and $\leq 10^9$ W/cm^3 × atm, respectively [1]. This pumping method made it possible to create many different high-pressure gas lasers (see, for example, [2]), some of which had rather significant output powers. For example, with electron-beam pumping of an excimer gas media based on the KrF ($\lambda = 248$ nm) molecule, an Aurora multichannel setup measured a laser energy of about 10 kJ over 500 ns [3]. When a chemical HF laser was initiated by an electron beam on H_2-F_2-O_2-NF_3 mixture, the output energy was 4.5 kJ over 50 ns [4]. A further increase in output performance is difficult mainly due to the complexity of pumping large volumes of active media uniformly.

Investigations of high-power pulsed lasers in the nanosecond range have mainly been directed toward creating a laser driver for facilities based on inertial confinement fusion (ICF). In order for the facility to be economically advantageous, the gain of the thermonuclear target must be about 100 [5]. According to calculations, this gain requires laser energy of 1–10 MJ [6]. At present, the energy (power) of the most powerful operating ICF laser in Russia is 30 kJ (120 TW) [7, 8]. Outside Russia, these values are 40 + 25 kJ (60 + 10 TW) [9, 10]; 30 kJ (~10 TW) [11]; and 150 kJ (15 TW) [12].

S.P. Melnikov et al., *Lasers with Nuclear Pumping*,
DOI 10.1007/978-3-319-08882-2_12

The last two lasers are separate sections of high-power facilities under construction: LMJ (240 channels) [11] and NIF (192 channels) [12]. According to calculations, these facilities should have roughly identical output performance, 1.8 MJ (500 TW) at $\lambda \approx 350$ nm. The commissioning of these facilities is planned for 2010–2012. Thus, building the laser driver needed for ICF facilities is a long and expensive task.

Using the γ-ray radiation from a nuclear explosion as the pump source provides unique opportunities to develop high-power lasers in the nanosecond range since it excites active gas media with volumes in the tens of cubic meters at specific power depositions up to ~10^9 W/cm^3 × atm. If a nuclear explosive device with a comparatively low energy release (about 0.5 kt of TNT) is used, laser pumping can be implemented with an output energy of 0.1–1 MJ [13]. These lasers may find their application, for example, in the solutions for certain ICF problems.

12.1 The First Experimental Studies

The first experiment on the pumping of a laser by γ-radiation from a nuclear explosion was presumably performed at VNIIEF in 1971. Liquid xenon emitted as Xe_2* ($\lambda = 172$ nm) excimer molecules were studied as an active media. This laser was pumped earlier with an electron beam by the staff at FIAN (Lebedev Physics Institute of the Academy of Science) [14].

Experiments on pumping two types of lasers with the γ-radiation from a nuclear explosion were carried out in the United States in 1973. This was done in a testing area in Nevada by the staff of LLNL [15] and LANL [16].

The LLNL experiments studied xenon gas at 6.25–18.75 atm and specific energy depositions of 4.5–15.3 J/cm^3. At xenon pressures higher than 9.4 atm, the gain of the spontaneous emission was recorded in the UV spectral range near 170 nm. In this case, the gain increased with the rise in pressure from 0.1 to 0.29 cm^{-1}. The light radiation had an obviously pronounced directivity, which is additional proof that the stimulated intense radiation was present in these experiments.

The staff at LANL checked the possibility of initiating a chemical HF-laser ($\lambda = 2.7$–3.0 μm) on SF_6-C_2H_6(8:1) mixture at 1.7 atm with the γ-radiation from a nuclear explosion. A 44 cm long laser cell 2.9 cm in diameter was exposed to a collimated γ-radiation for 12 ns duration. The exposed length of the cell was 10 cm, and the active volume was 65 cm^3. Lasing was reported in these experiments, and the laser pulse energy was 3.9 J ($\eta_l = 6.5$ %) at a 17 ns half-amplitude pulse duration.

Further studies of high-power gas lasers excited by the γ-radiation from a nuclear explosion were carried out at VNIIEF in 1977–1984. The laser setups examined next are single-pass, traveling-wave amplifiers.

Chemical HF-Lasers [17, 18]

The gas mixture SF_6-H_2 with a non-chain chemical reaction was used as the active media. This mixture was studied previously at VNIIEF with excitation by a relativistic electron beam at a 40 ns pulse duration [19, 20].

One of these experiments [17] used a 480-cm long cylindrical laser cell 11 cm in diameter without a laser cavity. The cell was filled with a SF_6-H_2 (9:1) mixture at 2.1 atm. The angle between the propagation direction of the γ-radiation and the optical axis of the cell was about 10°. Therefore, longitudinal excitation of the laser media was achieved in practice. Under these conditions, the divergence of the laser beam was determined by the geometric dimensions of the cell. To reduce the divergence, the section of the cell near the γ-source was comprised of closely placed channels, 60-cm long and 3-mm in diameter, which formed beam directivity in the remaining part of the cell.

The distribution of the specific energy deposition along the length of the cell is shown in Fig. 12.1. The average energy deposition by volume was about 1 J/cm^3 at a γ-pulse duration of 18 ns at the base. This experiment obtained the following parameters for the laser radiation: the energy density at the output of the cell was 4.4 J/cm^2; the half-amplitude pulse duration was 8 ns (Fig. 12.2); and the beam angular divergence was 7×10^{-3} rad.

The results of the experiment were analyzed based on a theoretical model developed by the authors [17]. This model is a one-dimensional transport equation for direct and reverse light waves at the vibrational-rotational transitions of the HF molecule ($\lambda = 2.7$–3.0 μm). It follows from the calculations that approximately 90 % of the radiation is in the forward direction corresponding to the direction of the initialization wave. Figure 12.2 shows the calculated shape of the laser pulse when energy is removed from the main volume of the cell as a result of the amplification of the narrow-band radiation from the former. Calculations also show that under experimental conditions, the energy of the narrow-band radiation from the cell is 20–25 % of the potential laser energy.

A significantly higher laser output, about 70 kJ, was obtained in an experiment with about 6 m^3 of active medium [18]. The active medium and the conditions for pumping it (average specific energy contribution and γ-pulse length) were the same

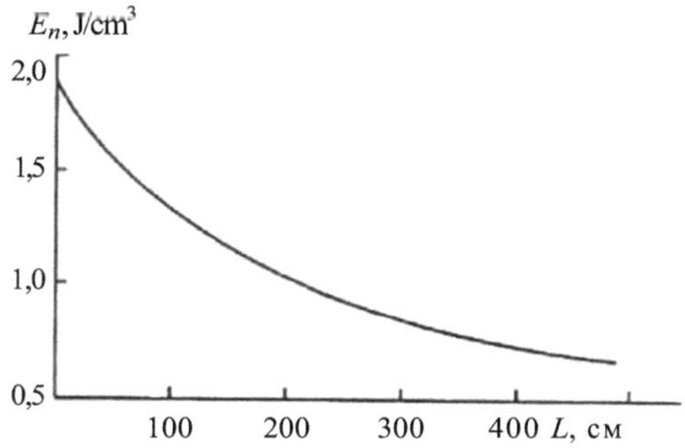

Fig. 12.1 The distribution of the specific energy deposition along the length of the laser cell for a SF_6-H_2 mixture at 2.1 atm [17]

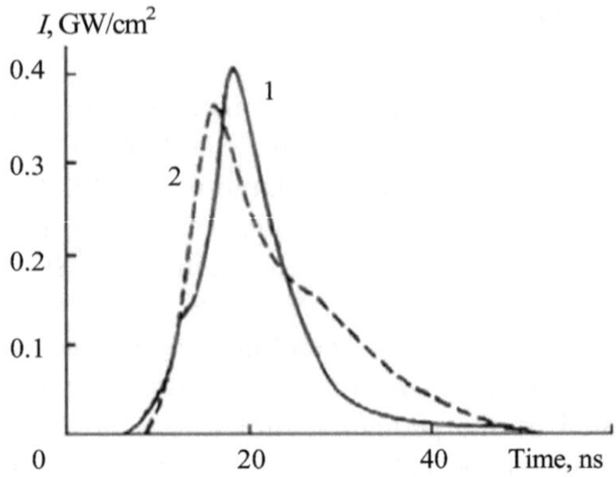

Fig. 12.2 Experimental (1) and calculated (2) laser pulses [17] (I is the laser power density at the output of the cell)

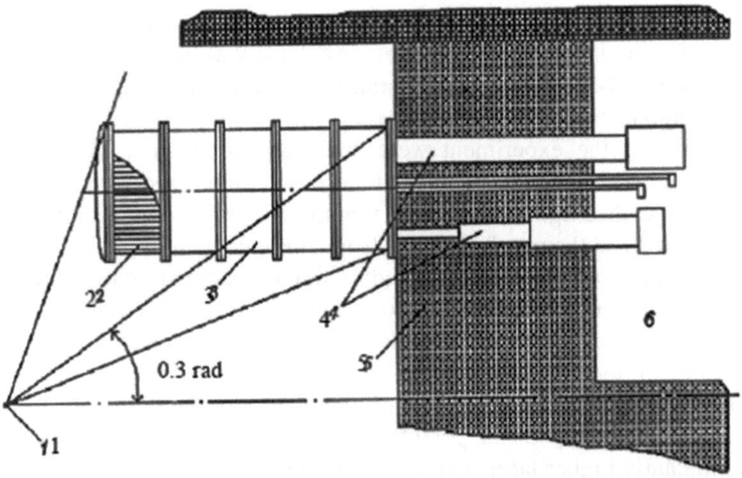

Fig. 12.3 Schematic of the experiment [18]: (1) pumping source; (2) former of beam directivity; (3) laser cell; (4) channels for withdrawal of laser radiation; (5) concrete shielding; (6) instrumentation compartment

as in the previous work [17]. The experiments whose schematics are shown in Fig. 12.3 used a cylindrical laser cell, 130 cm in diameter and 480 cm long, without a laser cavity. Laser radiation distribution in the beam cross-section was measured using bolometric wire-netting calorimeters.

As in the work [17], to reduce the angular divergence of the beam, a 60-cm long former of radiation directivity was used that consisted of elongated channels 3-mm in diameter. The inner volume of the main part of the cell was divided into long partitions to reduce energy loss due to amplification of the spontaneous emission in the transverse direction. The average energy density in the cross-section of the laser beam at the cell output was 5.1 J/cm^2, and the shape and duration of the laser pulse were about the same as in Fig. 12.2.

Excimer XeF-Lasers [13, 21–23]

These experiments used three various laser devices, which consisted of two volumes connected optically to each other by a small coupling aperture [22]. The first cylindrical volume, which was located near the γ-radiation source, formed the field of the radiation that was synchronized with the pumping wave. Subsequent radiation proceeded through the coupling aperture and entered the second (conical) volume and was amplified to the necessary energy. As a result, a radiation wave front, which was almost spherical, formed at the output of the conical amplifier, and the equivalent beam divergence was determined by the ratio of the diameter of the coupling aperture to the cone length.

The cylindrical part of these three laser devices was 100-cm long with a diameter of 10 cm. The conical portion of two from these devices was 2-m long, and the third device which had target chamber, was 14-m long and 0.8-m base diameter. As the active medium, a Ne-Xe-NF_3 (1,000:26.2:6.2) mixture was used at 5 atm. The dependence of the specific power deposition into the active medium on the distance to a coupling aperture 3 mm in diameter is shown in Fig. 12.4. The scheme of the laser device with a 30 l vacuum target chamber which was placed 12.5 m from the top of the cone is shown in Fig. 12.5. A spherical mirror 80-cm in diameter with a 2.1 m focal distance was situated at a distance of 15 m from the top of the cone. The mirror focused the laser radiation inside the chamber on the target through a quartz window 36-cm in diameter. The target was shielded from the direct influence of the γ-radiation by a 50-cm thick steel screen. The experiments used a target with a reversed corona [24], which is a thin-walled polyethylene cone 30-mm high with an open base and an apex angle of 30°.

The experiments measured laser parameters and the characteristics of the X-ray emitted by the plasma that formed when the target was irradiated with the laser radiation. The transmittance of the target chamber window at the laser wavelength

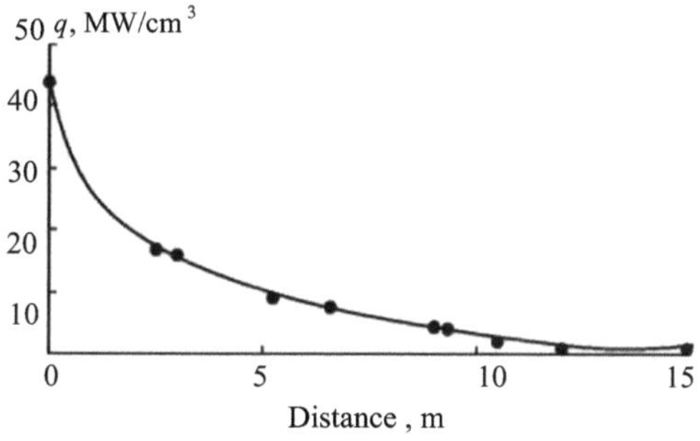

Fig. 12.4 The dependence of the specific power deposition of the active medium on the distance to the coupling aperture placed at the top of the cone [23]

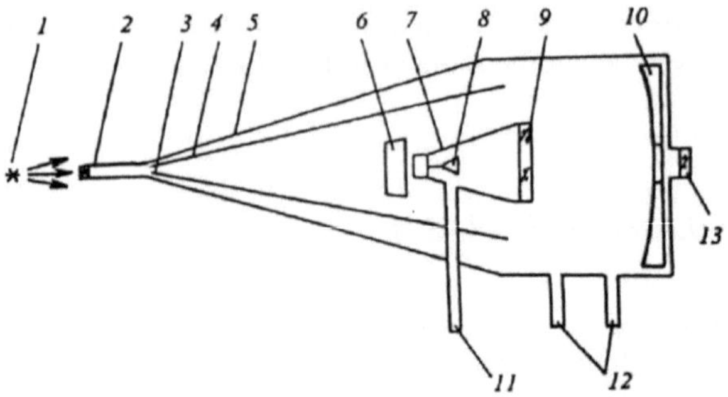

Fig. 12.5 A schematic of an experiment with a laser device where the laser radiation was focused on the target [22, 23]: (1) source of the γ-radiation; (2) cylindrical laser; (3) coupling aperture; (4) light-absorbing conical covering; (5) cone body; (6) steel shielding; (7) vacuum chamber; (8) target; (9) entrance window into the chamber; (10) focusing mirror; (11) vacuum channel for recording X-ray radiation; (12) optical diagnostic channels; (13) window of the frontal survey of the target

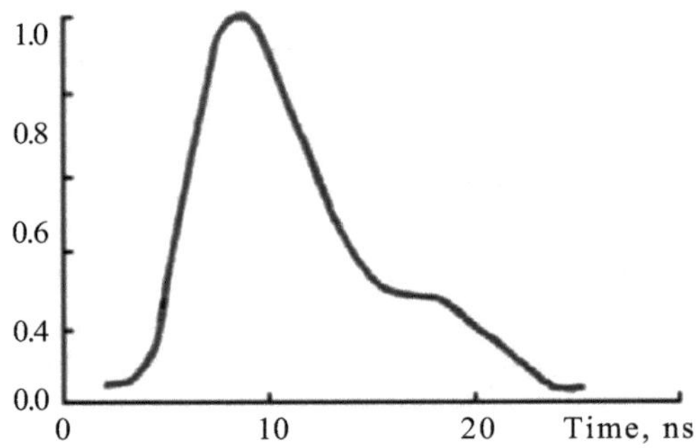

Fig. 12.6 An oscilloscope trace of a radiation pulse of a XeF-laser excited by the γ-radiation of a nuclear explosion [22]

can, in principle, be reduced by γ-radiation. Therefore, this parameter was controlled.

The laser device with a target chamber had the following laser characteristics [13, 21–23]: the energy (power) of the laser radiation was 700 J (1.4×10^{11} W), half-amplitude pulse duration of the laser radiation was 7 ns (Fig. 12.6), laser energy efficiency was 3 %, and the contrast of the output radiation was $\geq 10^4$. The experiments showed that the transparency of the quartz window did not change during a pump pulse. The characteristics of the recorded X-ray radiation agree with the calculated results for the laser energy at the target—about 1 kJ. This testifies that a spherical light wave formed in the conical amplifier since only in this case is it possible to transfer the laser energy to the target without losses using a focusing system.

The reported energy parameters were not the maximum. In subsequent experiments, a cylindrical XeF-laser with a larger active medium volume showed a laser energy of 6 kJ at an intrinsic efficiency of about 4 % [13].

Theoretical Modeling of Excimer Media

The features of experiments using nuclear explosive devices which relate to the destruction of the laser device require careful preparation, in particular, of a detailed simulation of the physical processes in the active laser media and preliminary calculation of laser characteristics. These calculations were performed for lasers based on the transitions of the excimer molecules: XeF [23] and KrF [25].

Calculations [23] of experimental conditions [21–23] used a kinetic model of the XeF laser, which determined the laser characteristics on two spectral lines: 351 and 353 nm. Resulting from the solution of the radiative transfer equation, the total power density on each line was determined as a function of the distance to the coupling aperture at the top of the cone (Fig. 12.7). The radiation power at the surface of the focusing mirror was $I\pi R^2 = 7.2 \times 10^{10}$ W ($R = 40$ cm). This is approximately two times less than the power obtained experimentally. The calculated maximum efficiency of the amplifier (2 %) was also lower than the experimental value (3 %).

A paper [25] analyzed the functioning of a laser device consisting of a cylindrical former and a conical amplifier with γ-ray, travelling-wave pumping. A gas medium based on KrF ($\lambda = 249$ nm) excimer molecules was considered as the active media. Special attention was paid to an analysis of the conditions for forming the directed radiation. Calculations showed that the efficiency of energy extraction from the conical amplifier as a function of its height was 20–50 %; the beam divergence at the output of the cone was $\theta \approx d_0/L = 10^{-4}$ rad (where d_0 is the diameter of the coupling aperture, and L is the length of the cone). These

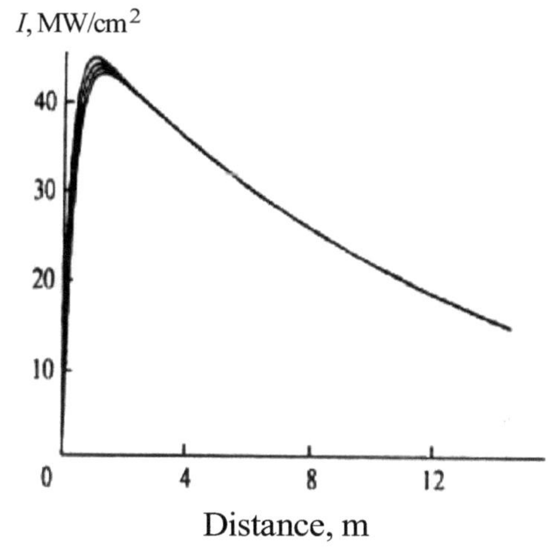

Fig. 12.7 Dependence of the total power density of the 351 and 352 nm lines on the distance to the coupling aperture. The initial power densities at the input of the conic amplifiers were 1, 10^2, 10^4, and 10^6 W/cm^2 (from *bottom* to *top*) [23]

calculations also showed that amplification of the spontaneous emission may influence the directivity of the useful radiation from the cone if the intensity of the spontaneous noise near the top of the cone is comparable to the saturation intensity of the active medium.

Scheme of an Experiment with a Focused Radiation on the Target

Paper [25] proposes a scheme for an experiment (Fig. 12.8), which focuses the beam of ten identical conical laser modules on a target using elliptical mirrors to initiate a thermonuclear reaction. Each conical module is separated from the vacuum light guide by a quartz plate. The axes of the all laser modules and light guides are arranged on two conical surfaces whose axes are aligned and pass through the source of the γ-radiation and target. The tops of the conical surfaces coincide with the source of the γ-radiation and target. Due to the identical geometry of the arrangement of all the schematic elements, radiation from the conical amplifiers falls on the target simultaneously.

Thus, the use of the most powerful energy sources—nuclear explosive devices for gas-laser pumping would solve one of the main problems of the ICF: to determine the energy level that laser drivers need to reach one or another target gain.

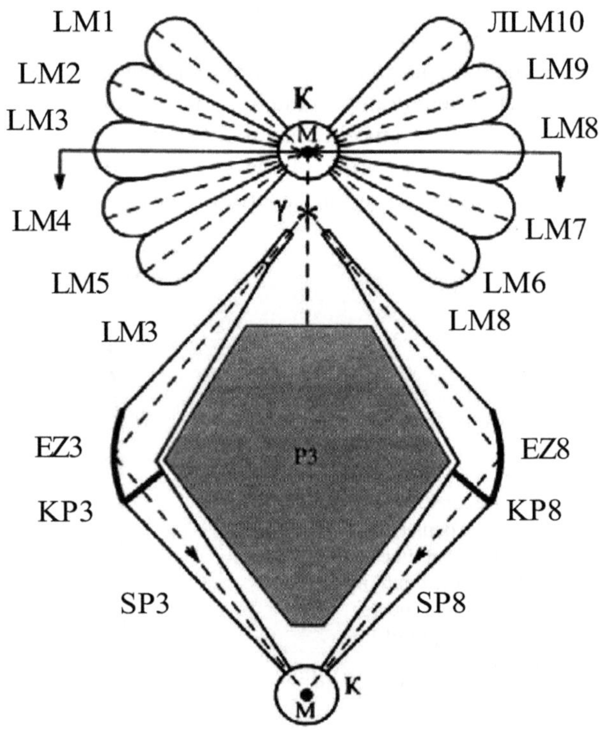

Fig. 12.8 A schematic of an experiment on the initialization of thermonuclear reactions in the target using the radiation of excimer lasers with γ-ray pumping [25]: LM1–LM10 are conical laser modules; EZ3, EZ8 are elliptical mirrors; KP3, KP8 are quartz plates; RZ is radiation shielding; SP3, SP8 are vacuum light guides; K is the chamber; M is the target

References

1. A.A. Kuznetsov, S.S. Sulakshin, O vozbuzhdenii gazovykh lazerov tyazhelymi uskorennymi zaryazhennymi chastitsami [On the excitation of gas lasers with heavy accelerated charged particles]. ZhTF [J. Tech. Phys.] **55**(5), 864–868 (1985)
2. G.A. Mesyats, V.V. Osipov, V.F. Tarasenko, Impulsnyye gazovyye lazery [Pulsed gas lasers] M: Nauka (1991)
3. L.A. Rosocha, P.S. Bowling, M.D. Burrows, M. Kang, J. Hanlon, J. McLeod, G.W. York, An overview of Aurora: A multikilojoule KrF laser system for inertial confinement fusion. Laser Part. Beams **4**(1), 55–70 (1986)
4. F. Kannari, H. Inagaki, M. Obara, High efficiency multikilojoule HF chemical lasers using an electron beam initiated low-pressure mixtures of $H_2/F_2/NF_3$ or $H_2/F_2/SF_6$. Appl. Phys. Lett. **48** (4), 266–268 (1986)
5. J.J. Duderstadt, G.A. Moses, *Inertial confinement fusion* (J. Wiley and Sons, New York, NY, 1982)
6. P.P. Dyachenko, A.V. Zrodnikov, Yu.A. Prokhorov, V.Ya. Pupko, S.S. Satin, V.I. Subbotin, N.G. Basov, V.B. Rozanov, Yu.V. Senatskiy, L.P. Feoktistov, A.A. Mak, S.B. Papernyy, V.A. Serebryakov, V.Ye. Yashin, Gibridnaya yaderno-termoyadernaya elektrostantsiya [Hybrid nuclear-thermonuclear power plant]. Preprint FEI [Preprint IPP], No. 2327, Obninsk (1993)
7. V.T. Punin, G.A. Kirillov, V.M. Murugov, V.I. Shemyakin, High power laser system ISKRA V. Proc. Int. Conf. LASERS'90, San Diego, 1990, pp. 819–826
8. V.I. Annenkov, V.A. Bagretsov, V.G. Bezuglov et al., Kvantovaya elektronika [Quant. Electron.] **18**(5), 536–537 (1991)
9. T.R. Boehly, D.L. Brown, R.S. Craxton et al., Initial performance results of the OMEGA laser system. Optics Commun. **133**(1–6), 495–596 (1997)
10. J.H. Kelly, L.J. Waxer, V. Bagnoud et al., OMEGA EP: high-energy petawatt capability for the OMEGA laser facility. J. Phys. IV France **133**, 75–80 (2006)
11. B. Bigot, Inertial fusion science in Europe. J. Phys. IV France **133**, 3–8 (2006)
12. E.I. Moses, G.H. Miller, R.L. Kauffman, The ICF status and plans in the United States. J. Phys. IV France **133**, 9–16 (2006)
13. Ye.K. Bonyushkin, R.I. Il'kaev, A.I. Pavlovski, A.P. Morovov, N.G. Basov, S.Yu. Gus'kov, V.B. Rozanov, N.V. Zmitrenko, About testing experiment with targets for gain based on using of the powerful pulsed lasers pumped by γ radiation of the underground nuclear explosion. Proc. Int. Conf. on Laser Interaction with Matter, 1994, Oxford, pp. 89–91
14. N.G. Basov, V.A. Danilychev, Yu.M. Popov, D.D. Khodkevich, Kvantovyy generator v vakuumnoy oblasti spectra pri vozbuzhdenii zhidkogo ksenona elektronnym puchkom [A quantum oscillator in the vacuum region of the spectrum with excitation of liquid xenon by an electron beam]. Pisma v ZhETF [Lett. J. Exp. Theor. Phys.] **12**(10), 473–474 (1970)
15. P.J. Ebert, L.J. Ferderber, H.A. Koehler, R.W. Kuckuck, D.L. Redhead, Amplified spontaneous emission in xenon pumped by gamma rays. IEEE J. Quant. Electron. **QE-10**(9), 736 (1974)
16. P.B. Lyons, J.S. Clarke, D.S. Metzger, Gamma initiated HF laser. IEEE J. Quant. Electron. **QE-10**(9), 736 (1974)
17. A.I. Pavlovskiy, Ye.K. Bonyushkin, V.V. Varaksin, G.S. Vinyarskiy, A.Ye. Lakhtikov, G.M. Mishchenko, A.P. Morovov, V.D. Urlin, Issledovaniye impulsnogo khimicheskogo HF-lasera s nakachkoy gamma-izlucheniyem [An investigation of a pulsed chemical HF-laser with gamma-emission pumping]. Doklady AN RF [Proceedings of the Academy of Science of the Russian Federation], 1993, vol. 331, no. 3, pp. 299–301
18. A.I. Pavlovskiy, Ye.K. Bonyushkin, V.V. Varaksin, G.S. Vinyarskiy, A.Ye. Lakhtikov, A.P. Morovov, G.M. Mishchenko, V.D. Urlin, Issledovaniye kharakteristik moshchnogo HF-lasera, vozbuzhdayemogo gamma-izlucheniyem yadernogo vzryva [An investigation of the characteristics of a high-power laser excited by the gamma-emission of a nuclear explosion]. Trudy 2-y mezhdunarodnoy konf. [Proceedings of the Second International Conference]

"Fizika yaderno-vozbuzhdayemoy plazmy i problemy lazerov s yadernoy nakachkoy" [Physics of Nuclear-Induced Plasma and Problems of Nuclear-Pumped Lasers]. Arzamas-16, 1995, vol. 2, pp. 214–218

19. L.Ye. Dovbysh, N.I. Zavada, A.T. Kazakevich, A.A. Karpikov, S.P. Melnikov, I.V. Podmoshenskiy, A.A. Sinyanskiy, Radiatsionno-khimicheskiy lazer na smesyakh SF_6+H_2 i CCl_2F_2+H_2 pri davleniyakh do 3 ata [Radiation-chemical SF_6+H_2 and CCl_2F_2+H_2 mixture lasers at pressures up to 3 ata. [sic atm]]. Pisma v ZhETF [Lett. J. Exp. Theor. Phys.] **20**(6), 407–410 (1974)
20. A.M. Voinov, L.Y. Dobvysh, A.T. Kazakevich, S.P. Melnikov, A.A. Sinyanskiy, Radiatsionno-khimicheskiy lazer vysokogo davleniya, initsiiruyemyy puchkom bystrykh elektronov iz koaksialnogo vysokovoltnogo dioda [A high-pressure radiation-chemical laser inititated by a high-velocity electron beam from a coaxial high-voltage diode]. Kvantovaya elektronika [Quant. Electron.] **4**(2), 426–427 (1977)
21. B.V. Alekhin, Ye.K. Bonyushkin, V.V. Varaksin, B.V. Lazhintsev, A.Ye. Lakhtikov, A.P. Morovov, G.F. Nasyrov, V.A. Nor-Arevyan, A.I. Pavlovskiy, R.A. Orlov, V.G. Rogachev, V.B. Shlyakhovoy, Eksimernyy lazer begyshchey volny s nakachkoy gamma-izlucheniyem [A traveling-wave excimer laser with gamma-emission pumping]. Trudy otraslevoy konf. [Proceedings of the Specialist Conference] "Fizika yaderno-vozbuzhdayemoy plazmy i problemy lazerov s yadernoy nakachkoy" [Physics of Nuclear-Induced Plasma and Problems of Nuclear-Pumped Lasers], Obninsk, 1993, vol. 3, pp. 272–273
22. B.V. Alekhin, Ye.K. Bonyushkin, V.V. Varaksin, B.V. Lazhintsev, A.Ye. Lakhtikov, A.P. Morovov, G.F. Nasyrov, V.A. Nor-Arevyan, A.I. Pavlovskiy, R.A. Orlov, V.G. Rogachev, V.B. Shlyakhovoy, Issledovaniye formirovatelya lazernogo izlucheniya s nakachkoy gamma-izlucheniyem yalernogo vzryva [Investigation of a laser emission former with pumping by the gamma-emission of a nuclear explosion]. Trudy 2-y mezhdunarodnoy konf. [Proceedings of the Second International Conference] "Fizika yaderno-vozbuzhdayemoy plazmy i problemy lazerov s yadernoy nakachkoy" [Physics of Nuclear-Induced Plasma and Problems of Nuclear-Pumped Lasers]. Arzamas-16, 1995, vol. 1, pp. 338–341
23. A.M. Boychenko, Ye.K. Bonyushkin, A.V. Karelin, B.V. Lazhintsev, A.Y. Lakhtikov, A.P. Morovov, S.I. Yakovlenko, Issledovaniye XeF-lasera s nakachkoy gamma-izlucheniyem yadernogo vzryva [An investigation of a XeF laser with pumping by the gamma-emission of a nuclear explosion]. Kvantovaya elektronika [Quant. Electron.] **23**(5), 420–422 (1996)
24. A.V. Bessarab, G.V. Dolgoleva, A.I. Zaretskiy, G.A. Kirillov, G.G. Kochemasov, V.M. Murugov, G.F. Nasyrov, V.G. Rogachev, Yu.A. Romanov, A.V. Sennik, S.A. Sukhanov, Issledovaniye termoyadernoy plazmy v mishenyakh s obrashchennoy korronoy [An investigation of thermonuclear plasma in targets with a reversed corona]. Doklady AN SSSR [Proc. Acad. Sci. USSR] **282**(4), 857–861 (1985)
25. B.V. Lazhintsev, V.A. Nor-Arevyan, Formirovaniye uzkonapravlennogo izlucheniya v eksimernom lazernom ustroystve s nakachkoy gamma-izlucheniyem [Forming of narrow-band beam emission in an excimer laser device with gamma-emission pumping]. Kvantovaya elektronika [Quant. Electron.] **32**(6), 557–561 (2002)

Chapter 13
Comments About Nuclear-Pumped Laser Research in the United States

13.1 Introduction

This book primarily discusses NPL research in Russia. The Russian program has been larger and remains active whereas work in the United States and elsewhere has dwindled. Shortly after the end of the Cold War, two circumstances combined to cause stoppage of NPL work. While the military need ceased, public opposition to nuclear power mounted following the Three Mile Island Nuclear reactor accident in the United States. Such concerns clearly carried over to a reactor pumped laser, i.e. to NPLs. Here we briefly review earlier work is the United States, prior to a build-up of effort as part of the "Star Wars" program as cold war tension grew. For completeness, the earlier history of NPL research in the United States is briefly reviewed here, largely adapted from earlier review articles [1]. Additionally, some content also appears in another previous review of NPLs [2] while some new material has been added, particularly concerning theory.

NPLs are driven by the reaction products of fission, fusion, or radioisotopes, providing the direct conversion of nuclear energy to directed optical energy. Aside from scientific aspects, research on NPLs was initially motivated because it has the potential to introduce many exciting new applications for fission reactors. This is especially true for systems where a nuclear energy source is a logical choice. The large volume made possible by using neutron excitation leads to ultra-large, steady-state power or intense pulsed-energy outputs that cannot be matched by other laser systems. Because nuclear energy is directly converted to excited states and subsequently to lasing, a relatively high efficiency (for traditional lasers) is achieved. Furthermore, the integrated reactor-laser system can be compact because no intermediate conversion equipment is needed. Consequently, researchers interested in military type laser began to take note of NPLs and a classified research program was mounted.

Chapter 1 of this book reviews work in the United States quite well from the Russian point of view, and this chapter provides a brief review from the American

S.P. Melnikov et al., *Lasers with Nuclear Pumping*,
DOI 10.1007/978-3-319-08882-2_13

perspective as I (George H. Miley) experienced it. Due to limitations on length, this chapter cannot do justice to the major effort done in the reactors at Sandia National Laboratory and earlier work at the Idaho National Laboratory [3–9]. These projects originated by Victor George at Lawrence Livermore National Laboratory under "Star Wars" funding and were classified (interested readers may now be able to find declassified versions of some reports). It appears that the project was aimed at KW-level demonstration results when funding stopped at the end of the Cold War. Thus, this chapter might be viewed as one from the perspective of academia. Admittedly, it is also weighted toward my work at the University of Illinois that my students, and former colleagues were involved in.

Reflections on early work at Illinois are briefly noted here to bring out some of the thinking and problems faced by early NPL researchers. In addition to my early work, other early U.S. pioneers in this field included Lloyd Herwig, David McArthur, Richard Schneider, Philip J. Ebert, Herbert Helmick, and Joe Verdeyen. Indeed, follow-on work by researchers noted in this chapter has been in a sense equally pioneering in this new unexplored field.

My experiments were uniquely enabled by access to the pulsed TRIGA reactor at the University of Illinois. These studies were enhanced by special facilities at the Illinois TRIGA reactor—a graphite thermal column, a horizontal "through-port," and a vertical access to the center core region. All three were used to great advantage at various times. Further, because I had earned a TRIGA reactor operator's license from the NRC, we could schedule laser experiments at convenient times, often at nights when the lab was relatively free from complications presented when other experiments were running. Also at these times the reactor experimental bay could be darkened to aid visual observations of the optics involved.

My interest in the field of NPLs began when I attended a faculty summer school at the Idaho National Reactor site in 1963. This frequently involved a one-hour bus ride out in the desert to the historic Fast Breeder Reactor Experiment. To pass the time, I read a book about the invention of the laser, which was discovered only a few years earlier, which inspired the thought, "Why not pump the laser with nuclear energy instead of electrical?" I then mastered the physics of laser threshold calculations from the book, and decided that the CO_2 laser with its long wavelength in the IR, hence low threshold, would be ideal for nuclear pumping. When I returned to the University of Illinois that fall, I discussed this with Prof. Joe Verdeyen, a well-known laser researcher in the Gaseous Electronics lab at the University of Illinois. Verdeyen had already done some pioneering work on the Transversely Excited Atmospheric Pressure (TEA) CO_2 Laser, and was extremely knowledgeable (he later went on to write a classic textbook on lasers, *Laser Electronics*, now in its third edition). He agreed with my assessment that CO_2 could easily be pumped in the TRIGA reactor, even with the low neutron fluxes available with steady state operation giving neutron fluxes of ~10^{14} n/cm^2 s. Pulsing would be "overkill." A boron coating on the laser tube would generate alpha particles from neutron-alpha reactions, and the charged alphas entering the CO_2 gas would deposit the energy needed for pumping the laser. Thus I teamed up with Verdeyen, along with several students, and began experiments. Unfortunately, this led to several years of frustration.

Though I initially thought I had invented the idea on that Idaho bus, I soon discovered that Lloyd Herwig at Northrop Grumman in California had already proposed the concept and had actually done experiments. It is hard to be first in this age of fast-moving technology! More importantly, the University of Illinois research team slowly learned two physics phenomena that represented significant challenges to the research. First, the nuclear radiations dissociated CO_2 molecules and other molecules involved in the lasing reaction. Because that effect was not included in my threshold calculations, it threw everything off. (Indeed, this has prevented nuclear pumping of CO_2 to this day. Amazingly, David MacArthur did succeed with CO, and that was the first experimental NPL in the United States.)

Next Illinois researchers discovered a new problem. The alpha heating of the gas gave a radial temperature profile that peaked near the wall, causing a thermal lensing effect that defocused the internal beams and threw the cavity optics off. Indeed, when using pulses, the effect was dynamic, such that the beam deflection was large and varied with time during the pulse. The effect was vividly illustrated when a cylindrical tube open to the air but with a boron coating on the tube's interior wall was used one night. A reference He-Ne laser beam was set up to pass through the boron coated tube located in the through beam port and projected on the building wall about 30 feet away. During the reactor pulse, the beam underwent huge spiral-circular gyrations on the wall, ultimately settling back to its original steady-state position. The effect of the gas lenses was thus abundantly obvious. The question was what to do about it to avoid this effect from spoiling the gain in the NPL cavity? Verdeyen proposed an over-focused cavity design (focal spots for internal beams were located at a point behind the cavity mirrors). That seemed to overcome the problem, although not without sacrificing some gain. (This effect was later studied in some detail by Maria Petra in her thesis work at Illinois, as discussed later.)

With this new knowledge my team continued trying to obtain lasing with CO_2, in desperation going to yet higher pulse sizes. Still, all attempts failed due to the dissociation effect already noted. We did achieve and publish papers on a radiation-enhanced, electrically pumped TEA CO_2 laser, i.e., this electrically driven laser output was increased by pulsing a version using a boron-coated tube. The effect could be attributed to the increased electrical conductivity created by the auxiliary ionization from the alpha particles. However, this was not the "pure" NPL we desired. So we next turned to study a possible He-Ne laser. The logic was that this medium would be less prone to dissociation interference, and again approximate threshold calculations predicted lasing could be achieved with TRIGA reactor pulses, maybe even with steady state operation. However, after over a year of frustrating results, it became evident that once again the threshold calculations were off, and achieving threshold was exceedingly challenging.

The reader may think that progress with NPL research was very slow. In contrast, research elsewhere on electrically pumped laser was making great strides. To understand the differences it must be realized that reactor-pumped laser experiments are quite different from normal lab bench laser research typical of electrically pumped lasers. The NPL is remotely located inside a tube deep within the reactor, making alignment adjustments difficult. Only a limited number of pulses could be obtained per day, and the reactor experiment schedule might only provide a limited number of days a month. Laser cavity components become activated, preventing immediate access. Also, radiation can damage optical components and detectors. Indeed, shielding is required around the detector array and in addition the laser beam must enter into the detector array without a straight path that radiation could follow. Thus, reflectors are needed to form a zigzag route. After a steep learning curve, my research group (and others elsewhere working in this field) became adept at overcoming these many obstacles. While their knowledge and technique had grown significantly, they had not yet achieved a NPL when the CO NPL was achieved by David MacArthur at Sandia National Laboratory (discussed later in this chapter). Despite this frustration, my group returned to work, even more determined to find new types of NPLs. The CO achievement was truly monumental, but due to the cooling requirements, this did not appear to be a practical result. New types of electrically pumped lasers were being reported quite frequently in those days, so why should not a variety of NPLs be found? This turned out to be the case, and, as seen from the following discussion, the University of Illinois was prolific in discovering new types of NPLs.

13.2 Comments About NPL Research in the United States

A brief historic outline of NPL developments to date in the United States is given in Table 13.1. Clearly, parallel developments were achieved in Russia, and a history of these developments is discussed in previous chapters. Also, the history of early developments in the United States has been detailed in several review articles [1, 2, 10–13], so only the highlights of these earlier studies will be given on here before turning to more recent developments.

The first demonstration of an NPL was in 1974, when a γ-ray-driven laser pumped by a thermonuclear explosion was reported [14, 15]. To distinguish the work focused on here from such experiments, some researchers have proposed the term reactor-pumped lasers (RPLs), but this discussion will continue to use the traditional NPL terminology. A similar concept using a fission reactor had been proposed as early as 1963 by Lloyd Herwig, who analyzed a CO_2 laser approach [16, 17]. Despite attempts by several groups to demonstrate pumping of gases such as CO_2, the first successful NPL was not achieved until 1975, when lasing of CO using a fast-burst reactor was reported at Sandia National Laboratories [7]. Since then, some 20 different experimental NPLs have been reported by groups at Sandia, Los Alamos National Laboratory (LANL), NASA Langley Research Center, and the Universities of Illinois, Missouri, and Florida.

Table 13.1 Highlights of NPL research in the United States

Dates	Event	Laboratory
1963	CO_2 NPL proposed	United Aircraft Labs
1974–1975	CO laser achieved	Sandia National Labs
1974–1975	3.51-μm Xe laser achieved	University of Florida/LANL
1975–1977	Low-threshold impurity lasers achieved	University of Illinois
1976	Neutron feedback ICF proposed	University of Illinois
1977	He-Hg laser (first visible NPL)	University of Illinois/Sandia
1977–1981	Extensive noble gas laser experiments/theory	NASA-Langley
1978	First nuclear pumped gain in excimer (XeF) reported	University of Illinois
1981	1-kW "box"-type He-Ar NPL operated	NASA-Langley
1983	First nuclear pumping of O_2 ($^1\Delta$)	University of Illinois
1984–1988	Gain measurements in XeF at high flux	Sandia National Labs
1987–1989	5-MJ NPL ICF target facility design	University of Illinois
1989	High-gain He-Ar-Xe (2.03 μm) studies	Sandia National Labs
1991	First nuclear-pumped flashlamp laser achieved ($XeBr-I_2$)	University of Illinois
1992	Visible $^3He-Ne-H_2$ NPL achieved	University of Illinois

Note: Table covers period up to 1993 as summarized; little work was done in the field after 1993, from several review articles [1, 2, 10–13]

Development Milestones

Initial studies of nuclear pumping concentrated on molecular gas lasers such as CO_2 because early theoretical predictions indicated lasing with low pump (neutron flux) thresholds of ~10^{10} n/cm^2 s [6, 7]. Indeed, experiments at the University of Illinois initiated during this period showed an enhanced output from an electrically excited CO_2 laser with nuclear pumping superimposed [11]. Experiments at the Sandia National Laboratories also demonstrated reactor-ionized electrical excitation of the CO_2 laser [6]. However, it was later found that lasing in CO_2 was prevented by dissociation of the molecule during irradiation. Thus, other NPLs were sought.

In their search for a NPL, Sandia workers led by David McArthur obtained, in early 1974, a small-signal gain and lasing in CO gas cooled to 77 °K [7]. This provided the first clear observation of lasing in the United States with pure fission-fragment excitation. Preliminary systems studies were then performed to study NPL scaling to large sizes and large powers with a source configuration that retained existing pulsed reactor technology [8, 18]. An early attempt at scaling of the cooled CO laser used a "folded path laser" apparatus, which lased at a power ~100 W.

Lasing was subsequently observed in rare gas mixtures and in CO gas mixtures at room temperature.

Shortly thereafter, a University of Florida team lead by Richard Schneider obtained pumping in a noble gas laser (He-Xe at 3.51 μm) in collaborative experiments at the Los Alamos Scientific Laboratory [19]. Subsequently, much was learned about the physics of nuclear pumping of noble gas NPLs through a series of studies by the NASA-Langley Research Center group led by Frank Hohl and Russell Deyoung (supported via Karl Heinz Thom in NASA headquarters). This work involved lasers with wavelengths generally in the near infrared. The NASA studies used a fast-burst reactor at the Army Aberdeen Test Laboratory and the ^{3}He (n, p)^{3}H reaction for pumping. Specifically, lasing was achieved in ^{3}He-Ar (1.27, 1.79 μm), ^{3}He-Xe (2.026 μm), ^{3}He-Kr (2.19, 2.52 μm), and ^{3}He-Cl (1.58 μm) using thermal neutron pulses of 200 ps full width and approximately 10^{16} n/cm^2 s [20]. These systems lased on atomic transitions where inversion occurred by collisional radiative recombination of the noble gas atomic ions. A record peak laser power of 3.7 W was achieved from a relatively small ^{3}He-Ar (1 % Ar) (1.79 μm) NPL at a total pressure of 4 atm. However, further increases in laser power were prevented by the small fraction of neutrons interacting with either ^{3}He or coated walls in the cell. Thus, a new laser geometry was devised to intercept more of the reactor neutrons. The laser volume was expanded to 40 × 30 × 3 cm by use of internal mirrors to create a "zigzag" type optical path through the excited ^{3}He-Ar media as shown in Fig. 13.1.

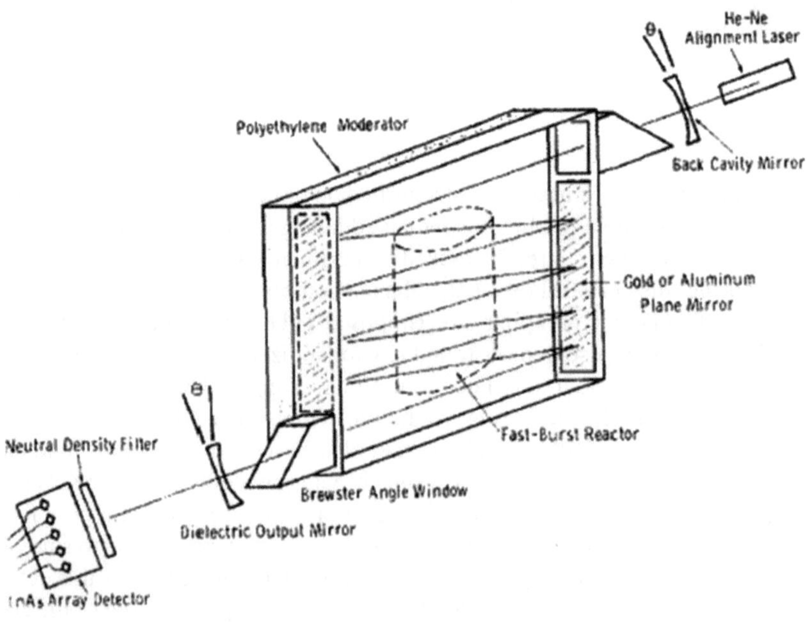

Fig. 13.1 Nuclear pumped multiple pass box laser [21]

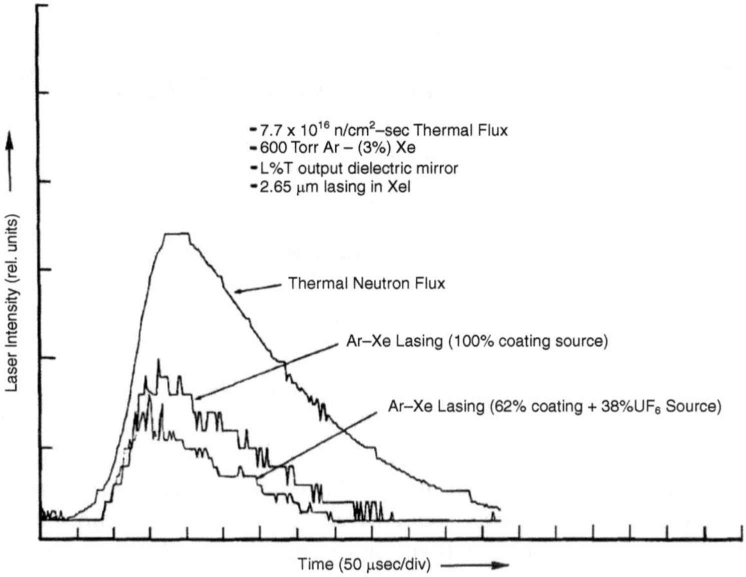

Fig. 13.2 Ar-Xe-$^{235}UF_6$ nuclear pumped lasing. *Note*: Table covers period up to 1993 as summarized from several review articles [1, 2, 10–13]; little work was done in the field after 1993

An output of 1.1 kW was achieved with a ^{3}He-Ar mixture at 2,300 Torr total pressure (0.74 % Ar) at 4.3×10^{16} thermal n/cm^2 s [21]. This system demonstrated the ability to scale the NPL power with increased laser gas volume. During this period, much progress was made in the theoretical interpretation of the pumping mechanism with stress on the secondary electron energy distributions created by the nuclear reaction products. Studies at the University of Illinois [22–24] employed both Monte Carlo and analytic solutions for a variety of gases, including the possible superposition of a weak electric field. This work is discussed in more detail in the theory section later in this chapter. Intensive numerical studies were also done at North Carolina State University [25]. These results defined the source term for reaction rate calculations. Next, attention turned to kinetic calculations for both noble gases [20] and excimers [26].

Another important pumping reaction considered at NASA Langley is the $^{235}UF_6$(n, ff)FF reaction, yielding 160 MeV of kinetic energy via the two highly charged fission fragments (FF_5). The importance of this reaction is the possibility of using a gaseous core reactor to combine the reactor and laser as one integral system. With this in mind, experiments were done in the U.S. Army Aberdeen Test Center Fast Burst Reactor to excite a mixture of $^{235}UF_6$-Ar-Xe (600 Torr). This produced lasing at 2.65 μm in Xe [27]. UF_6 gas is an effective laser quencher; thus, only small amounts could be added. A uranium coating, deposited from the UF_6 condensing on the internal laser cavity wall also produced fission fragments that excited the gas. Some results from this experiment are shown in Fig. 13.2.

With 3 Torr of UF_6 in Ar-Xe, 38 % of the energy deposited came from gaseous UF_6 and the remainder from the a ^{235}U coating formed on the cavity wall by UF_6

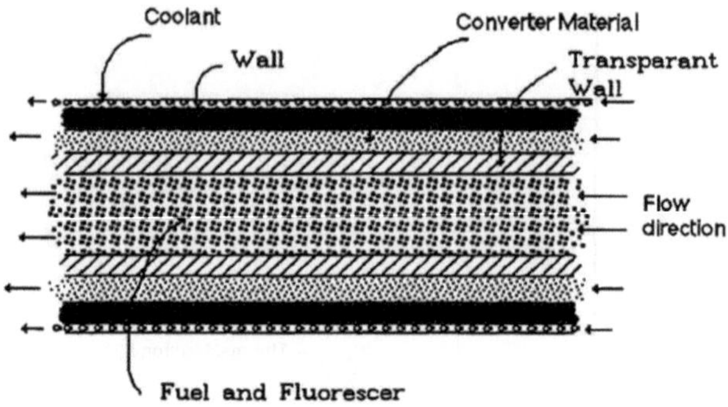

Fig. 13.3 Non-concentrating Aerosol Reactor Energy Conversion System (ARECS) module

decomposition. In summary, both ^{3}He and ^{235}UF$_6$ have been effective for pumping noble gas lasers. This coating reaction does not scale with volume as desired. Hence, if an ultra-compact fission core laser is desired, additional research is needed to find efficient systems for using UF$_6$ to achieve volumetric reactions.

One alternate approach that could avoid the quenching limits on the UF$_6$ concentration uses an aerosol of fissile microspheres in a fluorescer gas, making it an attractive approach to a higher-power nuclear-pumped flashlamp [28]. Thus work at the University of Missouri addressed an aerosol reactor concept called Aerosol Reactor Energy Conversion System (ARECS), illustrated in Fig. 13.3. The key component is a nuclear-driven flashlamp. Flashlamps do not require a long photon mean free path as do lasers. Thus, a media with relatively poor optical transport properties can be used to provide "pump" photons for lasers.

The ARECS fuel consists of fissile microspheres mixed with a fluorescer gas as an aerosol. The transport efficiency of the radiation from the nuclear fuel to the fluorescer medium strongly depends on the micropellet size distribution and the uniformity of pellet density. For example, as the radius of the micropellets increases, the transport efficiency decreases. While there have not been any experimental observations of a fissioning aerosol at high temperatures and flow, some key points are clear. A relatively high particle density with sizes averaging 2–10 μm in radius at temperatures of around 1,000 °K is required. Other experience with aerosols suggests that these conditions are achievable.

The fluorescer medium must channel the energy absorbed from the nuclear radiation to an excited state (versus gas heating). Irradiation of rare gas and rare gas halide mixtures can efficiently generate excimer fluorescence at relatively short wavelengths; thus, these gases are prime candidates for the fluorescer.

Several options are possible for coupling the light out of the fluorescer. One is to cause lasing directly in it. The other is to transport the light out via a "non-concentrating" cell geometry such as in Fig. 13.3 (light concentrating geometries are also possible). The coupling efficiency of a non-concentrating geometry can range from 30 to 80 % (the more complicated concentrating geometry could achieve even higher efficiencies).

The ARECS concept is potentially attractive for several applications including: photolytic lasers, photo-chemical production, and photoelectric or photo-electrochemical generation of electricity. These processes can be achieved at both high efficiency and high temperature. Consequently, they can serve as topping cycles for an even more efficient, integrated energy conversion system. To illustrate the potential performance, Boody and Prelas considered a photolytic laser system [29]. Such lasers are relatively efficient energy conversion systems, primarily due to resonance transitions. Boody and Prelas estimated an overall system efficiency of 0.03 using a photolytically pumped XeF* laser. This result compares well with other high power laser alternatives which have predicted system efficiencies in the 0.005–0.05 range.

Theoretical Studies in Support of U.S. NPL Research

During the earlier work on NPLs in the United States, a number of groups provided important insightful theoretical studies that clarified the physics of this new field. This work generally tackled one of the fundamental issues involved—the details of the source term created by the incoming radiation, the transfer of energy to the laser media, the reaction kinetics of the species in the lasing medium, and the optics associated with the laser, especially special effects such as thermal blooming encountered in such lasers. There is a complex matrix where these issues overlap with applications and a variety of possible laser media such as noble gases, excimers, molecular gases, liquids, etc. This matrix is far from full, so much work remains to be done to fully understand all of the possible routes that NPLs may take. The main goals are to obtain improved efficiencies, and ways to take advantage of the large energy but relatively low peak powers available with this form of pumping the laser.

Mark A. Kushner reported a number of important theoretical studies of nuclear-pumped lasers while at the University of Illinois. A summary of his early studies with NPLs is presented next in the section.

Nuclear-pumped lasers are characterized by high-threshold neutron fluxes and low gain. The closer the fissile gas can be coupled to the actual lasing process, the more efficient the laser will be. One approach is to seek laser systems such as chemical lasers that have low threshold power requirements. Kushner presented a model for a nuclear-pumped HF laser using a $^{235}UF_6$-H_2 gas mixture [30]. This model predicted a peak gain of about 50 %/m and a threshold neutron flux of $<10^{14}$/cm^2. He provided recommendations concerning the optimum use of this unique system. However, as discussed in other work by Wilson and DeYoung [31] and Maceda [43], this and other lasers requiring $^{235}UF_6$ are difficult to achieve experimentally, so the practicality of such lasers remains to be determined.

Kushner also sought other classes of lasers well suited to nuclear pumping. In study [32], he proposed study of a class of NPLs in which the laser levels are populated by negative ion–positive ion neutralization. He analyzed two particular systems, $^3He/N_2O/O_2$ and $^3He/N_2O/N_2$. Laser action was predicted in atomic oxygen and nitrogen as a result of O^- -O^+ and O^- -N^+ neutralizations. The source of negative ions is a dissociative attachment to N_2O. A threshold neutron flux of $\approx 10^{15}$ cm^{-2} s^{-1}, peak power of >500 W/cm^2, and a fission power to laser power conversion efficiency of ≈0.4 % were predicted. The electron density in the attachment process was evaluated, and the highly selective nature of the ion–ion neutralization in producing product-excited states.

Kushner studied the excitation source term along with average energies going into excitation and ionization [*W*-values] for nuclear pumping with fission fragments. This is particularly important to nuclear pumping because of the unique highly non-Maxwellian characteristic of the electron energy distribution created during the slowing down process. In study [33] Kushner and Moratz described a method for calculating excitation and ionization rates in a plasma generated by the slowing of fission fragments in a gaseous medium such as neon. The energy distribution of the fission fragments as they slow down and the excitation and ionization of neon due to collisions with the fission fragments was calculated. Effective *W* values for ionization and excitation of neon directly by the fission fragments of uranium were found to be about 71.6 and 11 eV, respectively. To illustrate the non-equilibrium characteristic of this process, they compared the source function for electrons produced by an e-beam to the source function of electrons produced by heavy ions which are shown to have had a lower average energy.

In another study [34], Kushner and colleagues presented a model for the heavy-ion pumping of a XeF ($B \rightarrow X$) laser by uranium fission fragments. The model is a self-consistent accounting of the generation and transport of the fission fragments through the fission foils, slowing of the fragments in the gas, evolution of the secondary-electron-source function and distribution, and the XeF laser plasma kinetics. By stimulating the same quantities for an e-beam-pumped plasma, direct comparisons can be made for laser performance. Kushner and colleagues found that the secondary-electron source generated by the e-beam was more energetic than that for direct ionization by fission fragments due to a more favorable mass ratio for momentum transfer collisions with orbital electrons. This difference in the electron-source functions significantly affects *W* values (energy/ion pair) and excitation fractions. The impact on laser performance, though, is not large due to the high efficiency of channeling deposited energy to the upper level in XeF lasers. For conditions typical of fission fragment excitation (power deposition 1–3 kW cm^{-3}, pulse length ≈200 μs), this model predicted 10–15 % higher gain with e-beam excitation than heavy-ion excitation.

Later, Kushner et al. pointed out that excimer lasers excited by electron or ion beams having energy deposition of 100 s J/1-atm over many microseconds experience a temperature rise of hundreds of degrees (K) [35]. The increase in gas temperature can significantly impact both the kinetics and spectroscopic parameters. In this study, Kushner and colleagues considered the high-temperature (≤900 K) plasma kinetics and absorption in He and Ne buffered gas mixtures for particle beam pumped XeF lasers. They found both gain and absorption depended differently on gas temperature in these mixtures (absorption decreased in He mixtures, but increased in Ne mixtures). The differences were attributed to a reduction in diatomic absorbing species with increasing temperature and differences in the temperature dependence of the optical absorption cross-sections for $NeXe^+$ and Xe_2^+.

Later, Kushner and colleagues discussed the effects of He addition on the performance of fission fragment excited Ar/Xe lasers [36]. This was viewed as an important laser system because both Russian and Sandia researchers were studying such systems experimentally. The intrinsic power efficiency of the atomic xenon laser depends upon the electron density because of the mixing of the laser levels by electron collisions while the electron density in high-pressure particle-beam excited plasmas increases with increasing gas temperature. Therefore, in order to reduce the amount of electron collisional mixing when operating at high-energy loadings (>100 s J/1-atm), mixtures having a high-heat capacity are required. In particle-beam excited Ar/Xe mixtures, which typically yielded the highest intrinsic laser efficiencies, increasing the gas pressure to increase the heat capacity was not always practical due to the high-stopping power of the gas mixture. For this reason, Kushner and colleagues investigated adding He to Ar/Xe mixtures for a fission-fragment excited atomic xenon laser. Adding He increased the heat capacity without appreciably perturbing the favorable kinetics resulting in efficient operation of the laser in Ar/Xe mixtures. They found that when adding He to Ar/Xe mixtures the dominant laser transition wavelength switched from 1.73 to 2.03 μm without significantly decreasing the efficiency. The laser pulse length also increased, an effect attributed to a lowering of both the electron temperature and gas temperatures.

Subsequent studies of the infrared atomic xenon laser ($5d \rightarrow 6p$) by Kushner et al. found it to be an attractive candidate for fission fragment excitation, providing low-power deposition (l–100 W cm^{-3}), long pulse lengths (1–10 ms), and high-energy deposition (100's J/l) [37]. The optical gain at 1.73 and 2.03 μm was measured in a reactor-excited xenon laser and yielded values exceeding 0.03–0.05 cm^{-1} at power depositions of less than 10 s W/cm^3. The gain was also found to rapidly terminate before the peak of the pump pulse for some experimental conditions. A computer model was developed to predict gain in fission-fragment-excited xenon lasers. It was found that the termination of gain was most likely attributable to gas heating which increased the electron density, leading to electron collision quenching. The specific dependence of gain on pump rate suggested that a reduced rate of recombination of molecular ions with increasing gas temperature was partly responsible for this behavior.

In another study pertaining to this system, Kushner and colleagues predicted gain of the 1.73 μm [$5d(3/2)_1$-$6p(5/2)_2$] and 2.03 μm [$5d(3/2)_1$-$6p(3/2)_1$] atomic xenon transitions for gas temperatures between 290 and 590 K [38]. Fission-fragment excitation was used to generate the gain in Ar/Xe, He/Ar/Xe, and Ne/Ar/Xe gas mixtures at a pump power of 8 W/cm^3. For constant gas density, the gain exhibited an approximate T^{-n}_{gas} dependence with n between 2 and 3. The primary reactions responsible for the temperature dependence of the gain were identified as dimer formation, dissociative recombination, and enhanced electron collisional mixing of the laser manifold due to an increased electron density.

In a following study [39], Kushner and Shon discussed excitation mechanisms and gain modeling of the high-pressure atomic Ar laser in He/Ar mixtures. The high-pressure (>0.5 atm) atomic Ar laser ($3d \rightarrow 4p$) oscillated on four infrared transitions (1.27–2.4 μm). Quasicontinuous oscillation on the 1.79 μm transition was obtained using electron beam and fission fragment excitation over a wide range of power deposition and gas pressure. In this regard, a computer model was developed to investigate excitation mechanisms of the Ar laser. Results from the model suggest that the upper laser level of the 1.79 μm transition $[Ar(3d[1/2]_1)]$ is dominantly populated by dissociative recombination of $HeAr^+$. In contrast, the dissociative recombination of $Ar^+{}_2$ was believed to predominantly produce Ar ($4s$) states. Electroionization from Ar metastables at moderate to high pump rates was likely to be responsible for the high efficiency of the Ar laser. Gain and laser oscillation were discussed and compared to experiments for He/Ar gas mixtures using various Ar mole fractions and total pressures. These results showed that the optimum Ar mole fractions in He/Ar mixtures are ~0.1–5 % for quasicontinuous pumping.

In a study aimed at understanding the pumping source term for coated wall type lasers, Miley and colleagues studied the transport of heavy charged particles produced by the ^{10}B (n, α) nuclear reaction using a mean-range straight-flight model [23]. The slowing down of these particles in a gas adjacent to the coating where they are born is described in terms of their flux energy spectrum, scalar flux, average energy, and energy-loss rate. These results were used in a plasma kinetics model which was then compared to measurements of metastable excited state densities in helium and neon plasmas created by the heavy charged particles. The resulting plasma kinetic model was shown to agree reasonably well with measurements involving ^{10}B-coated gas containers inserted in the Illinois pulsed TRIGA reactor. $Ne(1_{S3})$ and $Ne(1_{S5})$ densities ranging from 2×10^8 to 5×10^9 atom/cm^3 and 3×10^9 to 4×10^{10} atom/cm^3, respectively, were obtained in neon at 16.5 Torr with fluxes from 10^{12} to 10^{15} n/(cm^2 s). Maximum $He(2^1S)$ densities of about 10^{-4} atom/(cm^3 unit flux) were obtained in helium at pressures above 200 Torr. Such measurements, not previously available, were the first attempt to directly evaluate excitation processes in a plasma created in a nuclear reactor core.

In a related study, P.E. Theiss and Miley developed a method to evaluate the spatial dependence of the energy spectrum of charged particle currents entering a process fluid from a bordering solid fuel region [40]. This enabled calculation of spatially dependent ionization–excitation rates including the general case where the formation rate varies with particle energy. Non-dimensional plots permit application of the results to various fluid–fuel combinations. Also, with adjustment of two parameters, these results can also be used for various particles including fission fragments, betas, alphas, and protons. The example of excitation of helium by alpha irradiation is also presented in some detail in study [40].

Another concept that is important to future NPLs involves uranium-pumped nuclear plasmas. In a study focusing on this, Maceda and Miley developed a model for the atomic levels above ground state in neutral, U^0, and singly ionized, U^+, uranium based on identified atomic transitions [41]. They found 168 states in U^0

and 95 in U^+. A total of 1,581 atomic transitions were used to complete this process in this study. Atomic inverse lifetimes and line widths for the radiative transitions as well as the electron collisional cross sections were presented. They identified 285 transitions in U^0 and 278 transitions in U^+, and during the course of the research employed 1,581 atomic transitions. The final compilation of these data is given in study [41].

As one approach to the use of an uranium plasma for NPLs, Y.R. Shaban and Miley proposed a hybrid uranium-core-laser reactor (UCLR) for advanced space missions that could radiantly transfer energy to a propellant or alternately active laser action [42]. The propellant mode would be employed in the phases of the mission requiring a higher thrust. However, for the bulk of the travel, the propellant would be turned off and the ultrahigh specific impulse laser mode of operation would be employed. The concept was further developed, research and development issues were identified, and the different operational modes were discussed in study [42].

In another study concerning uranium plasma pumping, Maceda and Miley [43] computed the opacity and radiative-energy current due to line radiation was calculated for a U^{235} plasma with a temperature range 5,000–8,000 K. Also, a variation in the neutron flux of 2×10^{12} to 2×10^{16} neutrons/(cm^2 s) was considered. The plasma formed a cylinder with a diameter and height of 1 m. Because the electron states in uranium lie below 5 eV, recombination is the principal excitation mechanism. Inversions were found above 6,000 K and, at all temperatures, the line radiation at line center is greater than the corresponding blackbody radiation. An example of this is the 28,763–5,762-cm^{-1} transition in neutral uranium, where the Planck function at 5,000 K is 6.49×10^{-6} ergs/cm^2, and the calculated radiative-energy current is 1.492×10^{-4} ergs/cm^2. Negligible changes in the radiative-energy current were observed for changes in the neutron flux at a given temperature. As a benchmark, it is noted that the opacity at 5,000 K agrees well with earlier calculations done by Parks and colleagues [44], while recombinational excitation explains the variation in the opacity with temperature.

Beam profile control is an extremely important problem for NPLs due to the lensing effect caused by heating during pumping, especially for a wall coating type system. Thus several theoretical studies of this effect were performed at Illinois [45]. NPLs potentially offer an attractive method for high power laser applications such as a space power beaming. However, thermal gradients created by the pumping of the ion medium produce a dynamic "thermal blooming" (lensing effect due to temperature profile effects) which must be understood and controlled for accurate beam focusing. In paper [45], Petra and Miley performed basic experimental studies of thermal blooming under a variety of conditions presented. It was found that the laser pulse length could be severely restricted by thermal blooming, but some kinetic control is possible by changing the driving pulse shape. One promising approach found for reducing the effect was to employ a combined volume-wall pumping technique. Petra provided calculation of optimal coating and ^{3}He concentrations to achieve minimum thermal blooming.

In study [46], Petra and Miley considered the issue of control of the extracted beam quality of a NPL. One possible control method involved changing the spatial power deposition profile, which is a function of the gas pressure and the pumping mechanism. Thus, we proposed a combined ^{3}He and wall coating (e.g., B^{10} or U^{235}) technique. Then, the pump power deposition profile could be adjusted by varying the ^{3}He partial pressure. To examine this technique, measurements were performed with an in-core cell by imaging the radiation-induced fluorescence onto a linear diode array located some distance from the cell. Results from these measurements are presented in reference [46], and other related issues such as gas dynamics, index of refraction effects, and efficient beam extraction from a large-volume cavity are discussed.

In an effort to simplify calculational methods, R. H. Lo and Miley developed an integral balance technique for calculation of the electron energy distribution in a radiation-induced plasma [47]. Results predicted W-values reasonably well and compared favorably with more complicated Monte-Carlo calculations. As expected, results showed that the electron energy distribution was non-equilibrium with a high energy tail but lower average energy. This distribution differs from one using a normal electrical discharge and favors excitation over ionization. This then provides insights into how to seek NPLs which use a laser medium with energy levels receptive to such a distribution, illustrated in Fig. 13.4.

In addition to the source electron energy distribution, other reaction kinetics such as recombination play an important role in laser performance, as illustrated by a Monte Carlo simulation model for radiation-induced plasmas with nonlinear properties due to recombination [24]. This Monte Carlo model employed a piecewise-linearized predict–correct technique. Several variance reduction techniques were used, including antithetic variants. The resulting code was applied to the determination of the electron energy distribution for a noble-gas plasma created by alpha-particle irradiation. Results are presented in reference [21] for helium with an electron source rate from 10^{14} to 10^{18} electrons/(cm^3 s), initial energies from 70 to 1,500 eV, pressures from 10 to 760 Torr, and electric-field-to-pressure ratios from 0 to 10 V/(cm Torr). The low-energy portion of the distribution function approached a Maxwellian for zero electric field and Druyvesteyn's distribution with an applied electric field. However, above the ionization potential and extending to the source energy, a parabolic-shaped distribution (tail) occurred. This distribution is not too different from the one found in [47], although the effect of recombination on it is clearly discernable.

Due to pump source restrictions, even with a fast burst reactor NPLs typically exhibit relatively long (micro- to millisecond) pulse lengths with modest peak powers but with very high total energy [48]. These pump power restraints seriously limit the choice of laser media, favoring energy storage type lasers. One way to avoid this constraint is to employ a Nuclear-Driven Flashlamp (NDF) for the primary pumped element in the system. The fluorescence from this NDF can then be used for pumping a laser or for other high intensity light applications. The first experimental example of this approach was a ^{3}He-$XeBr_2$ NDF employed by Williams and Miley [49] to pump a small iodine laser. Study [50] discussed issues

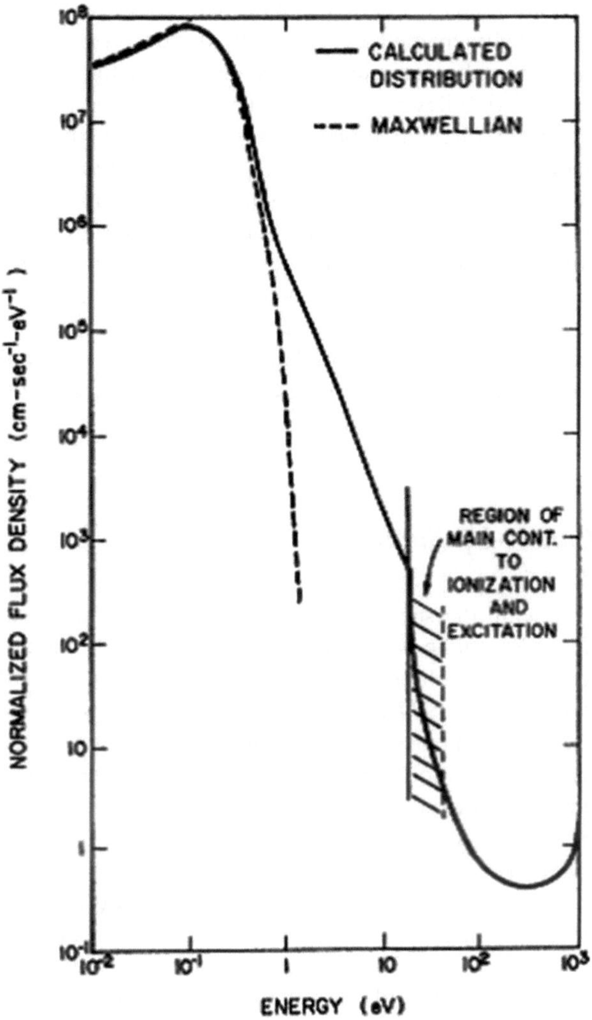

Fig. 13.4 Complete distributions for a 1-keV source in 10-Torr Helium, indicated by 5×10^{13} neutrons cm^{-2} s^{-1}. For comparison, a Maxwellian distribution is shown as a dashed curve [47]

involved in scaling such an NDF up to high power levels. Possible optimum configurations included use of microsphere or fiber pump elements dispersed in the NPF media. Analysis of such possibilities is presented along with consideration of special reflecting surface designs in references [49–51].

Another very important potential NPL laser involves the pumping of the iodine line. Miley reevaluated the existing data in study [52]. At the time, the chemically pumped O_2–I_2 transfer laser was gaining interest as a high-power laser for a variety of applications. Thus the nuclear-pumped version would be very significant and as already discussed seems to offer an electron energy distribution favorable to this system. Indeed, Garschadin had previously shown that an electrically pumped version faced obstacles due to competition between ionization and unwanted ionization and excitation. (After years of work, additives to combat this problem have finally enabled an electrically driven version, termed the "e-COIL" [53].) An interest in alternate pumping techniques for some situations remained, however. In study [54] Miley reviewed earlier work on the potential use of nuclear pumping, either for direct pumping of O_2 or for pumping of an excimer flashlamp in a photolytic iodine laser system. Such lasers are attractive for space power beaming systems where the compact, high-energy system achievable with a pulsed reactor is desired. For a lower pressure transfer laser system, direct nuclear pumping of O_2 to create $O_2(^1\Delta)$ is the simplest, yet efficient approach. For higher-pressure lasers, e.g., $\gtrsim 1$ atm, the use of a nuclear-pumped excimer appears to be attractive. In that case, a transfer laser can be developed through photolysis of O_3 to generate $O_2(^1\Delta)$, or the flashlamp can be used directly in a photolytic iodine laser system. The choice among these options largely depends on the specific application as this translates into system requirements.

Follow-Up Studies

Following the successes with noble gas pumping, various laboratories undertook new NPL studies with three main goals: development of shorter wavelength lasers, lower neutron thresholds, and improved efficiencies. A NPL with visible output was developed in 1977 by a University of Illinois/Sandia team using an He-Hg charge exchange scheme [20]. Even shorter wavelengths were predicted when Miley's team reported in 1978, for the first time, that gain had been measured with nuclear pumping of excimers [55]. Then, in the search for lower thresholds the University of Illinois group discovered a new class of NPLs where lasing occurs in a minority species (i.e., "impurity" lasers). The first such laser used small concentrations of N_2 in neon [56]. Subsequently, a low-threshold (0.04 W/cm^3 pump power) laser was developed using carbon from the dissociation of CO and CO_2 during irradiation [57]. Later, an ^{3}He-Ne laser with an even lower threshold was reported at the University of Florida, but the results are disputed [58, 59]. Later, gain measurements were reported on ^{3}He-Ne at low fluxes or order 10^8 n/cm^2 s [60].

Experimental Research During the 1980s

Work on NPLs narrowed to four laboratories in the 1980s: the Sandia National Laboratory, Idaho National Environmental Laboratory (INE), University of Illinois, and University of Missouri. The general thrust focused on increasing efficiencies through improved mating with the reactor system. NPL research at INE employed the TREAT pulsed reactor and has explored several possible laser systems. However, no results have been published to date. Sandia National Laboratory used the Sandia Pulsed Reactor III (SPR-III) for experiments at high excitation rates over small volumes, and the Sandia Annular Core Research Reactor (ACRR) at lower rates but larger volumes. The energy deposition rate in the laser medium with SPR-III was maximized to increase the range of lasers that could be studied [61, 62]. Laser excitation rates obtained were ~10 kW/cc with ^3He excitation [63] and −2 kW/cc with fission-fragment excitation [5]. Sensitive techniques were developed at ~0.35 μm for measuring small-signal gains as low as 0.5 %/pass (using a tunable probe laser) and laser cavity losses as low as 1 %/roundtrip. With XeF, a gain coefficient of 0.7 %/cm was observed at 0.351 μm with ^3He excitation at several atm [9]. Intermittent gains ~0.5 %/cm were also observed with the tunable probe laser for fission-fragment excitation of XeF gas mixtures using Ar buffer gas at ~1 atm. However, attempts at lasing with either type of excitation at ~2 kW/cc were unsuccessful [9]. Possible explanations include too low of a pumping rate, transient gas lasing, and a non-optimum gas mixture. Models of the entire fission-fragment excitation process have been developed in collaboration with University of Illinois researchers [63]. Measurements were also made of the index gradients at moderate to high pressures with uranium wall coatings [3].

More recently, studies were performed on the reactor-pumped atomic Xe laser at both 1.73 and 2.03 μm (first reported in [34]). Lasing was observed with the ACRR and SPR-III reactor facilities [37], and Rigrod experiments were performed to measure small-signal gain, saturation intensity, and non-saturable losses. Pump power densities ranged from 5 to 1,000 W/cm^3. Small-signal gains such as 0.7 %/cm were measured at 1.73 μm in Ar-Xe mixtures, with reported laser energy efficiencies ~3 % and power efficiencies ~5 %. Small-signal gains ~1–3 %/cm were measured at 2.03 μm in He-Ar-Xe mixtures, with energy efficiencies ~2.4 %. Non-saturable losses appeared to be small [64, 65]. Lasing was also observed at ~2 μm in He-Ar at the lower pump rates of the ACRR. Other Sandia experiments studied extraction of laser energy from a larger portion of the excited region through the use of unstable laser resonators and the development of specialized equipment for excitation of larger volumes of laser gas. Adaptive optic concepts have been intensely studied as to counter act "thermal blooming" as described in reference [35].

Studies at the University of Illinois pioneered the development of three new types of NPLs: (1) $O_2(^1\Delta)$-I_2; (2) nuclear-pumped flashlamp I_2; and (3) ^3He-Ne-H_2. These are briefly described next.

Measurements by using the University of Illinois TRIGA showed relatively efficient production of $O_2(^1\Delta)$ in low-pressure irradiations. A scaled-up design based upon these results would use ozone production at higher pressure followed by photodissociation with a nuclear-pumped flashlamp to form $O_2(^1\Delta)$ [66]. Direct lasing on the $O_2(^1\Delta)$ using a "π-pulse" technique was also proposed by Scully and colleagues [67]. This is based on the concept that a pulsed laser can efficiently extract energy from an inverted molecular medium, even though the transitions involved are highly forbidden, for example singlet delta oxygen which lives for 40 min in its excited state. This extraction is accomplished by injecting pulses that are strong enough to drive the molecules from the excited to the weakly coupled ground state. A well-known situation in which this occurs is provided by Rabi flopping of the molecule from the excited to the ground state via a so-called "π-pulse." Thus, for example, one envisions $O_2(^1\Delta)$, $NF(^1\Delta)$, or $S_2(^1\Delta)$ molecular medium produced by collision with fission fragments in the reactor and the gas injected into an optical cavity. A "π-pulse" produced by a dye laser is then injected into the inverted medium that "sweeps" the energy from the system. Further detailed discussion is given in study [68].

The use of a nuclear flashlamp to pump a laser [49] was first demonstrated experimentally at the University of Illinois by Williams and Miley [49–51]. The nuclear-pumped flash lamp used ^3He at 2,000 Torr with 0.5 Torr $CHBr_3$ and 200 Torr Xe. This mixture was in a hollow cylinder surrounding the laser cavity, which was filled with various pressures of C_3F_7I. Stimulated XeBr* fluorescence pumped an atomic iodine laser at 1.31 μm. This experiment was extremely important in demonstrating the potential for nuclear flashlamp pumping, which was discussed in a number of theoretical articles. While the efficiency of the flashlamp was less than anticipated, subsequent kinetic studies suggested modifications in the mixture ratio that could potentially result in improved operation [49].

A subsequent University of Illinois discovery involved the use of H_2 as a quenching agent to achieve lasing at 0.58 μm in a nuclear-pumped ^3He-Ne-H_2 mixture [50]. This NPL had a much lower pump threshold than reported earlier for ^3He-Ne-Ar mixtures (5 versus 120 W/cm^3). An efficiency of order of 1 % was anticipated. This laser then appears to be well suited for various space power beaming applications. Further details are given in the presentations [69, 70].

Reactor Systems

A variety of concepts for combining an NPL and fission reactor into a compact efficient NPL system have been considered [6, 10, 70–74]. For example, as previously mentioned, work at the University of Missouri involved an aerosol reactor concept called Aerosol Reactor Energy Conversion System (ARECS) proposed as a nuclear-driven flash lamp [75–77]. The ARECS concept is potentially attractive for several applications, including photolytic lasers, photochemical production, and photoelectric or photo-electrochemical generation of electricity. These processes can be achieved at both high efficiency and high temperature. Consequently, this approach could serve as a topping cycle for an integrated reactor energy conversion system.

Applications

The NPL allows a laser system with an extremely large excitation volume and a "built-in" energy source. This unique possibility could lead to a wide variety of applications that require high power sources. Here, two examples are discussed: space applications and laser fusion.

The power demand for future space missions will significantly increase in the twenty-first century, especially for such missions as establishing lunar and Mars

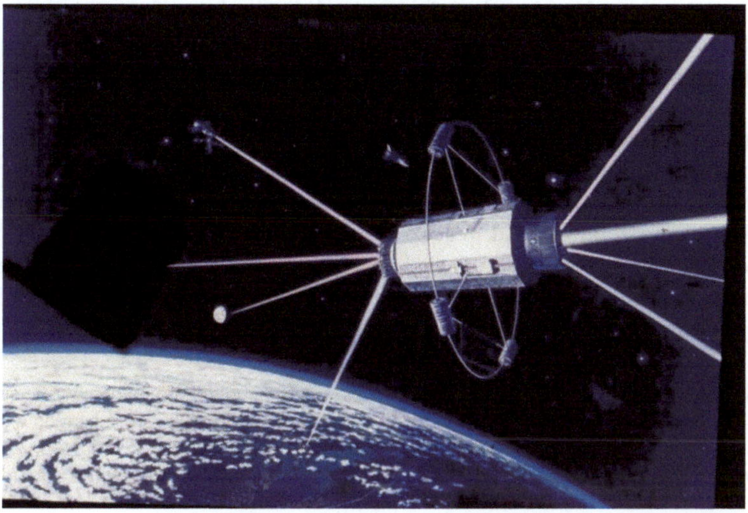

Fig. 13.5 Artist's concept of a space-based NPL system studied by NASA researchers [21]

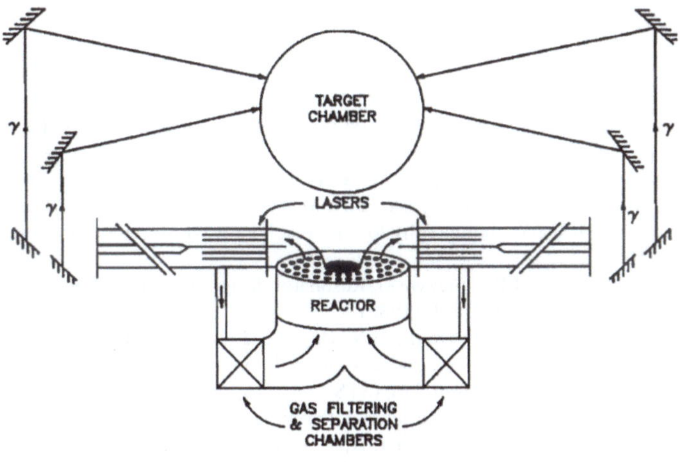

Fig. 13.6 Schematic diagram of the overall NPL system for use with the Laser Microfusion Facility (LMF) [71]

bases, creating space industrialization, and developing new propulsion systems. NASA-Langley and University of Illinois researchers have considered the use of NPLs to provide a cost-effective, enabling technology that could beam power in the form of a laser beam to remote users in space [78]. These users could then convert the laser energy into either electricity or propulsion. The space-based reactor power station could be placed in a high nuclear-safe orbit (as illustrated in Fig. 13.5).

The Laboratory Microfusion Facility (LMF), shown in Fig. 13.6, has been proposed to study inertial confinement fusion (ICF) targets with an MJ-scale laser to obtain reactor-grade gains. An advanced solid-state laser is the prime candidate as the driver for the LMF. However, researchers at the University of Illinois have developed a preliminary conceptual design for an alternate approach using NPLs [70, 71]. In this concept, a pulsed fission reactor is used to excite an $O_2(^1\Delta)$-I_2 NPL. An important potential advantage of the NPL driver is that projected costs are significantly lower than for a conventional solid-state laser. Another key advantage is that the LMF NPL experiment would provide a direct path to the development of a high-efficiency commercial reactor using the neutron feedback NPL driver approach [71]. In that case, the fission reactor would be eliminated with neutrons from the target implosion directly driving the NPL. This approach and a variant on it using D-^{3}He instead of conventional D-T targets are discussed in [78].

Acknowledgments As indicated by the references, many people have made significant contributions to NPL research in the United States. This chapter was largely adapted, with some additions (especially regarding theoretical advances) from a prior review in the *Journal of Laser and Particle Beams* [2], and a presentation at the ANS Conference on *Fifty Years with Nuclear Fission* authored by Miley, R. DeYoung, D. McArthur, and M. Prelas [1].

References

1. G.H. Miley, D. McArthur, M. Prelas, R. DeYoung, *Fission Reactor Pumped Lasers: History and Prospects*. Fifty years with nuclear fission (American Nuclear Society, La Grange Park, IL, 1989), p. 333
2. G.H. Miley, Overview of nuclear-pumped lasers. Laser Part. Beams **11**(3), 575 (1993)
3. D.R. Neal, W.C. Sweatt, J.R. Torczynski, D.A. McArthur, R.J. Gross, W.J. Alford, G.N. Hays, Application of high speed photography to time-resolved wavefront measurement. Proc. SPIE **832**, 5256 (1987)
4. D.R. Neal, W.C. Sweatt, J.R. Torczynski, Resonator design with and intracavity time-varying index gradient. Proc SPIE **965**, 130 (1988)
5. J.K. Rice, D.R. Neal, D.A. McArthur, G.N. Hays, W.J. Alford, D.E. Bodette, W.C. Sweatt, Reactor pumped laser research at Sandia National Laboratories, 1987, In Proceedings of the International Conference on Lasers ‘86, (STS Press, McLean, VA, p. 571)
6. D.A. McArthur, G.H. Miller, P.B. Tollefsrud, Pumping of high-pressure CO2 laser media via a fast-burst reactor and electrical sustainer. Appl. Phys. Lett. **23**, 303 (1973)
7. D.A. McArthur, P.B. Tollefsrud, Observation of laser action in CO gas excited only by fission fragments. Appl. Phys. Lett. **26**, 187 (1975)
8. D.A. McArthur, G.H. Miley, R. DeYoung, M. Prelas, Report SAND76-0584, (Sandia National Labs., Albuquerque, NM, 1977); reprinted November 1983
9. D.A. McArthur, G.N. Hays, D.R. Neal, J.K. Rice, Recent results from nuclear-pumped laser studies: gain and measurement in XeF, in *Laser Interaction and Related Plasma Phenomena*, ed. by H. Hora, G.H. Miley, vol. 8 (Plenum Press, New York, 1988), p. 75
10. K. Thom, R. Schneider, Nuclear pumped gas lasers, AIAA, **10**(4), (1972)
11. G.H. Miley, Review of nuclear pumped lasers, in *Laser Interaction and Related Plasma Phenomena*, ed. by H. Schwarz, H. Hora, vol. 4a (Plenum Press, New York, 1977), p. 181
12. R.T. Schneider, F. Hohl, Nuclear-pumped lasers, in *Advances in Nuclear Science and Technology*, ed. by J. Lewins, M. Becker (Plenum Publishing, New York, 1984), p. 123
13. G.H. Miley, Review of nuclear pumped lasers, in *Laser Interactions and Related Plasma Phenomena*, ed. by H. Hora, G.H. Miley, vol. 6 (Plenum Press, New York, 1984), p. 47

14. P.J. Ebert, L.J. Terderber, H.A. Koehler, R.W. Kuckkuck, D.L. Redhead, *Xenon Pumped by Gamma Rays* (IEEE IQEC-8, San Francisco, CA, 1974)
15. P.B. Lyons, J.S. Clarke, D.S. Metzger, *Gamma Initiated HF Laser* (IEEE IQEC-8, San Francisco, CA, 1974)
16. L. Herwig, *Report C-l 10053-5* (United Aircraft Labs, East Hartford, CT, 1964)
17. L. Herwig, Concepts for direct conversion of stored nuclear energy to laser beam power. Trans. Am. Nucl. Soc. **7**, 131 (1964)
18. T.R. Schmidt, D.A. McArthur, *Report SAND76-0139* (Sandia National Labs, Albuquerque, NM, 1976)
19. H.H. Helmick, J.L. Fuller, R.T. Schneider, Direct nuclear pumping of a helium-xenon laser. Appl. Phys. Lett. **26**, 327 (1975)
20. R.J. DeYoung, N.W. Jalufka, F. Hohl, Direct nuclear-pumped lasers using the $^3He(n,p)^3H$ reaction, 1978, In American Inst. Aeronautics and Astronautics, 16th Aerospace Sciences Meeting, p. 991
21. R.J. DeYoung, Kilowatt multiple-path ^3He-Ar nuclear-pumped laser. Appl. Phys. Lett. **38**, 297 (1981)
22. G.H. Miley, *Direct Conversion of Nuclear Radiation Energy* (American Nuclear Society and Atomic Energy Commission (AEC), Hinsdale, IL, 1971)
23. J.C. Guyot, G.H. Miley, J.T. Verdeyen, Application of a two-region heavy charged particle model to noble-gas plasmas induced by nuclear radiations. Nucl. Sci. Eng. **48**(4), 373–386 (1972)
24. B. Wang, G.H. Miley, Monte Carlo simulation of radiation-induced plasmas. Nucl. Sci. Eng. **52**, 130–141 (1973)
25. H.A. Hassan, J.E. Deese, Electron distribution function in a plasma generated by fission fragments. Phys. Fluids. **19**(12), 2005 (1976)
26. S.J.S. Nagalingam, G.H. Miley, Theoretical analysis of nuclear pumped XeF, In Proceedings of the International Conference on Lasers '80, New Orleans, LA, p. 699
27. R.J. DeYoung et al., Fission-fragment nuclear lasing of Ar(He)-Xe. Appl. Phys. Lett. **37**, 679 (1980)
28. M.A. Prelas, F.P. Boody, G.H. Miley, J.F. Kunze, Nuclear driven flashlamps. Laser Part. Beams **6**, 25 (1988)
29. F.P. Boody, M.A. Prelas, Photolytic dual-media nuclear pumping of excimer lasers. AIP Conf. Proc. **100**, 349–354 (1983)
30. M.J. Kushner, Characteristics of a UF_6-H_2/HF nuclear pumped laser. J. Appl. Phys. **51**(5), 2421–2428 (1980)
31. J.W. Wilson, R.J. DeYoung, Power deposition in volumetric $^{235}UF_6$-He fission-pumped nuclear lasers, J. Appl. Phys., **49**(3), 989–993
32. M.J. Kushner, Nuclear-reactor pumped lasers excited by ion-ion neutralization. J. Appl. Phys. **54**(1), 39–47 (1983)
33. T.J. Moratz, M.J. Kushner, Fission fragment pumping of a neon plasma. J. Appl. Phys. **63**(5), 1796–1798 (1988)
34. T.J. Moratz, T.D. Saunders, M.J. Kushner, Heavy-ion versus electron-beam excitation of an excimer laser. J. Appl. Phys. **64**(8), 3799–3810 (1988)
35. T.J. Moratz, T.D. Saunders, M.J. Kushner, High-temperature kinetics in He and Ne buffered XeF lasers: the effect on absorption. Appl. Phys. Lett. **54**(2), 102–104 (1989)
36. W.J. Alford, G.N. Hays, M. Ohwa, M.J. Kushner, The effects of He addition on the performance of the fission-fragment excited Ar/Xe atomic xenon laser. J. Appl. Phys. **69**(4), 1843–1848 (1991)
37. J.W. Shon, M.J. Kushner, G.A. Hebner, G.N. Hays, Predictions for gain in the fission-fragment-excited atomic xenon laser. J. Appl. Phys. **73**(6), 2686–2694 (1993)
38. G.A. Hebner, J.W. Shon, M.J. Kushner, Temperature dependent gain of the atomic xenon laser. Appl. Phys. Lett. **63**(21), 2872–2874 (1993)

39. J.W. Shon, M.J. Kushner, Excitation mechanisms and gain modeling of the high-pressure atomic Ar laser in He/Ar mixtures. J. Appl. Phys. **75**(4), 1883–1890 (1994)
40. G.H. Miley, P.E. Thiess, A unified approach to two-region ionization-excitation density calculations. Nucl. Appl. **6**, 434–451 (1969)
41. E.L. Maceda, G.H. Miley, An atomic model for neutral and singly ionized uranium. Nucl. Sci. Eng. **70**, 82–95 (1979)
42. Y.R. Shaban, G.H. Miley, Concept for a hybrid uranium gaseous-core laser reactor, (STAIF 2001, Albuquerque, NM, 2001). Accessed 10–14 Feb 2001
43. E.L. Maceda, G.H. Miley, Optical properties of uranium plasmas. J. Appl. Phys. **49**, 4369–4376 (1978)
44. G.L. Parks, J.C. Stewart, S. Peyton, NASA CR 72348 (unpublished, available from NTIS, Springfield, VA, 1968)
45. M. Petra, G.H. Miley, Thermal blooming in NPLs, SPIE Conference on High-Power Laser Ablation, (Santa Fe, NM, 1998) vol. 3343, pp. 242–253
46. M. Petra, G.H. Miley, Nuclear-powered laser beam profile control method. Laser Part. Beams **11**(3), 549–557 (1993)
47. R.H. Lo, G.H. Miley, Electron energy distribution in a helium plasma created by nuclear radiations. IEEE Trans. Plasma Sci. **PS-2**, 198–205 (1974)
48. G.H. Miley, M.A. Prelas, Nuclear-pumped flashlamp sources, in *High-Power Laser Ablation III*, ed. by Claude R. Phipps, Proceedings of SPIE, vol. 4065 (2000), pp. 728–742
49. W.H. Williams, G.H. Miley, Nuclear-induced XeBr* flourescence pumping of the atomic iodine laser, in *Proceedings of the Specialist Conference on Physics of Nuclear Induced Plasmas and Problems of Nuclear Pumped Lasers* (NPL-92) (Obninsk, Russia, 1992)
50. W.H. Williams, Nuclear induced XeBr fluorescence pumping of the atomic iodine laser, 1992 Ph.D. Thesis (University of Illinois, Urbana, IL)
51. G.H. Miley, W.H. Williams, Nuclear pumped stimulation of the atomic iodine laser, in *Proceedings of the 7th Interdisciplinary Laser Science Conference*, (Monterey, CA, 1991)
52. G.H. Miley, Nuclear pumping of the iodine laser – revisited, *SPIE Conference on High-Power Laser Ablation*, Santa Fe, NM, vol. 3343 (1998), pp. 692–700
53. D.L. Carroll, J.T. Verdeyen, D.M. King, B. Woodard, L. Skorski, J. Zimmerman, W.C. Solomon, Modeling of the ElectriCOIL system. IEEE J. Quant. Electron. **39**(9), 1150–1159 (2003)
54. G.H. Miley, Overview of nuclear-pumped lasers. Bull. APS **581**(24), 117 (1978)
55. M.A. Akerman, G.H. Miley, D.A. McArthur, A helium-mercury direct nuclear pumped laser. Appl. Phys. Lett. **30**, 409 (1977)
56. R.J. DeYoung, W.E. Wells, G.H. Miley, J.T. Verdeyen, Direct nuclear pumping of a Ne-N_2 laser. Appl. Phys. Lett. **28**, 519 (1976)
57. M.A. Prelas, Nuclear-pumping of the atomic carbon laser, Ph.D. Thesis (University of Illinois, Urbana, IL, 1979)
58. B.D. Carter, M.J. Rowe, R.T. Schneider, Nuclear-pumped cw lasing of the ^{3}He-Ne system. Appl. Phys. Lett. **36**, 115 (1980)
59. R.T. Schneider, B.D. Carter, M.J. Rowe, Reply to comments on the nuclear-pumped cw lasing of the ^{3}He-Ne system. Appl. Phys. **52**(11), 6980 (1981)
60. S. Soramoto, K. Mochiki, M. Nakazawa, A. Okumura, H. Hayashida, H. Nakamura, New neutron detection methods using laser technique, Nucl. Instrum. Meth. Phys. Res. Sect. A, **306**(3), 524
61. D.M. Minnema, D.A. McArthur, Reactor-pumped laser: kinetics effects on the SPRIII reactor, in *Proceedings of the Fast Burst Reactor Workshop*, (Albuquerque, NM, 1986), p. 57
62. D.M. Minnema, D.A. McArthur, Report SAND87-0098, (Sandia National Labs., Albuquerque, NM, 1987)
63. G.N. Hays, D.A. McArthur, D.R. Neal, J.R. Rice, Gain measurements near 351 nm in 3He/Xe/NF3 mixtures excited by fragments from the 3He(n, p)3H reaction. Appl. Phys. Lett. **49**, 363 (1986)

64. A.M. Voinov, L.E. Dovbysh, V.N. Krivonosov, S.P. Melnikov, L.V. Podmoshenskii, A.A. Sinyanskii, Low-threshold nuclear-pumped lasers using transitions of atomic xenon. Sov. Phys. Dokl. **24**, 189 (1979)
65. W.J. Alford, G.N. Hays, Measured laser parameters for reactor-pumped He/Ar/Xe and Ar/Xe lasers. J. Appl. Phys. **65**, 3760 (1990)
66. H.E. Elsayed-Ali, G.H. Miley, Kinetic model of photolytic $O_2(^1\Delta)$ generation. IEEE J. Quantum Eng. **20**, 977 (1984)
67. M.O. Scully, G. H. Miley, J. Meyer-ter-Vehn, On the nuclear-pumped π–pulse laser, in *Laser Interaction and Related Plasma Phenomena,* ed. by H. Hora and G. H. Miley, vol. 8, (1988) pp. 105–107
68. M.O. Scully, R.F. Shea, K. Kompa, Efficient energy extraction from highly forbidden transitions by means of high-power pulses. J. Opt. Soc. **B3**, 996 (1986)
69. Y.R. Shaban, G.H. Miley, Search for a visible wavelength nuclear pumped laser, In *Proceedings of the U.S.-Japan Seminar on Physics of High Power Laser Matter Interactions*, (Kyoto, Japan, 1992)
70. Y.R. Shaban, G.H. Miley, Practical, visible wavelength nuclear-pumped laser. Laser Part. Beams **11**, 559 (1993)
71. M.A. Prelas, F.P. Boody, M. Zediker, M. Rowe, A direct energy conversion technique based on aerosol core reactor concept, in *Proceedings of the IEEE International Conference on Plasma Science,* 1984, IEEE Publication 84CH1958-8, p. 38
72. G.H. Miley, Neutron feedback ICF, *Atomkern./Kerntech.* **45**, 14
73. G.H. Miley, A nuclear pumped laser for the laboratory microfusion facility. Laser Part. Beams **7**, 411 (1989)
74. S. Balog, H. Chung, M. Doczy, R. Duraski, M. Hutcheson, L.R. Eglezopoulos, D. Russ, Nuclear pumped laser for next step ICF driver, 1989 Fusion Tech. **15**, 383–390
75. G.H. Miley, 1992 A nuclear pumped flashlamp iodine laser, in *Proceedings of the U.S.-Japan Seminar on Physics of High Power Laser Matter Interactions*, (Kyoto, Japan)
76. M.A. Prelas, J.B. Romero, E.F. Pearson, A critical review of fusion systems for radiolytic conversion of inorganics to gaseous fuels. Nucl. Tech. Fusion **2**, 143 (1982)
77. A. Chung, M.A. Prelas, Charged particle spectra from B-10 and UO_2 slab and spherical sources, 1983, in *Proceedings of IEEE International Conference on Plasma Science,* IEEE Publication Number 83CH1847-3
78. G.H. Miley, Y. Shaban, M. Petra, R.J. DeYoung, Nuclear pumped laser system for power beaming, in *Proceedings of the Nuclear Technologies for Space Exploration,* (Jackson, WY, 1992)

MIX
Papier aus verantwortungsvollen Quellen
Paper from responsible sources
FSC® C105338

If you have any concerns about our products,
you can contact us on
ProductSafety@springernature.com

In case Publisher is established outside the EU,
the EU authorized representative is:
Springer Nature Customer Service Center GmbH
Europaplatz 3, 69115 Heidelberg, Germany

Printed by Libri Plureos GmbH
in Hamburg, Germany